Learning Approaches in Signal Processing

About the Series

A. G. Constantinides, Series Editor

Digital Signal Processing (DSP) is an area which either implicitly or explicitly forms the fundamental bedrock of a very wide range of applications and needs in our modern society. The broadening of the theoretical developments coupled with the accelerated incorporation of DSP to an ever wider range of applications highlights the need to have a rapid-reaction focus for the dissemination of the various advancements. The series is aimed at providing such a focus.

Pan Stanford Series on Digital Signal Processing — Volume 2

Learning Approaches in Signal Processing

edited by

Wan-Chi Siu, *PhD, DIC, Life-FIEEE*

Lap-Pui Chau, *PhD, FIEEE*

Liang Wang, *PhD, SrMIEEE*

Tieniu Tan, *PhD, DIC, FCAS, FREng, FTWAS, FIEEE, FIAPR*

PAN STANFORD PUBLISHING

Published by

Pan Stanford Publishing Pte. Ltd.
Penthouse Level, Suntec Tower 3
8 Temasek Boulevard
Singapore 038988

Email: editorial@panstanford.com
Web: www.panstanford.com

British Library Cataloguing-in-Publication Data
A catalogue record for this book is available from the British Library.

Learning Approaches in Signal Processing

ISBN 978-981-4800-50-1 (Hardcover)
ISBN 978-0-429-06114-1 (eBook)

Contents

PART III

IMAGING TECHNOLOGIES

Part V

Motions in Videos, Pose Recognition, and Human Activity Analysis

Part VII

Discussion on AI for Healthcare

Preface

This is the second volume in the Pan Stanford Series on Digital Signal Processing (DSP), which is published in honor of Emeritus Professor Anthony George Constantinides, a pioneer, a mentor, and a world leader in DSP. The editors were either his PhD students from Imperial College London or his grand PhD students, whilst, interesting enough, one of the Publishers was also a student of Professor Constantinides.

This series was triggered by a gathering of Professor Constantinides and many of his ex-PhD students in the Department of Electrical and Electronic Engineering, Imperial College London. Since 1960, Professor Constantinides has provided a wide range of contributions in DSP, including the publication of one of the very first textbooks on DSP, entitled *Introduction to Digital Filtering*, in 1975. Digital filters (in the form of kernels in convolutional neural networks with deep learning structures) and transform kernels are two most fundamental structures for modern signal processing techniques, such as stochastic signal processing, statistical inference, optimization, and machine learning, with a large variety of applications, including big data analysis, Internet of things, pattern recognition and object identification, tracking, security, smart city, bio-informatics and healthcare. This second book in the series comes timely in response to the development of these topics. It enriches readers with complete knowledge and innovative ideas on these fast-advancing technologies.

This book is divided into seven sections: (I) Tutorial and Overview of Learning Approaches, (II) Filter Design and Multirate Signal Processing, (III) Imaging Technologies, (IV) Biometrics and Health Applications, (V) Motions in Videos, Pose Recognition and Human Activity Analysis, (VI) Language Processing, Cooperative

Network and Communications, and (VII) Discussion on AI for Healthcare.

There are five chapters in Section I. Chapter 1 gives a tutorial on random tree and random forests for beginners, and it is expected to form useful background materials for supporting other chapters, such as Chapter 8 on super-resolution imaging and Chapter 12 on human behavior capture. This chapter contains detailed explanation of the basic concepts, and there are plenty of examples. Chapter 2 introduces some sparsity-based dictionary learning methods and their applications to image processing, including de-noising, super-resolution, and in-painting. It compares PSNR and visual quality of de-noised images using different dictionary learning methods and discusses various challenges associated with dictionary-based signal processing. Chapter 3 presents a comprehensive survey on group theory- and topology-based persistent homology, including mathematical backgrounds and pattern recognition applications. Several perspectives and directions for research of applying persistent homology are also discussed. Chapter 4 describes some low-rank matrix estimation models, and numerical techniques that are the cornerstones of more sophisticated algorithms. Applications in signal processing and machine learning are discussed to demonstrate the effectiveness of the models. Chapter 5, the last chapter in this section, initially gives a simplified step-by-step tutorial introduction to face recognition making use of eigenface/PCA (principal component analysis), which is a classical approach for beginners. The chapter consequently describes briefly the linear discriminant analysis (LDA), other modern subspace learning approaches, and deep learning networks for face recognition.

Section II contains two chapters on digital filtering, which is a fundamental element of most digital signal processing systems. Chapter 6 describes a specific class of Kalman filtering estimation, namely ensemble Kalman filter (EnKF). It has the capability to use the sample of the state covariance to approximate covariance matrix. The results can be used for weather forecasting. Chapter 7 starts with some very fundamental concepts of linear time invariant (LTI) systems and points out several problems in using conventional filter realization techniques for the realization of multirate signal processing systems. Subsequently, this chapter introduces new ways and useful tools for students to make realizations of multirate systems with appropriate programming and debugging techniques.

Section III contains two chapters on imaging technology. Chapter 8 gives a comprehensive introduction on conventional machine learning techniques and then the recent learning and deep learning approaches for image super-resolution. The core idea of each technique has been introduced in details; particularly, the use of random forests and convolutional neural networks has been thoroughly discussed. Chapter 9 presents two learning-based classification methods, namely discriminative dictionary learning and convolutional neural network, for non-contact three-dimensional measurement.

Section IV consists of two chapters relating to bioinformatics. DNA contains the genetic composition, which is extremely useful for the identification of living organism. However, DNS is an extremely long chain. With this constraint, Chapter 10 addresses the way to identify protein-coding regions reliably, with statistical features and learning approaches. Food recognition and analysis are effective identification techniques in various applications such as dietary logging and management. However, there are not many accurate food recognition systems that can satisfy mobile applications. Chapter 11 presents a novel compact neural network architecture to fulfill the challenging requirement.

There are three chapters in Section V, covering motions in videos, pose recognition, and human activity analysis. Chapter 12 starts with a short review on the randomized decision tree (RDT) algorithm. It then reports recent research works on this topic. A variant of the RDT algorithm that can be trained at a much lower computational complexity is discussed. Two applications are presented in this chapter to show the efficiency of the technique. Among all the sensors to capture hand gesture, electromyography (EMG) is considered a suitable one since it captures muscles' electrical signals. Chapter 13 presents the use of the deep learning method for electromyography-based gesture recognition. The chapter starts with data acquisition. It is then followed by data preparation, and the details of state-of-the-art neural network architectures are also discussed. Intrinsic synchrosqueezing coherence (ISC) is proposed in Chapter 14 for the analysis of coupled non-linear and non-stationary multivariate signals. Useful applications of the ISC algorithm to the quantification of inter-channel dependence in respiratory and heart rate variability (HRV) frequencies are also discussed in this chapter.

Section VI is on language processing, cooperative network and communications. Multitask cooperative networks are discussed in Chapter 15. There are tremendous applications for the diffusion adaptation of cooperative networks, such as communications, social networks, and modeling physiological system. In the case of brain application, the cooperative networks can be very promising, including brain–computer interface and Parkinson's problems. Diffusion adaptation may be combined with other signal processing techniques for various applications. Chapter 16 describes the concept of kernel memory, which offers versatile representations for spoken language processing. It provides a new idea of the connectionist's representation for spoken word recognition. Neural representation of syntactical structures and some illustrative examples of processing the embedded sentences are discussed. Furthermore, a holistic model of the spoken language processing is also proposed based on the kernel memory concept in this chapter.

Finally, Chapter 17 in Section VII gives a non-mathematical discussion on some recent advances in AI with special attention to healthcare. The chapter ends the book with an interesting highlight on machine learning in medicine, which can help diagnosis and personalized treatment.

This book has been written by a group of scholars, who were either PhD graduates of Imperial College London under the mentorship of Professor Constantinides, his grand PhD graduates, and friends and colleagues from Imperial College London. Hence, it follows the great tradition of Imperial College London to offer the most innovative ideas and practical applications. It is interesting to note that Imperial College London ranks 8th globally in the QS

World Rankings in 2018–19 and ranks 2nd as the most innovative universities in Europe for the third year running, in Reuters' latest (2018) innovation ranking.

Wan-Chi Siu
Lap-Pui Chau
Liang Wang
Tieniu Tan

Part I

Tutorial and Overview of Learning Approaches

Chapter 1

Introduction to Random Tree and Random Forests for Fast Signal Processing and Object Classification

Wan-Chi Siu, Xue-Fei Yang, Li-Wen Wang, Jun-Jie Huang, and Zhi-Song Liu

Center for Multimedia Signal Processing,
Department of Electronic and Information Engineering,
The Hong Kong Polytechnic University, Hong Kong
enwcsiu@polyu.edu.hk

Decision Tree is a learning tool suitable for resolving classification and data mining problems. The classification can simply be done by asking a series of simple questions about the feature space. Each time an answer is received, a follow-up question is then asked until the conclusion about the class label can be achieved. Related to some randomly selected features, the series of questions and answers are organized in a tree structure. Its major advantage is the ability to make fast decision with reference to an appropriately trained tree structure. Using more than one tree forms a forest structure, and in many cases multi forests are used. This is not a chapter on a comprehensive survey of the subject area, but it gives an introduction to the basic concepts of random tree and random forests which form the materials supporting several other chapters

Learning Approaches in Signal Processing
Edited by Wan-Chi Siu, Lap-Pui Chau, Liang Wang, and Tieniu Tan

ISBN 978-981-4800-50-1 (Hardcover), 978-0-429-06114-1 (eBook)
www.panstanford.com

in this book. Realization details of three practical examples are provided.

1.1 Introduction

Decision tree was first proposed by Breiman et al. [1] in 1984 and is now commonly used in classification and data mining. The idea of decision tree is to solve a complex problem by testing some simple questions which are organized hierarchically, partitioning the problem to a more specific region of the decision space according to the answers to the questions and making a decision when the problem reaches a region where the response is confident enough. For example, as shown in Fig. 1.1, we can identify the name of unknown animal by asking a series of questions, such as whether it moves, has breathing, needs battery, etc. When the number of candidate animals fulfilling the conditions reduces to one, you could identify this unknown animal.

As shown in Fig. 1.2, a binary decision tree is in a tree-structure where a node with two child nodes is called a non-leaf node and a node without a child node is called a leaf node. A non-leaf node is responsible for classification by asking a simple question and partitioning the training or the testing data into its left or right child node according to the answer to the question. At each leaf node, a prediction model (a classifier or a regressor) is learned using the arrived training data. The testing data are mapped to its desired form with the prediction models.

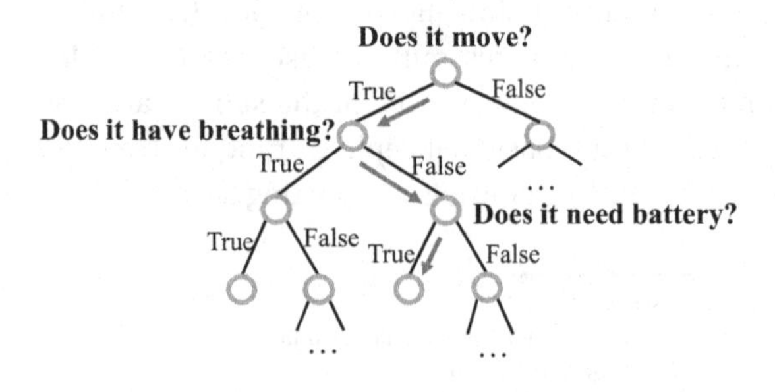

Figure 1.1 Decision tree.

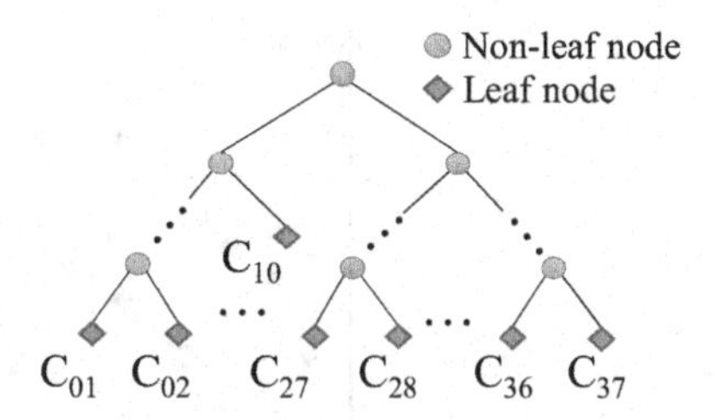

Figure 1.2 Node types.

1.1.1 Split Functions with Binary Tests

Each non-leaf node associates with a binary test θ which specifies the parameters of a testing condition to partition the training or the testing data. While the split function $h(\mathbf{x}, \theta)$ defines the testing condition and has binary outputs $\{0, 1\}$ (representing "true and false" or "left and right"). $\mathbf{x}$ is the input feature vector. The split function relates to the decision boundary type of the data separation including linear separation and nonlinear separation.

Linear Data Separation

There are two types of linear data separation split functions: axis-aligned hyperplane (see Fig. 1.3a) and general oriented hyperplane (see Fig. 1.3b). The general oriented hyperplane is a generalized version of axis-aligned hyperplane. The corresponding binary test θ of the general oriented hyperplane is defined as

$$\theta = (\phi, \mathbf{\Psi}, \tau), \tag{1.1}$$

where $\phi = \phi(\mathbf{x})$, which selects n (for $n > 1$) features from input feature vector $\mathbf{x}$, $\mathbf{\Psi} = (\psi_1, \psi_2, \ldots \psi_n)$ is a $n \times 1$ matrix and denotes the coefficients for a generic line in homogeneous coordinates, and τ is a threshold which makes the formulation more general, which also add one extra freedom in the feature domain.

The binary test parameters can describe a hyperplane which can separate the affine space into two half-spaces.

$$\phi(\mathbf{x}) \cdot \mathbf{\Psi} = \tau, \tag{1.2}$$

Or

$$\psi_1 x_1 + \psi_2 x_2 + \ldots + \psi_n x_n = \tau, \tag{1.3}$$

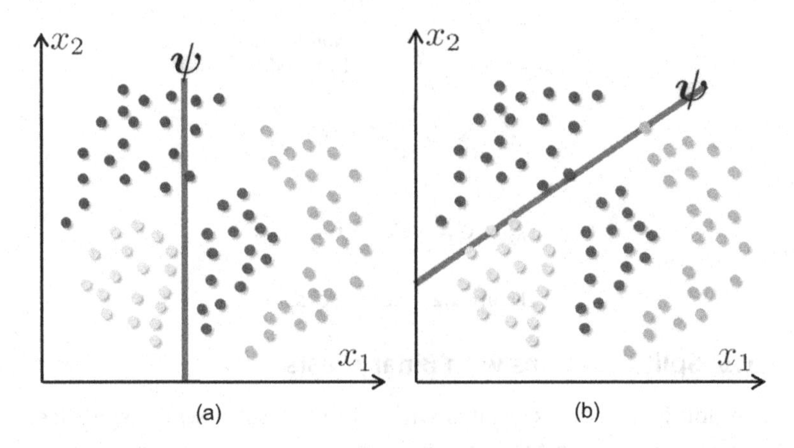

Figure 1.3 Split functions: (a) axis aligned separator, (b) general linear separator.

which is a plane dividing the n-dimensional plane into two halves, where $\mathbf{x} = (x_1, x_2, \ldots x_n)$.

If it is 2D, we have generally a line

$$ax_1 + bx_2 = \tau, \quad \text{for} \quad \psi_1 = a \quad \text{and} \quad \psi_2 = b. \tag{1.4}$$

If it is 1D, we have the axis-aligned hyperplane, generally

$$ax = \tau \text{ is a vertical line separator or} \tag{1.5}$$

$$by = \tau \text{ is a horizontal line separator.} \tag{1.6}$$

A split function is used to determine which half-space the input feature vector $\mathbf{x}$ belongs to and is defined as

$$h(\mathbf{x}, \theta) = \begin{cases} 0, & \text{if } \phi(\mathbf{x}) \cdot \mathbf{\Psi} < \tau \\ 1, & \text{otherwise.} \end{cases} \tag{1.7}$$

For axis-aligned hyperplane, $\phi(\mathbf{x})$ only select one feature ($n = 1$) from the input feature vector $\mathbf{x}$. The hyperplane is perpendicular or horizontal to the selected feature axis. The general oriented hyperplane has higher degree of freedom over the axis-aligned hyperplane. It is seen from Fig. 1.3 that the general oriented hyperplane can make the classification boundary adhere to the inherent structure in the data space, which can be better than the axis-aligned hyperplane. For the split function in Eq. 1.7, one can select more complex split functions such as SVM [2] and neural

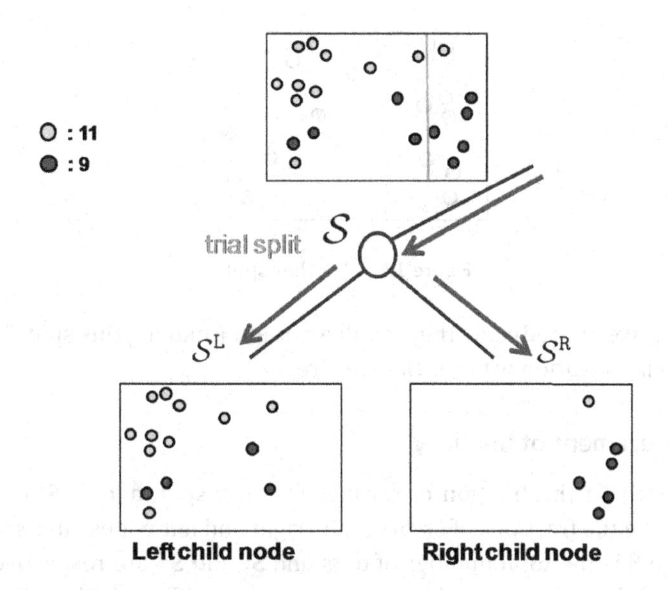

Figure 1.4 A simple split function: straight lines (example).

networks [3], etc., such that higher discrimination power could be obtained. However, the complexity increases and the advantage of the binary test is ruined. Let us use an example to illustrate the ideas up to now as clear as possible.

Example: Let us consider a dataset with 20 data points with a spatial distribution as shown in Fig. 1.4. Our objective is to learn a classifier that best separates these points.

Let us pass each of the 20 data points to the first node, and drop it either to the left or to the right child node according to a split function as shown in Fig. 1.4.

We can see that there are 11 yellow dots and 9 red dots entering into the root node. Let us start with a simple vertical separator as shown in Fig. 1.4. to separate the data. We count how many error dots we have before and after the split as shown below, and we also consider the majority to be the class type.

Total error reduction, R,

$$\begin{aligned} R &= \text{Error before the split} - \text{Total error after the split} \\ &= \text{Error before the split} - (\text{Error in the left} + \text{Error in the right}) \\ &= 9 - (4 \text{ red ones} + 1 \text{ yellow one}) = 4 \end{aligned}$$

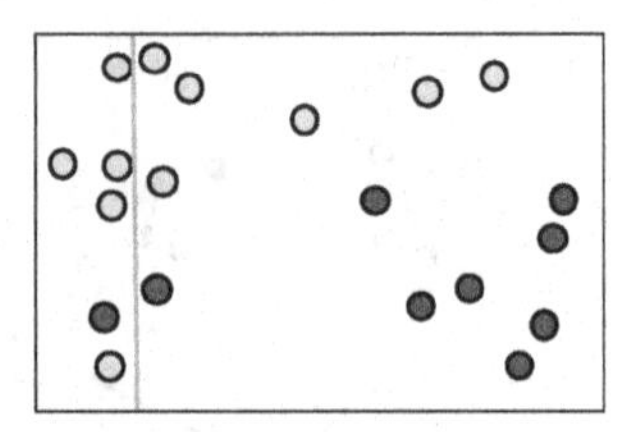

Figure 1.5 Another split.

Hence, we are reducing the overall error after making this split. The data classification is purer than before.

Measurement of Impurity

Let $E(\mathbf{S})$ be the fraction of error before the split, and $E(\mathbf{S}^R)$ and $E(\mathbf{S}^L)$ be the fractions of errors at the right and left nodes after split, where $\mathbf{S}$ is the total number of dots and $\mathbf{S}^L$ and $\mathbf{S}^R$ are respectively the number of dots at the Left Hand Side (LHS) and Right Hand Side (RHS) after split. Our objective is to reduce entropy in terms of information gain I (fractional error) after the split. If we consider the left hand side be the class for yellow dots and the right hand side be the class for red dots. There are two error dots in the left hand side and also two error dots in the right hand side. Hence, the total entropy I (fractional error reduced) becomes

$$I = E(\mathbf{S}) - [Err(\mathbf{S}^L) + Err(\mathbf{S}^R)] = E(\mathbf{S}) - \left[\frac{\mathbf{S}^L}{\mathbf{S}}E(\mathbf{S}^L) + \frac{\mathbf{S}^R}{\mathbf{S}}E(\mathbf{S}^R)\right] \tag{1.8}$$

Or

$$I = \frac{9}{20} - \left[\frac{14}{20} \times \frac{4}{14} + \frac{6}{20} \times \frac{1}{6}\right] = 0.2,$$

where $Err(\mathbf{S}^i)$ for $i = L$ or R is the fraction of error in the left or right child node.

There should be an infinite number of possible split functions. Let us see a few more examples. Figure 1.5 shows another possible split function. The total error reduction is

$$R = 9 - (1 + 6) = 2$$

This split is worse than the first one.

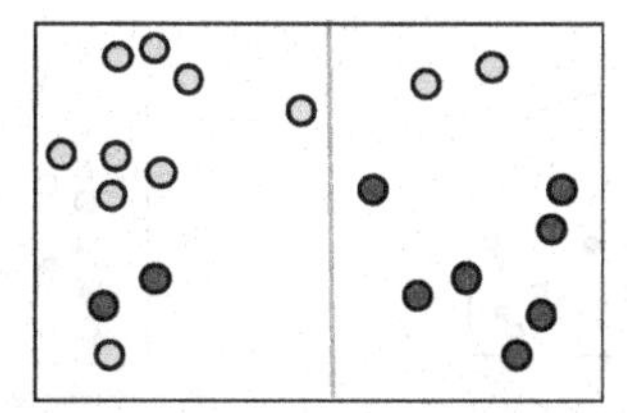

Figure 1.6 Another split.

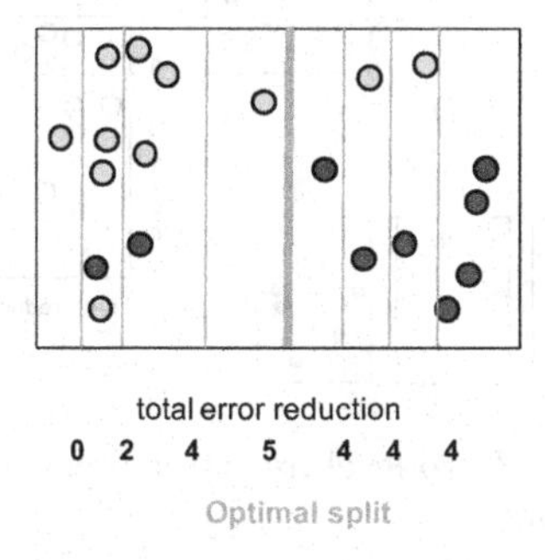

Figure 1.7 All possible splits.

Similarly, the split in Fig. 1.6 gives a total error reduction of

$$R = 9 - (2 + 2) = 5.$$

This is the best result we have so far (see Fig. 1.7). Hence, the basic process of finding the optimal split function for a node can be described as follows:

- Randomly generate one trial split and calculate its total error reduction.
- Randomly generate another trial split and calculate the error reduction.
- ...
- After generating an adequate number of trial splits, select the one that maximizes the total error reduction, which is the optimal split.
- Record this "Split function" according to this optimal split

The split can be continued to a lower level of nodes. However, we do not need to restrict ourselves to a single type of split function.

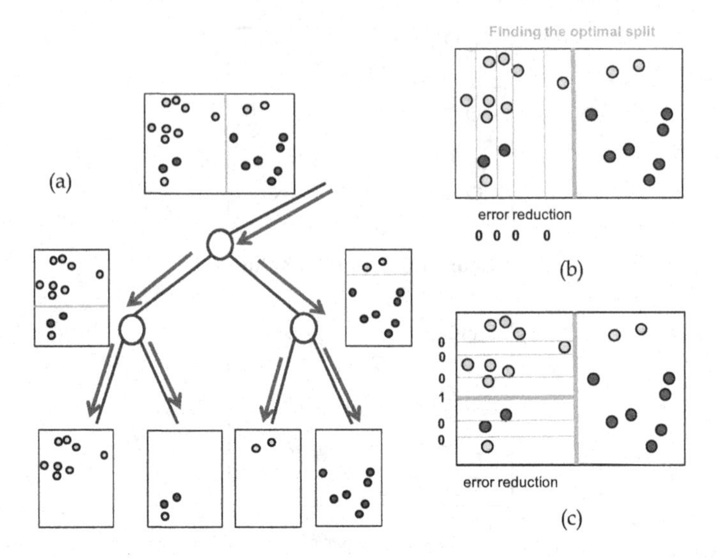

Figure 1.8 (a) Sample of two-level tree structure. (b) Vertical separators. (c) Horizontal Separators.

After the first level of grouping, let us start the second level of grouping (or split) as shown in Fig. 1.8a. All possible splits can be obtained using vertical separators as shown in Fig. 1.8b. However, we can find that none of the splits can get any entropy (error) reduction. This leads us to find other simple split functions, such as horizontal separators as shown in Fig. 1.8c. We can try all possible cases as indicated with horizontal lines. The reader may check that the thick horizontal blue line gives the optimal result, with the entropy reduction equal to

$$2 - (0 + 1) = 1.$$

Similarly, the right node can be split as shown, with an entropy reduction equal to $2 - (0 + 0) = 2$. Then the optimized splits can be obtained as shown in Fig. 1.9.

This is an over-simplified example for beginners, and it shows that

(i) learning (tests select the optimal separator) is required for finding the optimal split functions;

(ii) the first-level separator is a vertical line, $x = a$ (or $a = 5$) which implies that the split function can be written as if $x > 5 \rightarrow$ right node, otherwise left node;

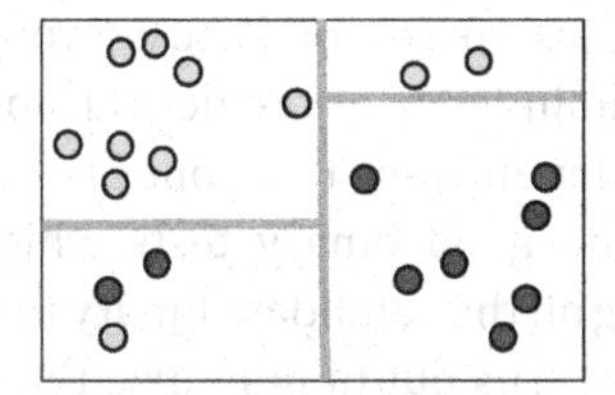

Figure 1.9 Optimal splits.

(iii) the feature is the location of a dot during training or during the recall (for classification). With this example in mind, let us introduce the random tree and random forests more systematically.

1.1.2 Definition and Structure of Random Trees

Training Sets

In many applications, like the regression problem, the training data is in the form of observed data **x** and ground-truth data **y** pairs (e.g. the LR-HR patch pairs in image up-sampling problem). Let us denote a pair of training data as $\{p_i = (x_i, y_i)\}$ and assume that there are l pairs of training data. The whole training dataset is denoted as $\mathbf{S_0} = \{p_i | i = 1, \ldots, l\}$. In supervised learning algorithms, the quantity and the quality of the training data could greatly affect the training results. Often learning algorithms themselves are very simple, but the complexity of a learning algorithm comes from the data. Thus, the training data should be carefully selected during preparation.

Energy Models

During training, a set of binary tests will be evaluated and the best one will be applied to split the training data and stored for testing. The goodness of each binary test is measured by energy models, including entropy and information gain.

Entropy

In information theory, entropy measures the average information (uncertainty) in the data. The decision tree gradually partitions the training data into a more specific region by the binary tests on the non-leaf nodes. This corresponds to entropy decrease for the training data. The training data at the deeper nodes in the decision tree have overall smaller entropy. Thus, one of objectives for decision tree training is to find good binary tests which can achieve the lowest entropy among all the candidate binary tests. The objective of decision tree training turns out to minimize the variance evaluated at the training data. Thus, the decision tree training tries to reduce the ambiguity among the training data and makes each training datum have a confident (i.e. with low variance) fitting value.

The discrete entropy of sample set $\mathbf{S}$ at a certain node can be written as Eq. 1.9.

$$H(\mathbf{S}) = -\sum_{c \in \mathbf{C}} p(c) \log(p(c)), \tag{1.9}$$

where $\mathbf{C}$ indicates all categories and $p(c)$ indicates the proportion (probability) of class c in $\mathbf{S}$.

Different from the discrete entropy, the entropy of a continuous distribution can be modeled as the fitting error of the training data.

$$H(\mathbf{S}) = \sum_{p_i \in \mathbf{S}} ||y - f(\mathbf{x}, \boldsymbol{\omega})||_2^2,$$

where $f(\mathbf{x}, \boldsymbol{\omega})$ is the fitting value, $\boldsymbol{\omega}$ is the set of model parameters.

This approach has similar reasoning as the discrete entropy approach. Because the fitting value $f(\mathbf{x}, \boldsymbol{\omega})$ is the average prediction value and the mean squared loss is related to variance of the estimated value. The difference is that using fitting error to evaluate entropy is more direct than discrete entropy and is without the constraint of statistical assumption.

Information Gain

Information gain $I(\mathbf{S}, \theta)$ which has been defined initially in Eq. 1.8 is used to measure the entropy (uncertainty) reduction achieved by partitioning the training data $\mathbf{S}$ into left and right child nodes $\mathbf{S}^L$ and $\mathbf{S}^R$ using a binary test θ.

$$\begin{aligned} \mathbf{S^L} &= \{\mathbf{P_k} | h(\mathbf{x}_k, \theta) = 0, \quad \mathbf{P_k} \in \mathbf{S}\} \\ \mathbf{S^R} &= \mathbf{S}/\mathbf{S^L} \end{aligned} \tag{1.10}$$

where θ is the binary test, h $(\mathbf{x}_k, \theta)$ is the split function, and k is the number of randomly selected binary tests, and $\mathbf{P_k}$ represents samples in **S**. The first equation says that if elements of $\mathbf{P_k}$ pass the binary test, the data will join the left child node, $\mathbf{S^L}$, and the second equation says the otherwise, which means to join right child node, $\mathbf{S^R}$. The information gain is

$$I(\mathbf{S}, \theta) = H(\mathbf{S}) - \sum_{n \in \{L,R\}} \frac{|\mathbf{S}^n|}{|\mathbf{S}|} H(\mathbf{S}^n), \tag{1.11}$$

where $|\mathbf{S}|$ is defined as the cardinality of training data **S**, and n is the number of child nodes. The information gain could be understood as the difference between the entropy at the parent node and the sum of the weighted entropies at the child nodes. If there are only two child nodes, we have

$$I(\mathbf{S}, \theta) = H(\mathbf{S}) - \left[\frac{|\mathbf{S}^L|}{|\mathbf{S}|} H(\mathbf{S}^L) + \frac{|\mathbf{S}^R|}{|\mathbf{S}|} H(\mathbf{S}^R)\right], \tag{1.12}$$

which is just Eq. 1.8, with a generalized condition.

Tree Training

During training, the root node is initialized by all the training data $\mathbf{S_0}$ and considered as an unprocessed non-leaf node. The aim of tree training is to construct a decision tree which could effectively classify training data into leaf nodes through hierarchical binary tests at the non-leaf nodes and make predictions with high generalization ability at leaf nodes. The three key elements of a decision tree to be learned during training are (1) optimal binary tests stored at the non-leaf nodes; (2) construction of the tree-structure; and (3) prediction model at the leaf nodes.

Optimal Binary Tests

For each unprocessed non-leaf node, a set of K candidate binary tests $\Theta = \{\theta_i | i = 1, \ldots, K\}$ will be generated to try to split the training data at this node. The corresponding information gain of each test is evaluated. The binary test which achieves the highest information gain will be selected as the optimal binary test θ^* at this non-leaf node.

$$\theta^* = \arg\max_{\theta_i \in \Theta} I(\mathbf{S}, \theta_i) \tag{1.13}$$

The number K determines the randomness during splitting the decision tree nodes and controls the training speed to construct the decision tree. A small K introduces higher randomness between decision trees and little tree correlation. A large K decreases the randomness and increases tree correlation.

In theory, K could be chosen as the number of all possible tests of the discrete variables, i.e. searching over the whole parameter space (just like our over-simplified example). The best binary test among all binary tests is selected at each non-leaf node and there is no randomness during training. For a single decision tree, maximal K could be beneficial for the quality of training result. However, for random forests (which is an ensemble of decision trees and will be discussed in the next Section) without bagging, choosing the maximal value of K will make every decision tree the same and highly correlated which diminishes the better generalization advantage of random forests. Besides, this strategy has very low efficiency, since the size of the parameter space could go extremely large when the dimensionality of the feature vector is large or the binary test parameters are very complex.

Randomized node optimization (RNO) is a more efficient approach which randomly generates a subset of all the possible binary tests. The extremely randomized trees [4] suggested that the default values of K should be $\sqrt{d}$ and d for classification problem and regression problem respectively, where d is the number of attributes of the observed feature vector. However, the suggested default values may not be the most suitable values for certain applications. One can choose an appropriate value of K through cross-validation adapting to a particular requirement.

Tree-Structure Construction

When the optimal binary test is found, two child nodes of this non-leaf node will be constructed and the training data at this non-leaf node will be exclusively partitioned into its child nodes according to the optimal binary test as in Eqs. 1.10 and 1.11.

There are three circumstances that a non-leaf node will not construct its child nodes and not partition its training data. (1) This non-leaf node reaches the pre-defined maximum tree depth, $D_{\max}$. (2) The information gain achieved by the optimal binary test is smaller than a threshold, I_T. (3) The number of training data at this non-leaf node is too few to learn a robust predication model. If any of these three conditions is met, this non-leaf node will be declared as a leaf node.

The maximum tree depth $D_{\max}$ restricts the decision tree size. The testing time of a decision tree is strongly related to $D_{\max}$ (or the average number of non-leaf nodes (binary tests) a testing datum will go through). The information gain threshold I_T ensures the quality of the decision tree. The negative information gain has no positive impact on the testing data prediction accuracy. Sometimes, the decision tree complexity (decision tree depth) is included into the information gain formulation which makes a tradeoff between the decision tree size and the information reduction. The minimum number of training data at a leaf node is closely connected with the testing prediction accuracy of the prediction model constructed by the training data at that node. When the number of training data is insufficient, the generalization ability would be severely declined. Under this consideration, a node will be declared as a leaf node when the training data are too few to split.

Leaf Prediction Models

Non-leaf node is responsible for classification, while leaf node is to make the prediction. The non-leaf node splits the data into its left or right child node according to the result of the split function with a binary test. When a testing datum reaches a leaf node, a prediction should be made with reference to the subset of training data at that node. It is assumed that the testing data reached have similar statistics of the training data reached. Thus, the desired result (value) of a testing sample is predictable.

For a classification problem, the predicted class label should be the class label that has the highest probability within the training data at that leaf node. For a regression problem, any regression model could be applied, such as linear regression model, Gaussian regression model, neural networks, etc.

Tree Testing

The testing algorithm of decision tree is very simple. The testing data $\mathbf{x}$ is passed to the learned decision tree. Starting from the root node, the testing data $\mathbf{x}$ is tested by the optimal binary tests on the non-leaf

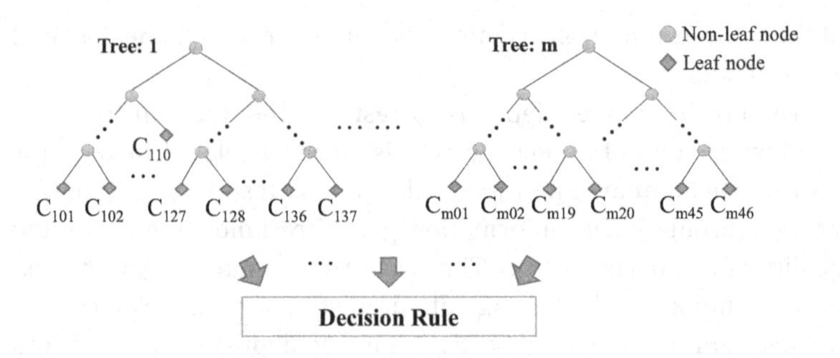

Figure 1.10 Random forest.

nodes and partitioned to the left or right child node until it reaches a leaf node. The prediction model (a classifier or a regressor) at the leaf node estimates the output value (a class label for a classification problem, or a model or continuous value for a regression problem) for the input testing data **x**.

1.1.3 Random Forests

The random forests approach [5] is an ensemble of decision trees which improves the generalization ability as shown in Fig. 1.10. The key idea is to combine prediction models from de-correlated decision trees to reduce prediction variance. The reason behind is that each prediction result by a single decision tree could have low bias and high variance. If all the decision trees are de-correlated, their average prediction results may achieve low bias and low variance.

Bagging

Breiman [5] proposed the bagged training which is a method to reduce overfitting and improve generalization capability. Each decision tree in the random forests is trained using a different randomly sampled subset of the whole training dataset. Bagging helps to prevent the decision trees overfitting to a particular training dataset and insert extra randomness into the decision trees which makes the decision trees de-correlated. Although bagging could accelerate the training speed (as only a subset of the training data is applied for training), the training efficiency of individual tree could decrease because not all training data are utilized for all decision trees.

Combining Decision Trees

During testing, a testing data $\mathbf{x}$ is passed to all the decision trees in the random forests. The testing data $\mathbf{x}$ will retrieve a prediction model (or result) from the reached leaf node for each decision tree. The prediction retrieving procedure could be done on parallel which makes random forests algorithm be able to achieve high runtime efficiency, because decision trees are independent from each other. The prediction models from all the decision trees in the random forests are usually averaged [5] to form a more robust model, or the majority decision is accepted for classification.

1.2 Object Classification

Handwritten digit recognition is a task to let the machine automatically distinguish digit characters 0 to 9 written by hands. It is widely used, say for example, in post offices in recognizing ZIP codes in order to alleviate the manual burden and to speed up the distribution of mails. With the advances in machine learning techniques, a well-learned model could achieve equal or even higher accuracy than human beings.

In this section, we introduce how random forests algorithm can resolve the handwritten digit recognition problem. We use images with 8×8 pixels to represent each digit, and the intensity of each pixel has been quantized into 16 levels for better illustration.

Before processing by the random forests, each image is represented into a row vector. As shown in Fig. 1.11, the left-top figure is the input 8×8 image that the intensity of each position value ranges from 0 to 15. Then, concatenating all the pixel intensities is made to form a 64-d row vector. As indicated in red rectangles in Fig. 1.11, pixel at position (2,4) of the 2-d image is indexed as the 20th in the row vector with a value 0.

1.2.1 Training Stage

Let us focus on the training process of the decision tree. First, we design a binary test that compares two pixel intensities of the input image at each non-leaf node. Second, we introduce the optimization

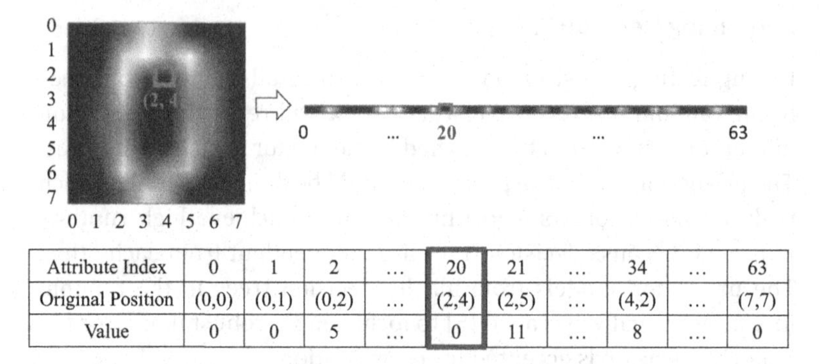

Attribute Index	0	1	2	...	20	21	...	34	...	63
Original Position	(0,0)	(0,1)	(0,2)	...	(2,4)	(2,5)	...	(4,2)	...	(7,7)
Value	0	0	5	...	0	11	...	8	...	0

Figure 1.11 Example of digit image and its corresponding feature vector.

process that select the most suitable test from a set of candidate tests. Then, the construction process is presented in details. Finally, we give the prediction process of leaf nodes and some experimental results.

Design of Binary Test

In Fig. 1.12b, there is a set of binary tests ("*BT-p*," where $p = 1, 2, \ldots m$) and m is the number of tests in the tree. Let $\mathbf{I} = \{I_1, I_2, \ldots, I_k\}$ denote the row vector of the image under question, and $I_1, I_2, \ldots, I_k$ denote the feature attributes of vector $\mathbf{I}$, which are actually the intensity value of the image, and k is the image size. The binary test we use in this example is a simple equation with two dimensions. Let us recall the linear data separation in Eq. 1.7. In this example, $\phi(\mathbf{I}) = (I_i, I_j)$ denotes two selected feature attributes and $\boldsymbol{\Psi} = (1, -1)^T$ denotes the weights assigned to the selected features. Hence, the binary test is $\phi(\mathbf{I}) \cdot \boldsymbol{\Psi} = (I_i, I_j) \cdot (1, -1)^T$ which implies a split function $I_i - I_j < \tau$. Then the function can be written as Eq. 1.14. Indices i and j are two randomly generated index numbers during training stage. I_i and I_j are the attribute values at the i-th and j-th entry of $\mathbf{I}$. τ denotes a threshold value. *Left* represents partitioning the incoming data into the left child node, while *Right* means to the right child nodes. Then this binary test can be interpreted as comparing the feature values at two different positions with respect to their respective vectors. If the first one is smaller (with a certain

(a) A sample feature vector (the original image is the handwriting digit "0")										
Attribute Index	0	1	2	…	20	21	…	34	…	63
Original Position	(0,0)	(0,1)	(0,2)	…	(2,4)	(2,5)	…	(4,2)	…	(7,7)
Value	0	0	5	…	0	11	…	8	…	0

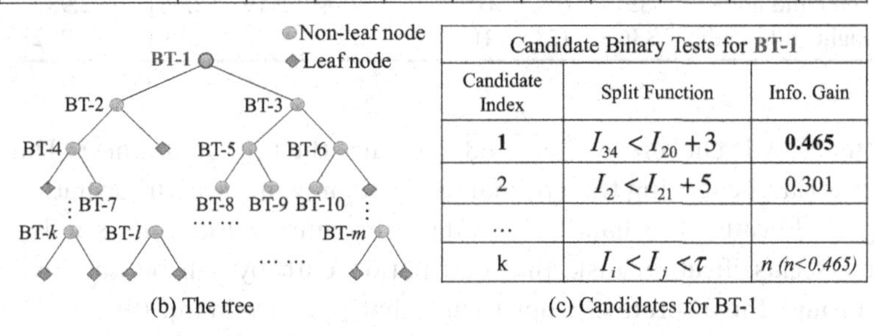

Candidate Binary Tests for BT-1		
Candidate Index	Split Function	Info. Gain
1	$I_{34} < I_{20} + 3$	**0.465**
2	$I_2 < I_{21} + 5$	0.301
…		
k	$I_i < I_j < \tau$	n (n<0.465)

(c) Candidates for BT-1

Figure 1.12 Design of binary test.

threshold), the sample goes to the left, and vice versa.

$$h(i, j, \tau) = \begin{cases} \text{Left} & \textit{if } I_i < I_j + \tau \\ \text{Right} & \text{otherwise} \end{cases} \quad (1.14)$$

Let us make use of Fig. 1.12 to illustrate how this binary test would work in separating data. As we have mentioned that i, j, and τ are the randomly selected parameters. For binary test "*BT-1*" in Fig. 1.12b, we randomly choose $i = 34$, $j = 20$, and $\tau = 3$. In other words, the binary test compares the 34th and 20th attribute values of the row vector of the input image. The feature values are shown in Fig. 1.12a, and we can obtain $I_{34} = 8$ and $I_{20} = 0$. Obviously, for this sample, the binary test "$8 < 0 + 3$" is not satisfied. Then this sample will enter into the right child node. Assume another draw has the attributes $I_{34} = 5$, $I_{20} = 7$ and $\tau = 3$. As the binary test "$5 < 7 + 3$" is satisfied, the sample will enter into the left child node. Similarly, all the samples in the training set **S** will join two groups separately after the processing of the binary test.

Optimal Binary Tests

As mentioned before, we need to select the optimal setting from K candidate binary tests (i, j, and τ in Eq. 1.14). Because the size of image is 8×8 (length of the feature vector is 64), there are a total of 64×63 combinations for the two-attribute (I_i and I_j) binary test.

Table 1.1 Distribution of samples of candidate binary test 1

Category	0	1	2	...	8	9	Total	Entropy
Root node	375	389	380		380	382	3,822	3.322
Left child node	39	374	370		254	319	2,161	2.937
Right child node	336	15	10		126	63	1,661	2.752

Hence, K is chosen as $\sqrt{64 \times 63}$ to ensure the full randomness. The criteria of choosing the optimal test is to maximize the information gain. Because the handwritten-digit-recognition example is a 10-class classification task, the information entropy can be obtained through the discrete entropy theory that is shown in Eq. 1.9.

First, we need to calculate the information entropies of all three nodes (root node and two child nodes). Table 1.1 shows the distribution of samples in the example after the processing of candidate binary test 1. The training set **S** at parent node (also root node in this case) contains 3,822 samples. After the separation of the binary test, 2,161 samples enter into the left child node, while the other 1,661 samples enter the right child node. For the root node, **S** is the set of the training samples at the node, $\mathbf{C} = \{0,1,2,\ldots,9\}$ denote the ten categories of our handwritten digit recognition task. From Table 1.1 we know that the number of samples for category "0" is 375, therefore $p_{(c=0)} = 375/3{,}822 = 0.098$. Then, based on the value of $p_{(c=0)}$, $-p_{(c=0)}\log(p_{(c=0)}) = -0.098{*}\log(0.098) = 0.032$. Similarly, we can obtain $-p_{(c=1)}\log(p_{(c=1)})$, $-p_{(c=2)}\log(p_{(c=2)})$, ..., $-p_{(c=9)}\log(p_{(c=9)})$. Next, by summarizing $-p_{(c)}\log(p_{(c)})$ for $c =$ 0,1, ..., 9, we can obtain the entropy $H(\mathbf{S})$ for the root note that is 3.322. For left and right child nodes, we can calculate the entropies in the same way that are 2.937 and 2.752 separately.

$$H(\mathbf{S}) = -\sum_{c \in C} p(c)\log(p(c)) = -\left[\frac{375}{3,822}\log\left(\frac{375}{3,822}\right) + \frac{389}{3,822}\log\left(\frac{389}{3,822}\right) + \ldots + \frac{382}{3,822}\log\left(\frac{382}{3,822}\right)\right] = 3.322$$

Second, we can calculate the information gain for this candidate binary test 1 (parameter is θ) through Eq. 1.12. $H(\mathbf{S})$ is the information entropy before the separation, which is actually the entropy at parent node (also root node in this case), therefore $H(\mathbf{S}) = 3.322$. Similarly, $H(\mathbf{S}^L)$ and $H(\mathbf{S}^R)$ denote the entropies at left and right child node, which are $H(\mathbf{S}^L) = 2.937$ and $H(\mathbf{S}^R) = 1.753$. $|\mathbf{S}|$, $|\mathbf{S}^L|$ and $|\mathbf{S}^R|$ are the cardinality of the training data in the parent, left-child and right-child nodes, which are 3822, 2161 and 1161 according to Table 1.1. Then, the information gain can be obtained by Eq. 1.12.

Up to now, we have obtained the information gain of the first candidate for binary test. Actually, there are 64 candidates for "*BT-1*" in Fig. 1.12b. Figure 1.12c shows the information gain of the candidates, and the maximum information gain (0.465) is obtained for the candidate $I_{34} < I_{20} + 3$. Therefore, this binary test is the optimized, and will be used at testing stage.

$$I(\mathbf{S}, \theta) = H(\mathbf{S}) - \left[\frac{|\mathbf{S}^L|}{|\mathbf{S}|} H(\mathbf{S}^L) + \frac{|\mathbf{S}^R|}{|\mathbf{S}|} H(\mathbf{S}^R)\right]$$

$$= 3.32 - \left[\frac{2,161}{3,822} \times 2.93 + \frac{1,661}{3,822} \times 2.75\right] = 0.465$$

Construction of the Tree Structure

After the optimization process at root node, we have designed the binary test "*BT-1*" as shown in Fig. 1.12b. The next step is to work on the optimization for the two child nodes "*BT-2*" and "*BT-3*."

Because we have obtained the optimized binary test of the root node, the training samples can be split into the left and right child nodes based on the test, "*BT-1*." Figure 1.13 shows the sample distribution of the root node and its two child nodes. The 3,822 samples are divided into two groups that 2,161 samples entered into the left child node, and the other 1,661 samples went to the right child node. Each color of the pie charts is a category. We can see that the separation at root node is actually a weak classification among all ten classes. Classes 1, 2, 3, 8, 9 occupy more percentage in the left child node, and classes 0, 4, 5, 6, 7 occupy more in the right one. Furthermore, if you view the combination of classes 1, 2, 3, 8, 9 as a big class, i.e. Big Class $\mathbf{A} = \{1, 2, 3, 8, 9\}$, and Big Class $\mathbf{B} = \{0, 4, 5, 6, 7\}$, then this is exactly the same as a two-class classification.

For the left child node with 2,161 samples, further classification is needed by adding another optimized binary test at this node,

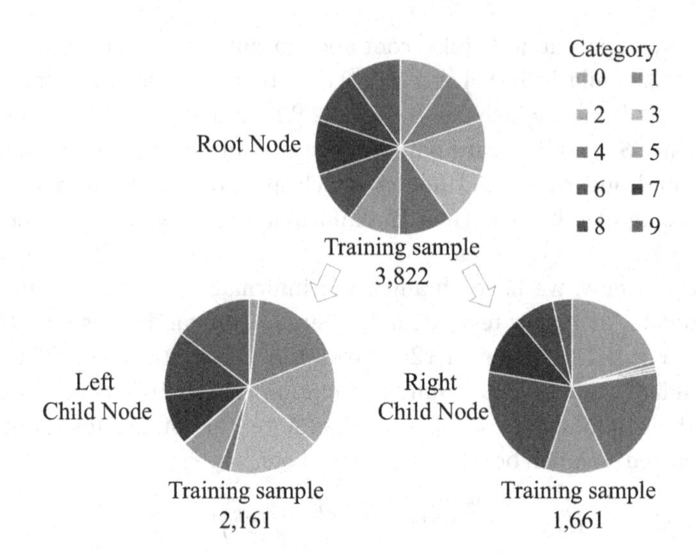

Figure 1.13 Example of data distribution in parent and child nodes.

which means the above optimization process will be used again to find a suitable test for the node. The process is the same as above. First, it randomly designs 64 candidate binary tests for "*BT-2*," and the information gains of these candidates are calculated at the same time. Second, selecting the optimized binary test for "*BT-2*" by finding the maximum information gain. "*BT-3*" should also be conducted the same process as "*BT-2*."

Next, the 1,161 samples of right child node will conduct "*BT-3*" that splits these samples into two child nodes. By this growing strategy, the binary tests "*BT-1*," "*BT-2*," ..., "BT-m" can be designed automatically. And the growing process will be terminated as we mentioned in the first section (e.g. it reaches the maximum depth, or the number of training sample is too little, etc.).

Prediction Model

During the training process, all training samples finally enter into leaf nodes (the node without child node). For each leaf node, it contains a subset which is formed by a set of data with similar attributes distribution. According to Bayes' Theorem, the class distribution of sample data fallen into this leaf reflects the class it

Table 1.2 Distribution of training samples at leaf node 1

Category	0	1	2	3	4	5	6	7	8	9	Total
Number	0	0	0	0	4	1	0	39	0	7	51
Probability	0	0	0	0	0.078	0.02	0	0.765	0	0.137	1

predicts. The class that has the highest posterior probability is the final predicted class, and that probability is the confidence score of this prediction.

Table 1.2 shows a leaf node (leaf node 1 in Fig. 1.14) which contains 51 training samples after training, where 39 of them are the "7" that the proportion is 0.765 (39/51). Note that in this leaf node seven samples are "9," four samples are "8" and one sample is "5." During testing stage, if a sample enters into this leaf node, it will be given the predicted as "7" with probability (let us also refer it loosely as confidence) 0.765.

1.2.2 Testing Stage

After the training process, we obtained a randomly trained decision tree that contains 66 leaf nodes as shown in Fig. 1.14. Each leaf node has a dominated category. For example, the dominated class of leaf node 1 is "7" with confidence 0.76, while for leaf node 27, the dominated class is "0" with confidence 1.0. Although leaf node 1 and 2 have the same dominated class, the higher confidence with 0.94 in leaf node 2 suggests a more reliable prediction at the testing stage.

Let us recall the example in Fig. 1.14, let a sample enter into the root node; hence, the binary test "*BT-1*" is conducted. The sample satisfies the binary test, and can then be sent to the left child node and binary test "*BT-2*" can be conducted subsequently. The red arrows indicate the decision path that the sample finally entered into leaf node 1 after conducting a set of binary tests ("*BT-1*" → "*BT-2*" → "*BT-4*" → "*BT-l*" → "*leaf node1*"). Finally, the leaf node made the prediction that belongs to category "7" with confidence 0.76.

Similarly, we can conduct some tests on handwritten digit recognition dataset. The training dataset contains 3,823 samples and the testing dataset contains 1,797 samples. The distribution of the datasets is shown at Table 1.3.

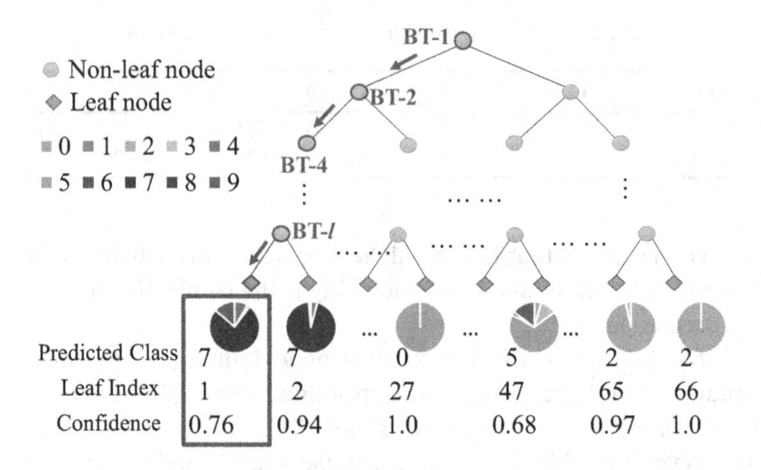

Figure 1.14 Distribution of leaf nodes.

As we mentioned, each non-leaf node is a weak classifier that is based on a split test. The depth of the tree indicates the number of split tests before a testing sample enters the leaf node. We investigated the performances of different depths of trees. The result is shown in Fig. 1.15a. The horizontal axis is the depth of the tree, and the depth of trees ranges from one to ten. The vertical axis is the average accuracy of a ten-tree decision forest. When the depth is small, the classification performance was bad. It is because the number of split tests at non-leaf node is insufficient for the classification. With the increase of depth, the classification accuracy became higher, which suggests the importance of sufficient number of split tests before making prediction.

Random forests are formed by a set of randomly trained decision trees. Because each tree is trained randomly, the prediction of different trees is independent with each other. Each decision tree can be regarded as a weak classifier, the combination of these trees can obtain a better generalization ability. Figure 1.15b shows the

Table 1.3 Distribution of handwritten digit recognition dataset

Category	0	1	2	3	4	5	6	7	8	9	Total
Training	375	389	380	389	387	376	377	387	380	382	3,823
Testing	178	182	177	183	181	182	181	179	174	180	1,797

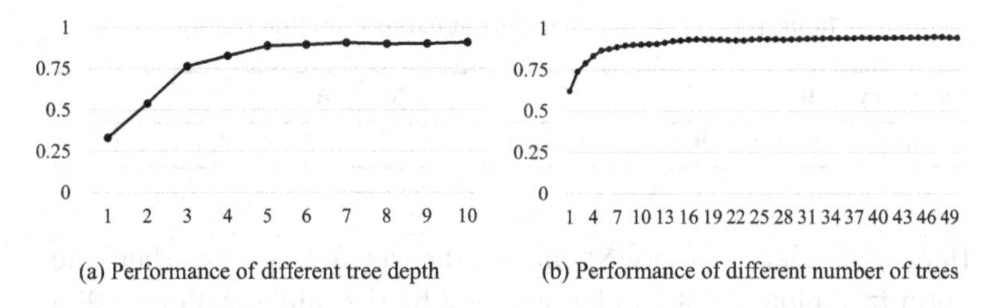

(a) Performance of different tree depth (b) Performance of different number of trees

Figure 1.15 Performance of different settings.

performance of different sizes of random forests. The horizontal axis is the number of trees, and the vertical axis is the classification accuracy. The maximum depth of these trees is set to ten. For random tree classifier (only one tree), the accuracy was 0.619. When using more trees, the classification accuracy increased significantly, and when the tree's number reaches ten (accuracy was 0.901). Then, the trend of the growth becomes saturated and has a score of 0.942 when tree number is 50.

Table 1.4 shows the accuracy of different categories. The random forests classifier consists of ten randomly trained decision trees that the depth of each tree is ten. Because the samples of our dataset are low resolution images (8×8, 16-level), it is difficult to recognize even for our human. However, the random forests algorithm has achieved over 0.8 accuracy for most categories, which suggests the outstanding classification ability.

We may consider classification and regression tree (CART) structure proposed by Breiman et al. [1] is the original decision tree algorithm. CART approach iterates all the possible splits of each attribute at non-leaf node, and then select the binary test with Gini index, which is the summation of purities of two child nodes as shown in Eq. 1.15. Similar to entropy in the information theory in Eq. 1.9, Eq. 1.15 defines a measurement of purity of the data,

$$Impurity(\mathbf{S}) = \sum_{c \in \mathbf{C}} \sum_{c \neq c'} p(c)p(c'), \tag{1.15}$$

where $p(c)$ denotes the proportion of class c in set $\mathbf{S}$. Class c' represents classes that are different from class c.

Therefore, $Impurity(\mathbf{S})$ actually is the probability acquiring two different classes for randomly selecting two samples from set $\mathbf{S}$.

Table 1.4 Accuracy of different categories (ten trees)

Category	0	1	2	3	4	5	6	7	8	9
Accuracy	0.994	0.808	0.915	0.951	0.917	0.973	0.95	0.872	0.782	0.844

Hence, smaller *Impurity*(**S**) means the set **S** is purer. Then the optimized binary test can be obtained by the minimization of Gini index.

$$Gini\_index(\mathbf{S}, \theta) = \left[\frac{|\mathbf{S}^L|}{|\mathbf{S}|} Impurity(\mathbf{S}^L) + \frac{|\mathbf{S}^R|}{|\mathbf{S}|} Impurity(\mathbf{S}^R)\right] \quad (1.16)$$

In our example, the candidate tests pool is formed by $I_i < \tau$ (i = 0, 1, 2, ..., 63; τ = *0, 1, 2, ..., 15*). Then the candidate test with the minimum Gini index is then selected. Obviously, the binary tests in random-tree approach contain more randomness than the original decision tree approach. More randomness can help to reduce the overfitting risk. We implemented the original CART tree for this handwritten-digit-recognition task by OpenCV 3.0 package. The experimental result shows our random forests approach described above achieves better performance (accuracy is 0.942) compared with the CART tree method (accuracy is 0.851).

1.3 Random Tree for Shadow Detection

Vehicle detection is a core function of any driving assistant system. The detection process always needs high computing cost to face various challenges that are caused by different environmental conditions. However, the driving assistant system needs to provide real-time safety guarantee, which means the detection process should be as fast as possible.

Let us briefly mention a promising structure for vehicle detection systems [6]. The structure can be roughly described as a two-stage detection process: the hypothesis generation stage and the hypothesis verification stage. The hypothesis generation stage searches for the whole frame, trying to locate all candidate regions where a vehicle might exist, and then generate a proper region of interest for further exploration. Fast detection is required at this

(a) Position of bumper area

(b) Example of shadow patches under various illumination conditions

Figure 1.16 Shadow patches.

stage, so many of the hypothesis generation methods exploit simple features that describe local properties of the vehicle.

Shadow underneath the vehicle is one of the possible features to be used due to its simplicity. As shown in Fig. 1.16a, there are two observations of the property of a vehicle: (1) The bumper area is right inside two road lanes. (2) The bumper area has very sharp contrast that is distinguishable even in very difficult situations. Based on these properties, we design a shadow-patch detection process that tries to use a small rectangular area right inside the road lanes to estimate the existence of vehicles. It is a pre-process at an early detection stage, acting as a fast detection of any possible vehicle locations, and then followed by a more accurate vehicle detection module.

Shadow-patch detection reduces the computation cost by generating a set of hypothesis areas which allows a fast and simple test to quickly check whether there is a vehicle. It is actually a binary classification between bumper area and non-bumper area. As we mentioned, the shadow-patch detection process is not the

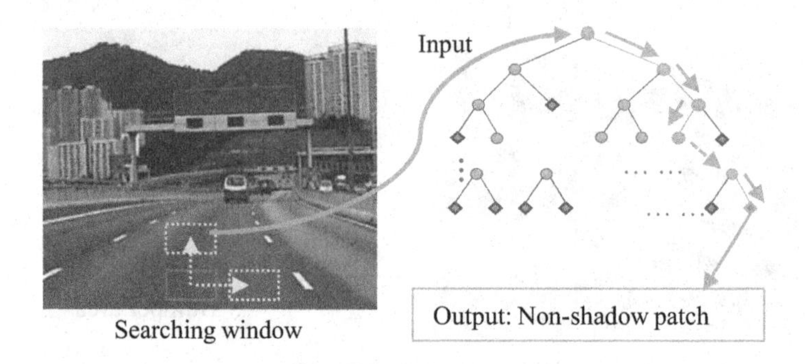

Figure 1.17 Searching for shadow area.

final process in vehicle detection, so it is very important that the classification algorithm could perform fast, that is why we would like to choose a random trees based method.

Let us focus on utilizing the random tree algorithm to search for a shadow area from an input image. As shown in Fig. 1.17, first, a 25×10 searching window is used to generate a set of patches from the input image as shown. The feature vector we used in this example is also the concatenated intensity values of the image patch. Similar to section 1.2 on Object Classification, each patch is preprocessed to a row vector (we use grey channel only, the intensity ranges from 0 to 255). Therefore, each patch is described by a 250-d feature vector. Second, the feature vector is sent into the random tree classifier to check whether the image patch belongs to a shadow or non-shadow area. Obviously, the classification between shadow and non-shadow of the image patches is the key part of the whole detection process. The training details of the random tree classifier are shown below.

1.3.1 Training Stage

The first step to train a random tree is to build a training dataset for our shadow detection task. Then the training process in this example is similar as the previous example. The binary test at each non-leaf node compares two pixel intensities and the optimization makes use of information gain. Details are as follows:

Pre-Training Process

We need to prepare some training data for the random tree. First, we annotated the bumper shadow from 263 images. Second, using the 25×10 searching window at the annotated shadow area to generate a set of patches which are labeled as shadow patch (+ve sample). We also randomly generated some patches from background area, that are labeled as non-shadow patches (−ve sample). The ratio between +ve and -ve patches is about 1:1.5 to avoid sample imbalance problem of the training process.

Design of Binary Test

Similarly, we adapt the binary test that is presented in Eq. 1.14. In this example, $\mathbf{I}$ is the 250-d feature vector of each input patch (size: 25×10). The binary test randomly selects two feature attributes (I_i and I_j) and a threshold τ. The randomly selected feature entries (i and j) are actually a pixel pair of the image patch (as shown in Fig. 1.18). Illumination often influences the intensity of the image. Strong light leads to brighter images, while dim light makes the image darker. The size of shadow patches area is small with only 25×10 pixels, which means the illumination at this small area is evenly distributed. The split function can be written as $I_i - I_j < \tau$, where $I_i - I_j$ removes the changes caused by different illumination conditions. Hence, the split test is robust to different illumination conditions.

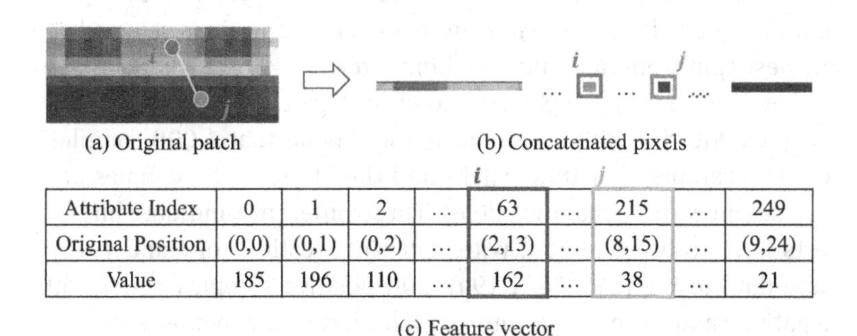

Attribute Index	0	1	2	...	63	...	215	...	249
Original Position	(0,0)	(0,1)	(0,2)	...	(2,13)	...	(8,15)	...	(9,24)
Value	185	196	110	...	162	...	38	...	21

Figure 1.18 A pixel pair of the shadow patch.

The binary test does not necessarily need to be the differences of pixel pairs. You may choose other features such as color difference or HOG, or even single pixel intensity. However, we found that using intensity difference is very useful and it performs well for this task. The approach is simple and it characterizes the shadow pattern well since the shadow and the bright areas have obvious intensity differences. Suppose we have a tree of n depth levels, theoretically the decision tree utilizes $2n$ possible pattern combinations for the prediction of each input sample. This is more than enough to represent the various types of shadow patches if n is deep enough. (Please note that due to certain termination criteria, some paths might terminate early before reaching the n-th depth, so empirically, the number of tested features is equal or smaller than $2n$).

Optimal Binary Tests

For the image patch with 250-d feature vector, the vector of it contains 250 attribute values. It means the combination of index i and j are $250 \times (250–1)$. In other word, there are $250 \times (250–1)$ split functions that can be chose at each non leaf node. Essentially, the learning process is to find the optimal binary test for each non-leaf node from the $250 \times (250–1)$ candidate tests. To ensure enough randomness of the decision tree, only $\sqrt{250 \times (250 - 1)}$ candidate tests are evaluated for finding the optimal one.

Similar as Section 1.2, we use the information gain to choose the suitable test at each node. As we mentioned before, a large information gain indicates the data becomes purer after the split, and thus the split with the maximum information gain is selected as the best split from the candidate binary tests.

Let us give an example. As shown in Fig. 1.19, there are 3,000 samples entering into a certain non-leaf node that 1,000 samples are +ve samples (shadow patch) and the other 2,000 samples are −ve samples (non-shadow patch). Three different candidate binary tests have to be generated with different locations of i and j. For candidate test 1 (see Fig. 1.19a), 300 positive samples and 1,200 negative samples entered into the left child node, while the other 700 positive and 800 negative samples entered into the right child node. Similarly, we use the discrete entropy as shown in Eq. 1.9.

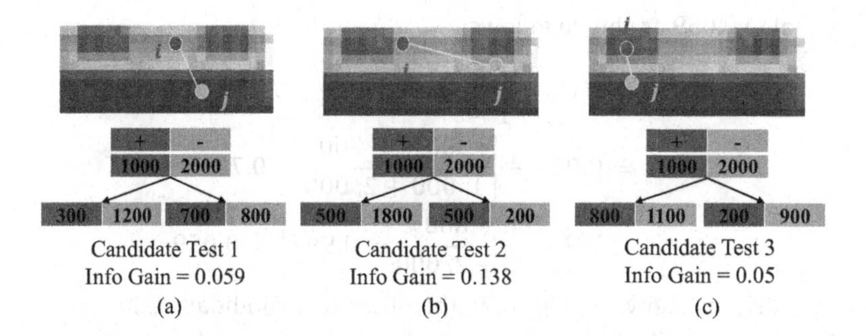

Figure 1.19 Different split candidates.

The training set **S** contains two classes ("+ve" and "−ve") in this example, therefore $C = \{\text{"+ve," "−ve"}\}$. The information entropies of the parent $H(\mathbf{S})$, left-child $H(\mathbf{S}^L)$ and right-child $H(\mathbf{S}^R)$ nodes can be obtained as follows:

$$\begin{aligned} H(\mathbf{S}) &= -\sum_{c \in \mathbf{C}} p(c)\log(p(c)) \\ &= -\left[\frac{1,000}{1,000+2,000}\log\left(\frac{1,000}{1,000+2,000}\right) \right. \\ &\quad \left. + \frac{2,000}{1,000+2,000}\log\left(\frac{2,000}{1,000+2,000}\right)\right] = 0.918 \end{aligned}$$

$$\begin{aligned} H(\mathbf{S}^L) &= -\sum_{c \in \mathbf{C}} p(c)\log(p(c)) \\ &= -\left[\frac{300}{300+1,200}\log\left(\frac{300}{300+1,200}\right) \right. \\ &\quad \left. + \frac{1,200}{300+1,200}\log\left(\frac{1,200}{300+1,200}\right)\right] = 0.722 \end{aligned}$$

$$\begin{aligned} H(\mathbf{S}^R) &= -\sum_{c \in \mathbf{C}} p(c)\log(p(c)) \\ &= -\left[\frac{200}{200+900}\log\left(\frac{200}{200+900}\right) \right. \\ &\quad \left. + \frac{900}{200+900}\log\left(\frac{900}{200+900}\right)\right] = 0.997 \end{aligned}$$

Also, based on the information gain theory as shown in Eq. 1.12, the information gain of the candidate test 1 can be calculated, which is equal to 0.059 as shown follows:

$$I(\mathbf{S},\theta) = H(\mathbf{S}) - \left[\frac{|\mathbf{S}^L|}{|\mathbf{S}|}H(S^L) + \frac{|\mathbf{S}^R|}{|\mathbf{S}|}H(\mathbf{S}^R)\right]$$
$$= 0.918 - \left[\frac{300 + 1,200}{1,000 + 2,000} \times 0.722 + \frac{700 + 800}{1,000 + 2,000} \times 0.997\right] = 0.059$$

Similarly, as shown in Fig. 1.19, the other two candidate tests lead to gains of 0.138 and 0.05 separately. Obviously, candidate split 2 showed the largest information gain which means the data can be purer after using this split function. Then, for this parent node, we use the candidate test 2 from the 3-candidate pool (actually there are 250 candidate tests in practice) as the final binary test. Similarly, we can work on the optimization for "*BT-1*," "*BT-2*," ..., "*BT-m*" as shown in Fig. 1.13, and the tree stops growing when meeting the termination conditions.

Prediction Model

A leaf node is the end of the tree structure and cannot be further split into child nodes. Each leaf node stores a prediction model determined by the data arriving at that node. The advantage of decision tree classifier is its fast speed, since only several binary tests are computed during the testing stage. Each binary split can be viewed as a weak classifier that tries to classify the input sample as shadow patch or non-shadow patch. However, the prediction is made purely based on the posterior probability of the training data at leaf node, and the classifications at non-leaf nodes along the path have not been explicitly used. In this section, we will introduce a novel prediction process that takes into account the classification of each split along the path.

Weak Classifier: Figure 1.20 shows an example of a binary split, where in the parent node 40% of training samples are negative and 60% of training data are positive. The split then separates these training data into left and right child nodes. The data distribution in the left child node becomes 20%:80%, where the data distribution is 90%:10% in the right child node. We then can conclude that more

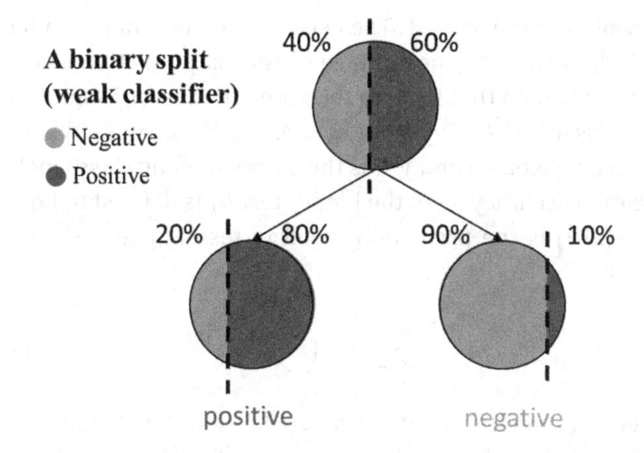

Figure 1.20 An example of binary test.

positive samples have been distributed to the left, and similarly, more negative samples have been distributed to the right. We can thus regard this binary split as a weak classifier where the path to the left represents a positive classification, and the path to the right represents a negative classification. Furthermore, we can even assign a confidence score to this weak classifier according to how well it separates the training samples. One simple way to assign this confidence scores to use the percentage of the majority class in the child node along the path. Hence, for example the classification result of the left path as shown in Fig. 1.20 is positive, and the confidence of that prediction is 0.8.

***Final Prediction Model*:** Let h_j represent the binary test for non-leaf node S_j at the j-th depth level. Let C_j (where $C_j \in \{0,1\}$ for binary classification) represent the classification result from test h_j, and the overall prediction model C^* would be $C^* = Vote(C_1, C_2, \ldots, C_j, \ldots)$. If we adopt the majority vote, as expressed in Eq. 1.17, where c represents the class label, and $c \in \{0, 1\}$ for binary classification, $\delta(\cdot) = 1$ if its argument is true and 0 otherwise.

$$C^* = \arg\max_c \sum_j \delta(C_j = c) \tag{1.17}$$

In addition to the prediction model, we also propose a confidence measure for the classification result, by evaluating the classification accuracy of each binary test. The classification accuracy of a binary test h_j is defined as the percentage of correct prediction evaluated on the training data that fall into the parent node. For example, for a training subset $\{(x^{(k)}, y^{(k)})|k = 1, 2, \ldots M\}$ of M examples, where $x^{(k)}$ is the k-th image patch and $y^{(k)}$ is the corresponding class label. The classification accuracy α_j of the binary test h_j is defined in Eq. 1.18, where $C_j(x^{(k)})$ is the prediction result of testing patch $x^{(k)}$ by the binary test h_j.

$$\alpha_j = \frac{1}{M}\sum_{k=1}^{M} \delta(C_j(x^{(k)}) = y^{(k)}) \tag{1.18}$$

The overall prediction confidence $\alpha(C^*)$ of the j-th node is the summation of the classification accuracy of each binary test that provides the correct prediction result in Eq. 1.19.

$$\alpha(C^*) = \sum_j (\alpha_j, if(C_j = C^*)) \tag{1.19}$$

Let us give an example. As shown in Fig. 1.21, assume that there is a sample that entered the leaf node L after passing the split tests S_1, S_2, S_3 and S_4. For split test S_1, the predicted class C_1 is "–" (negative category) with confidence score F_1 equals to p_1^- which is the proportion of negative samples during the training stage. However, split test S_2 gave a "+" (positive category) result with confidence score p_2^+, which is the proportion of positive samples at training stage. Then, the sample passed tests S_3 and S_4, and both of them predicted it as a negative sample. Therefore, the final prediction combines all the decisions of tests S_1, S_2, S_3 and S_4, and makes the final decision that the category of this sample is negative, and the confidence can be obtained through Eq. 1.18, which is $(p_1^- + p_3^- + p_4^-)/4$.

The overall confidence is affected by (1) the majority prediction from all binary tests, and (2) the number of binary tests giving that prediction. The idea can be illustrated in Fig. 1.22. The left path shows the classification of testing shadow patches with low confidence. Nodes 0 and 4 classified the patch as negative, while nodes 1, 9, and 16 predicted them as positive. The right path shows the classification of a high-confidence patch. All nodes along the path classify the patch as positive.

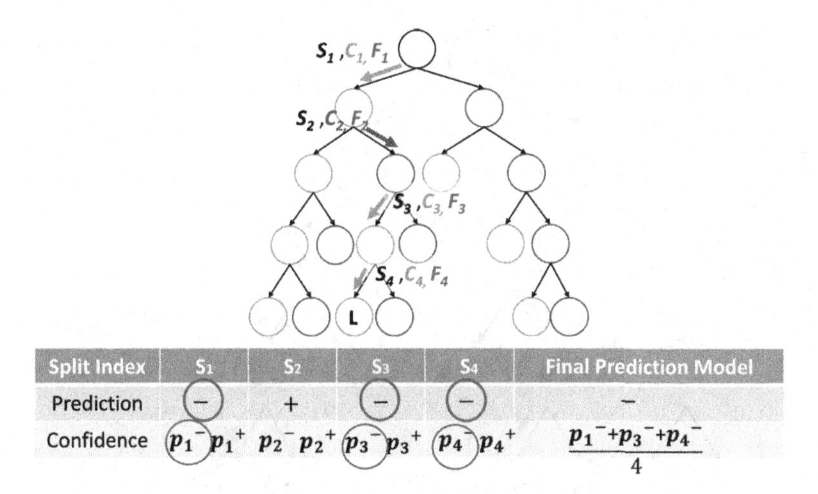

Split Index	S1	S2	S3	S4	Final Prediction Model
Prediction	−	+	−	−	−
Confidence	$p_1^- \; p_1^+$	$p_2^- \; p_2^+$	$p_3^- \; p_3^+$	$p_4^- \; p_4^+$	$\frac{p_1^- + p_3^- + p_4^-}{4}$

Figure 1.21 Example of prediction model.

One benefit of using the final confidence is to determine how much we could trust this classification. A high confidence indicates most of the weak classifiers have classified the testing patch into the same class, hence a high probability of correct classification. A low confidence indicates that for some tests the patch is classified as positive, while negative for others; hence, it is more likely to be a difficult case and indicates a lower probability of being classified correctly. For example, most patches that go through the path $0 \rightarrow 1 \rightarrow 4 \rightarrow 9 \rightarrow 16 \rightarrow 22$ are shadow patches under extreme illumination conditions, and are more difficult to recognize. The patches that go through the path $0 \rightarrow 2 \rightarrow 6 \rightarrow 12 \rightarrow 18 \rightarrow 24$ can be easily recognized because of their discriminative appearances.

1.3.2 Testing Stage and Experimental Results

As we mentioned before, we extracted and annotated 263 frames that contain many conditions with ground truth bounding box of the bumper shadow area. A large number of positive shadow patch and negative non-shadow patches (some examples are shown in Fig. 1.23) can be then generated (as shown in Table 1.5).

Let us compare the random tree based method (Decision Tree) with other approaches. The traditional non-learning based method

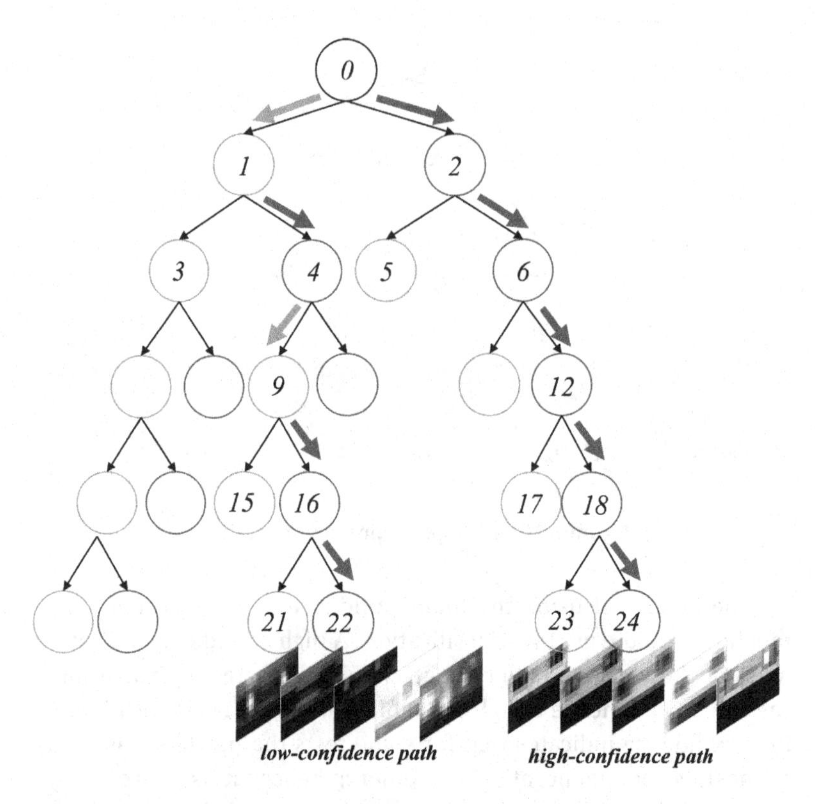

Figure 1.22 Classifying shadow patches.

(Traditional Approach) adopts some hand-engineered features that are based on color, edge information and so on, which is actually a weak detector in the driving assistant system of our group. The SVM classifier implemented by OpenCV package uses HOG features that is a popular approach in the field. The training dataset is the same for SVM classifier and for Modified DT ap-

Table 1.5 Number of annotated frame and extracted positive and negative samples for training and testing datasets

Dataset	Number of frames	Number of +ve samples	Number of -ve samples
Training	140	7,700	10,980
Testing	123	6,215	43,307

Table 1.6 Performance of (1) HOG and SVM classifier, (2) Traditional non-learning based classifier (Traditional Approach), (3) random tree based method (Decision Tree)

Method	Precision	Recall	F1 score	Time (ms) per patch
HOG+SVM	0.35	0.97	0.52	1.2
Traditional Approach	0.88	0.61	0.72	0.397
Decision Tree	0.64	0.98	0.78	0.028

proach. Table 1.6 shows the performance comparison among these classifiers.

To compare the performances of different approach, we use recall and precision indicators that are common used in object detection filed, and the calculation is shown below.

$$Recall = \frac{\text{Correctly recognized shadow patch}}{\text{All shadow patches in the dataset}},$$

$$Precision = \frac{\text{Correctly recognized shadow patch}}{\text{All patches estimated as shadow patches}}$$

Recall rate is closely related to the missing detection. In practice, we prefer as less missing detection as possible, which is also a larger recall rate. Similarly, higher precision value means less wrong detections. F1 score is an overall measure of a classifier by combining the precision and recall rate, and the calculation is

$$F1 = (2 \times precision \times recall)/(precision + recall).$$

As shown in Table 1.6, our Decision Tree approach shows the highest recall rate, and also a quite satisfying precision rate. In other words, the method has the lowest missing rate and a satisfying false alarm rate in operation, which is acceptable since the shadow recognition is only the first stage in a vehicle detection system. Note that our Decision Tree uses low-level feature (intensity) only, while the Traditional Approach utilizes a set of effective features. Our experimental results show that the Decision Tree approach has the best performance among the compared methods with the largest *F1* score. Besides, the Decision Tree approach requires the least computation time among the three algorithms, which is 42 times faster than the SVM classifier with the HOG feature.

Figure 1.23 Examples of non-shadow patches.

1.4 Random Forests for Light Rail Vehicle Detection

Traffic accidents endanger human safety. However, most of the accidents were caused by the subjective factors of drivers, like distraction, improper driving speed, and so on. Light rail vehicle (we further refer it to as train, for simplicity) is a type of vehicle that plays an important role in city development. It also has the inherent collision hazard. But these accidents can be avoided if a warning signal is released by a driving assistant system (DAS) in advance.

Obviously, the vehicle detection part is the core of the DAS systems. We take train for example to present a vehicle detection procedure that is based on random forests algorithm. The method we presented is actually a part of the Driving Assistant System (CMS-V111) in our research group [7]. The architecture is shown in Fig. 1.24. First, different sizes of searching windows are used to generate a set of patches. Second, Feature descriptor is operated to extract feature vectors from these image patches. Next, the feature vectors are sent to Random forests classifier to distinguish vehicles from backgrounds.

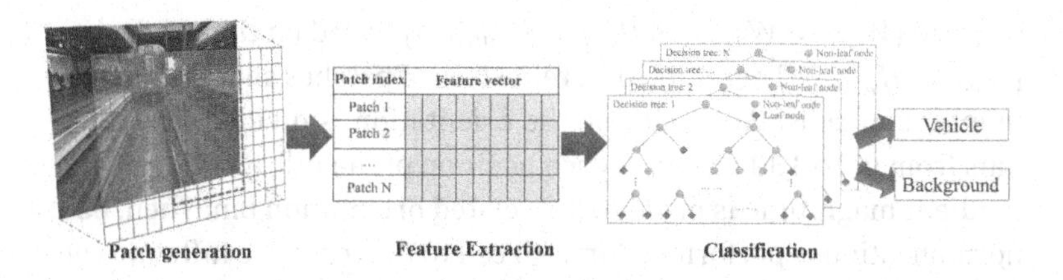

Figure 1.24 Architecture of vehicle detection.

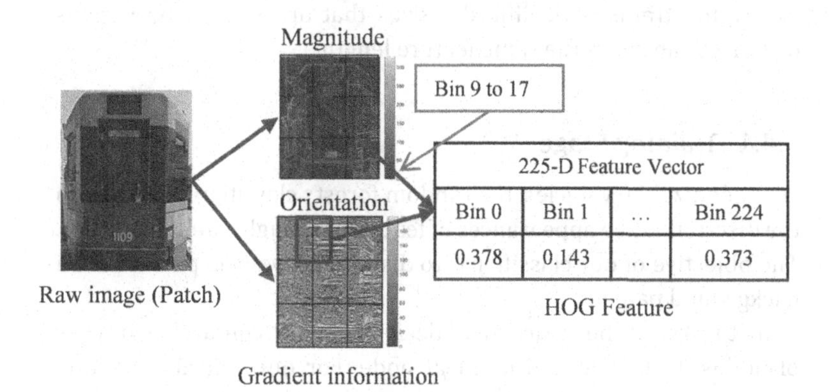

Figure 1.25 HOG feature extraction.

Histogram of Oriented Gradient (HOG) is based on gradient information and is robust to illumination and weather conditions. In our design, the original HOG feature descriptor is modified to realize a real-time and different-scale-adapt in scheme.

$$Grad_{mag}(x, y) = \sqrt{Grad_{hor}^2(x, y) + Grad_{ver}^2(x, y)} \tag{1.20}$$

$$Grad_{ori}(x, y) = \tan^{-1}\frac{Grad_{hor}(x, y)}{Grad_{ver}(x, y)} \tag{1.21}$$

At each position (x, y) of a patch, the typical 1-D Sobel operator [-1, 0, 1] is first used to calculate the horizontal gradient $Grad_{hor}(x, y)$ and vertical gradient $Grad_{ver}(x, y)$. Then the magnitude $Grad_{mag}(x, y)$ and orientation $Grad_{ori}(x, y)$ can be calculated by the Eqs. 1.20 and 1.21. Next, the patch area is divided into 5 × 5 cells with overlaps (as shown in Fig. 1.25). Different from the general HOG process that the size of each cell is fixed, we set the cell size ($W_{cell} = W_{patch}/3, H_{cell} = H_{patch}/3$) based on the patch size (W_{patch}, H_{patch}). Therefore, different sizes of patches share the same numbers of cells. For each cell, the orientation is divided into nine bins from 0° to 180°. Then at each position of the cell, a voting on its gradient magnitude is made to the related orientation bin. Then, L2-normalization is performed for each cell to decrease the influence of illumination. Finally, concatenating all bins' values is made to form a 225-D (25 cells $\times$ 9 bins/cell) vector. Therefore, by using this feature descriptor, trains with different sizes that are formed by various distances can share the same feature length.

1.4.1 Training Stage

To recognize the vehicles, the random forests classifier should learn the discriminative appearances of features through training process. The objective of our classifier is to distinguish vehicle patches from background patches.

Let us recall the structure of decision tree. It contains two types of nodes: leaf node and non-leaf node. For our vehicle detection task, the non-leaf node at our decision tree tests whether the input sample belongs to vehicle or background, while the leaf node gives the prediction. The training process is shown below.

Pre-training Process

To recognize the vehicles from the video frames, we need to face the challenge of size variation. In other words, the size in the image varies because of these different intervals. Therefore, we use different sizes of sliding windows to generate a set of patches from the video frame. The aspect ratio of searching windows is set to 0.8 that is the same to the vehicle's. There are 20 sizes of searching windows whose heights vary from 70 to 480 pixels. Enough training samples are essential for any machine learning algorithm. We manually labeled all trains as ground truth ones. Let us define Intersection over Union (IoU), as shown in Eq. 1.22.

$$IoU = \frac{Area\ of\ Overlap}{Area\ of\ Union} \tag{1.22}$$

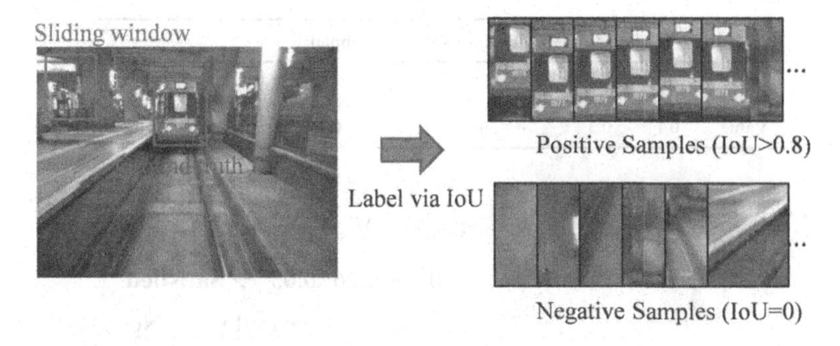

Figure 1.26 Training sample generation.

where "Area of Overlap" means overlapped area of the searching window and ground-truth vehicle area, and "Area of Union" represents the total area of searing window and. ground-truth vehicle area. The larger value represents the higher proportion of the overlapped area.

As shown in Fig. 1.26, patches that cover trains with the right proportion (IoU value is larger than 0.8) are regarded as positive (+ve) patches, while other patches that only contain background are negative (−ve) patches. Our training dataset is formed by 3,249 video images captured that contains various conditions. By using this strategy, 527,066 +ve patches and 1,805,691 −ve patches were obtained.

Design of Binary Test

Our vehicle-background classification is conducted after HOG feature extraction process. Because the patch we designed has 25 cells and each cell forms a 9-bin histogram, the length of the feature vector is 225 (= 25 × 9). For binary test in each non-leaf node, it randomly selects two attributes from the 225-d feature vectors. Similar to the previous works, the binary test we designed is to compare the value of these two feature attributes (see Eq. 1.23). Let $\mathbf{V} = \{V_0, V_1, ..., V_k\}$ denote the HOG vector of the patch under question. and $V_0, V_1, ..., V_k$ denote the attributes of vector **V**, and k is the size of the vector ($k = 255$ in this example). Indices i and j denote the randomly chosen positions of the vector **V** during

HOG Feature Vector (a sample)										
Index	0	1	...	**25**	...	**120**	...	223	224
Value	0.1	0.3		**0.13**		**0.26**		0.06	0.17

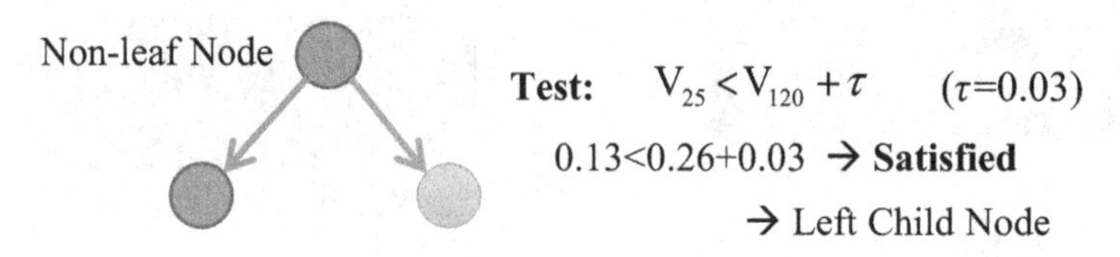

Figure 1.27 An example of split test.

training stage. τ is a threshold value. *Left* and *Right* denote the left and right child nodes separately.

$$h(i, j, \tau) = \begin{cases} \text{Left} & \text{if } V_i < V_j + \tau \\ \text{Right} & \text{otherwise} \end{cases} \quad (1.23)$$

Let us show an example to clarify the idea. Assume that a non-leaf node randomly chooses 25th and 120th feature attributes as V_i and V_j separately, and the threshold value τ is also randomly set to 0.03. Each sample that reaches this node will check the relationship between V_{25} and $(V_{120} + 0.03)$ which is based on the function $h(i, j, \tau)$. As shown in Fig. 1.27, there is a sample (feature vector) entering the non-leaf node. Its 25th and 120th feature attributes are 0.13 and 0.26 separately. Obviously, it satisfies the split function at the non-leaf node which is $0.13 < 0.26 + 0.03$. Therefore, the sample should go to the left child node.

Optimal Binary Tests

As we mentioned before, the number of randomly selected tests at each node depends on the length of feature vector to ensure sufficient randomness; hence, there are 225 candidate binary tests at each non-leaf node in the example. Similar to section 1.2 on Object Classification, we make use of the information gain (Eq. 1.12) to select the suitable binary tests (with the maximum information gain) from the candidate pool.

Let us show an example. As shown in Table 1.7, a non-leaf node contains 1,000 vehicle patches and 2,000 background patches. The

Table 1.7 Candidate binary test 1

	#Vehicle	#Background	Entropy
Non-leaf Node	1,000	2,000	0.918
Left Child Node	800	300	0.845
Right Child Node	200	1700	0.485

Information Gain: 0.301

information entropy can be obtained by the Eq. 1.9, that is, 0.918.

$$\begin{aligned} H(\mathbf{S}) &= -\sum_{c \in \mathbf{C}} p(c) \log(p(c)) \\ &= -\left[\frac{1,000}{1,000+2,000} \log\left(\frac{1,000}{1,000+2,000}\right) \right. \\ &\quad \left. + \frac{2,000}{1,000+2,000} \log\left(\frac{2,000}{1,000+2,000}\right)\right] \\ &= 0.918, \end{aligned}$$

where $\mathbf{C} = \{$*"vehicle," "background"*$\}$ denotes the two classes in this vehicle-background-classification task.

Then, a split function (Candidate binary test 1) leads to the separation of the samples, where the left child node contains 800 vehicle patches and 300 background patches, and the right child node contains 200 vehicle patches and 1700 background patches. Similarly, entropies of the leaf and right node can be obtained separately as 0.845 and 0.485. Then, the information gain of this split function can be calculated by Eq. 1.12, and the information gain of this candidate test is 0.301.

$$\begin{aligned} I(\mathbf{S}, \theta) &= H(\mathbf{S}) - \left[\frac{|\mathbf{S}^L|}{|\mathbf{S}|} H(\mathbf{S}^L) + \frac{|\mathbf{S}^R|}{|\mathbf{S}|} H(\mathbf{S}^R)\right] \\ &= 0.918 - \frac{800+300}{1000+2000} \times 0.845 - \frac{200+1700}{1000+2000} \times 0.485 \\ &= 0.301 \end{aligned}$$

Similarly, information gains of all the candidate binary tests can be obtained and then the suitable binary test can be found. The same tree growing process in section 1.2 on Object Classification can be used again until meeting the terminated conditions.

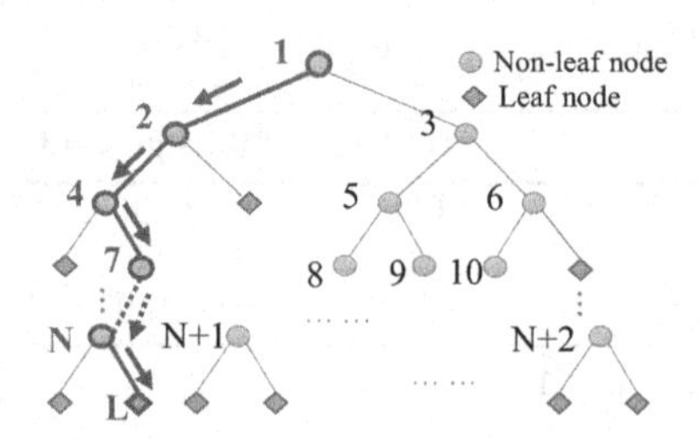

Figure 1.28 Path of arriving leaf node *L*.

1.4.1.1 Prediction Model

As we mentioned before, for each leaf node, it contains a subset which is formed by a set of data with similar attributes distribution. For example, Fig. 1.28 shows a tree structure. All the samples that arrived leaf node L have passed tests 1, 2, 4, 7, ..., N. Then the perdition can be made based on the posterior probability model.

Let us show the distribution results of leaf nodes of a tree in Table 1.8. For leaf node 1, 107 vehicle patches entered and no background patch arrived at this node during the training process. It is obvious that "*Vehicle*" is the predominant at this leaf. The proportion of vehicle patches is 100%. Hence, if a sample enters this leaf node, the leaf node will give the prediction: its category is vehicle and probability (or confidence) is 1.00.

Combination of Random Trees

As we mentioned before, each decision tree is trained randomly. The combination of those weak classifiers should lead to a better classification ability. Table 1.9 shows an example that a random

Table 1.8 Sample distribution of leaf nodes

Leaf Index	#Vehicle	#Background	Dominant class	Probability
1	107	0	Vehicle	1.00
2	34	97	Background	0.74
3	133	24	Vehicle	0.85
...				
N	0	455	Background	1.00

Table 1.9 Example of a five-tree random forest

Tree index	Leaf index	Used attributes $\{i, j\}$	Predicted class	Probability
1	652	{21,45}, {35,14},…, {92,135}	*Vehicle*	1.00
2	154	{214,45}, {32,15},…, {25,77}	*Vehicle*	0.97
3	784	{36,56}, {51,11},…, {21,47}	*Vehicle*	0.99
4	1,241	{121,35}, {84,49},…, {8,221}	*Background*	0.52
5	21	{41,98}, {101,54},…, {40,68}	*Vehicle*	1.00
$P_{Vehicle} = (1 + 0.97 + 0.99 + (1 - 0.52) + 1)/5 = \mathbf{0.88}$				

forest is formed by five randomly trained decision trees. We input a testing sample to the random forest. The first tree makes the prediction that it belongs to "*Vehicle*" with probability 1.0. This decision is made by the 652th leaf node. Before given the prediction, the sets of feature attributes {21th, 45th}, {35th, 14th} …, {92th, 135th} are used in the corresponding decision path. The second tree makes decision based on the 214th, 45th, 32th, …, 77th feature attributes that is different from the first tree. In other words, the decisions made by the two trees are in depended with each other. Then, the prediction of the 5-tree forests can be obtained. The estimation of this sample is vehicle with probability 0.88.

Table 1.10 shows the performance of random forests with different number of trees. The testing dataset was formed by 72,597 vehicle and 245,780 background patches that were extracted from 487 images. Similar as the definition of recall and precision in 1.3 on Shadow Detection, a larger recall rate means less missing detection in practice. In our experiment, the recall rate was 0.903 when only one tree was used. By using more trees, the recall rate increases

Table 1.10 Performance of random forests with different number of trees (time is classification time of one patch)

Tree number	Accuracy	Precision	Recall	F1-score	Time (us)
1	0.969	0.96	0.903	0.935	0.56
3	0.987	0.981	0.962	0.974	1.28
5	0.99	0.985	0.971	0.981	1.68
10	0.992	0.993	0.971	0.981	2.63
15	0.992	0.995	0.968	0.98	3.41
20	0.991	0.995	0.964	0.977	4.11
30	0.991	0.995	0.967	0.979	5.44
50	0.992	0.996	0.97	0.981	8.66

significantly, and a very high recall rate is obtained when the number of trees is larger than 5. A higher precision means the less wrong detections. A precision of 0.96 is good when using one decision tree, but it rises to over 0.99 when we use more than five trees. It is obvious that the classification accuracy and F1 score of random forests are higher than using one decision tree, which shows the overall classification ability becomes better. In practice, we always use 10 trees in order to balance the classification performance and computational cost.

1.4.2 Further Experimental Results

Let us compare other machine learning algorithms like SVM, decision tree, etc. with the random forests approach. We designed a patch-based experiment. A total of 303,445 patches' HOG vectors were extracted from video sequences. The experimental results are shown in Table 1.11. It is shown that the accuracy of random forests was the highest, which means the classification ability of random forests outperformed the SVM and decision tree in our task. Our random forests classifier showed the highest recall rate which suggests the less missing detection. For overall performance indicator "F1 score," our random forests classifier overtakes other two classifiers and reaches a 98.2% F1 score at our testing dataset. The speed of our random forests classifier is very fast requiring 2.53 us for each prediction. Because these trees are independent with

Table 1.11 Performance of different classifier (time is classification time of one patch)

Classifier	Accuracy	Recall	Precision	F1 score	Time (us)
SVM	0.981	0.919	1.000	0.958	415.16
Decision Tree	0.973	0.896	0.992	0.942	0.84
Random Forests	0.992	0.971	0.993	0.982	2.53

each other, the prediction can be realized in parallel making our random forests even faster.

1.5 Review of Some Related Works

Decision tree is a powerful machine learning tool for both classification and regression tasks [1–5, 8–11]. As we said, this is not a chapter on a comprehensive review of the random tree and random forests. However, if the reader is interested in more details of the topic, he/she can refer to references [1–5, 12–14]. Breiman et al. first proposed decision tree algorithm in 1984 [1] and then introduced the bagged training process to reduce overfitting problem [5]. Quinlan introduced ID3 [13, 14] and C4.5 Rule [12] for the implementation of decision tree. Yao et al. [2] and Rota Bulo et al. [3] selected more complex split functions like SVM and neural networks. Geurts et al. [4] introduced the extremely randomized trees. Good review papers include [5, 15, 16].

Handwritten digit recognition is a fundamental task for image classification. LeCun et al. [17] proposed a MNIST dataset and implemented a number of algorithms to face the recognition challenge, like K-nearest neighbors, SVM, etc. The examples in this chapter on shadow detection and train recognition are parts of a driving assistant system that is designed in our research group [6, 7, 18–20]. Shadow detection [6] is performed inside a region-of-interest area that determined by our previous study on railway detection [18], and followed by vehicle detection [7, 19] and distance estimation [20]. There are related works, like [21, 22], utilizing the property that shadow areas are usually darker than non-shadow areas. Cheon et al. [23] took similar approach, but heavily relying on the mean and standard deviation of road pixels. Vehicle detection is challenging because of different appearance of vehicles caused by various environment conditions. Good feature extraction and selection method is crucially important for reliable vehicle recognition [24]. Researches in [25, 26] prefer to use light-insensitive features because of the robustness. Machine learning algorithms can handle tough cases at the expense of large computational power, and are widely used in vehicle detection field [27].

1.6 Conclusion

In this chapter, we first give some easy-to-understand examples to show some basic concepts of decision tree. For non-leaf node, the randomly designed binary test is considered to work as a weak classifier that can be optimized based on information theory. The tree structure is constructed by the progressive binary tests. A test sample can then finally enter into a leaf node that is responsible for prediction. The extremely randomized trees can then form a set of decision forests for better classification ability.

We have also provided three examples: object classification, shadow detection and train recognition. In the example of object classification, the random tree is utilized to classify the handwriting digitals into ten classes. To achieve the goal, we have to design a binary test that randomly chooses two pixel intensities at each non-leaf node, and the information gain is used to optimize the tests. The prediction at leaf node is based on the posteriori estimation. Similarly, the example on shadow detection also adopts the binary test that makes the comparison between two pixels. As we mentioned in 1.3, the binary test brings robustness to resolve different illumination conditions. We also give the definition of confidence score that is based on decision path, which is an essential way for reliable prediction. Finally, the random forests algorithm is applied to more challenging task—vehicle detection. In the example, HOG features and different sizes of searching windows are used to face environment-variance and distance-variance challenge. The randomly trained decision trees are independent with each other. We combine the prediction of these trees, and the overall classification performance can be improved substantially.

One of the objectives of this chapter is to recall the basic random decision trees techniques to support other chapters in this book. For example, Chapter 8 makes use of the random forests for image sup-resolution [28–31], the performance of which matches very well with deep learning using CNN (Convolutional Neural Networks) with the speed advantage of being possibly used in real-time applications. Chapter 12 makes use of the random trees structures to detect human activities and for monitoring. These are very useful applications, for which the random decision trees/forests make the algorithms fast enough for practical uses.

References

1. Leo Breiman, Jerome Friedman, RA Olshen and Charles J Stone, *Classification and Regression Trees*, Taylor & Francis, 1984.
2. Bangpeng Yao, Aditya Khosla and Li Fei-Fei, Combining randomization and discrimination for fine-grained image categorization, *Proceedings of IEEE Conference on Computer Vision and Pattern Recognition (CVPR)* pp. 1577–1584, 2011, Colorado Springs.
3. Samuel Rota Bulo and Peter Kontschieder, Neural decision forests for semantic image labelling, *Proceedings of IEEE Conference on Computer Vision and Pattern Recognition (CVPR)*, pp. 81–88, 2014, Columbus, OH.
4. Pierre Geurts, Damien Ernst and Louis Wehenkel, Extremely randomized trees, *Machine Learning*, vol. 63, no. 1, pp. 3–42, 2006.
5. Leo Breiman, Random forests, *Machine Learning*, vol. 45, no. 1, pp. 5–32, 2001.
6. Xue-Fei Yang and Wan-Chi Siu, Vehicle detection under tough conditions using prioritized feature extraction with shadow recognition, *Proceedings of International Conference on Digital Signal Processing (DSP)*, pp. 1–5, 2017, London, United Kingdom.
7. Li-Wen Wang, Xue-Fei Yang and Wan-Chi Siu, Learning approach with random forests on vehicle detection, *Proceedings of International Conference on Digital Signal Processing (DSP)* (Accepted), 2018, Shanghai, China.

8. Bochuan Du, Wan-Chi Siu and Xue-Fei Yang, Fast CU partition strategy for HEVC intra-frame coding using learning approach via random forests, *Proceedings of Signal and Information Processing Association Annual Summit and Conference (APSIPA)*, pp. 1085–1090, 2015, Hong Kong, China.
9. Jun-Jie Huang and Wan-Chi Siu, Practical application of random forests for super-resolution imaging, *Proceedings of IEEE International Symposium on Circuits and Systems (ISCAS)*, pp. 2161–2164, 2015, Lisbon Portugal.
10. Jun-Jie Huang and Wan-Chi Siu, Fast image interpolation with decision tree, *Proceedings of International Conference on Acoustics, Speech and Signal Processing (ICASSP)*, pp. 1221–1225, 2015, Brisbane, Australia.
11. Jun-Jie Huang, Wan-Chi Siu and Tian-Rui Liu, Fast Image Interpolation via Random Forests, *IEEE Transactions on Image Processing*, vol. 24, no. 10, pp. 3232–3245, 2015.
12. Ross J Quinlan, *C4.5: Programs for Machine Learning*, Morgan Kaufmann Publishers Inc., 1993.
13. J. Ross Quinlan, Discovering rules by induction from large collections of examples, *Expert Systems in the Micro Electronics Age*, Edinburgh University Press, 1979.
14. J. Ross Quinlan, Induction of decision trees, *Machine Learning*, vol. 1, no. 1, pp. 81–106, 1986.
15. Antonio Criminisi, Jamie Shotton and Ender Konukoglu, Decision forests: A unified framework for classification, regression, density estimation, manifold learning and semi-supervised learning, *Foundations and Trends® in Computer Graphics and Vision*, vol. 7, no. 2–3, pp. 81–227, 2012.
16. Vladimir Svetnik, Andy Liaw, Christopher Tong, J Christopher Culberson, Robert P Sheridan and Bradley P Feuston, Random forest: A classification and regression tool for compound classification and QSAR modeling, *Journal of Chemical Information and Computer Sciences*, vol. 43, no. 6, pp. 1947–1958, 2003.
17. Yann LeCun, Léon Bottou, Yoshua Bengio and Patrick Haffner, Gradient-based learning applied to document recognition, *Proceedings of the IEEE*, vol. 86, no. 11, pp. 2278–2324, 1998.
18. Hao Wu and Wan-Chi Siu, Real time railway extraction by angle alignment measure, *Proceedings of IEEE International Conference on Image Processing (ICIP)*, pp. 4560–4564, 2015, Quebec City, QC.

19. Chup-Chung Wong, Wan-Chi Siu, Stuart Barnes and Paul Jennings, Low relative speed moving vehicle detection using motion vectors and generic line features, *Proceedings of IEEE International Conference on Consumer Electronics (ICCE)*, pp. 208–209, 2015, Hangzhou, China.
20. Hoi-Kok Cheung, Wan-Chi Siu, Steven Lee, Lawrence Poon and Chiu-Shing Ng, Accurate distance estimation using camera orientation compensation technique for vehicle driver assistance system, *Proceedings of IEEE International Conference on Consumer Electronics (ICCE)*, pp. 227–228, 2012, Las Vegas.
21. Ying-Li Tian, Max Lu and Arun Hampapur, Robust and efficient foreground analysis for real-time video surveillance, *Proceedings of IEEE Conference on Computer Vision and Pattern Recognition (CVPR)*, vol. 1, pp. 1182–1187, 2005, San Diego, CA.
22. Jia-Bin Huang and Chu-Song Chen, Moving cast shadow detection using physics-based features, *Proceedings of IEEE Conference on Computer Vision and Pattern Recognition (CVPR)*, pp. 2310–2317, 2009, Miami, FL.
23. Minkyu Cheon, Wonju Lee, Changyong Yoon and Mignon Park, Vision-based vehicle detection system with consideration of the detecting location, *IEEE Transactions on Intelligent Transportation Systems*, vol. 13, no. 3, pp. 1243–1252, 2012.
24. Xuezhi Wen, Ling Shao, Wei Fang and Yu Xue, Efficient Feature Selection and Classification for Vehicle Detection, *IEEE Transactions on Circuits and Systems for Video Technology*, vol. 25, no. 3, pp. 508–517, 2015.
25. Hossein Tehrani Niknejad, Akihiro Takeuchi, Seiichi Mita and David McAllester, On-road multivehicle tracking using deformable object model and particle filter with improved likelihood estimation, *IEEE Transactions on Intelligent Transportation Systems*, vol. 13, no. 2, pp. 748–758, 2012.
26. Jun-Wei Hsieh, Li-Chih Chen, and Duan-Yu Chen, Symmetrical surf and its applications to vehicle detection and vehicle make and model recognition, *IEEE Transactions on Intelligent Transportation Systems*, vol. 15, no. 1, pp. 6–20, 2014.
27. Sayanan Sivaraman and Mohan M Trivedi, Active learning for on-road vehicle detection: A comparative study, *Machine Vision and Applications*, vol. 25, no. 3, pp. 599–611, 2014.
28. Zhi-Song Liu and Wan-Chi Siu, Cascaded random forests for fast image super-resolution, *Proceedings of IEEE International Conference on Image Processing (ICIP)*, pp. 2531–2535, 2018, Athens, Greece.

29. Zhi-Song Liu, Wan-Chi Siu and Yui-Lam Chan, Fast image super-resolution via randomized multi-split forests, *Proceedings of IEEE International Symposium on Circuits and Systems (ISCAS)*, pp. 1–4, 2017, Baltimore, MD, USA.
30. Jun-Jie Huang and Wan-Chi Siu, Learning hierarchical decision trees for single-image super-resolution, *IEEE Transactions on Circuits & Systems for Video Technology*, vol. 27, no. 5, pp. 937–950, 2017.
31. Yu-Zhu Zhang, Wan-Chi Siu, Zhi-Song Liu and Ngai-Fong Law, Learning via decision trees approach for Video super-resolution, *Proceedings of International Conference on Computational Science and Computational Intelligence (CSCI'17)*, pp. 558–562, 2017, Las Vegas, USA.

Chapter 2

Sparsity Based Dictionary Learning Techniques

Raju Ranjan,[a] Sumana Gupta,[b] and K. S. Venkatesh[b]

[a]*School of Technology, Pandit Deendayal Petroleum University, Gandhinagar, Gujarat, India*

[b]*Indian Institute of Technology Kanpur, Uttar Pradesh, India*

rajuran@gmail.com, sumana@iitk.ac.in

Model selection in signal processing plays a crucial role to achieve the underlying objective. Use of arbitrary or generalized model without exploiting data prior results in poor performance. In recent times, non-parametric methods such as dictionary-based approach have been proposed for applications such as image de-noising, in-painting, de-mosaicking, and compression. In this chapter, we discuss regularized sparsity prior-based dictionary learning algorithm named Regularized K Times Sum of Optimally Weighted Vectors. We have mathematically formulated and derived atom update expression for different priors. Dictionary learning algorithm that considers a smoothing regularizer on dictionary atom has been discussed for image de-noising. It has the advantage of having closed form expression for atom update. The regularized dictionary learning algorithm has been shown to achieve overall

Learning Approaches in Signal Processing
Edited by Wan-Chi Siu, Lap-Pui Chau, Liang Wang, and Tieniu Tan

ISBN 978-981-4800-50-1 (Hardcover), 978-0-429-06114-1 (eBook)
www.panstanford.com

better performance compared to some of the existing dictionary learning techniques for image de-noising.

2.1 Introduction

Digital signal processing has almost replaced processing signals in analog domain due to availability of cheap computing units. Capturing natural signal using sensor followed by digitization results in a representation that is usually not suitable for many image processing and computer vision tasks. There exist a vast set of image processing and computer vision tasks ranging from image de-noising, in-painting, compression, object detection, image retrieval, and many more that require analyzing the signal in a suitable transform space. For example, image compression is efficiently achieved using Discrete Cosine Transform (DCT) and Wavelet Transform (WT). Object detection and image retrieval require the extraction of feature that is robust to noise, artifacts, and distortions present in an image. Filtering of digital signal is efficiently performed using Discrete Fourier Transform (DFT) and Fast Fourier Transform (FFT). Model that efficiently captures features salient to underlying application is of utmost importance in signal processing. Model should be flexible enough to allow formulation of prior that enhances system performance without losing mathematical tractability. A model that is appropriate for a particular application might not work efficiently for a different problem. Examples of generative data model techniques are Gaussian mixture model, Non-Negative Matrix Factorization (NNMF), and Principal Component Analysis (PCA). Perceptron, Support Vector Machine (SVM) [1, 2], and Neural Network (NN) [3] can be categorized under the paradigm of discriminative modeling techniques. Dictionary-based signal modeling and processing has been gaining popularity in recent times. The idea of dictionary has been adopted for several tasks in image processing and computer vision. There are a set of predefined dictionaries based on a mathematical model of the data such as DFT, DCT, Short Time Fourier Transform [4], Gabor [5], wavelet, wavelet packet [6], and so on. Efficiency of such analytic dictionaries has been questioned in recent times for its limited expressiveness. A straightforward extension aiming to exploit 1-D and 2-D mathematical models for constructing dictionary atom in higher dimension constrains its expressiveness for a range of natural phenomenon.

To capture and represent natural phenomenon of a specific class of signal, data driven trained dictionary has been proposed for various signal processing applications. A trivial but useful model is self-similarity model that collects a set of images from a class. Exemplar-based image in-painting uses such model with significant performance. Learning the model from the data itself or adapting it to the data without relying on predefined mathematical structures has opened a new paradigm of dictionary-based image processing techniques. This relates to dictionary learning where the image is modeled as being represented via a learned dictionary. Learning a dictionary to perform its best on a training set results in the seminal work of [7–12], and others.

Another important trend that has drawn much attention recently in signal processing is data models based on sparsity prior. Using this prior several state-of-the-art results have been produced in the field of image processing and computer vision. Widely used JPEG [13] standard is one such example that exploits sparsity for the task of compression. Such prior assumes natural signal has sparse representation in some domain often called sparse-land. Model that uses sparsity prior tries to describe signals as linear weighted combinations of a few atoms from some over-complete dictionary. It can also be thought of as a signal from a class that lies in the union of subspaces.

Dictionary that gives sparse representation for a class of signal is often essential for efficient processing of signal. In recent times, non-parametric methods such as dictionary-based approach have been proposed for applications such as image de-noising, in-painting, de-mosaicking, and compression. The existing dictionary learning algorithms such as Method of Optimal Direction (MOD) [8–10], Union of Orthonormal Bases [14, 15], Generalized Principal Component Analysis (GPCA) [12], and K Times Singular Value Decomposition (KSVD) [11] have been widely accepted in computer vision community.

In this article, we discuss algorithms for sparsity prior-based dictionary learning along with various priors on dictionary atoms. We have mathematically formulated and derived atom update expression for different priors. Dictionary learning algorithm that considers a smoothing regularizer on dictionary atom has been discussed for image de-noising. The regularized sparsity prior-based dictionary learning algorithm is named Regularized K Times Sum of Optimally Weighted Vectors (RKSOWV). In learning a discriminative dictionary for classification, a penalty on dictionary atom learning has been considered. The expression for atom update has the advantage of having closed form expression. The RKSOWV has also been compared with Over-complete Discrete Cosine Transform (ODCT), Method of Optimal Directions (MOD) [8–10], global K Times Singular Value Decomposition (KSVD), adaptive KSVD [11], and sparse KSVD (SKSVD) [16]. Dictionary learned using RKSOWV is shown to give marginally better de-noising performance in terms of PSNR compared to adaptive KSVD [11], and sparse KSVD (SKSVD) [16]. The RKSOWV gives PSNR as good as KSVD [11] for low σ values such as 10, 20, and 25. For higher σ values (50, 75, and 100), its PSNR is similar to SKSVD [16].

2.2 Brief History of Dictionary-Based Signal Processing

In signal representation, dictionary plays an important role in analyzing the underlying signal. A dictionary is usually a set of elementary signals that has the desirable property of orthogonality. This property helps to obtain the representation coefficients as an inner product of signal and dictionary atoms. In the case of a non-orthogonal dictionary, signal is multiplied with dictionary inverse to obtain the representation coefficients. Though the property of dictionary being orthogonal or independent provides better mathematical tractability, the drawback of such dictionaries is their limited expressiveness. To overcome this limitation, a newer set of over-complete dictionaries with number of atoms greater than the dimension of the signal was developed. Such over-complete dictionaries that contain redundant atoms are found to give state-of-the-art results in many areas of signal processing [17–22].

Dictionary learning encompasses a large subset of data driven learning problems such as Karhunen–Loeve Transform [6, 23] or PCA [24], Independent Component Analysis, NNMF, Factor Analysis, K-means, KLine, and many others.

For a given set of training vectors, dictionary learning problem is mathematically formulated as

$$\min_{D,A} \|Y - DA\|_F^2 \qquad (2.1)$$
$$\text{With a set of constraints on D and A.}$$

where, Y, D, and A represent training data, dictionary, and representation matrices, respectively. In following subsections, it is shown how the choice of a particular set of constraint results in data modeling suitable for an application.

2.2.1 K-means and KLine

The constraints $\|A_i\|_0 = 1, \forall i$ and $\|A_i\|_2^2 = 1, \forall i$ ensure that a vector is quantized to the nearest centroid d_k. Greedy algorithms such as Generalized Lloyd algorithm (GLA) [25] is used for K-means. The relation between K-means and sparse representation has been studied in detail in [8, 26, 27]. Relaxing the constraint results in representing a given vector by nearest line. Constraint $\|A_i\|_0 = 1, \forall i$ insures this and resulting model is termed as KLine.

A natural generalization would be to relax sparsity from 1 to L. By relaxing the constraint, we obtain the following mathematical relation.

$$\min_{D,A} \|Y - DA\|_F^2 \qquad (2.2)$$
$$\text{S.t } \|A_i\|_0 \leq L, \forall i.$$

Using such constraint results in modeling data as union of sub-spaces. Interestingly, this relaxation of constraint allows modeling of a diverse set of natural signals and leads to state-of-the-art results in signal processing.

2.2.2 Method of Optimal Directions

MOD was introduced by Engan et al. in 1999 [8–10] and was one of the first methods to implement what is known today as a sparsification process. Similar to other training methods, MOD alternates sparse coding and dictionary update steps. In first stage, representation corresponding to each training vector is obtained using any standard technique such as Orthogonal Matching Pursuit (OMP) [28] and Focal Under-determined System Solver (FOCUSS). The residual error is estimated as $E = \|Y - DA\|_F^2$, where Y is a matrix containing all training vectors, A is a representation matrix and E is error matrix. Next step updates dictionary D by taking derivative of E with respect to D. Taking derivative and solving for D results in updating D by $YA^T(AA^T)^{-1}$. Dictionary update and representation matrix estimation is repeated alternatively until a predefined convergence criterion is satisfied. MOD typically requires only a few iterations to converge. Method suffers from relatively high complexity of matrix inversion.

2.2.3 Union of Orthobases

Union of orthobases represents one of the first attempts at training a structured over-complete dictionary. The model suggests training a dictionary D which is concatenation of K orthogonal bases $[D_1, D_2 \ldots D_N]$ where, D_i are unitary matrices. Sparse coding over this dictionary can be performed efficiently through a Block Coordinate Relaxation (BCR) technique [15]. For a comprehensive study readers are advised to refer to [14, 15]. Drawback with union of orthobases is relative restrictiveness in signal expression. It does not perform as well for more flexible structures in practice.

2.2.4 Generalized Principal Component Analysis

GPCA was introduced in 2005 by Vidal, Ma, and Sastry [12]. GPCA is basically an extension of original PCA formulation, which approximates a set of examples by a low dimensional subspace. In the GPCA setting, the set of examples is modeled as the union of several low dimensional subspaces of variable dimensionality. GPCA viewpoint differs from sparsity model as each example in GPCA setting is represented using only one of the subspaces. Atoms from different subspaces cannot jointly represent a signal and has advantage of limiting over-expressiveness of dictionary. Disadvantage is, dictionary structure may be too restrictive for more complex natural signals.

2.2.5 K Times Singular Value Decomposition

KSVD [11] is state-of-the-art dictionary learning technique proposed by Michal Aharon et al. It follows two similar alternative steps for given training feature vectors such as estimating representation with respect to dictionary and updating dictionary for estimated representation matrix. First, residual matrix $\mathrm{E} = \|\mathrm{Y} - \mathrm{DA}\|_{\mathrm{F}}^{2}$ is estimated. Next, it updates dictionary to reduce residual error E. As name suggests, KSVD updates each of the K atoms sequentially. To update an atom, it finds out associated training vectors. For residual error vector corresponding to associated training vectors matrix, it forms a rank one approximation problem and uses SVD to update atom. Eigen vector corresponding to most dominant singular value is proposed as optimal solution. Sparsity-based data modeling using dictionary learning algorithm KSVD has been shown to give state-of-the-art results in various image processing tasks such as image de-noising [17], single image super-resolution [18], image in-painting [19], image de-mosaicking [20] etc. A comprehensive study is reported in [21, 22].

2.2.6 Efficient Implementation of the K Times Singular Value Decomposition Algorithm

An efficient implementation of the KSVD (EKSVD) has been suggested in [29]. The author has proposed to use an approximate version for atom update rather than using exact SVD. Implementation in [29] uses a single iteration of alternate optimization over the atom d and the coefficients row α^{H} (α^{H} is corresponding row of sparse representation matrix A corresponding to d) given by

$$d = \frac{E\alpha^{H^T}}{\|E\alpha^{H^T}\|_2}$$
$$\alpha^{H^T} = E^T d. \qquad (2.3)$$

Such method eliminates the need to explicitly compute the matrix E and saves both time and memory. Ref. [29] also propose to further speed up the sparse coding step using Batch-OMP.

2.2.7 Sparse K Times Singular Value Decomposition

Ref. [16] highlights some of the disadvantages such as lack of regularity and inefficiency of the results associated with the explicit dictionary learning techniques. It tries to retain the advantage of both analytic and adaptive dictionaries. A dictionary called sparse KSVD (SKSVD) [16] whose atoms themselves have sparse representation over a base dictionary have been shown to provide better de-noising results than KSVD [11]. ODCT has been taken as the initial base dictionary.

2.2.8 Challenges

Matrix inversion being a costly operation $\{O(N^3)$ for a matrix of size $(N \times N)\}$, updating D by $YA^T(AA^T)^{-1}$ limits the usage of MOD [8–10] to lower dimension A, therefore restricting the number of training vectors and number of atoms in dictionary that defines over-completeness. KSVD [11] reformulates dictionary update problem and updates dictionary atom one at a time avoiding the limitation of MOD [8–10] with better convergence and performance. KSVD [11] uses Singular Value Decomposition (SVD) to learn an atom. SVD of an $m \times n$ matrix has computational complexity of $O(km^2n + k'n^3)$ (k and k' are constants). Calculation of most significant singular value and corresponding singular vector (using truncated SVD) is indeed sufficient for KSVD [11]; nevertheless it is still a computationally expensive operation. This burdens computing resource for large number of training vectors. Atom update expression using KSVD [11] does not have closed form solution. SVD has closed form expression for 2×2 matrix. For an arbitrary sized matrix SVD uses diagonalization or an iterative algorithm to find singular vectors [30, 31]. Using SVD for dictionary atom update also makes it non-trivial to introduce regularization term for application specific dictionary learning. Dictionary learning is an evolving area of signal processing, many attempts such as described in Sections 2.2.6 and 2.2.7 have been made to address mentioned issues. On the same line, regularized dictionary learning technique RKSOWV and DKSOWV are derived by performing mathematical reformulation of the dictionary atom update that generalizes Lloyd's algorithm used for K-means. It has the closed form expression for atom update. A closed form expression for RKSOWV that consider a smoothing regularizer on dictionary atom has been derived and shown to achieve overall better PSNR results for image de-noising application compared to adaptive KSVD [11], and sparse KSVD (SKSVD) [16].

2.3 Regularized K Times Sum of Optimally Weighted Vectors

In this section, we extend Eq. (2.1) to allow prior if any about atoms in dictionary. In case of KSVD, a trivial and simplistic approach would be to select eigen vector out of available set that also comply with prior. Another approach could be to choose eigen vector corresponding to most dominant singular value (as proposed for KSVD) and project it to a set of vectors satisfying prior. However, compared to these suggested approaches, we derive closed form expression for atom update for a range of prior. Mathematical formulation of such dictionary learning with a prior is as follows:

$$\begin{gathered} \min_{D,A} \|Y - DA\|_F^2 + \lambda \|OD\|_F^2 \\ S.t \|A_i\|_0 \leq L, \forall i. \end{gathered} \tag{2.4}$$

In Eq. (2.4), O is an operator matrix that operates on each atom of dictionary D. λ is weighting parameter.

We derive the expression to update dictionary atom efficiently as follows. Sparse representation matrix A is estimated for training vectors Y. The mathematical formulation is given by

$$\begin{aligned} &\min_{A} \|Y - DA\|_F^2 \\ &\text{S.t. } \|A_i\|_0 \leq L, \forall i. \end{aligned} \tag{2.5}$$

Equation (2.5) can be equivalently written for each of the training vectors as

$$\begin{aligned} &\min_{\alpha} \|y - D\alpha\|_2 \\ &\text{S.t. } \|\alpha\|_0 \leq L. \end{aligned} \tag{2.6}$$

Sparse coding method OMP is used to estimate the sparse representation. In the next step, we update the dictionary. Residual error matrix $E = \|Y - DA\|_F^2$ is estimated and D is updated for given Y and A. Updating of dictionary is formulated as

$$\min_{D} \left\{ \|Y - DA\|_F^2 + \lambda \|OD\|_F^2 \right\} \tag{2.7}$$

Equation (2.7) can be reformulated and equivalently written as shown below:

$$\begin{aligned} &\min_{D} \left\{ \left\| Y - \sum_{j=1}^{K} d_j \alpha_j^H \right\|_F^2 + \lambda \|OD\|_F^2 \right\} \\ &= \min_{D} \left\{ \left\| \left(Y - \sum_{j \neq k} d_j \alpha_j^H \right) - d_k \alpha_k^H \right\|_F^2 + \lambda \|OD\|_F^2 \right\} \end{aligned}$$

Here, α_k^H is k_{th} row of sparse representation matrix A corresponding to d_k. It is to be noted that sparse representation of i_{th} training vector y is i_{th} column of A.

The algorithm updates K dictionary atoms one at a time sequentially (assuming remaining $K - 1$ atoms are known) to reduce residual error E. A training vector does not use every atom from dictionary. To update an atom d_k, associated training vectors $Y_k \subseteq Y$ to an atom are separated. For residual error corresponding to Y_k, it forms an optimization problem and solves for an atom.

$$\min_{d_k}\left\{\left\|E_k - d_k\alpha_k^H\right\|_F^2 + \lambda\|Od_k\|_2^2\right\}$$

$$= \min_{d_k}\left\{\left\|\left[e_{k1}\ e_{k2}\ \ldots\ e_{kP}\right] - \left[\alpha_k^H(1)\ \alpha_k^H(2)\ \ldots\ \alpha_k^H(P)\right]\cdot d_k\right\|_F^2 \ldots\right.$$

$$\left. + \lambda\|Od_k\|_2^2\right\}$$

where, $\alpha_k^H(1), \alpha_k^H(2), \ldots\ \alpha_k^H(P)$ are P scalar sparse coefficients.

As representation matrix A is sparse, there are zeros in α_k^H that do not affect minimization problem. Pruning such scalar values α_{kz}^H and respective error vectors e_{kz} results in minimization problem corresponding to k_{th} atom,

$$\min_{d_k}\left\{\left\|e_{knz1} - \alpha_{knz}^H(1)\cdot d_k\right\|_2^2 + \left\|e_{knz2} - \alpha_{knz}^H(2)\cdot d_k\right\|_2^2 + \ldots\right.$$

$$\left. + \left\|e_{knzM} - \alpha_{knz}^H(M)\cdot d_k\right\|_2^2 + \lambda\|Od_k\|_2^2\right\}$$

where, M is the number of nonzero elements in α_k^H. On differentiating above equation with respect to d_k and solving for d_k, a closed form expression (2.3) is obtained,

$$d_k = \left[\left\{\alpha_{knz}^H(1)^2 + \alpha_{knz}^H(2)^2 + \cdots + \alpha_{knz}^H(M)^2\right\}\cdot I + \lambda OO^T\right]^{-1}$$

$$\left\{\alpha_{knz}^H(1)\cdot e_{knz1} + \alpha_{knz}^H(2)\cdot e_{knz2} + \cdots + \alpha_{knz}^H(M)\cdot e_{knzM}\right\}$$

Atom update expression is sum of optimally weighted contributing training vectors with regularizer term and hence we term it as Regularized K Times Sum of Optimally Weighted Vectors (RKSOWV). It seems intuitive to obtain such expression for learning the basis of thin subspace of associated vectors to an atom. It is an iterative method that estimates sparse representation and updates dictionary in alternate steps until convergence. Dictionary D is initialized using one of the following methods: (1) Random selection of vectors from training set, (2) Patches at uniform spatial interval in image converted to vectors, (3) ODCT, (4) Random vectors from some distribution, and (5) Application specific structured dictionary such as Wavelets, Contourlets, Bandelets etc. Steps of this dictionary learning algorithm are given in Fig. 2.1. Atom update expression is derived in Fig. 2.2.

1: Input: Training Vectors $Y = \{y_i\}_1^N$, Sparsity = L and Number of iteration = T.
2: Task: To learn dictionary D.
3: Dictionary learning problem formulation:

$$\min_{D,A} \|Y - DA\|_F^2 + \lambda \|OD\|_F^2$$
$$\text{S.t. } \|A_i\|_0 \leq L, \forall i.$$

4: Initialization: Initialize D by randomly choosing K vectors from $Y = \{y_i\}_1^N$.
5: **for** t = 1, 2, ... T **do**
6: Sparse coding:
7: Find A:

$$\min_{\alpha_i} \|y_i - D\alpha_i\|_2$$
$$\text{S.t. } \|\alpha_i\|_0 \leq L \text{ for } i = 1, 2 \ldots N.$$

8: Dictionary update:
9: **for** k = 1, 2, ... K **do**
10: Find residual error matrix $E = Y - \sum_{j \neq k} d_j \alpha_j^H$.
11: Find non-zero coefficients α_{knz}^H out of α_k^H and corresponding $E_{knz} = \{e_c\}_{c=1}^M$.
12: Optimization problem: $\min_{d_k} \{\|E_{knz} - d_k \alpha_{knz}^H\|_F^2 + \lambda \|Od_k\|_2^2\}$.
13: Update:

$$d_k = \left[\left\{\alpha_{knz}^H(1)^2 + \alpha_{knz}^H(2)^2 + \cdots + \alpha_{knz}^H(M)^2\right\} \cdot I + \lambda OO^T\right]^{-1}$$
$$\left\{\alpha_{knz}^H(1) \cdot e_{knz1} + \alpha_{knz}^H(2) \cdot e_{knz2} + \cdots + \alpha_{knz}^H(M) \cdot e_{knzM}\right\}.$$

14: **end for**
15: **end for**

Algorithm 2.1 RKSOWV.

1: Input: Error matrix E_{knz}, α^{H}_{knz} and Operator matrix O.
2: Task: To learn dictionary atom d_k.
3: Optimization problem for d_k:

$$\min_{d_k} \{\|E_{knz} - d_k\alpha^{H}_{knz}\|_F^2 + \lambda\|Od_k\|_2^2\}.$$

$$= \min_{d_k} \left\{ \left\| (e_{knz1}, e_{knz2} \ldots e_{knzM}) - [d_k \cdot \{\alpha^{H}_{knz}(1)\alpha^{H}_{knz}(2) \ldots \alpha^{H}_{knz}(M)\}] \right\|_F^2 + \lambda\|Od_k\|_2^2 \right\}.$$

$$= \min_{d_k} \left\{ \left\| (e_{knz1}, e_{knz2} \ldots e_{knzM}) - [\{\alpha^{H}_{knz}(1), \alpha^{H}_{knz}(2) \ldots \alpha^{H}_{knz}(M)\} \cdot d_k] \right\|_F^2 + \lambda\|Od_k\|_2^2 \right\}.$$

$$= \min_{d_k} \left\{ \left\|e_{knz1} - \alpha^{H}_{knz}(1) \cdot d_k\right\|_2^2 + \left\|e_{knz2} - \alpha^{H}_{knz}(2) \cdot d_k\right\|_2^2 \cdots + \left\|e_{knzM} - \alpha^{H}_{knz}(M) \cdot d_k\right\|_2^2 + \lambda\|Od_k\|_2^2 \right\}.$$

4: Differentiating with respect to d_k and equating to zero gives expression for d_k:

$$d_k = \left[\left\{\alpha^{H}_{knz}(1)^2 + \alpha^{H}_{knz}(2)^2 + \cdots + \alpha^{H}_{knz}(M)^2\right\} \cdot I + \lambda OO^T \right]^{-1} \left\{\alpha^{H}_{knz}(1) \cdot e_{knz1} + \alpha^{H}_{knz}(2) \cdot e_{knz2} + \cdots + \alpha^{H}_{knz}(M) \cdot e_{knzM}\right\}.$$

Algorithm 2.2 Dictionary atom update using RKSOWV.

RKSOWV derivation is mathematically tractable and atom update expression has a closed form. The constraint $\lambda = 0$ on dictionary update expression (2.2) gives expression for regular dictionary atom update. Additional constraints $\|A_i\|_0 = 1, \forall i$ and $\|\alpha_i\|_2^2 = 1, \forall i$ on dictionary update expression (2.2) gives centroids of K-Means as shown below.

$$d_k = \frac{e_{knz1} + e_{knz2} + \ldots e_{knzM}}{M} \tag{2.8}$$

The constraint $\|\alpha_i\|_0 = 1, \forall i$ reduces (2.2) to KLine. It is apparent that expression (2.3) for updating dictionary atom is a generalized solution of K-Means and KLine.

2.4 Dictionary Atom with Smoothness Constraint

In this section, a smoothing criterion on dictionary atoms is considered for image de-noising task. Smoothness prior on dictionary atoms itself make sense as natural images are smooth. Underlying dictionary bases have to be aligned towards smoothness. JPEG compression considers nearly smooth DCT bases for extracting information. Mathematical formulation of such dictionary learning with smoothness prior is as follows:

$$\min_{D,A} \|Y - DA\|_F^2 + \lambda\{\|O_h D\|_F^2 + \|O_v D\|_F^2\}$$
$$S.t\|A_i\|_0 \leq L, \forall i. \tag{2.9}$$

In Eq. (2.9), O_h and O_v are horizontal and vertical filtering operator matrix, respectively, as shown below.

$$O_h = \begin{pmatrix} -1 & 1 & 0\,0 \ldots & 0 \\ 0 & -1 & 1\,0 \ldots & 0 \\ 0 & 0 & -1\,1 \ldots & 0 \\ \vdots & \vdots & \vdots\,0 \;\ddots & \vdots \\ 0 & 0 & 0\,0 \ldots & -1 \end{pmatrix}$$

$$O_v = \begin{pmatrix} -1 & 0 & 0\,0 \ldots & 0 \\ 1 & -1 & 0\,0 \ldots & 0 \\ 0 & 1 & -1\,0 \ldots & 0 \\ \vdots & \vdots & \vdots\,0 \;\ddots & \vdots \\ 0 & 0 & 0\,0 \ldots & -1 \end{pmatrix}$$

O_h and O_v force a smoothness prior on dictionary atoms. λ is weighting parameter that decides amount of introduced smoothness. The atom update expression that considers smoothing criterion on dictionary atoms and used for de-noising is given below,

$$d_k = \left[\left\{ \alpha_{knz}^{H}(1)^2 + \alpha_{knz}^{H}(2)^2 + \cdots + \alpha_{knz}^{H}(M)^2 \right\} \cdot I + \lambda\{(O_h O_h{}^T) \right.$$

$$\left. \cdots + (O_v O_v{}^T)\} \right]^{-1}$$

$$\left\{ \alpha_{knz}^{H}(1) \cdot e_{knz1} + \alpha_{knz}^{H}(2) \cdot e_{knz2} + \cdots + \alpha_{knz}^{H}(M) \cdot e_{knzM} \right\}.$$

The denominator of d_k for RKSOWV is a well structured tridiagonal matrix and computational complexity of finding d_k is of similar order as EKSVD [29]. Dictionary learned using such an expression is used for image de-noising in the next section.

2.5 Gray Image De-Noising

In this section, we apply RKSOWV to image processing tasks such as gray image de-noising and compare results with other state-of-the-art dictionary-based algorithms. De-noising is one of the most fundamental inverse problems in signal processing. Image de-noising problem for additive noise is modeled as,

$$Y = X + \mathcal{N}. \tag{2.10}$$

where, Y is a noisy image, $\mathcal{N}$ is an additive zero-mean, white, and homogeneous Gaussian noise, with standard deviation σ, and X is the desired image.

Sparsity prior-based image de-noising has drawn much attention in recent decades. State-of-the-art results have been obtained for image de-noising using sparsity-based dictionary learning [17, 19]. Gray image de-noising using KSVD-based dictionary [17] is formulated as follows:

$$\{\hat{A}_{ij}, \hat{X}\} = \min_{A_{i,j},X} \quad \lambda\|X - Y\|_2^2 + \sum_{ij} \mu_{ij}\|A_{ij}\|_0 + \sum_{ij} \|DA_{ij} - R_{ij}X\|_2^2 \tag{2.11}$$

Here, A is sparse representation matrix. λ and μ are constants. The R_{ij} is a matrix that extracts the (ij) block from the image.

The de-noised image is obtained by Eq. (2.12).

$$\hat{X} = \left(\lambda I + \sum_{ij} R_{ij}^T R_{ij} \right)^{-1} \left(\lambda Y + \sum_{ij} R_{ij}^T D \hat{A}_{ij} \right) \quad (2.12)$$

We use Eq. (2.12) as suggested in [17] for gray image de-noising and compare regularized KSOWV (RKSOWV) with ODCT, MOD [8–10], global KSVD (GKSVD), adaptive KSVD [11], and sparse KSVD (SKSVD) [16].

In ODCT, we obtain atoms analytically by over sampling cosine waveforms for a set of frequencies. Global KSVD is learned on 250000 patches extracted from a dataset of 1381 images of McGill Calibrated Color Image Database [32]. This dataset contains color images from a set of diverse classes such as flowers, animals, foliage, textures, fruits, landscapes, winter, man made, and shadows. Images are transformed to gray for extracting training patches. Global KSVD are learned for sparsity constraints of 8 and 200 iterations, respectively. MOD [8–10], KSVD [11], SKSVD [16], and RKSOWV learn dictionary from patches present in the noisy image itself. Atom sparsity for SKSVD [16] has been selected to be 6 as suggested in [16]. The size of patch is 9×9 for ODCT, MOD [8–10], GKSVD, adaptive KSVD [11], SKSVD [16], and RKSOWV. Dictionary over-completeness factor for all dictionary based approach is 4. Dictionaries for MOD [8–10], adaptive KSVD [11], SKSVD [16], and RKSOWV are learned for 20 iterations. Images selected for testing dictionary based de-noising are (1) Lena, (2) Barbara, (3) Boats, (4) Finger print, (5) House, and (6) Peppers. Noise $\mathcal{N}$ of standard deviation $\sigma = 10, 20, 25, 50, 75, \& 100$ are added to simulate a set of noisy images.

PSNR of de-noised images using RKSOWV are compared to ODCT, MOD [8–10], GKSVD, KSVD [11], and SKSVD [16] in Table 2.1.

The layout of each cell of PSNR values of de-noised images in Table 2.1 are ordered as: (From top to bottom and left to right) ODCT, MOD [8–10], GKSVD, KSVD [11], SKSVD [16], and RKSOWV. As seen from Table 2.1, overall better de-noising results in terms of PSNR is obtained using RKSOWV. The RKSOWV performs as good as KSVD [11] for low σ values such as 10, 20, and 25. For higher σ values (50, 75, and 100), its PSNR is similar to SKSVD [16]. Table 2.2 shows standard deviation of de-noised image PSNR using KSVD [11]

Table 2.1 PSNR of de-noised gray images. Entries of each cell are ordered as: (From top to bottom and left to right) ODCT, MOD [8–10], GKSVD, KSVD [11], SKSVD [16], and RKSOWV

σ/PSNR	Lena		Barb		Boats		Fgrpt		House		Peppers		Average	
	35.88	35.91	34.13	34.18	34.01	33.98	32.35	32.36	38.37	38.59	36.05	36.04	35.13	35.18
10/28.14	36.05	36.06	33.41	34.38	34.18	34.22	32.57	32.37	38.60	38.90	36.27	36.28	35.18	**35.37**
	36.01	36.04	34.24	34.37	34.18	34.21	32.36	32.38	38.90	38.90	35.74	36.27	35.24	**35.36**
	32.16	32.37	29.97	30.42	30.01	30.14	28.08	28.38	35.29	35.04	32.68	32.79	31.36	31.52
20/22.10	32.57	32.64	29.40	30.77	30.42	30.53	28.49	28.39	35.48	35.78	33.23	33.26	31.60	**31.90**
	32.54	32.63	30.56	30.75	30.41	30.53	28.30	28.40	35.87	35.75	33.12	33.27	31.80	**31.89**
	30.98	31.28	28.63	28.98	28.83	29.04	26.74	27.11	34.25	34.21	31.66	31.86	30.18	30.42
25/20.17	31.47	31.53	28.12	29.52	29.33	29.47	27.21	27.21	34.52	34.83	32.27	32.33	30.49	**30.81**
	31.46	31.52	29.36	29.53	29.36	29.47	27.11	27.21	34.89	34.84	32.20	32.33	30.73	**30.82**
	27.30	27.36	24.59	24.81	25.37	25.45	22.00	22.02	30.43	29.79	28.22	28.27	26.32	26.29
50/14.15	27.85	27.82	24.27	25.38	25.88	25.91	22.93	23.36	30.62	30.72	28.98	28.95	26.76	27.02
	27.87	27.86	25.36	25.37	25.94	25.96	23.35	23.40	30.79	30.75	28.97	28.99	**27.05**	**27.06**
	25.50	25.47	22.73	22.71	23.68	23.70	18.99	18.96	27.57	27.42	26.14	26.08	24.10	24.05
75/10.63	25.89	25.74	22.58	22.89	23.98	23.88	19.63	19.81	27.67	27.66	26.60	26.49	24.39	24.41
	25.80	25.81	23.01	22.92	23.95	23.96	19.85	19.89	27.74	27.69	26.54	26.54	**24.48**	**24.47**
	24.44	24.47	21.75	21.69	22.67	22.66	17.64	17.78	26.18	26.18	24.86	24.84	22.92	22.94
100/08.12	24.62	24.49	21.71	21.67	22.83	22.68	18.00	18.01	26.20	26.13	25.06	24.90	23.07	22.98
	24.57	24.58	21.79	21.69	22.75	22.78	18.07	18.14	26.24	26.19	25.01	24.99	**23.07**	**23.06**

Table 2.2 Standard deviation of de-noised image PSNR using KSVD [11] and RKSOWV

σ	**10**		**20**		**25**		**50**		**75**		**100**		**Legend**	
PSNR	0.0122	0.0084	0.0158	0.0130	0.0084	0.0089	0.0167	0.0167	0.0447	0.0482	0.0071	0.0045	KSVD [11]	RKSOWV

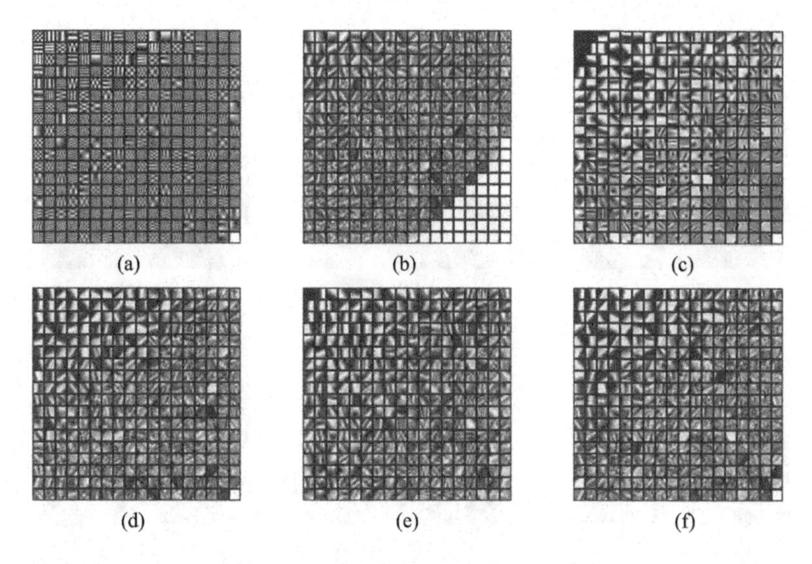

Figure 2.1 Dictionary atoms learned using patches of gray Lena image contaminated with noise of standard deviation 25: (a) ODCT, (b) MOD [8–10], c) GKSVD, (d) KSVD [11], (e) SKSVD [16], (f) RKSOWV.

and RKSOWV. Standard deviation has been calculated over five de-noising experiments.

Figure 2.1 shows dictionaries learned using ODCT, MOD [8–10], GKSVD, KSVD [11], SKSVD [16], and RKSOWV, respectively.

De-noised Lena images using ODCT, MOD [8–10], GKSVD, KSVD [11], SKSVD [16], and RKSOWV are shown in Fig. 2.2.

Figures 2.3 and 2.4 show the zoomed parts of noisy image contaminated with additive zero-mean, white, and homogeneous Gaussian noise, de-noised Lena using SKSVD [16] and RKSOWV for noise standard deviation of 25 and 50, respectively.

2.6 Discriminative K Times Sum of Optimally Weighted Vectors

For classification task, dictionary atoms of one class should be able to discriminate from other data classes. Discrimination is incorporated by penalizing dictionary atom that is closer to feature

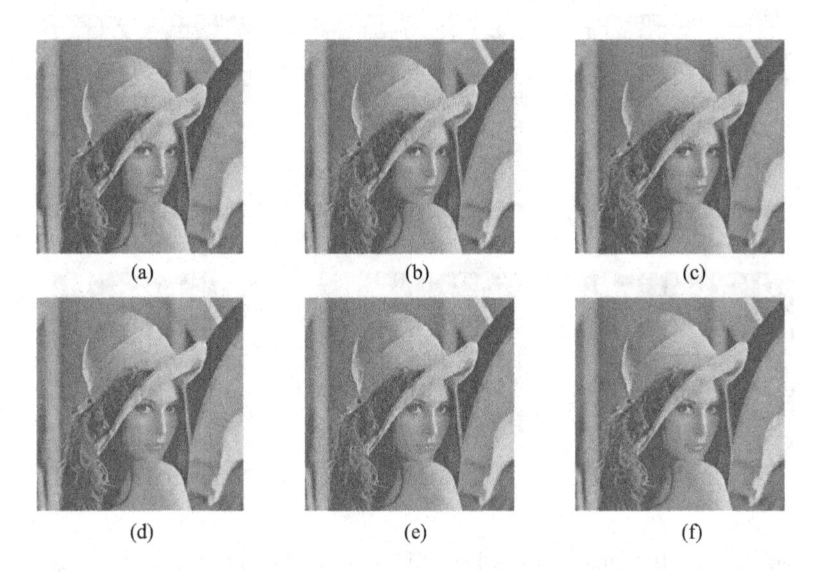

Figure 2.2 De-noised Lena image using different algorithms (Standard deviation of noise added to Lena image = 25): (a) ODCT, (b) MOD [8–10], (c) GKSVD, (d) KSVD [11], (e) SKSVD [16], (f) RKSOWV.

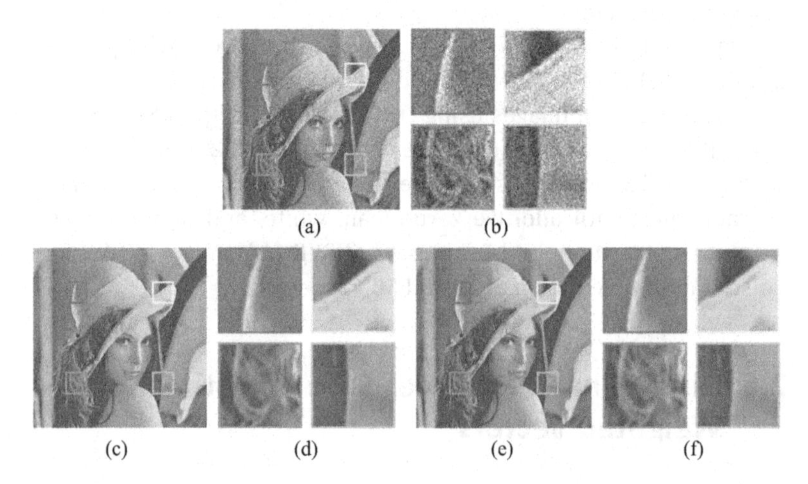

Figure 2.3 De-noised Lena image (Standard deviation of noise added to Lena image = 25): (a) Noisy, (b) Zoomed, (c) SKSVD [16], (d) Zoomed, (e) RKSOWV, and (f) Zoomed.

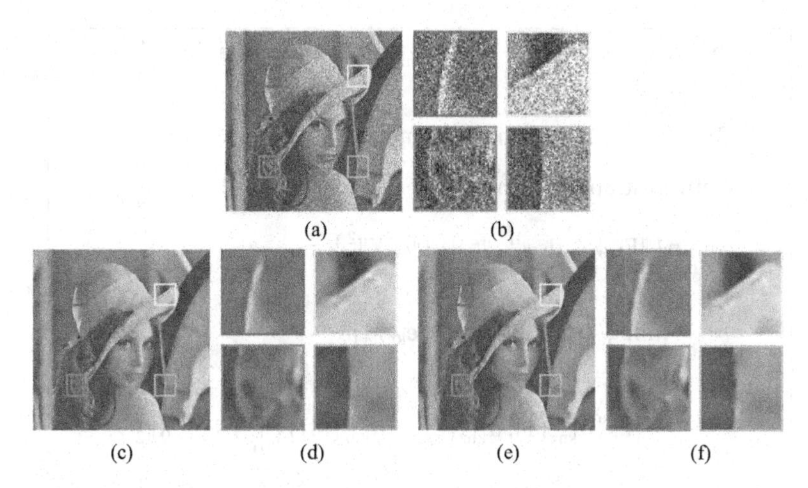

Figure 2.4 De-noised Lena image (Standard deviation of noise added to Lena image = 50): (a) Noisy, (b) Zoomed, (c) SKSVD [16], (d) Zoomed, (e) RKSOWV, and (f) Zoomed.

vectors of other classes. Mathematically, it is formulated as following dictionary learning problem.

$$\min_{D,A} \left\{ \|Y - DA\|_F^2 + \lambda \|D^T X\|_2^2 \right\} \quad \text{S.t. } \|A_i\|_0 \le L, \forall i. \tag{2.13}$$

Y and X are matrices containing training vectors from the given class and remaining classes, respectively. In Eq. (2.13), D is dictionary, A is sparse representation matrix, L is sparsity constraint, and λ is weight. Above dictionary learning problem is solved by estimating D and A alternately for a number of iterations. Balance between class specific characteristic and distinction from rest of the class is maintained using weighting factor λ that is found using validation. Discriminative dictionary atom update is performed using Eq. (2.14).

$$d_k = \frac{\alpha_{knz}^H(1) \cdot e_{knz1} + \alpha_{knz}^H(2) \cdot e_{knz2} + \cdots + \alpha_{knz}^H(M) \cdot e_{knzM}}{\{\alpha_{knz}^H(1)^2 + \alpha_{knz}^H(2)^2 + \cdots + \alpha_{knz}^H(M)^2\} \cdot I + \lambda XX^T} \tag{2.14}$$

Atom update algorithm for discriminative dictionary is given in Fig. 2.3.

1: Input: Error matrix E_{knz}, α^{H}_{knz}, Rest class training Vectors - $X = \{x_i\}_1^N$.

2: Task: To learn dictionary atom d_k.

3: Optimization problem for d_k:

$$\min_{d_k}\{\|E_{knz} - d_k\alpha^{H}_{knz}\|_F^2 + \lambda\|d_k^T X\|_2^2\}.$$

$$= \min_{d_k}\left\{\left\|(e_{knz1}, e_{knz2} \ldots e_{knzM}) - \left[d_k\left\{\alpha^{H}_{knz}(1), \alpha^{H}_{knz}(2) \ldots \alpha^{H}_{knz}(M)\right\}\right]\right\|_F^2 + \lambda\|d_k^T X\|_2^2\right\}.$$

$$= \min_{d_k}\left\{\left\|(e_{knz1}, e_{knz2} \ldots e_{knzM}) - \{\alpha^{H}_{knz}(1), \alpha^{H}_{knz}(2) \ldots \alpha^{H}_{knz}(M)\} \cdot d_k]\right\|_F^2 + \lambda\|d_k^T X\|_2^2\right\}.$$

$$= \min_{d_k}\left\{\|e_{knz1} - \alpha^{H}_{knz}(1) \cdot d_k\|_2^2 + \|e_{knz2} - \alpha^{H}_{knz}(2) \cdot d_k\|_2^2 \cdots + \|e_{knzM} - \alpha^{H}_{knz}(M) \cdot d_k\|_2^2 + \lambda\|d_k^T X\|_2^2\right\}.$$

4: Differentiating with respect to d_k and equating to zero gives expression for d_k:

$$d_k = \left[\left\{\alpha^{H}_{knz}(1)^2 + \alpha^{H}_{knz}(2)^2 + \ldots \alpha^{H}_{knz}(M)^2\right\} \cdot I + \lambda XX^T\right]^{-1} \left\{\alpha^{H}_{knz}(1) \cdot e_{knz1} + \alpha^{H}_{knz}(2) \cdot e_{knz2} + \ldots \alpha^{H}_{knz}(M) \cdot e_{knzM}\right\}.$$

Algorithm 2.3 Dictionary atom update using Discriminative K Times Sum of Optimally Weighted Vectors.

Atom update expression is essentially transformation of sum of optimally weighted associated training vectors to an atom. Transformation matrix is decided by both class of data vectors

(See denominator in 2.14). Dictionary atom learned using DKSOWV maintains similarity with training vectors with unnoticeable visual distinction. The atom update expression of DKSOWV has a closed form.

2.7 Fast RKSOWV

RKSOWV can be used to take advantage of parallel computing resources. Aforementioned dictionary learning algorithms are all executed serially. Fast execution of dictionary learning algorithm can be beneficial to a large number of real time applications. An approximate version of RKSOWV algorithm that is parallelized is discussed here. Consider following dictionary atom update formulation.

$$\min_{d_k}\left\{\|(Y-\sum_{i\neq k} d_i\alpha_i^T)-d_k\alpha_k^T\|_F^2+\lambda\|Od_k\|_2^2\right\} \tag{2.15}$$

Equation (2.15) is applied K times sequentially to update each dictionary atom in a dictionary update iteration. Invoking K such equation solvers (2.15) at a time gives parallelized speedup version of RKSOWV. By updating dictionary atom in parallel, estimated error is not updated as in case of original sequential dictionary atom learning. Parallel dictionary atom update setting can significantly improve dictionary learning time depending upon the extent of parallelization, compared to serial execution.

2.8 Conclusion

We have discussed regularized sparsity prior-based dictionary learning algorithm named RKSOWV (Regularized K Times Sum of Optimally Weighted Vectors). A smoothing regularizer on dictionary atom has been considered for image de-noising. RKSOWV has a closed form expression for updating the dictionary atom. Dictionary atom learned using RKSOWV preserves both structure and color that are present in training vectors. RKSOWV has been shown to achieve overall superior PSNR for image de-noising application compared to adaptive KSVD [11], and sparse KSVD (SKSVD) [16]. The RKSOWV gives PSNR as good as KSVD [11] for low σ values such as 10, 20, and 25. For higher σ values (50, 75, and 100), its PSNR is similar to SKSVD [16]. Dictionary learning is fundamentally a generative approach. For learning a discriminative dictionary for the purpose of classification, a penalty on dictionary atom learning has been imposed and a closed form expression has been obtained.

References

1. C. Cortes and V. Vapnik, Support-vector networks, *Machine Learning*, vol. 20, no. 3, pp. 273–297, 1995.
2. B. E. Boser, I. M. Guyon, and V. N. Vapnik, A training algorithm for optimal margin classifiers, in *Proceedings of the Fifth Annual Workshop on Computational Learning Theory*. ACM, 1992, pp. 144–152.
3. C. M. Bishop, *Neural Networks for Pattern Recognition*. Oxford University Press, 1995.
4. J. B. Allen and L. R. Rabiner, A unified approach to short-time Fourier analysis and synthesis, *Proceedings of the IEEE*, vol. 65, no. 11, pp. 1558–1564, 1977.
5. D. Gabor, Theory of communication. part 1: The analysis of information, *Electrical Engineers-Part III: Radio and Communication Engineering, Journal of the Institution of*, vol. 93, no. 26, pp. 429–441, 1946.
6. S. Mallat, *A Wavelet Tour of Signal Processing: The Sparse Way*. Academic press, 2008.
7. B. A. Olshausen et al., Emergence of simple-cell receptive field properties by learning a sparse code for natural images, *Nature*, vol. 381, no. 6583, pp. 607–609, 1996.
8. K. Engan, S. O. Aase, and J. H. Husøy, Multi-frame compression: Theory and design, *Signal Processing*, vol. 80, no. 10, pp. 2121–2140, 2000.
9. K. Engan, S. O. Aase, and J. H. Husoy, Method of optimal directions for frame design, in *Acoustics, Speech, and Signal Processing, 1999. Proceedings, 1999 IEEE International Conference on*, vol. 5. IEEE, 1999, pp. 2443–2446.
10. K. Engan, B. D. Rao, and K. Kreutz-Delgado, Frame design using focuss with method of optimal directions (mod), in *Proc. NORSIG*, vol. 99, 1999, p. 9.

11. M. Aharon, M. Elad, and A. Bruckstein, *k*-svd: An algorithm for designing overcomplete dictionaries for sparse representation, *IEEE Transactions on Signal Processing*, vol. 54, no. 11, pp. 4311–4322, 2006.
12. R. Vidal, Y. Ma, and S. Sastry, Generalized principal component analysis (gpca), *IEEE Transactions on Pattern Analysis and Machine Intelligence*, vol. 27, no. 12, pp. 1945–1959, 2005.
13. W. B. Pennebaker and J. L. Mitchell, *JPEG: Still Image Data Compression Standard*. Springer Science & Business Media, 1992.
14. S. Lesage, R. Gribonval, F. Bimbot, and L. Benaroya, Learning unions of orthonormal bases with thresholded singular value decomposition, in *Acoustics, Speech, and Signal Processing, 2005. Proceedings (ICASSP'05). IEEE International Conference on*, vol. 5. IEEE, 2005, pp. v–293.
15. S. Sardy, A. Bruce, and P. Tseng, Block coordinate relaxation methods for nonparametric signal denoising with wavelet dictionaries, Citeseer, Tech. Rep., 1998.
16. R. Rubinstein, M. Zibulevsky, and M. Elad, Double sparsity: Learning sparse dictionaries for sparse signal approximation, *IEEE Transactions on Signal Processing*, vol. 58, no. 3, pp. 1553–1564, 2010.
17. M. Elad and M. Aharon, Image denoising via sparse and redundant representations over learned dictionaries, *IEEE Transactions on Image Processing*, vol. 15, no. 12, pp. 3736–3745, 2006.
18. J. Yang, J. Wright, T. S. Huang, and Y. Ma, Image super-resolution via sparse representation, *IEEE transactions on Image Processing*, vol. 19, no. 11, pp. 2861–2873, 2010.
19. J. Mairal, M. Elad, and G. Sapiro, Sparse representation for color image restoration, *IEEE Transactions on Image Processing*, vol. 17, no. 1, pp. 53–69, 2008.
20. X. Wu, D. Gao, G. Shi, and D. Liu, Color demosaicking with sparse representations, in *Image Processing (ICIP), 2010 17th IEEE International Conference on*. IEEE, 2010, pp. 1645–1648.
21. R. Rubinstein, A. M. Bruckstein, and M. Elad, Dictionaries for sparse representation modeling, *Proceedings of the IEEE*, vol. 98, no. 6, pp. 1045–1057, 2010.
22. I. Tosic and P. Frossard, Dictionary learning, *IEEE Signal Processing Magazine*, vol. 28, no. 2, pp. 27–38, 2011.
23. A. K. Jain, *Fundamentals of Digital Image Processing*. Prentice-Hall, Inc., 1989.
24. I. Jolliffe, *Principal Component Analysis*. Wiley Online Library, 2002.

25. A. Gersho and R. M. Gray, *Vector Quantization and Signal Compression*. Springer Science & Business Media, 2012, vol. 159.
26. K. Kreutz-Delgado, J. F. Murray, B. D. Rao, K. Engan, T.-W. Lee, and T. J. Sejnowski, Dictionary learning algorithms for sparse representation, *Neural Computation*, vol. 15, no. 2, pp. 349–396, 2003.
27. E. W. Cheney and A. K. Cline, Topics in sparse approximation, 2004.
28. Y. C. Pati, R. Rezaiifar, and P. Krishnaprasad, Orthogonal matching pursuit: Recursive function approximation with applications to wavelet decomposition, in *Signals, Systems and Computers, 1993. 1993 Conference Record of The Twenty-Seventh Asilomar Conference on*. 1993, pp. 40–44.
29. R. Rubinstein, M. Zibulevsky, and M. Elad, Efficient implementation of the k-svd algorithm using batch orthogonal matching pursuit, *Cs Technion*, vol. 40, no. 8, pp. 1–15, 2008.
30. G. Golub and W. Kahan, Calculating the singular values and pseudo-inverse of a matrix, *Journal of the Society for Industrial and Applied Mathematics, Series B: Numerical Analysis*, vol. 2, no. 2, pp. 205–224, 1965.
31. G. H. Golub and C. Reinsch, Singular value decomposition and least squares solutions, *Numerische mathematik*, vol. 14, no. 5, pp. 403–420, 1970.
32. A. Olmos et al., A biologically inspired algorithm for the recovery of shading and reflectance images, *Perception*, vol. 33, no. 12, pp. 1463–1473, 2004.

Chapter 3

A Comprehensive Survey of Persistent Homology for Pattern Recognition

Zhen Zhou,[a] Yongzhen Huang,[a] Rocio Gonzalez-Diaz,[b] Liang Wang,[a] and Tieniu Tan[a]

[a]*Institute of Automation, Chinese Academy of Sciences, China*
[b]*Department of Applied Math, School of Computer Engineering, University of Seville, Spain*
zzhou@nlpr.ia.ac.cn, tnt@nlpr.ia.ac.cn

Persistent homology, one of the most important theories in algebraic topology, has attracted much attention in a growing number of studies such as clustering, shape analysis, natural image statistics and object recognition, showing its great potential in the tasks of pattern recognition. However, persistent homology is defined based on the knowledge of both the group theory and topology, which is not popular in the community of pattern recognition and not easy to get access for non-specialists. In this chapter, we make such a timely survey for pattern recognition, in which (1) mathematical backgrounds are extensively described; (2) various applications with persistent homology are introduced; (3) their relations are exploited and concluded as an appropriate taxonomy; (4) and finally open directions are discussed. It is believed that this work will benefit both beginners and practitioners in the area

Learning Approaches in Signal Processing
Edited by Wan-Chi Siu, Lap-Pui Chau, Liang Wang, and Tieniu Tan

ISBN 978-981-4800-50-1 (Hardcover), 978-0-429-06114-1 (eBook)
www.panstanford.com

of pattern recognition, and thus promote the future development of this field.

3.1 Introduction

3.1.1 Motivation

One of the most common properties of various data in pattern recognition is the shape, and the shape matters. However, the shape can appear with uncertain appearances, e.g., the shapes of a person in different poses. The most fundamental features of an arbitrary shape are the number of connected components, the number of holes and its higher-dimensional counterparts. These factors define topological invariants, which may be as simple as integers but represent complicated structures. This is just the place where topology comes into play for pattern recognition. In a nutshell, topology is a mathematical branch studying the properties of shapes that are preserved under continuous deformations, and provides a robust measure between deformed shapes.

Among various branches of topology, algebraic topology may be the most promising one to design efficient algorithms for describing topological invariants, and persistent homology (Edelsbrunner et al., 2000) is one of the most important theories in algebraic topology (Hatcher, 2002). It is widely used in topological data analysis (Li et al., 2014), which is a new area dealing with two main problems (Ghrist, 2008): Inferring a high-dimensional structure from low-dimensional representations, and assembling discrete points into a global structure. It has already achieved great success in discovering information hidden in many complex data sets (Emmett et al., 2014; Wildani and Sharpee, 2014; Xia et al., 2015). It is also worth trying to solve some problems in pattern recognition by persistent homology because

- most data in pattern recognition, e.g., images, videos and speeches waves, are inherently discrete and their global structures have not been explored by topological methods;
- features extracted from images, videos and speeches are generally of high dimension, which is hard to explore by traditional methods.

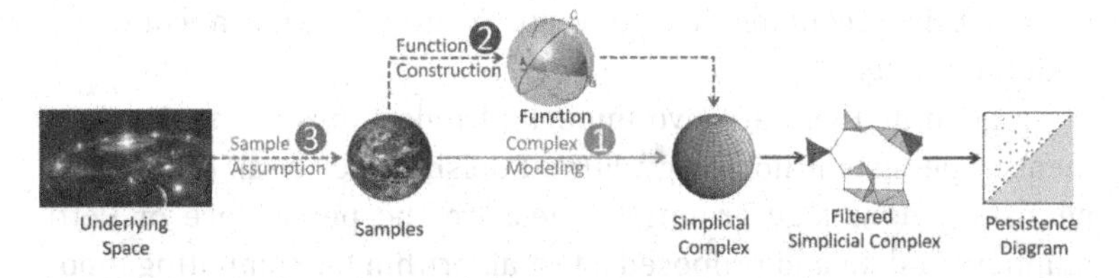

Figure 3.1 The pipeline of applying persistent homology and its three major research issues indicated by the numbers. See Section 3.2 for the relevant concepts and Section 3.3 for more detailed explanation of this pipeline.

Although persistent homology may be very useful in pattern recognition, there exist some difficulties in practice, which motivates us to write such a survey here.

- It is not easy to understand the theory of persistent homology, especially for those who lack related knowledge of the group theory. Therefore, we first present a comprehensive introduction of persistent homology.
- At the same time, the theory of persistent homology and its development are out of the study of pattern recognition. Therefore, to give a better understand how it works in solving related tasks of pattern recognition, we summarize its pipeline and provide a taxonomy of existing works according to its research issues (Fig. 3.1).
- Moreover, for both the beginners and practitioners in the area of pattern recognition who are interested in persistent homology, a summary of promising perspectives and open directions is needed, which is presented at the end of this chapter.

We believe that this work will benefit both beginners and practitioners in the field of pattern recognition.

3.1.2 Persistent Homology

In this part, we briefly depict the development of persistent homology. The pipeline of applying persistent homology into a specific task is shown in Fig. 3.1, which provides an intuitive expression. Since persistent homology is somewhat mathematically heavy, it is recommended to read Section 3.2 for a complete understanding.

In general, there are two important milestones in the development of persistent homology. First, Edelsbrunner et al. (2000) use persistent homology groups to measure the persistence of Betti numbers of data and proposed a fast algorithm for computing topological persistence in $\mathbb{R}^3$. They also divided these persistences as topological features or noise by simplification. Second, Zomorodian and Carlsson (2005) present a generalized method for computing persistence homology of spaces in an arbitrary dimension. This enables us to compute the topology of high-dimensional data. Afterward a growing number of works are proposed to apply persistent homology for real tasks (Carlsson et al., 2008; Chazal et al., 2009b, 2013b; Hu and Yin, 2013; Lamar-León et al., 2012; Li et al., 2014).

Persistent homology has achieved success in several applications, like neuroscience (Singh et al., 2008), bioinformatics (Kasson et al., 2007), shape classification (Chazal et al., 2009b), and sensor networks (De Silva and Ghrist, 2007). Moreover, recent pioneering works incorporate persistent homology with pattern recognition (Chazal et al., 2013b; Hu and Yin, 2013; Lamar-León et al., 2012; Li et al., 2014; Zhu, 2013). These works demonstrate that persistent homology alone or as complementary structural information can not only improve the performance of other methods, but also provide a new insight into the given problems.

3.1.3 Previous Work

The most related work of this chapter is probably the one proposed by Freedman and Chen (2009). However, there are several significant differences between them.

- Although both of them involve algebraic topology, there are six years since their paper was published, and it is necessary to make a timely survey introducing the latest development. A lot of new works have been established during these years (see Table 3.1 for details).

Table 3.1 Summary of the taxonomy of issues when applying persistent homology

Taxonomy	Topics	Applications
Complex Modeling	Vietoris–Rips complex, Čech complex, Delanunay complex, etc.	Natural image statistics (Lee et al., 2003), (Carlsson et al., 2008), (Adams and Carlsson, 2009) Shape analysis (Carlsson et al., 2004), (Collins et al., 2004), (Memoli, 2007), (Bronstein et al., 2009), (Skraba et al., 2010) Natural language processing (Zhu, 2013), Human identification (Lamar-León et al., 2012) Hand gesture recognition (Hu and Yin, 2013)
Function Construction	Density estimator, Heat kernel signature, convolution	Clustering and Segmentation (Chazal and Oudot, 2008), (Skraba et al., 2010), (Chazal et al., 2011), (Chazal et al., 2013b) Object recognition (Li et al., 2014), Image diffusion (Chen and Edelsbrunner, 2011)
Sample Assumption	Approximation, Converge, Stability	Manifold estimation, (Adler et al., 2010), (Genovese et al., 2012), Homology estimation, (Niyogi et al., 2008), (Balakrishnan et al., 2012) Persistence diagram estimation (De Vito et al., 2013), (Mileyko et al., 2011), (Bubenik, 2012)

- The major concerns are different. They focus on applications in computer vision, while we depict applications in a wider range. In particular, we not only introduce those appearing in their paper (e.g., shape analysis, natural image statistics, image denoising and segmentation), but also concern more topics in pattern recognition (e.g., natural language processing, clustering, manifold estimation).
- The way to introduce the development of persistent homology is different. They presented four applications in computer vision one by one, while we propose a taxonomy of existing works according to the general pipeline of applying persistent homology (Section 3.3). Such kind of taxonomy is more reasonable because the pipeline of applying persistent homology among various applications is similar.
- Moreover, we summarize several perspectives and open directions of persistent homology for future research, while these were not involved in their paper.

The main ideas of this chapter are summarized as follows.

- Propose a proper taxonomy of the existing works about applying persistent homology in pattern recognition. To the best of knowledge, this is the first taxonomy for the existing works in pattern recognition by using persistent homology. The taxonomy is built according to the pipeline of persistent homology, which is reasonable, i.e., the proposed taxonomy is consistent with the use of persistent homology in most literatures.
- Provide a comprehensive survey on the background of persistent homology and its recent progress in pattern recognition. This survey can help the beginners and the practitioners to quickly get familiar with this research field, and also understand its recent development in pattern recognition.
- Summarize several perspectives and open directions of persistent homology for future research.

The rest of this chapter is clearly organized, as shown in Fig. 3.2. In order to reveal main research issues when applying persistent homology (as shown in Fig. 3.1), we first introduce

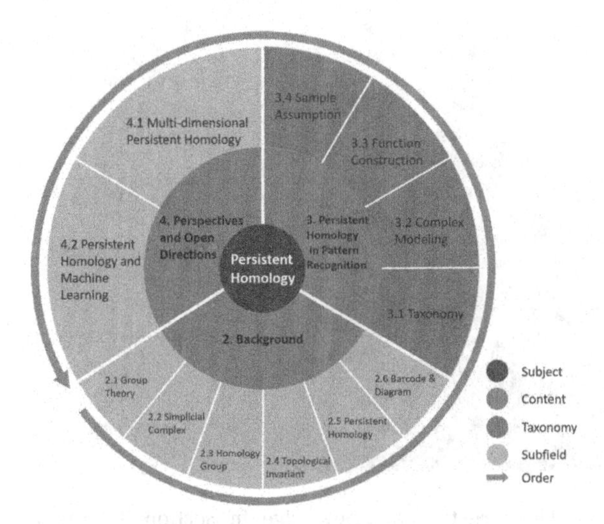

Figure 3.2 The organization of the rest of this chapter.

some background knowledge. Moreover, for the sake of a systematic understanding of persistent homology, it is necessary to further depict a comprehensive introduction of basic concepts in Section 3.2, i.e., from the fundamental group theory, then homology groups, and finally persistent homology, as organized in Fig. 3.3. Subsequently, three common issues involved in the proposed taxonomy are summarized in Section 3.3. The corresponding existing works are stated along each issue. Section 3.4 gives some perspectives and open directions for future research. We conclude the chapter in Section 3.5.

3.2 Background

In this section, we focus on the preliminary of persistent homology. It is related to abstract algebra and general topology. The structure of knowledge in this part is displayed in Fig. 3.3. First we will introduce the basis of group theory and simplicial complex. Simplicial complex is the discrete topological space that we are going to deal with while the group theory is the corresponding algebraic tool. The group theory helps to transfer the problem of counting the numbers

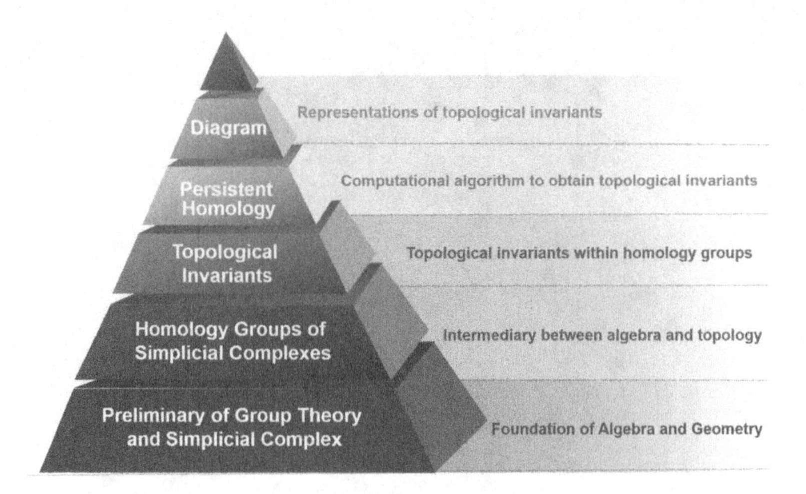

Figure 3.3 The structure of knowledge in Section 3.2 to understand persistent homology (from bottom to top).

of components and holes from a geometric issue to an algebraic one, which is convenient for formulation and inference. The result by using the group theory to describe simplicial complex is the homology group of simplicial complex. By exploring these homology groups, we can obtain the topological invariant, i.e., the number of components, holes and their counterparts. However, these single numbers are not very informative. For example, one square and one circle have the same number of components and holes, which makes them indistinguishable with respect to topological invariants (i.e., invariants under continuous deformation). This is where persistent homology enters the story. It tracks topological invariants at different spatial resolutions, which provides a way to differentiate objects that are topologically equivalent (i.e., one object can be continuously deformed to the other). In order to depict what persistent homology has done, we visualize the output of persistent homology by a barcode or a persistence diagram. Accordingly we can easily define metrics on these diagrams and measure the similarity of two objects in terms of their persistent homologies. Unless especially mentioned, the mathematical concepts in this section are derived from Armstrong (1983); Hatcher (2002); Lang (2005); Munkres (1984).

3.2.1 Preliminary of Group Theory

Definition 3.2.1.1. A **group** $< G, \star >$ is a set together with a binary operator $\star$ such that the following properties hold.

(1) Closure: If x and y are elements in G, then $x \star y$ is also in G.
(2) Associativity: For all x, y, z in G, $(x \star y) \star z = x \star (y \star z)$.
(3) Identity: There exists an identity element e in G such that $e \star x = x \star e = x$ for all x in G.
(4) Inverse: If x is an element in G, then there exists an inverse element y in G such that $x \star y = y \star x = e$.

If $\star$ represents the multiplication $\times$ (the addition $+$), then $e = 1$ ($e = 0$) and G is a multiplicative group (additive group). If $\star$ also satisfies being commutative, i.e., $x \star y = y \star x$ for all x, y in G, G is called Abelian group.

We give several examples of groups. $< \mathbb{Z}, + >$ and $< \mathbb{R}, + >$ are additive groups (also Abelian groups) with identity 0 and $-x$ as the inverse of x. However, $< \mathbb{R}, \times >$ is not a multiplicative group because 0 does not satisfy the inverse condition. Therefore, $< \mathbb{R}\backslash\{0\}, \times >$ and $< \mathbb{R}^+, \times >$ are multiplicative groups (also Abelian groups) with identity 1 and $\frac{1}{x}$ as the inverse of x. Sometimes, we just call G a group when the operator is explicit.

Definition 3.2.1.2. A **subgroup** $< H, \star >$ of $< G, \star >$ refers to a subset H of G and is itself a group.

The identity element of G constitutes a subgroup, which is called the trivial subgroup. $< \mathbb{R}^+, \times >$ is a subgroup of $< \mathbb{R}\backslash\{0\}, \times >$.

Definition 3.2.1.3. Let $< G, \star >$, $< G', * >$ be groups. A **homomorphism** $f : G \to G'$ of G into G' is a map satisfying the following property: $\forall x, y \in G$, we have $f(x \star y) = f(x) * f(y)$.

That is to say, the relation of the binary operator could remain unchanged after the homomorphism. For example, the map $f : x \mapsto e^x$ is a group homomorphism of $< \mathbb{R}, + >$ into $< \mathbb{R}^+, \times >$.

Definition 3.2.1.4. A group homomorphism $f : G \to G'$ is an **isomorphism** if it is injective and surjective.

An important property of isomorphisms is that if a property of group G is totally defined in terms of the group operation, then every other group will also have this property as long as it is isomorphic to G.

Definition 3.2.1.5. Let $f : G \to G'$ be a group homomorphism between two groups. The **kernel** of f consists of all elements $x \in G$ such that $f(x) = e'$ where e' is the identity element of G', i.e., $\ker f = \{x \in G | f(x) = e'\}$.

The kernel of f is obviously a subset of G. Moreover, f just "collapses" this subset of G into the identity element of G'. We will see later that this idea is also used in the quotient group.

Theorem 3.2.1.1. *The kernel of a homomorphism $f : G \to G'$ is a subgroup of G (Clark (1984, p. 49)).*

This theorem is useful when we define a group of cycles (Definition 3.2.3.5).

Definition 3.2.1.6. Let $< H, \star >$ be a subgroup of a group $< G, \star >$. a is an element of G. The set $a \star H = \{a \star x, x \in H\}$ is called a (left) **coset** of H in G.

Similarly, we can define a right coset $H \star a$. If group G is an Abelian group, i.e., it is commutative, $a \star H$ and $H \star a$ will be the same.

Definition 3.2.1.7. Let $< H, \star >$ be a subgroup of a group $< G, \star >$ satisfying $x \star H = H \star x$ for all $x \in G$. The **quotient group** (or factor group) is the collection of cosets of H denoted as G/H, i.e., $G/H = \{a \star H | a \in G\}$. The binary operator is defined as follows, $(a \star H)(b \star H) = a \star b \star H$ for all $a, b \in G$. The identity element is H (Corollary 4.6 of Munkres (1984)).

Definition 3.2.1.8. An equivalence relation in a set S refers to a measure between pairs of its elements, written $x \sim y$. The relation must satisfy the following properties. (1) $x \sim x$ for all $x \in S$. (2) If $x \sim y$ then $y \sim x$. (3) If $x \sim y$ and $y \sim z$ then $x \sim z$.

Thus a set S can be partitioned into several subsets $\{S_i\}$. Elements in each subset are equivalent to each other while elements of different subsets are not equivalent. Each subset is called an equivalence class of S. An equivalence class S_i can be represented by any one of its elements, written as $S_i : [x], x \in S_i$. x is called a representative. For example, $\mathbb{Z}$ can be divided into two equivalence classes, i.e., odd integers ([1]) and even integers ([2]), by modulo 2 operation, which is an equivalence relation of $\mathbb{Z}$. The coset is also an equivalence relation of the quotient group, thus leading to a partition of the quotient group.

Now we go back to the quotient group. Accordingly, G/H can be separated into some equivalence classes, which implies that we collapse G into some classes by the coset of H. The advantage is that instead of studying G, we can just explore the properties of its equivalence classes. This property will help to classify topological spaces in the remaining sections.

3.2.2 Simplicial Complex

It should be first pointed out that the class of spaces we are dealing with is the class that can be built from such "building blocks" as line segments, triangles, tetrahedrons, and their higher-dimensional analogs called simplices in general by "gluing them together" along their faces (Munkres, 1984). The reason for defining such a discrete space is that, in a nutshell, it is simpler to understand and compute.

In this part, our introduction will be focused on the geometric realization of simplicial complex, which is easier to understand, compared with abstract simplicial complex which is derived from the set theory. These two concepts are essentially telling the same thing but using two different "languages".

Two points are said to be geometrically independent if they are distinct. The same is true for three non-collinear points, four non-coplanar points, etc. For the same reason, we can define the concept of being geometrically independent on a point set. After that, we can present the definition of a simplex.

Definition 3.2.2.1. Let $\{a_0, \cdots, a_n\}$ be a geometrically independent set in $\mathbb{R}^N$. A **k-simplex** σ spanned by $a_0, \cdots, a_n$ is the set of all points

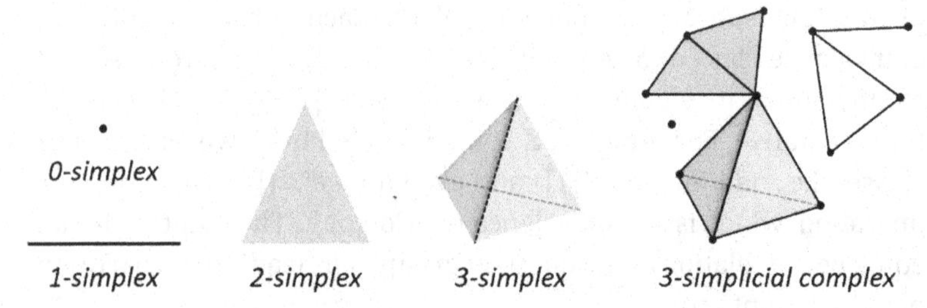

Figure 3.4 An illustration of the k-simplex for $k \leq 3$ and a simplicial complex. A 3-simplicial complex is constructed by properly "gluing together" points (0-simplex), line segments (1-simplices), triangles (2-simplices), tetrahedrons (3-simplices) and their faces.

x of $\mathbb{R}^N$ such that,

$$x = \sum_{i=0}^{n} t_i a_i, \quad \text{where} \quad \sum_{i=0}^{n} t_i = 1 \quad \text{and} \quad \forall t_i \geq 0. \tag{3.1}$$

The above definition implies that a k-simplex is the smallest convex hull of $k + 1$ geometrically independent points. In low dimensions, one can picture it easily, namely, a single point is a 0-simplex, and a line segment is an 1-simplex. In a similar way, a 2-simplex is a triangle and a 3-simplex is a tetrahedron, and so on (Fig. 3.4).

Definition 3.2.2.2. The convex hull of any nonempty subset of $k + 1$ points that define a k-simplex is called a face of the simplex. In particular, the convex hull of a subset of size $d + 1$ (of the $k + 1$ defined points) is a d-simplex, called a d-face of the k-simplex.

For example, the faces of a tetrahedron (a 3-simplex) include a tetrahedron itself (a 3-face), four triangles (2-faces), six edges (1-faces), and four points (0-faces). The sole k-face of the whole k-simplex is itself. Note faces are simplices themselves.

Definition 3.2.2.3. A simplicial complex $\mathcal{K}$ is a set of simplices that satisfies the following conditions.
(1) Any face of a simplex of $\mathcal{K}$ is also in $\mathcal{K}$.
(2) The intersection of any two simplices $\sigma_1, \sigma_2 \in \mathcal{K}$ is a face of both σ_1 and σ_2.

Definition 3.2.2.4. If a subset $\mathcal{K}_i \subseteq \mathcal{K}$ is itself a simplicial complex, then it is a **subcomplex** of $\mathcal{K}$.

A simplicial complex usually consists of a set of simplices, i.e., points, line segments, triangles and their higher-dimensional analogs, together with all their faces. Moreover, a simplicial complex can be constructed from other simplicial complexes by "gluing them together" as long as the glued one is itself a simplicial complex (Fig. 3.4, right).

Definition 3.2.2.5. The dimension of $\mathcal{K}$, denoted as $dim\mathcal{K}$, is the largest dimension of any simplex in $\mathcal{K}$, namely, the dimension of a k-simplicial complex is equal to k.

There are several ways to build a simplicial complex. The most commonly used is the Vietoris–Rips complex. Another similar one is the Čech complex, which is an approximation of the Vietoris–Rips complex with higher efficiency. The following concepts are introduced from Ghrist (2008).

Definition 3.2.2.6. The **Vietoris–Rips complex** $\mathcal{R}_\varepsilon$ is the simplicial complex whose p-simplices correspond to unordered $(p+1)$-tuples of points that are pairwise within a distance ε.

Definition 3.2.2.7. The **Čech complex** $\mathcal{C}_\varepsilon$ is the simplicial complex whose p-simplices correspond to unordered $(p+1)$-tuples of points whose closed $\frac{\varepsilon}{2}$-balls have nonempty intersection.

Here the closed ball is the space inside a sphere and its surface with a radius of $\frac{\varepsilon}{2}$. The top of Fig. 3.8 is a toy example of the Vietoris–Rips complex with different values of ε. It is obvious that the requirement of whether a p-simplex can be built in a Čech complex is more rigorous than that in a Vietoris–Rips complex. Only if the closed $\frac{\varepsilon}{2}$-balls of all three points have nonempty intersection, these three points can be seen as a 2-simplex added into the Čech complex. But for the Vietoris–Rips complex, these three points only need to be pairwise within the distance ε. In other words, a Čech complex can be 'approximated' by Rips complexes. This fact is used in computational applications, since working with Rips complexes is much more efficient than with Čech complexes (Adler et al., 2010).

3.2.3 Homology Groups of Simplicial Complexes

The discrete space aforementioned can be regarded as the underlying space of a simplicial complex. Before we introduce (simplicial) homology groups, we wish to briefly mention that singular homology groups, which are not the emphasis in this chapter, are defined for arbitrary spaces as a generalization of simplicial homology groups, while (simplicial) homology groups are defined on a simplicial complex. Singular homology groups are convenient for theoretical analysis. However, it can be proven that the simplicial homology of a given space does not depend on its triangulation, i.e., the way it is subdivided in points, edges, triangles, tetrahedrons and so on to form a simplicial complex.

Now we are ready to define (simplicial) homology groups of a simplicial complex by the chain group. Rather than defined on formal sum of simplices or oriented simplices which needs some extra concepts, we present this part using a similar manner as in Zhu (2013), which is easy to follow.

Definition 3.2.3.1. A p-chain is a formal linear combination of p-simplices. It can be interpreted as a subset of p-simplices if the ground of coefficients is $\mathbb{Z}_2$.

We take a 3-simplex (a tetrahedron) as an example. Then any subset of its four triangles (2-simplices) comprises a 2-chain of this tetrahedron, thus the total number of different 2-chains will be 2^4. For the same reason, there are 2^6 1-chains (six edges) and 2^4 0-chains (four points). Note that the empty set can be interpreted as the zero p-chain for any p.

We can also redefine p-chain by the operator $+_2$ which represents the binary addition, i.e., even times of repeated items will be canceled. We still use the tetrahedron as an example. Suppose its four triangles are denoted by $T = \{T_i, i \in I = 1, 2, 3, 4\}$. Then its 2-chains can be expressed as T_i, $T_i +_2 T_j$, $T_i +_2 T_j +_2 T_m$, $T_i +_2 T_j +_2 T_m +_2 T_n$ and $\emptyset$ where $i, j, m, n \in I$ are not equal between pairs. Note that the empty set can be written as $\emptyset = T_i +_2 T_i$. Therefore, it is easy to conclude that $< T, +_2 >$ is a group as described in the next theorem. In this way, we often say that such groups are defined over the field of $\mathbb{Z}_2$ (i.e., mod 2). Unless otherwise stated, the addition in the following all refers to $+_2$.

Theorem 3.2.3.1. *The set of p-chains of a simplicial complex forms a p-chain group $C_p(\mathcal{K})$ (Dey (2013, p. 20)).*

If $p < 0$ or $p > dim\mathcal{K}, C_p(\mathcal{K})$ denotes the trivial group. The above definition builds the relationship between the group theory and the simplicial complex. In order to connect p-chains of $\mathcal{K}$ with different values of p, we need the boundary operator.

Definition 3.2.3.2. The **boundary** of a p-simplex is the set of its $(p-1)$-faces.

Namely, the boundary of a tetrahedron is the set of its four triangles while the one of a triangle is its three edges. Obviously the boundary of a line is its two endpoints. The boundary of a single point is empty. Hence we can conclude that the boundary of a p-simplex is the $(p-1)$-chain which is the $+_2$ sum of all the $(p-1)$-faces of the p-simplex.

Definition 3.2.3.3. We define a group homomorphism $\partial_p : C_p(\mathcal{K}) \rightarrow C_{p-1}(\mathcal{K})$, called the **boundary operator**, which is the $+_2$ sum of the boundaries of the simplices of the p-chain $C_p(\mathcal{K})$.

So the boundary of the sum of any two p-chains is equal to the sum of their boundaries, i.e., $\partial_p(c1 + c2) = \partial_p(c1) + \partial_p(c2)$, $c1$, $c2 \in C_p(\mathcal{K})$. Take Fig. 3.5 as an illustration. Just like the example of the empty set aforementioned, faces shared by an even number of p-simplices in the chain will be canceled out.

There is an important conclusion about the boundary operator, which leads us to homology groups.

Theorem 3.2.3.2. *The boundary of a boundary is always zero, i.e., $\partial_p \circ \partial_{p+1} = 0$ (Dey (2013, p. 21)).*

Definition 3.2.3.4. The image of $\partial_{p+1} : C_{p+1}(\mathcal{K}) \rightarrow C_p(\mathcal{K})$ is called the group of p-boundaries and denoted $B_p(\mathcal{K})$.

Definition 3.2.3.5. The kernel of $\partial_p : C_p(\mathcal{K}) \rightarrow C_{p-1}(\mathcal{K})$ is a group of p-cycles and denoted $Z_p(\mathcal{K})$. Thus a p-cycle is a p-chain with the empty boundary, i.e., $\partial_p c = 0$, $c \in C_p(\mathcal{K})$.

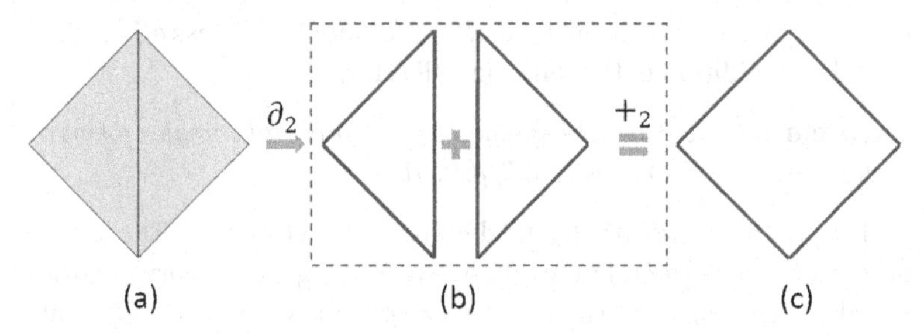

Figure 3.5 Faces shared by an even number of p-simplices in the chain will be canceled out by $+_2$ addition (Zhu, 2013). (a) Two triangles, equivalent to two 2-simplices or two 2-chains, are placed together with a duplicate edge. (b) The separate boundaries of the two triangles, which are also two 1-chains. (c) The boundary of the sum of two 2-chains is equal to the sum of their own boundaries.

The kernel of ∂_p is naturally a group by Theorem 3.2.1.1. To depict the concept of p-cycle, take Fig. 3.5c as an example. This is the border of a rhombus, i.e., an 1-chain. Its boundary is the sum of the boundaries of its four edges while the boundary of an edge is its two endpoints. Thus each point of the rhombus appears twice, which makes them being canceled out when $+_2$ adding up. Therefore, the boundary of the border of a rhombus is empty, i.e., the border of the rhombus is an 1-cycle.

According to Theorem 3.2.3.2, $\forall c \in C_{p+1}(\mathcal{K})$, $\partial_p(\partial_{p+1}c) = 0$, i.e., the boundary of a (p+1)-chain is a p-cycle. This implies that the image of ∂_{p+1}, denoted as $\operatorname{im}\partial_{p+1}$, is a subset of $\ker\partial_p$, i.e., $B_p(\mathcal{K}) \subset Z_p(\mathcal{K})$. Figure 3.6 summarizes the relationship of $C_p(\mathcal{K})$, $Z_p(\mathcal{K})$, and $B_p(\mathcal{K})$ under the boundary operator.

Now we are ready to define homology groups.

Definition 3.2.3.6. The p-th **homology group** of $\mathcal{K}$ is defined as the quotient group

$$H_p(\mathcal{K}) = Z_p(\mathcal{K})/B_p(\mathcal{K}) = \{z + B_p(\mathcal{K}), z \in Z_p(\mathcal{K})\}. \tag{3.2}$$

Moreover, we can simplify this form by using representatives,

$$H_p(\mathcal{K}) = \{[z], z \in H_p(\mathcal{K})\}, \quad \text{where} \quad [z] = z + B_p(\mathcal{K}). \tag{3.3}$$

[z] is a homology class represented by z. Homology groups are naturally Albelian groups. In the next section, we will explain that homology groups are topological invariants.

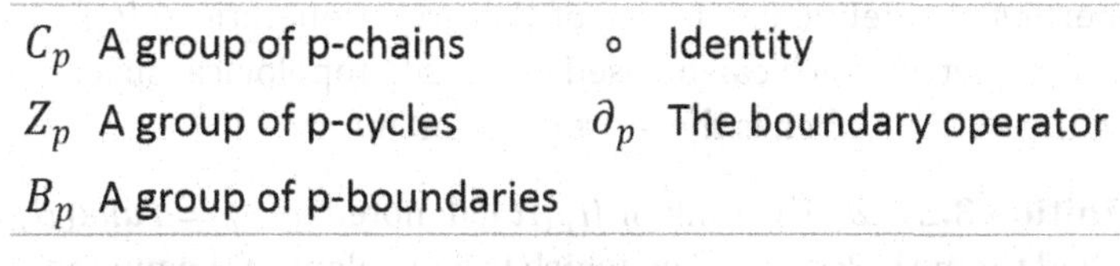

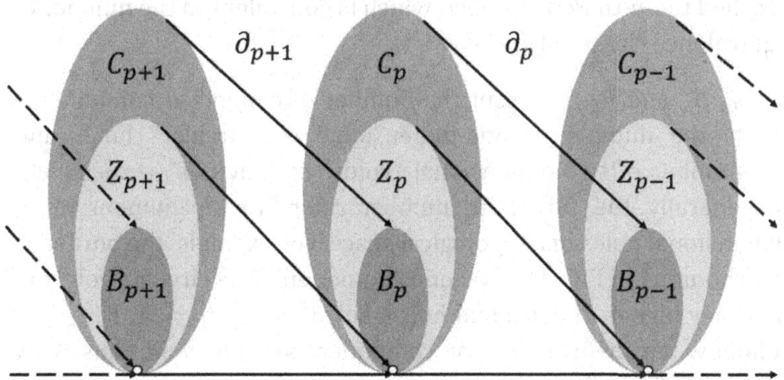

Figure 3.6 The relationship of chain, cycle and boundary groups under the boundary operator. Reprinted with permission from Zomorodian and Carlsson (2005). Copyright (2005) Springer Nature.

3.2.4 Topological Invariant

Definition 3.2.4.1. A function $f : X \to Y$ between two topological spaces X and Y is called *homeomorphism* if it has the following properties: (1) f is a bijection (one-to-one and onto), (2) f is continuous, and (3) the inverse function f^{-1} is continuous (f is an open mapping).

A function with the above properties is sometimes called bicontinuous. If such a function exists, we say X and Y are homeomorphic. The homeomorphisms form an equivalence relation on the class of all topological spaces. The resulting equivalence classes are called homeomorphism classes.

Theorem 3.2.4.1. *If $f : \mathcal{X} \to \mathcal{Y}$ is a homeomorphism, then $f_\# : H_p(\mathcal{X}) \to H_p(\mathcal{Y})$ is an isomorphism. Therefore, $H_p(\mathcal{X})$ is a topological invariant of $\mathcal{X}$ (Hatcher (2002)[Corollary 2.11]).*

As we have mentioned in Definition 3.2.1.4 that two groups will have the same properties as long as they are isomorphism and the properties are defined in terms of the group operation. $H_p(\mathcal{K})$ is such a property, which can be used to classify topological spaces. Its rank, the Betti number, is the most common metric.

Definition 3.2.4.2. The rank of $H_p(\mathcal{X})$, denoted as $\beta_p = rank(H_p)$, is called the p-th *Betti number*, which is equivalent to the number of equivalence classes of $H_p(\mathcal{X})$.

β_0, β_1, and β_2 represent the number of connected components of $\mathcal{X}$, the number of one-dimensional or "circular" holes and the number of two-dimensional "voids" or "cavities", respectively. In generally, the p-th Betti number refers to the number of p-dimensional holes of a topological space. For example, the border of the rhombus in Fig. 3.5c has one component, one circular hole and no other higher-dimensional holes, i.e., $\beta_0 = 1, \beta_1 = 1, \beta_{>1} = 0$. A hollow tetrahedron has one component and one void, thus $\beta_0 = 1, \beta_2 = 1, \beta_{>1} = 0$.

Another commonly used topological invariant is the Euler characteristic, which is derived from Betti numbers.

Definition 3.2.4.3. If β_k is the k-th Betti number of the topological space X, then the *Euler characteristic* is

$$\chi(X) = \sum_{k=0}^{N} (-1)^k \beta_k, \tag{3.4}$$

where N is the dimension of X.

The Euler characteristic behaves well with respect to many basic operations on topological spaces (Kalyanswamy, 2009). For example, a useful concept called Euler integration, which is originated from the Euler characteristic, is proposed to help solving a number of problems, for example, counting the total number of observable targets (e.g., persons, vehicles, landmarks) in a region using local counts performed by a network of sensors (Adler et al., 2007; Baryshnikov and Ghrist, 2009, 2010).

3.2.5 Persistent Homology

Persistent homology, as introduced by Edelsbrunner et al. (2000) and refined by Zomorodian and Carlsson (2005), is a method for computing topological invariants of a space at different spatial resolutions.

Now we recall the boundary operator of p-chain. It is easy to obtain the following chain structure,

$$\cdots \xrightarrow{\partial_{p+1}} C_p(\mathcal{K}) \xrightarrow{\partial_p} C_{p-1}(\mathcal{K}) \xrightarrow{\partial_{p-1}} \cdots \tag{3.5}$$

Persistent homology tracks homology classes along such chain structure. We will demonstrate this later. First, we import the concept of filtration, which induces the chain structure of homology groups.

Definition 3.2.5.1. An ordered set of the cells of K, $\{\sigma_1, \sigma_2, \cdots, \sigma_m\}$, is called a **filter**, such that if σ_i is a proper face of σ_j ('proper' refers to $\sigma_i \neq \sigma_j$) then $i < j$.

Definition 3.2.5.2. The nested sequence of subcomplex sets $\emptyset = \mathcal{K}_0 \subset \mathcal{K}_1 \subset \cdots \subset \mathcal{K}_m = \mathcal{K}$ is called a **filtration** with the constraint that a simplex enters the sequence not earlier than its faces.

We can use a filter function $f : \mathcal{K} \mapsto \mathbb{R}$ to build a filtration, which assigns each simplex in $\mathcal{K}$ a real value. Each value of a simplex is no smaller than that of its faces. Let $f_i = f(\sigma_i)$ and $f_0 = -\infty$. Then $\mathcal{K}_i$ is defined as $f^{-1}(-\infty, f_i]$.

It is easy to obtain an inclusion map from Definition 3.2.5.2.

$$\mathcal{K}_i \hookrightarrow \mathcal{K}_j, 0 \leq i < j \leq m. \tag{3.6}$$

This also induces a group homomorphism of the corresponding homology groups (Freedman and Chen, 2009).

$$F_p^{i,j} : H_p(\mathcal{K}_i) \to H_p(\mathcal{K}_j), 0 \leq i < j \leq m. \tag{3.7}$$

For example, suppose one topological trait (i.e., one homological class) first appears in $H_p(\mathcal{K}_i)$, and it is merged into another existing component in $H_p(\mathcal{K}_j)$. Then the interval of $f_j - f_i$ is called the **lifetime** of this component. And we state that this component is born at f_i and dead at f_j, or equivalently it persists from f_i to f_j. Without loss of generality, when we mention the topological trait of $\mathcal{K}$ or a homology class, it refers to its components ($p = 0$), holes ($p = 1$) and their higher-dimensional counterparts ($p > 1$).

Now we explain how to compute the birth and death time of a topological trait by Fig. 3.7. $H_p(\mathcal{K}_i)$ consists of two parts, one is

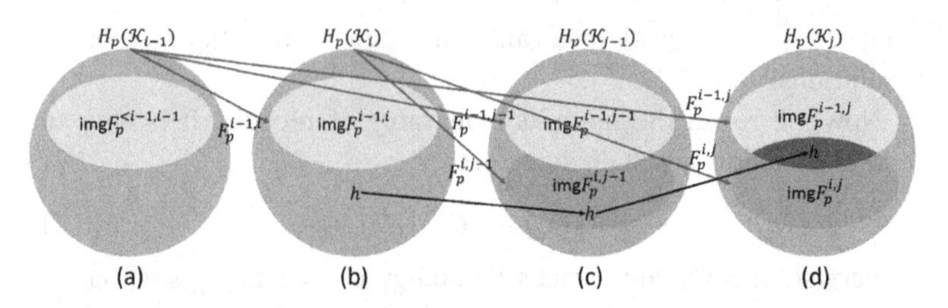

Figure 3.7 A homology class, h, is born at the time f_i if $h \in H_d(\mathcal{K}_i)$ but $h \notin \text{im}(F_d^{i-1,i})$. Given h is born at f_i, h dies at time f_j if $F_d^{i,j-1}(h) \notin \text{im}(F_d^{i-1,j-1})$ but $F_d^{i,j}(h) \in \text{im}(F_d^{i-1,j})$ (Freedman and Chen, 2009). See the context for more details.

the image of $F_p^{i-1,i}$ and the other is the rest. We should notice that all traits that survive at f_i are all in im $F_p^{i-1,i}$. Thus the elements in $H_p(\mathcal{K}_i) \backslash \text{im } F_p^{i-1,i}$ represent new traits. If $h \in H_p(\mathcal{K}_i) \backslash \text{im } F_p^{i-1,i}$, then its birth time is f_i. As to its death time, for example, say f_j, then it should satisfy the following two conditions, h still survives in $H_p(\mathcal{K}_{j-1})$ but merged in $H_p(\mathcal{K}_j)$. We illustrate this in Fig. 3.7. The first condition states that $F_p^{i,j-1}(h)$ should lie in the green region of im $F_p^{i,j-1}$, which means that it is not merged by the traits represented by im $F_p^{i-1,j-1}$. The second one states that $F_p^{i,j}(h)$ should be in the red region, the intersection of im $F_p^{i,j}$ and im $F_p^{i-1,j}$, which implies that h now is merged into the traits represented by im $F_p^{i-1,j}$.

For a fixed value of p, persistent homology tracks Betti numbers along $H_p(\mathcal{K}_i)$ for each value of i. For each value of i, we can compute the homology groups $H_p(\mathcal{K}_i)$, thus obtaining the Betti number of each $\mathcal{K}_i$. Through these maps in Equation 3.7 we know the dynamic changes of each trait, e.g., when it appears and how long it survives.

3.2.6 Barcode and Diagram

In this section, we concern two issues. One is the visualization tools recording the birth and lifetime of a topological trait, including the persistence diagram and barcode. The other is the metric between two persistence diagrams, which is used to distinguish one diagram from another and should be selected such that persistence diagrams are stable with respect to the metric.

Definition 3.2.6.1. Recall that $f : \mathcal{K} \mapsto \mathbb{R}$ and $\mathcal{K}_i = f^{-1}(-\infty, f_i]$. The p-th **persistence diagram** $\text{Dgm} f$ is a collection of points in the extended plane $\overline{\mathbb{R}}^2$, where $\overline{\mathbb{R}} = \mathbb{R} \cup \{-\infty, +\infty\}$, in which each point (b, d) corresponds to a topological trait (homology class) that is born at $H_p(\mathcal{K}_b)$ and dead at $H_p(\mathcal{K}_d)$. Note $b < d$, thus points only exist in the upper half plane.

Take Fig. 3.10 as an example. When $death = +\infty$ ($birth = 0$), we just place this point to the roof (floor). The lifetime of a topological trait is $death\text{-}birth$, which can be displayed by its vertical or horizontal distance from the diagonal. Points in a persistence diagram can be roughly divided into two categories, "important" and "unimportant". Points close to the diagonal are unimportant because their lifetimes are short which implies they may be noise while points far away from the diagonal persist for a long time indicating that they stand for important topological traits.

Definition 3.2.6.2. A **persistence barcode** is a collection of intervals, each of which uniquely represents a point in the corresponding persistence diagram. It is displayed by a plane whose horizontal axis corresponds to the time and vertical axis represents an (arbitrary) ordering of these intervals (Ghrist, 2008).

Take Figs. 3.8 and 3.9 as examples. Similarly, short intervals correspond to those unimportant points in the persistence diagram.

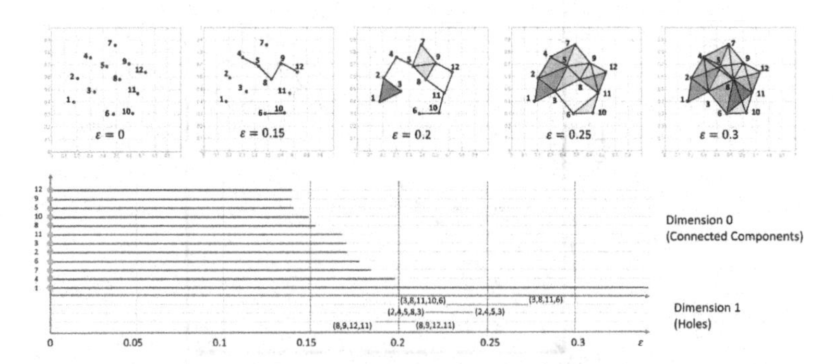

Figure 3.8 An illustration of the barcode of the Vietoris–Rips complex. The rank of the persistent homology group equals the number of intervals in the barcode intersecting the (dashed) line $\varepsilon = \varepsilon_i$, which also can be seen from the snapshots of Vietoris–Rips complex with specific ε.

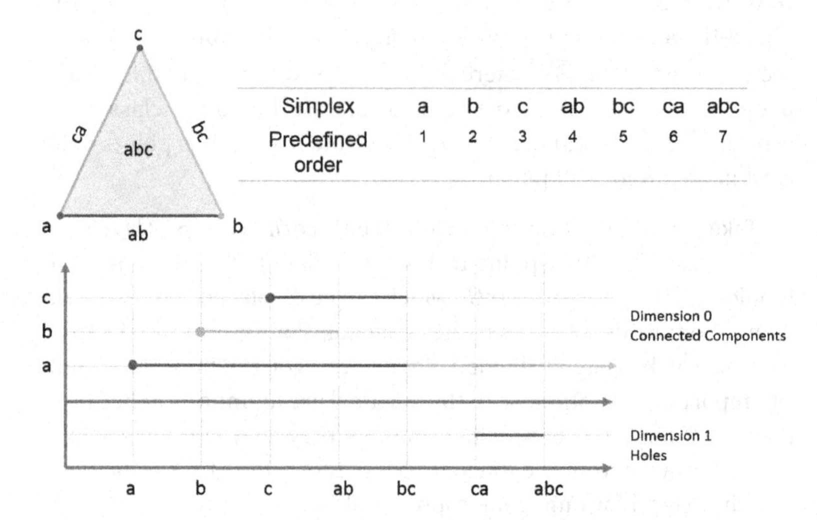

Figure 3.9 An example of a barcode of the simplicial complex shown on top-left. Each part of the triangle has been sorted previously. Then they appear in order. The green line in dimension 1 represents the hole before the face *abc* appears.

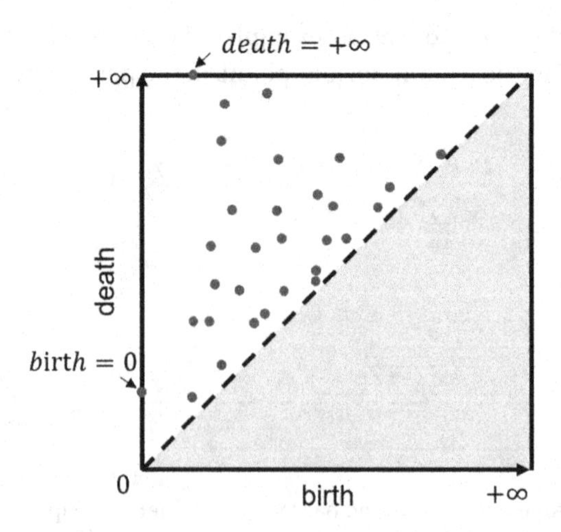

Figure 3.10 An example of the persistence diagram.

The information delivered by the persistence diagram and barcode of the same filtration is exactly the same.

Consider a persistence diagram to be a set of points, thus the metric between two diagrams becomes the problem of point set matching, which is well studied. Here we introduce two of the most commonly used, the Bottleneck distance and Hausdorff distance.

Definition 3.2.6.3. The Bottleneck distance between two persistence diagrams $\mathrm{Dgm}\, f_1$ and $\mathrm{Dgm}\, f_2$ is defined as follows.

$$d_B(\mathrm{Dgm}\, f_1, \mathrm{Dgm}\, f_2) = \min_{\varphi} \max_{x \in A} \|x - \varphi(x)\|_\infty, \tag{3.8}$$

where $\varphi : \mathrm{Dgm}\, f_1 \to \mathrm{Dgm}\, f_2$ is a bijection map and ranges over all possible such maps.

The bottleneck distance requires exploring all possible bijection maps, which makes it inefficient. However, it can be proven that the persistence diagram is stable with respect to this metric, as illustrated in Fig. 3.11.

Definition 3.2.6.4. The directed Hausdorff distance between two persistence diagrams $\mathrm{Dgm}\, f_1$ and $\mathrm{Dgm}\, f_2$ is

$$d_h(\mathrm{Dgm}\, f_1, \mathrm{Dgm}\, f_2) = \max_{x \in \mathrm{Dgm}\, f_1} \min_{y \in \mathrm{Dgm}\, f_2} \|x - y\|. \tag{3.9}$$

The corresponding Hausdorff distance is

$$\begin{aligned} &d_H(\mathrm{Dgm}\, f_1, \mathrm{Dgm}\, f_2) = \\ &\max\{d_h(\mathrm{Dgm}\, f_1, \mathrm{Dgm}\, f_2), d_h(\mathrm{Dgm}\, f_2, \mathrm{Dgm}\, f_1)\}. \end{aligned} \tag{3.10}$$

Technically the maximum and minimum operator should be the supremum and infimum, because the numbers of points in $\mathrm{Dgm}\, f_1$ and $\mathrm{Dgm}\, f_2$ are not necessarily the same. The Hausdorff distance is the largest minimum distance while the Bottleneck distance is the smallest maximum distance. In Section 3.3, we will introduce some applications based on the Hausdorff distance.

Remark 3.2.1. So far, we have introduced all necessary basics of algebraic topology and persistent homology. There are several available software programs that are open to researchers. JavaPlex (Tausz et al., 2011) is a well-documented and flexible java library which can be used in Matlab. Dionysus Morozov (2002) is a C++

library with Python bindings for some parts while Perseus (Nanda, 2007) is another C++ library but is used in Matlab. PHAT (Bauer et al., 2014) is a partially parallel C++ library by using OpenMP (oep, 2008), thus faster. The Gudhi library (Maria et al., 2014) is a generic C++ library for high-dimensional topological data analysis.

3.3 Persistent Homology in Pattern Recognition

In this part, we first propose a taxonomy of existing works according to the pipeline of persistent homology. Three issues in this taxonomy will be described. Applications of persistent homology in pattern recognition are depicted along with each issue.

3.3.1 Taxonomy

Theoretically speaking, there are three steps to get a topological representation (persistence diagram), i.e., (1) construct the simplicial complex, (2) then build the filtration, (3) finally compute the Betti numbers and obtain the persistence diagram. The first two steps may be done at the same time, for example, when we build the Vietoris–Rips complex we can get the filtered complex simultaneously. The pipeline of persistent homology is visualized in Fig. 3.1. There are usually three common research issues as indicated by the numbers in Fig. 3.1, where we call them "complex modeling", "function construction" and "sample assumption", respectively.

Most of the existing works can be classified according to which issue it studies. This taxonomy is original and can help pattern recognition researchers to apply persistent homology in a "non-blind" way. This taxonomy is stable because currently the pipeline of applying persistent homology into applications is constant. Moreover, with more studies on persistent homology in pattern recognition, this taxonomy is well organized and logical, compared with displaying existing works one by one.

Complex modeling is introduced first (Section 3.3.2) because it may be the most common problem when applying persistent homology. It directly models the data, transforming it to the format that persistent homology can handle. Another way to model data is to construct a function which also contains the structural information of the given data. We depict this issue next (Section 3.3.3). There is an important assumption before we use persistent homology, namely, the given data is assumed as samples from the underlying space. However, whether these samples can reveal the whole structure of the underlying space has not been explored. This is the third issue (Section 3.3.4).

3.3.2 Complex Modeling

This issue concerns about how to build a reasonable complex for the given data and the application. This is a tough task because usually we do not have ideal data: our data contain noise, outliers and sometimes can be incomplete. This is the first serious problem for applying persistent homology in real world applications.

The existing complexes, to name a few, include CW, cubical and simplicial complexes. Simplicial complexes have been introduced in Section 3.2, for example, the Vietoris–Rips complex and the Čech Complex. A cubical complex consists of squares, cubes and their higher-dimensional counterparts, which are called cubical cells. Generally, a cubical complex $\mathcal{Q}$ is a collection of cubical cells in some space satisfying that if a cell belongs to $c\mathcal{Q}$ then so do all its faces.

A CW complex comprises cells (of perhaps varying dimension) and each cell is a space homeomorphic to an open ball. These cells are glued together under additional constrains (Hatcher, 2002, p. 5). Due to the variability of cells and the glue operation, the CW complex can be used in wider applications and be proved over time to be the most natural class of spaces for algebraic topology.

Natural Image Statistics In order to decide which complex to use, we need to consider the given data. There are three types of data, i.e., the point cloud, functions and shapes (Edelsbrunner, 2012). We will talk about functions in next section. Here point cloud refers to a set of points that we do not know its underlying structure in advance. The goal is to explore and understand the overall shape of the point cloud. There are a lot of works in this topic (Adams and Carlsson, 2009; Carlsson et al., 2008; de Silva and Ghrist, 2007; Lee et al., 2012; Singh et al., 2008), we just name a few of them. To develop statistically better image representations, it is necessary to understand how natural image data is distributed in high-dimensional spaces (Carlsson et al., 2008; Lee et al., 2003). Adams and Carlsson (2009) deal with this problem by exploring the distribution of image patches. Each patch is an $n \times n$ square in the image and is regarded as a point in a high-dimensional space. Hence, all image patches compose a point cloud. In order to deal with noises, they calculate the D-norm of each patch and select the top ones. The core subset is implemented to remove outliers. Afterward they build a family of Vietoris–Rips simplicial complexes on the processed data nested by the parameter ε to describe its topology. It is found that optical image patches and larger range image patches are respectively clustered in a similar geometric way, which implies that the two image patches may have a strong connection.

Natural Language Processing Zhu (2013) proposes to use persistent homology to analyze the structure of documents. The document is divided into small pieces and each piece is represented by a point, thus producing a point cloud. It is expected that these points should be organized in a geometric structure. Persistent homology is accordingly applied to reveal these geometries. Vietoris–Rips complex is used to describe the document. Finally it is found that the pieces of a document are with a nice structure (e.g., a circle) if the document is with a conclusion paragraph that corresponds to the introduction paragraph.

Human Identification Shapes are subsets of ambient space that satisfy regularity conditions of one kind or another (Edelsbrunner, 2012), for example, the boundaries of 2D objects and the surfaces of 3D objects. In this case, there is another common complex called the Delaunay complex, which is defined as the Delaunay triangulation of the shape. Shape analysis by persistent homology is a hot topic in topological data analysis. One primary goal is to recognize and distinguish different shapes (Bronstein et al., 2009; Carlsson et al., 2004; Collins et al., 2004; Memoli, 2007; Skraba et al., 2010). Lamar-León et al. (2012) build a triangulated simplicial complex for silhouette-based human identification. A sequence of silhouettes is glued together through their gravity centers, forming a space-time shape (see Fig. 1 in (Blank et al., 2005)). Based on the space-time shape, they build the cubical complex according to the rule: from a point, building a cube (with length 1) if there exists another corresponding 7 points. By subdividing each face of the cube into two triangles and remaining those around the surface, they obtain the triangulated simplicial complex. Moreover, they build multiple functions according to different directions and fuse them in the score level, which leads to a better performance. The experimental performance shows its advantage over the methods used in Chen et al. (2009); Goffredo et al. (2010); Yu et al. (2006) with respect to single-view gait identification.

Hand Gesture Recognition The drawback of the Delaunay complex is its high computational cost to implement triangularization and sometimes the Vietoris–Rips complex is a popular alternative. Hu and Yin (2013) apply persistent homology to hand gesture recognition, which studies the holes between the hand posture and its convex hull (see Figs. 8 and 12 in Hu and Yin (2013)). We can also equivalently regard these holes (colored regions) as independent components. By applying morphological erosion operations with multiple scales, they obtain a matrix (see Figs. 2 and 3 in (Hu and Yin, 2013)), which records the changes of these components by persistent homology. The columns of the matrix represent the number of holes under each erosion scale. In order to be rotation invariant, the first column of the matrix always represents the holes which has the longest life-span. The same procedure can be also applied to arm posture recognition. Through plenty of experiments, this representation shows that it outperforms the traditional ones, e.g., Modified Census Transform (Just et al., 2006), Histograms of Oriented Gradients (Song et al., 2011) and Shape Context (Belongie et al., 2000).

Remark 3.3.1. Overall, if we want to apply persistent homology to real world applications, the first problem is that how to build a reasonable complex (or function) based on the given data. There are several kinds of complexes and one can propose new complexes according to Definition 3.2.2.3. Therefore, a serious problem is that what kind of complexes is suitable to solve the given problem. There are no general rules but a little experience. Empirical experience implies that for irregular structural data, e.g., the point cloud, the Rips complex is always worth trying. For regular structural data, for example, shapes or a sequence of human silhouettes (Lamar-León et al., 2012), the cubical complex and the Delaunay complex can be considered. Next section will introduce how to avoid explicitly modeling the complex, i.e., by defining a reasonable function.

3.3.3 Function Construction

Function construction is about building a (scalar valued) function from the given data, i.e., $f : \mathcal{K} \mapsto \mathbb{R}$, so that we can explore the topology of this function instead of the original data. The advantage of this shift lies in that it is independent of data dimension. Moreover, it is especially useful when the complex modeling is hard to achieve, for example, the texture data. However, we can build a reasonable function which represents the structure behind the texture data.

First we explain how to explore the topology of a function and then introduce some existing works in this aspect. Figure 3.11a is a function curve f (blue) of the underlying space. The red points are the samples and the red curve $\widehat{f}$ is the approximation of f defined on the given data. Next, we test the evolution of topology of super-level sets $\widehat{f}^{-1}([\alpha, +\infty])$ as α decreases. For example, as Fig. 3.11b shows, when $\alpha = \alpha_1$, it meets the first point, thus finding the first component. Then it meets another component as α decreases to α_2 while the first component is still "alive". Hence, we obtain a line segment (red) in Fig. 3.11b, which represents that the first

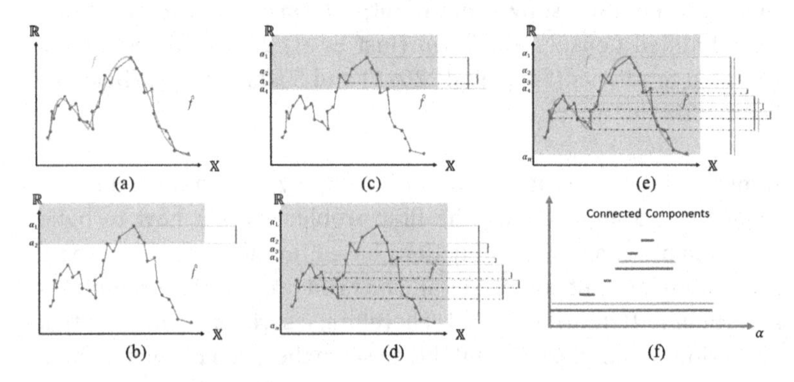

Figure 3.11 An illustration of Function Construction. (a) The real function f and its sampling $\widehat{f}$. (b–d) depict the evolution of topology of super-level sets $\widehat{f}^{-1}([\alpha, +\infty])$ as α decreases. (e) The results of f is also placed here. The line segments on the right of these subfigures form a barcode displayed in (f). More details please see in the context.

component survives for that much time. However, as α decreases to α_3, the second component is merged into the first one, i.e., it has a short lifetime as the short red line segment in Fig. 3.11,c. As α decreases to $-\infty$, we can get the whole information as shown in Fig. 3.11,d. Applying the same process we obtain the result of f as shown in Fig. 3.11e. A barcode consists of these line segments as shown in Fig. 3.11f. By removing the short line segments, barcodes of f and $\widehat{f}$ are the same. This implies that this approach is stable with respect to small deformation in approximation methods.

The common goal of function based persistent homology is to approximate the persistence diagram of an unknown scalar field from its values at a finite set of sample points (Chazal et al., 2011). An important issue is the stability property of persistence diagram under the perturbations of sampling methods, which has been theoretically proved in Chazal et al. (2009a); Cohen-Steiner et al. (2007). We will depict this stability topic in detail in next section.

Persistence-based Clustering and Segmentation We introduce several works by exploring functions defined on the data. Persistence-based clustering and segmentation are the typical examples (Chazal et al., 2011, 2013b; Chazal and Oudot, 2008; Skraba et al., 2010). It is assumed that a data set $\mathbb{X}$ is comprised of infinite samples $\mathbb{S}_n$ drawn from an unknown density function f. Clustering then becomes a problem of understanding the structure of the underlying density function f or $\widehat{f}$ estimated from the samples, e.g., estimated by a Gaussian kernel or k-nearest neighbors (Chazal et al., 2011, 2013b). More precisely, the goal consists of detecting the local peaks of $\widehat{f}$ in order to use them as cluster centers, and to classify the data set. Only pairwise distances between data points are required, which are used to construct the neighborhood graph. This means that $\mathbb{X}$ can be any metric space. Roughly speaking, the stability property states that if f is perturbed by no more than a $\varepsilon > 0$ (i.e., $\widehat{f}$), then the bottleneck distance between the diagrams of f and $\widehat{f}$ is limited, namely, the points on the persistence diagrams will also be perturbed by no more than ε.

The algorithm has three major steps. The first one is to construct the neighborhood graph and estimate $\widehat{f}$ by the finite sample points. Different neighborhood graphs can be selected, e.g., k-nearest neighbor graph and δ-Rips graph (1-skeleton of the Rips complex) (Chazal et al., 2013b). Meanwhile, various methods can be used to estimate $\widehat{f}$, for example, the local density used in Rodriguez and Laio (2014). The second is to find the peaks (local maximum) of $\widehat{f}$ as cluster centers. This is the so-called mode-seeking, so this persistence-based method can also be regarded as a variant of Mean-Shift (Comaniciu and Meer, 2002). As it is known to all, the original Mean-Shift algorithm produces many unstable clusters, i.e., over-segmentation. This is where persistent homology comes into play through the last step, i.e., computing the persistence diagram in order to merge small clusters, which are treated as noises. The goal of the third step is to regain some stability, which is ensured by Spatial Stability Theorem (Theorem 4.9 in Oudot and Skraba (2009)).

Object Recognition Li et al. (2014), inspired by the classical bag-of-features model (Lazebnik et al., 2006) and persistence-based shape analysis (Carlsson et al., 2004; Chazal et al., 2009b), apply persistent homology to recognition tasks. The used functions include heat kernel signature (Sun et al., 2009), wave kernel signature (Aubry et al., 2011) and scale invariant heat kernel signature (Kokkinos et al., 2012). The heat kernel $k_t(x, y)$, defined on the temporal and spatial domain $\mathbb{R}^+ \times \mathcal{M} \times \mathcal{M}$, can be understood as the amount of heat transferred from x to y within time t provided a unit heat at x. Therefore, $k_t(x, \cdot)$ represents the heat transferred from x to the other part. The heat kernel has many nice properties, i.e., symmetric, isometric invariant, informative (fully characterizes shapes up to isometry), multi-scale, and stable under perturbations of the underlying manifold (Sun et al., 2009). Heat kernel signature is described as $HKS(x, t) = k_t(x, x)$, which restricts the signature to the temporal domain to reduce the complexity and keep all information of $k_t(x, \cdot)_{t>0}$.

Moreover, they also consider combining multiple functions together to improve the performance. As these functions are built in different ways, it is not a good choice to directly combine them together. As a consequence, metric learning between multiple functions is used, which learns the importance of each function from labeled data. Meanwhile, superior performance by combining this persistence-based approach and the bag-of-features model, is achieved in 3D shape retrieval, hand gesture recognition and texture image classification.

Image Diffusion Image diffusion is a technique aiming at reducing image noise without removing significant parts of the image content, typically edges, lines or other details that are important for the interpretation of the image (Perona and Malik, 1990; Tsiotsios and Petrou, 2013). Chen and Edelsbrunner (2011) propose using persistent homology to quantify the effect of diffusion on critical points in an image. Interpreting an image as a real valued function on a compact subset of the Euclidean plane, they get its scale-space by diffusion, i.e., an image generates a parameterized family of successively more and more blurred images based on a diffusion process (Lindeberg, 1993b; Weickert, 1998). They idealize the effect of diffusion by convolving the function with an isotropic Gaussian kernel, which generated a scale space of 1-parameter family of functions on the $\mathbb{R}^2$. In general, diffusion washes out details, so the number of critical points is expected to decrease. Nevertheless, there are counterexamples indicating that the critical points grow in number. However, the experimental evidence implies that this phenomenon rarely happens. Previously, Lindeberg (1993a) proved that the expected number of critical points for $n = 1$-dimensional images and for random noise is constant $\frac{const}{t^{n/2}}$. But this theorem has not been proved to be correct for $n > 1$. Therefore, Chen and Edelsbrunner rationalize this theorem to $n \geq 1$ with weaker assumptions, which states as follows (Chen and Edelsbrunner, 2011)[Main Theorem].

Theorem 3.3.3.1. *Let $\Omega \subseteq \mathbb{R}^n$ be compact, $f : \Omega \to \mathbb{R}$ a function and $g_t : \mathbb{R}^n \to \mathbb{R}$ a Gaussian kernel with scale t such that $f_t = f * g_t$ is tame for all $t \geq 0$. Then*

$$\|Dgm(f_t)\|_p \leq \frac{const}{t^{n/2}} \tag{3.11}$$

for all $p > \frac{1}{2}(2n+1+\sqrt{4n^2+1})$, and the exponent on the right hand side of the inequality is best possible. $\|Dgm(f_t)\|_p$ is defined as

$$\|Dgm(f_t)\|_p = \left[\sum_{u \in Dgm(f_t)} \left|death(u) - birth(u)\right|\right]^{\frac{1}{p}}. \tag{3.12}$$

$death(u)$ and $birth(u)$ represent the coordinates of a point in the persistence diagram. This norm goes rapidly to zero as time goes to infinity. This property helps to prove Theorem 3.3.3.1.

Remark 3.3.2. In this section, we first depict the overall process of obtaining persistence diagrams from a function. Afterward we introduce three kinds of works, i.e., persistence-based clustering and segmentation, object recognition and image diffusion. There is another crucial issue before applying persistent homology by constructing functions, which we ignore in this part for sake of the content coherence, i.e., whether the defined function can represent the original data. In the following section, we concentrate on this topic.

3.3.4 Sample Assumption

Three Goals In the previous section we actually ignore the fact that the given data is sampled from the underlying topological space, which should be investigated systematically. It is reasonable to infer that these data are derived from an underlying space/distribution with random noise. We assume that these samples can approximate the structure of the underlying space. We explore this assumption in the following three aspects: 1) under what kind of situations we can accept this assumption, 2) how closely these samples can approximate the inherent space, 3) the stability of these approximation methods. These three aspects can be further formulated as Goal 3.3.4.1, Goal 3.3.4.2 and Goal 3.3.4.3, respectively.

Goal 3.3.4.1. Given points $S_n = \{x_1, x_2, \cdots, x_n\}$, each of which is independent and identically distributed (i.i.d.) of a function ρ. $\mathbb{M}$ is the support of ρ, which is a manifold. We define $U_n = \bigcup_{x \in S_n} B_\varepsilon(x)$. The goal is to make sure that for some $n > n_0$ with a probability more than $1 - \delta$, we have

$$homology(U_n) = homology(\mathbb{M}). \tag{3.13}$$

Goal 3.3.4.2. Our goal is to make sure that U_n and $\mathbb{M}$ are as close as possible with respect to persistence diagram, which can be formulated as follows,

$$\mathcal{P}\Big(d(\mathrm{Dgm}(U_n), \mathrm{Dgm}(\mathbb{M}))\Big) \to 1. \tag{3.14}$$

where $d(\cdot, \cdot)$ is a distance measure, and $\mathcal{P}(\cdot)$ is a probability measure. With the increase of n, we expect this probability can quickly converge to one.

Goal 3.3.4.3. Let f and g be two functions defined as $f, g : S_n \to \mathbb{R}$. Then we try to figure out the difference between topologies induced by f and g, i.e., the relationship between $\| f - g \|$ and $\| topology(f) - topology(g) \|$.

Answer to Goal 3.3.4.1 To answer the question in Goal 3.3.4.1, we have the Nerve Theorem (Adler et al., 2010).

Theorem 3.3.4.1. *Suppose that the intersections $\bigcap_{x \in S'_n} B_\varepsilon(x)$ are either empty or contractible for any subset S'_n of S_n. Then the Čech complex C_ε is homotopy equivalent to $\bigcup_{x \in S_n} B_\varepsilon(x)$. In particular, if $\mathcal{M}$ is a finite-dimensional normed linear space, or a compact Riemannian manifold with convexity radius greater than ε, and if $B_\varepsilon(x)_{x \in S_n}$ is a cover of the space $\mathbb{M}$, then C_ε is homotopy equivalent to $\mathbb{M}$.*

This provides a computational way to approximate the underlying space. Moreover, the computation of the Čech complex can be intractable when facing a large data set. Therefore, the Rips complex, α-shape and witness complex De Silva and Carlsson (2004) are proposed to reduce computing time and resources.

Answer to Goal 3.3.4.2 Goal 3.3.4.2 is to ensure that with the increase of n, the probability that persistence diagrams of U_n and $\mathbb{M}$ is the same can converge to one (Balakrishnan et al., 2012; Genovese et al., 2012; Niyogi et al., 2008). Moreover, the convergence rate is expected to be as fast as possible. The following two theorems are derived from Chazal et al. (2014), which solves the problem of the convergence and gives its upper and lower bounds.

Theorem 3.3.4.2. *Let μ denote a probability measure on it equipped with its Borel algebra. $\mathbb{X}_\mu$ denotes the support of μ, namely, the smallest closed set with probability one. $\widehat{\mathbb{X}}_n = \{X_1, \cdots, X_n\}$ is a set of points drawn i.i.d. from μ. Provided $a > 0$ and $b > 0$, if $\forall x \in \mathbb{X}_\mu$ and $\forall r > 0$, $\mu(B(x, r)) \geq \min(1, ar^b)\}$, then we say μ satisfies the (a, b)-standard assumption on $\mathbb{M}$. Then when $n \to \infty$,*

$$\mathcal{P}\Big((d_B(Dgm(\mathbb{X}_\mu), Dgm(\widehat{\mathbb{X}}_n))) \leq C\big(\frac{\log n}{n}\big)^{1/b} \Big) \to 1. \tag{3.15}$$

where the constant C only depends on a and b.

Theorem 3.3.4.3. *Let $(\mathbb{M}, \rho)$ denote a metric space and $\mathbb{P} = \{\mu$ on $\mathbb{M} | \mu$ satisfies (a, b)-standard assumption$\}$. Then the upper bound of*

$\mathcal{P}$ is

$$\sup_{\mu\in\mathbb{P}} \mathcal{E}\left[d_B(Dgm(\mathbb{X}_\mu), Dgm(\widehat{\mathbb{X}}_n))\right] \leq C_1\left(\frac{\log n}{n}\right)^{1/b}, \tag{3.16}$$

where $\mathcal{E}$ stands for mathematical expectation and the constant C_1 only depends on a and b. d_B is the bottleneck distance. Assume x is a non isolated point in $\mathbb{M}$. Consider any sequence $\{x_n\} \in (\mathbb{M}\backslash\{x\})^{\mathbb{N}}$ such that $\rho(x, x_n) \leq (an)^{-1/b}$, then for any estimation $\widehat{Dgm}_n$ of $Dgm(\mathbb{X}_\mu)$, we have

$$\lim_{n\to\infty} \inf \rho(x, x_n)^{-1} \sup_{\mu\in\mathbb{P}} \mathcal{E}\left[d_B(Dgm(\mathbb{X}_\mu), \widehat{Dgm}_n)\right] \geq C_2 \tag{3.17}$$

where C_2 is an absolute constant.

Through Equation 3.15, we can believe with more confidence that the two persistence diagrams are the same when n increases its value. Equation 3.16 and Equation 3.17 give the upper and lower bounds for the rate of convergence of persistence diagram estimation, respectively. It can be concluded that the difference between the estimator and the origin becomes smaller and smaller with the increase of n.

Answer to Goal 3.3.4.3 The Goal 3.3.4.3 can be answered by the bottleneck stability theorem, which has been studied in many literatures (Chazal et al., 2009b, 2012, 2013a; Cohen-Steiner et al., 2007; Edelsbrunner, 2013; Edelsbrunner and Harer, 2008). The following is the common statement of this theorem.

Theorem 3.3.4.4. *Let $\mathbb{X}$ be a topological space with tame functions $f, g : \mathbb{X} \to \mathbb{R}$. Then for each dimension p the bottleneck distance between the dimension p persistence diagrams is bounded from above by the difference between the functions,*

$$d_B\left(Dgm_p(f), Dgm_p(g)\right) \leq \|f - g\|_\infty. \tag{3.18}$$

We refer the reader to Cohen-Steiner et al. (2007) and to Chazal et al. (2012) for proofs of this theorem.

Manifold, Homology, and Persistence Diagram Estimation Generally speaking, the above three goals are studies in the fields of manifold estimation, homology estimation and persistence diagram estimation, which can be seen as three ways to approximate the underlying space. Persistence diagram estimation is strongly connected to the better known problem of measure support estimation (De Vito et al., 2013). Some recent attempts exploring the space of persistence diagrams from a statistical view have been reported (Bubenik, 2012; Mileyko et al., 2011). The goal of manifold estimation is about estimating smooth manifolds from finite samples (Adler et al., 2010, Chapter 5.1), which aims at nonlinear dimensionality reduction (Balasubramanian and Schwartz, 2002; Saul and Roweis, 2000). The estimated manifold $\widehat{\mathbb{M}}$ and the original one $\mathbb{M}$ should share the same topological properties. Furthermore, Genovese et al. (2012) provide a closeness guarantee between $\widehat{\mathbb{M}}$ and $\mathbb{M}$ with respect to the Hausdorff distance. Homology estimation refers to estimate the homology groups of a manifold. The reason lies in that homology groups of a manifold are important topological summaries of the space. Niyogi et al. (2008) and Balakrishnan et al. (2012) show that if the parameters λ for the density estimator and ϵ for the balls at each point are chosen properly, then an estimation of the homology of $\mathbb{M}$ can be reached with high probability. Moreover, they provide an upper bound on how many examples needed to be drawn in order to reliably reconstruct the homology of $\mathbb{M}$. In terms of the speed issue, generally speaking, persistence diagram estimation is more time-consuming compared with homology estimation while manifold estimation is the most efficient (Chazal et al., 2014).

Stable Topological Features At last, we use the previously mentioned persistence-based clustering and segmentation (Chazal et al., 2013b) as an example to show how to obtain stable topological features. As suggested in Fig. 2 in Chazal et al. (2013b), (c) is the persistence diagram of points in (a). A point near the diagonal of the persistence diagram implies that its birth time and death time are close such that we believe it corresponds to noise (unstable clusters). Intuitively, we use a threshold (as indicated in the dotted line) to distinguish the real clusters and the noisy ones. From the figure, we can also conclude that even though the sample points are of small perturbations, two clear clusters are also stable. Therefore, this method is robust to noise.

In a nutshell, by applying persistent homology, we always assume that the given data are sampled from the underlying space. In this section, we explain how samples affect the approximation and under what kind of conditions they can reveal the real topological structures of the underlying space.

Remark 3.3.3. We conclude this section in Table 3.1. We have presented the taxonomy of issues during applying persistent homology to real world problems. First we need to process the give data in order to build the reasonable complex, or we can instead build a well designed function defined on the given data and explore the topologies of the function. However, before we evaluate the performance obtained from the constructed function, we need to consider whether the given data can reveal the structure of the underlying space. The corresponding applications are also introduced along each issue.

3.4 Perspectives and Open Directions

In this part, we present several perspectives and open directions for applying persistent homology.

3.4.1 Multi-Dimensional Persistent Homology

This chapter restricts the filter function to be $f : \mathcal{K} \mapsto \mathbb{R}$. However, it is natural to wonder if multiple filter functions is more informative, or similarly, $f : \mathcal{K} \mapsto \mathbb{R}^n$, $n > 1$. In many applications of topology, we need to study a family of spaces parameterized along multiple geometric dimensions as shown in Fig. 3.12. This is the so-called multi-dimensional or multi-filtration persistent homology. This change, from scalar-valued functions to vector-valued functions, leads to a totally different theoretical analysis. It has been proved that there are no similar complete discrete invariants (persistence intervals) for multi-dimensional persistent homology (Carlsson and Zomorodian, 2009).

In order to obtain a representation like persistence diagram, Carlsson and Zomorodian (2009) propose the rank invariant, a discrete invariant for the robust estimation of Betti numbers in a multi-filtration case, and prove its completeness in one dimension. Afterward, they present a polynomial time algorithm for

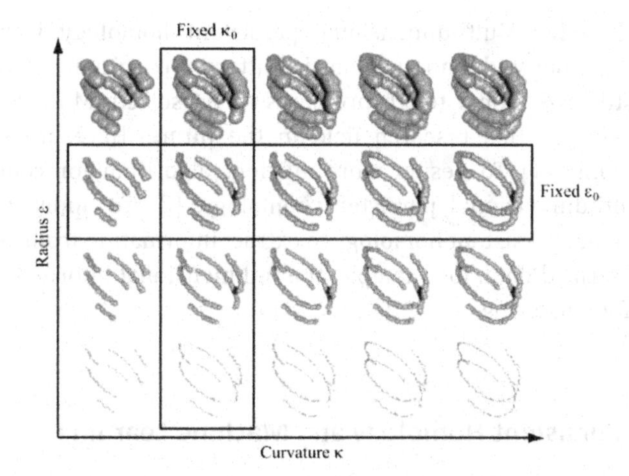

Figure 3.12 A bifiltration parameterized along curvature κ and radius ϵ. Without multi-dimensional persistent homology, we can only apply persistent homology to a filtration, i.e., either fix ϵ or κ. Reprinted with permission from Carlsson and Zomorodian (2009). Copyright (2009) Springer Nature.

computing multi-dimensional persistence by recasting this problem into computational algebraic geometry (Carlsson et al., 2009). Cerri et al. (2009) later prove that multi-dimensional rank invariants are stable with respect to function perturbations, i.e., small changes of the function imply small changes of the rank invariants under the constructed distance. Except for the rank invariant, Cerri and Landi (2013) try to generalize the concept of persistence diagram to a vector-valued continuous function meeting the requirements that the persistence diagram should satisfy:

- The persistence diagram can be defined via multiplicities obtained from persistent Betti numbers.
- It is allowed to completely reconstruct persistent Betti numbers.
- The coordinates of its off-diagonal points are homological critical values.

It makes sense to expect the proposed new concept, called persistence space, should also possess the above properties. This is a good way to follow for designing new features for multi-dimensional persistent homology.

Remark 3.4.1. Multi-dimensional persistent homology is more natural and has wider applications in pattern recognition. However, there still exist some tough problems to be solved. More efforts are needed in this research field in the future. Here are some suggestions: (1) To design more efficient algorithm for computing multi-dimensional persistent homology. (2) To apply multi-dimensional persistent homology to extract information from multi-parameterized data, for example, the 0-dimensional information is useful for clustering.

3.4.2 Persistent Homology and Machine Learning

Recently persistent homology has drawn attention from multiple fields, e.g., machine learning icm (International Conference on Machine Learning Workshop, 2014). We are of great interest in this connection since machine learning is of great importance for pattern recognition. We point out several promising directions in this aspect.

Developing Statistical Methods One of recent works in persistent homology has concentrated on developing statistical methods for data analysis because of the requirement of quantifying the statistical significance of topological features detected with persistent homology (Balakrishnan et al., 2012; Bubenik et al., 2007; Fasy et al., 2013, 2014; Mukherjee, 2013; Skraba et al., 2010).

One common question by developing statistical methods for persistent homology (Mileyko et al., 2011) is that given persistence diagrams from one hundred realizations of point cloud data obtained from one geometric object, what is the average diagram and how much do these diagrams vary? Bubenik (2013) deals with this problem by a new representation called persistence landscapes by mapping the persistence diagram into a designed function, which is easy to be combined with statistical tools. It has been demonstrated to be useful in random geometric complexes, random clique complexes and Gaussian random fields. Mileyko et al. (2011) show that the space of persistence diagrams allows us to define probabilistic measures, e.g., means, variances and conditional probabilities. After that, Turner et al. (2014) develop an algorithm to compute the Fréchet means in this space.

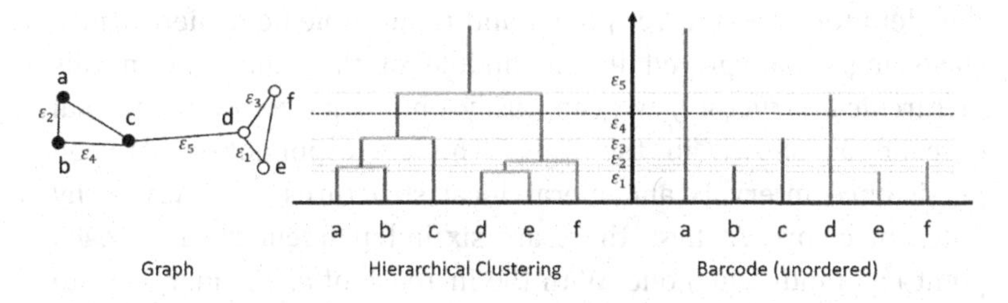

Figure 3.13 An example to explain the relation between the zero-dimensional barcode and hierarchical clustering. See context for more details.

In machine learning, one of the common tasks is to infer a model's parameters from its observations. We can consider solving this problem by using persistent homology. Emmett et al. (2014) perform parametric inference using statistical information defined on the computed topological invariants. In this way, they distinguish biological events occurring at different genetic scales.

Interpreting as Unsupervised Learning Persistent homology can also work for unsupervised learning in pattern recognition. We mention the regression and clustering here.

The goal of regression is to find the best fitting function and its parameters for a set of points. A simple case is the linear regression, which is a line or hyperplane. The problem is that provided that we have no prior knowledge about the given data, it is not easy to decide what kind of functions to use. For example, it will definitely fail if the data distributes like a circle while we choose to build a linear model. In this case, persistent homology can play as a data pr-processing method to give a rough prior about the structure. With this prior in mind, we can determine the complexity of the used function.

By using the Vietoris–Rips complex, the computed zero-dimensional barcode by persistent homology can be easily described as a way of hierarchical clustering, which is explained by a toy example in Fig. 3.13. There are six points divided into two clusters, represented by the solid and hollow points, respectively. ε_i is the distance between its connected endpoints and $\varepsilon_1 < \cdots < \varepsilon_5$. With the definition of the Vietoris–Rips complex, two points are connected if they are within the distance of ε. We follow this definition to connect points and then implement hierarchical clustering, as displayed in the middle of the figure. From this hierarchical structure, we can easily obtain its zero-dimensional barcode as shown in the right. The correspondence between persistence intervals and hierarchical structures is indicated by different colors. At first, there are six independent clusters, each point is an individual one. With the increase of ε, e is merged into d at ε_1 and b is merged into a at ε_2. Similarly, f becomes a part of d at ε_3 and c as a part of a at ε_4. Finally, d and a are connected as a whole at ε_5. By drawing a horizontal dash line in dark when $\varepsilon_4 < \varepsilon < \varepsilon_5$, we can obtain the correct two clusters from both the hierarchical tree and the zero-dimensional barcode. It can be concluded that we can obtain different level categories by the barcode as well as the hierarchical tree.

Compared with the hierarchical clustering, persistent homology has the following advantages.

- There is no need to decide where to cut the hierarchical tree to obtain the final clusters. By transforming the barcode to a persistence diagram, we can observe that some points are far away from the diagonal. Each of these points represents a cluster.
- By computing higher-dimensional barcodes, more structural information can be obtained, for example, we can detect the hole formed by a,b and c by the one-dimensional barcode.

There are a few prior works in this way. Niyogi et al. (2011) make a topological view of clustering, which may be interpreted as trying to find the number of connected components of an underlying geometrically structured probability distribution. They estimated the homology of the underlying manifold with high confidence from noisy data if the variance of the Gaussian noise is small in a certain sense.

Connecting with Deep Learning Deep learning is now a very popular topic in many fields and it has achieved great success in many applications. One of current hot research issues is to understand what a neural network is really doing. An interesting attempt is by Olah (2014), who states the following theorem.

Theorem 3.4.2.1. *Suppose that the activation is tanh or sigmoid but not ReLU. Then layers with N inputs and N outputs are homeomorphisms, if the weight matrix W is non-singular.*

Note that we also assume that there is no pooling operation and dropout technique. This theorem states that such neural network implements continuous deformation to the input data in order to make it linearly separable. Networks may differ in structures, but as long as they satisfy the above requirement, they all transform the input continuously and the transformed data are topologically equivalent. This helps better understand what the network does to the input and how the network varies.

Remark 3.4.2. Persistent homology and machine learning are two significant theories in their fields. And recent works have shown the potential to combine them together. We describe three topics in the above and give some suggestions for those who are interested. (1) It is worth exploring the structure of the space of model parameters in order to obtain better models. (2) The information computed by persistent homology can not only be used for unsupervised learning (e.g., regression and clustering), but also be complementary to other models (Li et al., 2014). (3) Persistent Homology and deep learning can be incorporated in many ways, especially with respect to large scale data (Kibardin, 2015).

3.5 Conclusion

In this chapter, we have first provided a comprehensive study of persistent homology and its preliminary concepts, and then proposed an elegant taxonomy of existing works, which contains three research issues. At last, we summarize several open research directions.

Persistent homology has beautiful mathematical formulations and theories. It deals with data from shape, point cloud, and functions with the output of the lifespan of topological features at different dimensions. These topological features are used as topological representations for different applications. Persistent homology and its practice in pattern recognition are of great potential. More efforts should be invested to explore its connection with machine learning and multi-dimensional persistent homology, which will possibly produce great value for persistent homology in pattern recognition.

References

(2008). The OpenMP website, http://openmp.org.

(International Conference on Machine Learning Workshop, 2014). Topological methods for machine learning, http://topology.cs.wisc.edu/.

Adams, H. and Carlsson, G. (2009). On the nonlinear statistics of range image patches, *SIAM Journal on Imaging Sciences* **2**, 1, pp. 110–117.

Adler, R. J., Bobrowski, O., Borman, M. S., Subag, E. and Weinberger, S. (2010). Persistent homology for random fields and complexes, in *Borrowing Strength: Theory Powering Applications–A Festschrift for Lawrence D. Brown* (Institute of Mathematical Statistics), pp. 124–143.

Adler, R. J., Samorodnitsky, G. and Taylor, J. E. (2007). Excursion sets of stable random fields, *arXiv preprint arXiv:0712.4276*.

Armstrong, M. A. (1983). Basic topology. undergraduate texts in mathematics, *Springer-Verlag, Berlin-Heidelberg-New York* **8**, p. 8.

Aubry, M., Schlickewei, U. and Cremers, D. (2011). The wave kernel signature: A quantum mechanical approach to shape analysis, in *Computer Vision Workshops (ICCV Workshops), 2011 IEEE International Conference on* (IEEE), pp. 1626–1633.

Balakrishnan, S., Rinaldo, A., Sheehy, D., Singh, A. and Wasserman, L. A. (2012). Minimax rates for homology inference, in *Proceedings of the Fifteenth International Conference on Artificial Intelligence and Statistics (AISTATS-12)*, Vol. 22, pp. 64–72.

Balasubramanian, M. and Schwartz, E. L. (2002). The isomap algorithm and topological stability, *Science* **295**, 5552, pp. 7–7.

Baryshnikov, Y. and Ghrist, R. (2009). Target enumeration via Euler characteristic integrals, *SIAM Journal on Applied Mathematics* **70**, 3, pp. 825–844.

Baryshnikov, Y. and Ghrist, R. (2010). Euler integration over definable functions, *Proceedings of the National Academy of Sciences* **107**, 21, pp. 9525–9530.

Bauer, U., Kerber, M., Reininghaus, J. and Wagner, H. (2014). Phat–persistent homology algorithms toolbox, in *Mathematical Software–ICMS 2014* (Springer), pp. 137–143.

Belongie, S., Malik, J. and Puzicha, J. (2000). Shape context: A new descriptor for shape matching and object recognition, in *NIPS*, Vol. 2, p. 3.

Blank, M., Gorelick, L., Shechtman, E., Irani, M. and Basri, R. (2005). Actions as space-time shapes, in *Computer Vision, 2005. ICCV 2005. Tenth IEEE International Conference on*, Vol. 2 (IEEE), pp. 1395–1402.

Bronstein, A. M., Bronstein, M. M. and Kimmel, R. (2009). Topology-invariant similarity of nonrigid shapes, *International Journal of Computer Vision* **81**, 3, pp. 281–301.

Bubenik, P. (2012). Statistical topology using persistence landscapes, *arXiv preprint arXiv:1207.6437*.

Bubenik, P. (2013). Statistical topological data analysis using persistence landscapes, *arXiv preprint arXiv:1207.6437*.

Bubenik, P., Kim, P. T. et al. (2007). A statistical approach to persistent homology, *Homology, Homotopy and Applications* **9**, 2, pp. 337–362.

Carlsson, G., Ishkhanov, T., De Silva, V. and Zomorodian, A. (2008). On the local behavior of spaces of natural images, *International Journal of Computer Vision* **76**, 1, pp. 1–12.

Carlsson, G., Singh, G. and Zomorodian, A. (2009). Computing multidimensional persistence, in *Algorithms and Computation* (Springer), pp. 730–739.

Carlsson, G. and Zomorodian, A. (2009). The theory of multidimensional persistence, *Discrete & Computational Geometry* **42**, 1, pp. 71–93.

Carlsson, G., Zomorodian, A., Collins, A. and Guibas, L. (2004). Persistence barcodes for shapes, in *Proceedings of the 2004 Eurographics/ACM SIGGRAPH symposium on Geometry processing* (ACM), pp. 124–135.

Cerri, A., Di Fabio, B., Ferri, M., Frosini, P. and Landi, C. (2009). Multidimensional persistent homology is stable, *arXiv preprint arXiv:0908.0064*.

Cerri, A. and Landi, C. (2013). The persistence space in multidimensional persistent homology, in *Discrete Geometry for Computer Imagery* (Springer), pp. 180–191.

Chazal, F., Cohen-Steiner, D., Glisse, M., Guibas, L. J. and Oudot, S. Y. (2009a). Proximity of persistence modules and their diagrams, in *Proceedings of the twenty-fifth Annual Symposium on Computational Geometry* (ACM), pp. 237–246.

Chazal, F., Cohen-Steiner, D., Guibas, L. J., Mémoli, F. and Oudot, S. Y. (2009b). Gromov-hausdorff stable signatures for shapes using persistence, in *Computer Graphics Forum*, Vol. 28 (Wiley Online Library), pp. 1393–1403.

Chazal, F., de Silva, V., Glisse, M. and Oudot, S. (2012). The structure and stability of persistence modules, *arXiv preprint arXiv:1207.3674*.

Chazal, F., De Silva, V. and Oudot, S. (2013a). Persistence stability for geometric complexes, *Geometriae Dedicata*, pp. 1–22.

Chazal, F., Glisse, M., Labruère, C. and Michel, B. (2014). Convergence rates for persistence diagram estimation in topological data analysis, in *Proceedings of The 31st International Conference on Machine Learning*, pp. 163–171.

Chazal, F., Guibas, L. J., Oudot, S. Y. and Skraba, P. (2011). Scalar field analysis over point cloud data, *Discrete & Computational Geometry* **46**, 4, pp. 743–775.

Chazal, F., Guibas, L. J., Oudot, S. Y. and Skraba, P. (2013b). Persistence-based clustering in riemannian manifolds, *Journal of the ACM (JACM)* **60**, 6, p. 41.

Chazal, F. and Oudot, S. Y. (2008). Towards persistence-based reconstruction in euclidean spaces, in *Proceedings of the twenty-fourth Annual Symposium on Computational Geometry* (ACM), pp. 232–241.

Chen, C. and Edelsbrunner, H. (2011). Diffusion runs low on persistence fast, in *Computer Vision (ICCV), 2011 IEEE International Conference on* (IEEE), pp. 423–430.

Chen, C., Liang, J., Zhao, H., Hu, H. and Tian, J. (2009). Frame difference energy image for gait recognition with incomplete silhouettes, *Pattern Recognition Letters* **30**, 11, pp. 977–984.

Clark, A. (1984). Elements of abstract algebra.

Cohen-Steiner, D., H. and Harer, J. (2007). Stability of persistence diagrams, *Discrete & Computational Geometry* **37**, 1, pp. 103–120.

Collins, A., Zomorodian, A., Carlsson, G. and Guibas, L. J. (2004). A barcode shape descriptor for curve point cloud data, *Computers & Graphics* **28**, 6, pp. 881–894.

Comaniciu, D. and Meer, P. (2002). Mean shift: A robust approach toward feature space analysis, *Pattern Analysis and Machine Intelligence, IEEE Transactions on* **24**, 5, pp. 603–619.

De Silva, V. and Carlsson, G. (2004). Topological estimation using witness complexes, in *Proceedings of the First Eurographics conference on Point-Based Graphics* (Eurographics Association), pp. 157–166.

de Silva, V. and Ghrist, R. (2007). Coverage in sensor networks via persistent homology, *Algebraic and Geometric Topology* **7**, pp. 339–358, doi:10.2140/agt.2007.7.339.

De Silva, V. and Ghrist, R. (2007). Homological sensor networks, *Notices of the American mathematical society* **54**, 1.

De Vito, E., Rosasco, L. and Toigo, A. (2013). Learning sets with separating kernels, *Applied and Computational Harmonic Analysis.*

Dey, T. K. (2013). Chains, boundaries, homology groups, betti numbers [notes on munkres book], http://web.cse.ohio-state.edu/~tamaldey/course/CTDA/homology.pdf.

Edelsbrunner, H. (2012). Persistent homology: theory and practice, in *Proceedings of the European Congress of Mathematics.*

Edelsbrunner, H. (2013). Persistent homology in image processing, in *Graph-Based Representations in Pattern Recognition* (Springer), pp. 182–183.

Edelsbrunner, H. and Harer, J. (2008). Persistent homology-a survey, *Contemporary mathematics* **453**, pp. 257–282.

Edelsbrunner, H., Letscher, D. and Zomorodian, A. (2000). Topological persistence and simplification, in *Foundations of Computer Science, 2000. Proceedings. 41st Annual Symposium on* (IEEE), pp. 454–463.

Emmett, K., Rosenbloom, D., Camara, P. and Rabadan, R. (2014). Parametric inference using persistence diagrams: A case study in population genetics, *arXiv preprint arXiv:1406.4582.*

Fasy, B. T., Lecci, F., Rinaldo, A., Wasserman, L., Balakrishnan, S. and Singh, A. (2013). Statistical inference for persistent homology: Confidence sets for persistence diagrams, *arXiv preprint arXiv:1303.7117.*

Fasy, B. T., Lecci, F., Rinaldo, A., Wasserman, L., Balakrishnan, S., Singh, A. et al. (2014). Confidence sets for persistence diagrams, *The Annals of Statistics* **42**, 6, pp. 2301–2339.

Freedman, D. and Chen, C. (2009). Algebraic topology for computer vision, *Computer Vision. Nova Science.*

Genovese, C. R., Perone-Pacifico, M., Verdinelli, I., Wasserman, L. et al. (2012). Manifold estimation and singular deconvolution under hausdorff loss, *The Annals of Statistics* **40**, 2, pp. 941–963.

Ghrist, R. (2008). Barcodes: the persistent topology of data, *Bulletin of the American Mathematical Society* **45**, 1, pp. 61–75.

Goffredo, M., Bouchrika, I., Carter, J. N. and Nixon, M. S. (2010). Self-calibrating view-invariant gait biometrics, *Systems, Man, and Cybernetics, Part B: Cybernetics, IEEE Transactions on* **40**, 4, pp. 997–1008.

Hatcher, A. (2002). Algebraic topology, (Cambridge University Press).

Hu, K. and Yin, L. (2013). Multi-scale topological features for hand posture representation and analysis, in *Computer Vision (ICCV), 2013 IEEE International Conference on* (IEEE), pp. 1928–1935.

Just, A., Rodriguez, Y. and Marcel, S. (2006). Hand posture classification and recognition using the modified census transform, in *Automatic Face and Gesture Recognition, 2006. FGR 2006. 7th International Conference on* (IEEE), pp. 351–356.

Kalyanswamy, S. (2009). Euler characteristic, *Math 4530 Introduction to Topology, Essay. Cornell University, Ithaca.*

Kasson, P. M., Zomorodian, A., Park, S., Singhal, N., Guibas, L. J. and Pande, V. S. (2007). Persistent voids: a new structural metric for membrane fusion, *Bioinformatics* **23**, 14, pp. 1753–1759.

Kibardin, E. (2015). 6 crazy things deep learning and topological data analysis can do with your data, https://www.linkedin.com/pulse/6-crazy-things-deep-learning-topological-data-can-do-your-kibardin.

Kokkinos, I., Bronstein, M., Yuille, A. et al. (2012). Dense scale invariant descriptors for images and surfaces, *INRIA*.

Lamar-León, J., García-Reyes, E. B. and Gonzalez-Diaz, R. (2012). Human gait identification using persistent homology, in *Progress in Pattern Recognition, Image Analysis, Computer Vision, and Applications* (Springer), pp. 244–251.

Lang, S. (2005). Undergraduate algebra, (Springer).

Lazebnik, S., Schmid, C. and Ponce, J. (2006). Beyond bags of features: Spatial pyramid matching for recognizing natural scene categories, in *Computer Vision and Pattern Recognition, 2006 IEEE Computer Society Conference on*, Vol. 2 (IEEE), pp. 2169–2178.

Lee, A. B., Pedersen, K. S. and Mumford, D. (2003). The nonlinear statistics of high-contrast patches in natural images, *International Journal of Computer Vision* **54**, 1–3, pp. 83–103.

Lee, H., Kang, H., Chung, M. K., nyun Kim, B. and Lee, D. S. (2012). Persistent brain network homology from the perspective of dendrogram, in *IEEE T. Med. Imaging*.

Li, C., Ovsjanikov, M. and Chazal, F. (2014). Persistence-based structural recognition, in *Proceedings of the IEEE Conference on Computer Vision and Pattern Recognition*, pp. 1995–2002.

Lindeberg, T. (1993a). Effective scale: A natural unit for measuring scale-space lifetime, *Pattern Analysis and Machine Intelligence, IEEE Transactions on* **15**, 10, pp. 1068–1074.

Lindeberg, T. (1993b). Scale-space theory in computer vision, (Springer).

Maria, C., Boissonnat, J.-D., Glisse, M. and Yvinec, M. (2014). The gudhi library: Simplicial complexes and persistent homology, in *Mathematical Software–ICMS 2014* (Springer), pp. 167–174.

Memoli, F. (2007). On the use of gromov-hausdorff distances for shape comparison, in *Eurographics Symposium on Point-based Graphics* (The Eurographics Association), pp. 81–90.

Mileyko, Y., Mukherjee, S. and Harer, J. (2011). Probability measures on the space of persistence diagrams, *Inverse Problems* **27**, 12, p. 124007.

Morozov, D. (2002). A practical guide to persistent homology, Software availabel at http://www.mrzv.org/software/dionysus/.

Mukherjee, S. (2013). Statistical learning and bayesian inference and topology.

Munkres, J. R. (1984). Elements of algebraic topology, (Addison-Wesley Reading).

Nanda., V. (2007). Perseus, the persistent homology software. Software availabel at http://www.sas.upenn.edu/~vnanda/perseus/.

Niyogi, P., Smale, S. and Weinberger, S. (2008). Finding the homology of submanifolds with high confidence from random samples, *Discrete & Computational Geometry* **39**, 1–3, pp. 419–441.

Niyogi, P., Smale, S. and Weinberger, S. (2011). A topological view of unsupervised learning from noisy data, *SIAM Journal on Computing* **40**, 3, pp. 646–663.

Olah, C. (2014). Neural networks, manifolds, and topology, http://colah.github.io/posts/2014-03-NN-Manifolds-Topology/.

Oudot, F. C. G. and Skraba, P. (2009). Persistence-based clustering in riemannian manifolds, *INRIA*.

Perona, P. and Malik, J. (1990). Scale-space and edge detection using anisotropic diffusion, *Pattern Analysis and Machine Intelligence, IEEE Transactions on* **12**, 7, pp. 629–639.

Rodriguez, A. and Laio, A. (2014). Clustering by fast search and find of density peaks, *Science* **344**, 6191, pp. 1492–1496.

Saul, L. K. and Roweis, S. T. (2000). An introduction to locally linear embedding, unpublished. Available at: http://www.cs.toronto.edu/~roweis/lle/publications. html.

Singh, G., Memoli, F., Ishkhanov, T., Sapiro, G., Carlsson, G. and Ringach, D. L. (2008). Topological analysis of population activity in visual cortex, *Journal of Vision* **8**, 8, p. 11.

Skraba, P., Ovsjanikov, M., Chazal, F. and Guibas, L. (2010). Persistence-based segmentation of deformable shapes, in *Computer Vision and Pattern Recognition Workshops (CVPRW), 2010 IEEE Computer Society Conference on* (IEEE), pp. 45–52.

Song, Y., Demirdjian, D. and Davis, R. (2011). Tracking body and hands for gesture recognition: Natops aircraft handling signals database, in *Automatic Face & Gesture Recognition and Workshops (FG 2011), 2011 IEEE International Conference on* (IEEE), pp. 500–506.

Sun, J., Ovsjanikov, M. and Guibas, L. (2009). A concise and provably informative multi-scale signature based on heat diffusion, in *Computer Graphics Forum*, Vol. 28 (Wiley Online Library), pp. 1383–1392.

Tausz, A., Vejdemo-Johansson, M. and Adams, H. (2011). Javaplex: A research software package for persistent (co)homology, Software available at http://javaplex.github.io/.

Tsiotsios, C. and Petrou, M. (2013). On the choice of the parameters for anisotropic diffusion in image processing, *Pattern recognition* **46**, 5, pp. 1369–1381.

Turner, K., Mileyko, Y., Mukherjee, S. and Harer, J. (2014). Fréchet means for distributions of persistence diagrams, *Discrete & Computational Geometry* **52**, 1, pp. 44–70.

Weickert, J. (1998). Anisotropic diffusion in image processing, (Teubner, Stuttgart, Germany).

Wildani, A. and Sharpee, T. O. (2014). Persistent homology for characterizing stimuli response in the primary visual cortex, in *Proceedings of the International Conference of Machine Learning (ICML) Workshop on Topology*.

Xia, K., Zhao, Z. and Wei, G.-W. (2015). Multiresolution persistent homology for excessively large biomolecular datasets, *The Journal of chemical physics* **143**, 13, p. 134103.

Yu, S., Tan, D. and Tan, T. (2006). A framework for evaluating the effect of view angle, clothing and carrying condition on gait recognition, in *Pattern Recognition, 2006. ICPR 2006. 18th International Conference on*, Vol. 4 (IEEE), pp. 441–444.

Zhu, X. (2013). Persistent homology: an introduction and a new text representation for natural language processing, in *Proceedings of the Twenty-Third International Joint Conference on Artificial Intelligence* (AAAI Press), pp. 1953–1959.

Zomorodian, A. and Carlsson, G. (2005). Computing persistent homology, *Discrete & Computational Geometry* **33**, 2, pp. 249–274.

Chapter 4

Low-Rank Matrix Estimation and Its Applications in Signal Processing and Machine Learning

Aimin Jiang,[a] Hon Keung Kwan,[b] and Yanping Zhu[c]

[a] *College of Internet of Things Engineering, Hohai University, Changzhou, Jiangsu 213022, China*

[b] *Department of Electrical and Computer Engineering, University of Windsor, Windsor, Ontario N9B 3P4, Canada*

[c] *School of Information Science and Engineering, Changzhou University, Changzhou, Jiangsu 213164, China*

jiangam@hhuc.edu.cn, kwan1@uwindsor.ca, zhuyanping@cczu.edu.cn

Low-rank structures play an important role in signal processing and machine learning [1–6], with various applications ranging from digital filter designs and medical imaging to dimensionality reduction and sensor network localization. In these applications, high-dimensional data can be approximately modeled as lying in a low-dimensional subspace or manifold. Under this assumption, a variety of data processing tasks (e.g., noisy data filtering, missing data interpolation, principle components learning, etc.) can be successfully accomplished. Furthermore, low-rank properties also result in significant reduction in computation and storage, that is extremely important in big data scenarios and leads to a plethora of

Learning Approaches in Signal Processing
Edited by Wan-Chi Siu, Lap-Pui Chau, Liang Wang, and Tieniu Tan

ISBN 978-981-4800-50-1 (Hardcover), 978-0-429-06114-1 (eBook)
www.panstanford.com

recent progress in low-rank modeling techniques and computational efficient numerical algorithms.

In this chapter, we focus on some basic low-rank models, which are widely used in the scenario of signal processing and machine learning. We also review some numerical techniques that are cornerstones of more sophisticated algorithms developed to tackle practical problems exploiting low-rank structures. Some practical problems are introduced later to demonstrate the power and effectiveness of low-rank modeling. Throughout this chapter, we use boldface capital letters (such as $\mathbf{A}$) to denote matrices, and boldface lowercase letters (such as $\mathbf{a}$) to denote vectors. The (i, j)th (ith) entry of matrix $\mathbf{A}$ ($\mathbf{a}$) is represented by A_{ij} (a_i), and $\mathbf{A}_{i_1:i_2,j_1:j_2}$ denotes a submatrix (or subvector) of $\mathbf{A}$. The transpose and the conjugate transpose of a matrix or vector are indicated by the superscript ^{T} and ^{H}, respectively. For a vector $\mathbf{a}$, its l_p norm is defined by $\|\mathbf{a}\|_p := \left(\sum_i a_i^p\right)^{1/p}$. Some important examples of ℓ_p norm used later are $\|\mathbf{a}\|_2 = \sqrt{\mathbf{a}^T\mathbf{a}} = \sqrt{\sum_i a_i^2}$ and $\|\mathbf{a}\|_1 = \sum_i |a_i|$. For a matrix $\mathbf{A}$, $\mathrm{Tr}(\mathbf{A})$, $\mathrm{rank}(\mathbf{A})$, and $\|\mathbf{A}\|_F$ stand for, respectively, the trace (i.e., $\sum_i A_{ii}$), the rank and the Frobenius norm (i.e., $\sqrt{\mathrm{Tr}\left(\mathbf{A}^T\mathbf{A}\right)}$) of $\mathbf{A}$. For a square matrix, $\mathrm{diag}(\mathbf{A})$ is used to retrieve diagonal elements of $\mathbf{A}$. If $\mathbf{A}$ is symmetric, $\mathbf{A} \succeq 0$ implies that it is a positive semidefinite (PSD) matrix. Finally, $\mathbf{1}$ ($\mathbf{0}$), and $\mathbf{I}$ to denote vectors of ones (zeros), and an identity matrix, respectively. Their dimensions can be clear in the context.

4.1 Low-Rank Matrix Estimation Models

Let $\mathbf{X} \in \mathbb{R}^{n\times m}$ be a matrix-form signal of interest. The singular value decomposition (SVD) is denoted by

$$\mathbf{X} = \mathbf{U}\Sigma\mathbf{V} = \sum_{i=1}^{\min\{n,m\}} \sigma_i \mathbf{u}_i \mathbf{v}_i^T \tag{4.1}$$

where $\mathbf{U} = [\mathbf{u}_1\ \mathbf{u}_2\ \ldots\ \mathbf{u}_n]$, $\mathbf{V} = [\mathbf{v}_1\ \mathbf{v}_2\ \ldots\ \mathbf{v}_m]$, and singular values are supposed to be organized in a non-ascending order, that is, $\sigma_1 \geq \sigma_2 \geq \ldots$. Using the SVD, the nuclear norm of $\mathbf{X}$ is defined by $\|\mathbf{X}\|_* = \sum_i \sigma_i$. If a large number of singular values are close zeros, $\mathbf{X}$ has an approximate low-rank representation. The best rank-r approximation of $\hat{\mathbf{X}}$ is given by

$$\hat{\mathbf{X}} = \arg\min_{\text{rank}(\mathbf{G}) \leq r} \|\mathbf{X} - \mathbf{G}\|_F, \tag{4.2}$$

whose optimal solution is guaranteed by the Eckart–Young–Mirsky theorem

$$\hat{\mathbf{X}} = \sum_{i=1}^{r} \sigma_i \mathbf{u}_i \mathbf{v}_i^T. \tag{4.3}$$

Based on the low-rank assumption, the approximation error of (4.2) should be sufficiently small for some $r \ll \min\{n, m\}$.

In many practical applications, one is not able to directly observe (even noisy) $\mathbf{X}$, but rather indirect measurements or incomplete entries of it. To recover $\mathbf{X}$, the low-rank assumption can also be exploited. For instance, if only a subset of entries of $\mathbf{X}$ are observed, then the corresponding problem can be cast as

$$\hat{\mathbf{X}} = \arg\min_{\mathbf{X} \in \mathbb{R}^{n \times m}} \text{rank}(\mathbf{X}), \quad \text{s.t.} \quad \mathbf{x} = \mathbf{M} \circ \mathbf{X} \tag{4.4}$$

where $\mathbf{M}$ is a binary mask matrix with one only on entry (i, j) if X_{ij} is provided in $\mathbf{x}$, $\circ$ denotes the Hadamard (or elementwise) product. If only noisy observations of $\mathbf{X}$ are available, the constraint in (4.4) can also be replaced by $\|\mathbf{x} - \mathbf{M} \circ \mathbf{X}\|_2 \leq \varepsilon$, leading to the following optimization problem

$$\hat{\mathbf{X}} = \arg\min_{\mathbf{X} \in \mathbb{R}^{n \times m}} \text{rank}(\mathbf{X}), \quad \text{s.t.} \quad \|\mathbf{x} - \mathbf{M} \circ \mathbf{X}\|_2 \leq \varepsilon. \tag{4.5}$$

Sometimes, (4.5) can also be formulated as, if the upper bound of the rank of $\mathbf{X}$ is given,

$$\hat{\mathbf{X}} = \arg\min_{\mathbf{X} \in \mathbb{R}^{n \times m}} \|\mathbf{x} - \mathbf{M} \circ \mathbf{X}\|_2, \quad \text{s.t.} \quad \text{rank}(\mathbf{X}) \leq r. \tag{4.6}$$

In the above formulation, however, the rank is a nonconvex function of $\mathbf{X}$, and (4.5) [also (4.6)] is generally NP-hard. To attain tractable solutions, convex relaxation can be applied to reformulate (4.5) [also (4.6)] as a convex optimization problem [7]. Note that rank $(\mathbf{X})$ is essentially equal to the number of nonzero singular values of $\mathbf{X}$. Inspired by the l_1 norm used as the tightest convex surrogate of the sparsity of a vector, one can replace rank $(\mathbf{X})$ by $\|\mathbf{X}\|_*$, resulting in

$$\hat{\mathbf{X}} = \arg\min_{\mathbf{X} \in \mathbb{R}^{n \times m}} \|\mathbf{X}\|_*, \quad \text{s.t.} \quad \|\mathbf{x} - \mathbf{M} \circ \mathbf{X}\|_2 \leq \varepsilon \tag{4.7}$$

or a regularized optimization problem

$$\hat{\mathbf{X}} = \arg\min_{\mathbf{X} \in \mathbb{R}^{n \times m}} \|\mathbf{x} - \mathbf{M} \circ \mathbf{X}\|_2 + \gamma \|\mathbf{X}\|_* . \tag{4.8}$$

A special case of low-rank models aforementioned deals with symmetric and PSD matrices (e.g., covariance matrix of statistical signals). In this situation, low-rank property is defined by the eigenvalue decomposition (EVD), i.e.,

$$\mathbf{X} = \mathbf{U}\Sigma\mathbf{U}^T = \sum_{i=1}^{n} \lambda_i \mathbf{u}_i \mathbf{u}_i^T , \tag{4.9}$$

where $\{\lambda_i\}$ are nonnegative eigenvalues of $\mathbf{X}$. Since now $\|\mathbf{X}\|_* = \mathrm{Tr}(\mathbf{X}) = \sum_i \lambda_i$, the nuclear norm is correspondingly replaced by the trace in (4.7) and (4.8).

Using the nuclear norm or the trace to replace the original rank function generally yields semidefinite programming (SDP) problems [8–10]. In principle, they can be solved by second-order-based interior-point methods, which adopt Newton's method to minimize a log-barrier penalty function and, thus, are stable and accurate. When dealing with SDP problems involving large-scale matrices, however, interior-point methods face many computational issues, which motivates recent advances of first-order-based numerical algorithms, such as singular value thresholding [11], the augmented Lagrangian multiplier methods [12], and Frank-Wolfe method [13]. Another strategy of efficiently solving low-rank matrix estimation problems is to exploit the low-rank factorization $\mathbf{X} = \mathbf{L}\mathbf{R}^T$ where $\mathbf{L} \in \mathbb{R}^{n \times r}$ and $\mathbf{R} \in \mathbb{R}^{m \times r}$ [14–17], such that numerical algorithms work directly on $\mathbf{L}$ and $\mathbf{R}$, instead of handling the whole matrix $\mathbf{X}$. For a symmetric matrix, computational costs can be further reduced. In this chapter, we focus on basic low-rank matrix estimation models and also their applications in signal processing and machine learning. Readers interested to computationally efficient numerical methods are referred to some wonderful review papers [1–6, 9]. In the subsequent section, we shall present some practical applications of low-rank matrix estimation models in the fields of signal processing and machine learning.

4.2 Applications of Low-Rank Models in Signal Processing and Machine Learning

4.2.1 Infinite Impulse Response (IIR) Digital Filter Design

4.2.1.1 Design model

A digital filter is a tool, implemented by software or hardware, used to extract useful information and remove undesired signal components from input signals. Basically, there are two types of digital filters, i.e., finite impulse response (FIR) and IIR [18, 19]. Compared with an FIR digital filter, an IIR digital filter has better approximation capability under the same specifications. But its design problem faces more challenging due to its nonconvex property. Due to the presence of the denominator in its rational transfer function, an IIR digital filter design problem (involving both magnitude and phase approximation) cannot be readily formulated as an equivalent convex optimization problem. Furthermore, for stability, all the poles of an IIR digital filter must be constrained within a stability domain, which, however, is generally nonconvex.

The most prevalent design strategy is to employ sequential procedures to gradually approach an optimal design. To cope with the nonconvexity essence, the Levy (L) linearized function [20] and the Sanathanan–Koerner (SK) iterative technique [21], which we shall collectively call the L-SK, are widely employed by many design methods [22–27]. In each iteration, the denominator of an approximation error is replaced by its counterpart obtained in the previous iteration, resulting in a convex approximation error to be minimized. The primal drawback of the L-SK design approaches is that their convergence cannot be guaranteed. Taylor series approximation [28] or more sophisticated numerical techniques (e.g., Gauss–Newton method [29]) can also be used to achieve a convexized objective function. But it could be trapped in the neighborhood of locally optimal solutions. In practice, the performance of the design approaches relies on many factors, such as initial points, step size, and stability conditions adopted. They generally have no ways to detect if a globally optimal solution is obtained or not. In this subsection, however, we shall show that, given a frequency-domain approximation error, the IIR digital filter design problem in the minimax sense can be relaxed to an SDP problem. The globally optimal solution can possibly be obtained by bisection searching procedure.

The transfer function of an IIR digital filter is defined by

$$H(z) = \frac{P(z)}{Q(z)} = \frac{\sum_{n=0}^{N} p_n z^{-n}}{1 + \sum_{m=1}^{M} q_m z^{-m}} = \frac{\mathbf{p}^T \varphi_N(z)}{1 + \mathbf{q}^T \varphi_M(z)}, \tag{4.10}$$

where

$$\mathbf{p} = [p_0 \ p_1 \ \ldots \ p_N]^T, \tag{4.11}$$

$$\mathbf{q} = [q_1 \ q_2 \ \ldots \ q_M]^T, \tag{4.12}$$

$$\varphi_N(z) = [1 \ z^{-1} \ \ldots \ z^{-N}]^T, \tag{4.13}$$

$$\varphi_M(z) = [z^{-1} \ z^{-2} \ \ldots \ z^{-M}]^T. \tag{4.14}$$

Given an ideal frequency response $D(\omega)$ over $[0, \pi)$ and a nonnegative weighting function $W(\omega)$, the minimax design problem of an IIR digital filter can be formulated as

$$\min_{\delta, \mathbf{x}} \ \delta, \quad \text{s.t.} \quad W(\omega) \left| H(e^{j\omega}) - D(\omega) \right| \le \delta \text{ for } \omega \in \Omega_I, \tag{4.15}$$

where $\mathbf{x} = [\mathbf{q}^T \ \mathbf{p}^T]^T$ and Ω_I denotes the union of all frequency bands of interest. In practice, the constraint of (4.15) can be implemented on a grid of frequency points sampled over Ω_I. It is worth noting that there exists in the above problem an implicit stability constraint imposed on $\mathbf{q}$, such that all the poles lie inside the unit circle. To proceed, we first introduce

$$\mathbf{X} = \mathbf{x}\mathbf{x}^T = \begin{bmatrix} \mathbf{X}_q & \mathbf{X}_{q,p} \\ \mathbf{X}_{q,p}^T & \mathbf{X}_p \end{bmatrix}, \tag{4.16}$$

where $\mathbf{X}_q \in \mathbb{R}^{M \times M}$, $\mathbf{X}_p \in \mathbb{R}^{(N+1)\times(N+1)}$, and $\mathbf{X}_{q,p} \in \mathbb{R}^{M \times (N+1)}$. Then, the constraint of (4.15) can be reformulated as

$$\begin{aligned}
&\left| D(\omega) Q(e^{j\omega}) - P(e^{j\omega}) \right|^2 \\
&= |D(\omega)|^2 + 2\mathrm{Re}\left\{ D(\omega)\mathbf{c}^H(\omega) \right\} \mathbf{x} + \mathbf{x}^T \mathbf{A}(\omega)\mathbf{x} \\
&= |D(\omega)|^2 + 2\mathrm{Re}\left\{ D(\omega)\mathbf{c}^H(\omega) \right\} \mathbf{x} + \mathrm{Tr}\left(\mathbf{X}\mathbf{A}(\omega) \right) \le \frac{\delta^2}{W^2(\omega)} \cdot \left| Q(e^{j\omega}) \right|^2 \\
&= \frac{\delta^2}{W^2(\omega)} \cdot \left[1 + 2\mathrm{Re}\left\{ \varphi_M^H(e^{j\omega}) \right\} \mathbf{q} + \mathbf{q}^T \mathbf{B}(\omega)\mathbf{q} \right] \\
&= \frac{\delta^2}{W^2(\omega)} \cdot \left[1 + 2\mathrm{Re}\left\{ \varphi_M^H(e^{j\omega}) \right\} \mathbf{q} + \mathrm{Tr}\left(\mathbf{X}_q \mathbf{B}(\omega) \right) \right]
\end{aligned} \tag{4.17}$$

where Re$\{\cdot\}$ is used to retrieve the real part of a complex variable, and

$$\mathbf{c}(\omega) = \begin{bmatrix} D(\omega)\varphi_M(e^{j\omega}) \\ -\varphi_N(e^{j\omega}) \end{bmatrix}, \tag{4.18}$$

$$\mathbf{A}(\omega) = \text{Re}\left\{\mathbf{c}(\omega)\mathbf{c}^H(\omega)\right\}, \tag{4.19}$$

$$\mathbf{B}(\omega) = \text{Re}\left\{\varphi_M(e^{j\omega})\varphi_M^H(e^{j\omega})\right\}. \tag{4.20}$$

Note that, given δ, (4.17) becomes a linear inequality constraint with respect to $\mathbf{x}$ and $\mathbf{X}$, which are related by (4.16). The remaining obstacle is that matrix equality constraint (4.16) is nonconvex. To the end of constructing a convex optimization problem, we relax (4.16) to $\mathbf{X} \succeq \mathbf{x}\mathbf{x}^T$, which is equivalent to

$$\mathbf{Z} = \begin{bmatrix} 1 & \mathbf{x}^T \\ \mathbf{x} & \mathbf{X} \end{bmatrix} \succeq 0. \tag{4.21}$$

Now, we have the following feasibility problem:

$$\text{find} \quad \mathbf{Z} \tag{4.22a}$$

$$\text{s.t.} \quad |D(\omega)|^2 + 2\text{Re}\left\{D(\omega)\mathbf{c}^H(\omega)\right\}\mathbf{x} + \text{Tr}\left(\mathbf{X}\mathbf{A}(\omega)\right)$$
$$\le \frac{\delta^2}{W^2(\omega)} \cdot \left[1 + 2\text{Re}\left\{\varphi_M^H(e^{j\omega})\right\}\mathbf{q} + \text{Tr}\left(\mathbf{X}_q\mathbf{B}(\omega)\right)\right], \quad \forall\omega \in \Omega_I, \tag{4.22b}$$

$$\mathbf{Z} = \begin{bmatrix} 1 & \mathbf{x}^T \\ \mathbf{x} & \mathbf{X} \end{bmatrix} = \begin{bmatrix} 1 & \mathbf{q}^T & \mathbf{p}^T \\ \mathbf{q} & \mathbf{X}_q & \mathbf{X}_{q,p} \\ \mathbf{p} & \mathbf{X}_{q,p}^T & \mathbf{X}_p \end{bmatrix} \succeq 0. \tag{4.22c}$$

In practice, the minimum approximation error is unknown. But one can find it by the bisection searching procedure. In each iteration, (4.22) is modified to

$$\min_{z, \mathbf{Z} \succeq 0} \quad z \tag{4.23a}$$

$$\text{s.t.} \quad |D(\omega)|^2 + 2\text{Re}\left\{D(\omega)\mathbf{c}^H(\omega)\right\}\mathbf{x} + \text{Tr}\left(\mathbf{X}\mathbf{A}(\omega)\right)$$
$$\le \left(\frac{\delta^{(k)}}{W(\omega)}\right)^2 \cdot \left[1 + 2\text{Re}\left\{\varphi_M^H(e^{j\omega})\right\}\mathbf{q} + \text{Tr}\left(\mathbf{X}_q\mathbf{B}(\omega)\right)\right] + z,$$
$$\forall\omega \in \Omega_I, \tag{4.23b}$$

$$\mathbf{Z} = \begin{bmatrix} 1 & \mathbf{x}^T \\ \mathbf{x} & \mathbf{X} \end{bmatrix} = \begin{bmatrix} 1 & \mathbf{q}^T & \mathbf{p}^T \\ \mathbf{q} & \mathbf{X}_q & \mathbf{X}_{q,p} \\ \mathbf{p} & \mathbf{X}_{q,p}^T & \mathbf{X}_p \end{bmatrix}. \tag{4.23c}$$

Three important remarks about the optimal solution $\left(z^{(k)}, \mathbf{Z}^{(k)}\right)$ to (4.23) are given below.

(1) If $z^{(k)} > 0$, one can conclude that, for the given $\delta^{(k)}$, there is no feasible solution satisfying the relaxed constraints and, thus, also the original nonconvex constraints in (4.15). In this situation, one should increase $\delta^{(k)}$ to achieve a feasible solution in the next iteration.
(2) A negative $z^{(k)}$ implies that there exists a feasible solution to the relaxed problem (4.23). This, however, does not mean that, for a given $\delta^{(k)}$, one can definitely find a feasible solution to the original problem. More sophisticated techniques have to be employed to check the feasibility of the original problem. For example, to achieve a low-rank solution, $\mathrm{Tr}\,(\mathbf{Z})$ can be incorporated as a regularization term in the objective function of (4.23) [30].
(3) If $z^{(k)} < 0$ and $\mathrm{rank}\left(\mathbf{Z}^{(k)}\right) = 1$, the obtained $\left(\mathbf{p}^{(k)}, \mathbf{q}^{(k)}\right)$ is an feasible solution to the original problem. Furthermore, if in all the iterations, one always achieves rank-1 solutions and the final output of the bisection searching procedure satisfies the stability condition, the globally optimal solution to (4.15) is attained.

4.2.1.2 Design example

Note that Remark 3 indicates that the optimality of the obtained solution can be verified, even though (4.15) is a nonconvex optimization problem. This can be demonstrated by the following low-pass filter design example. The ideal frequency response is given by

$$D(\omega) = \begin{cases} e^{-j12\omega}, & 0 \leq \omega \leq 0.4\pi, \\ 0, & 0.56\pi \leq \omega < \pi. \end{cases} \tag{4.24}$$

Filter order are chosen as $N = 15$ and $M = 4$. Problem (4.23) is solved by SeDuMi [31], a well-known MATLAB toolbox. Although no stability constraints are used, a stable design is still obtained by the SDP relaxation approach. Magnitude response and passband group delay of the obtained IIR digital filter are depicted in Fig. 4.1. The approximation error is shown in Fig. 4.2. Simulation results

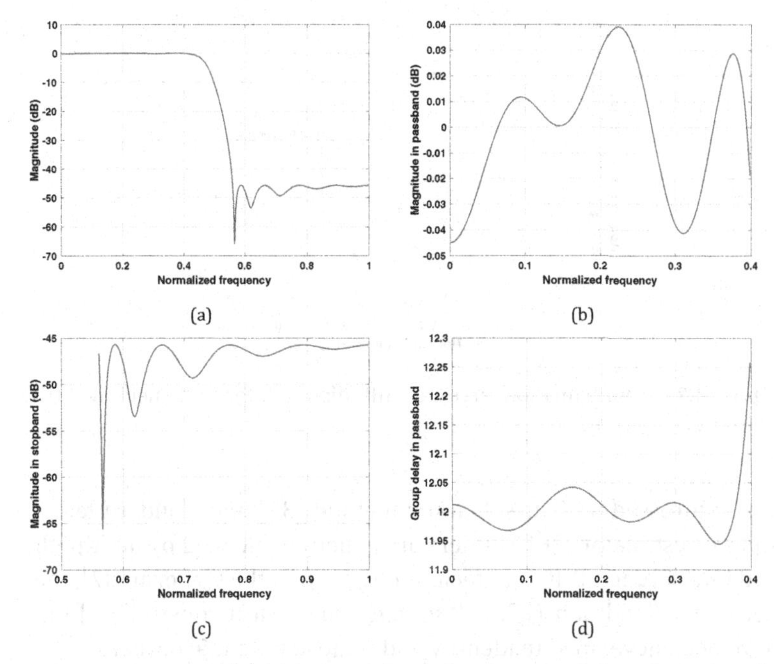

Figure 4.1 Magnitude response and passband group delay of IIR digital filter designed by SDP relaxation algorithm.

reveal that, after 13 iterations, the SDP relaxation design approach terminates the iterative procedure. Rank-1 solutions are always obtained in each iteration. Thus, the final output of the bisection searching procedure can be claimed to be the optimal design of the original problem. Using the same specifications, the L-SK design algorithm [23] is also employed to design an IIR digital filter whose approximation error is also shown in Fig. 4.2.

4.2.2 Time-Difference-of-Arrival (TDOA) Estimation

4.2.2.1 TDOA estimation model

TDOA measurements are widely used in various applications of sensor networks, e.g., source localization [32] and tracking [33]. They can be obtained by general cross-correlation (GCC) [34], but adversely affected by noise, outliers, and even missing data. When only considering additive Gaussian noise, TDOA estimation can be

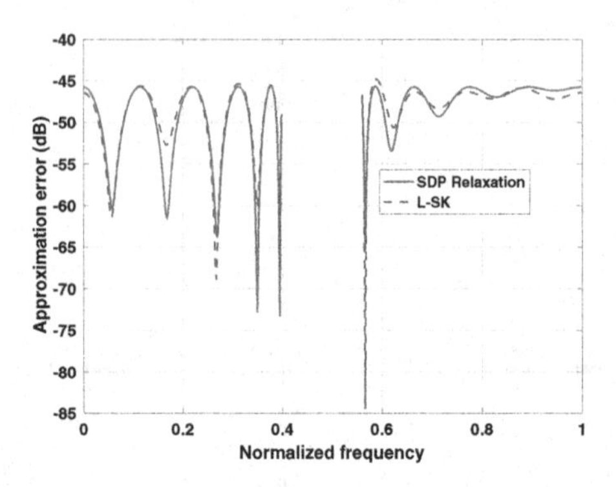

Figure 4.2 Approximation error of IIR digital filter designed by SDP relaxation algorithm.

well addressed by Gauss–Markov method [35] and standard least-squares estimator [36]. Outliers are generally induced by multipath and interference. Different techniques, such as the ℓ_p-norm [37], the Zero-Sum Condition (ZSC) [38], and geometrical constraints [39], have been developed to identify and remove potential outliers.

Missing data of TDOA measurements are induced by communication failure or severe non-line-of-sight (NLOS) propagation. Recent studies show that low-rank property of a TDOA matrix can be exploited to handle missing data along with additive noise and outliers [40]. For a network with n sensors, the (i, j)th element of a TDOA matrix represents the TDOA value between sensors i and j with respect to a common reference source. Thus, it is a skew-symmetric matrix and has rank 2, which can be exploited to improve the denoising performance and the accuracy of TDOA matrix completion. In this subsection, we shall demonstrate how to exploit the low rank essence of the Gram matrix of the TDOA matrix to enhance estimation accuracy. The resulting algorithm does not rely on any extra prior information (e.g., the exact number of outliers, statistical assumptions of noise and outliers, sensors' positions).

Suppose that, in a sensor network, one source and n sensors are located at $\mathbf{s}_0 \in \mathbb{R}^d$ and $\mathbf{s}_i \in \mathbb{R}^d, i = 1, \ldots, n$, respectively. The TDOA value for a pair of sensors is computed by $a_{ij} = d_i - d_j$ where $d_i = \|\mathbf{s}_i - \mathbf{s}_0\|_2$. Then, the TDOA matrix is defined by $\mathbf{A} = [a_{ij}] \in \mathbb{R}^{n \times n}$. In practice, TDOA measurements are contaminated by noise and some elements of the TDOA matrix cannot be reliably obtained, yielding noisy and incomplete TODA matrix $\tilde{\mathbf{A}} = [\tilde{a}_{ij}]$.

For noiseless TDOAs, a TDOA matrix can be written as [40]

$$\mathbf{A} = \phi(\mathbf{x}) = \mathbf{x} \cdot \mathbf{1}^T - \mathbf{1} \cdot \mathbf{x}^T, \tag{4.25}$$

where $\mathbf{x} \in \mathbb{R}^n$. Eq. (4.25) indicates that rank($\mathbf{A}$) $\leq$ 2. Once TDOAs are contaminated by noise and outliers, the rank of noisy TDOA matrix $\tilde{\mathbf{A}}$ is generally larger than 2. By exploiting low-rank essence of noiseless $\mathbf{A}$, closed-form solutions can be obtained [40]. But if $\tilde{\mathbf{A}}$ is incomplete, the estimation problem becomes more complicated. To achieve an accurate estimation, we further introduce

$$\mathbf{Y} = \frac{1}{n}\mathbf{A}\mathbf{A}^T = \Upsilon(\mathbf{X}), \tag{4.26}$$

where, by defining $\mathbf{E} = \frac{1}{n}\mathbf{1} \cdot \mathbf{1}^T$,

$$\mathbf{X} = \mathbf{x}\mathbf{x}^T, \tag{4.27}$$

$$\Upsilon(\mathbf{X}) = \mathbf{X} + \mathrm{Tr}\{\mathbf{X}\} \cdot \mathbf{E} - (\mathbf{X} \cdot \mathbf{E} + \mathbf{E} \cdot \mathbf{X}). \tag{4.28}$$

Obviously, rank($\mathbf{Y}$) = rank($\mathbf{A}$) $\leq$ 2. Similar to (4.16), (4.27) defines a nonconvex constraint. Moreover, rank constraint leads to computational intractability. To overcome these obstacles, we adopt SDP relaxation technique to reformulate (4.27) as $\mathbf{Z} = \Omega(\mathbf{x}, \mathbf{X}) = \begin{bmatrix} 1 & \mathbf{x}^T \\ \mathbf{x} & \mathbf{X} \end{bmatrix} \succeq 0$ and replace rank($\mathbf{Y}$) by Tr ($\mathbf{Y}$), yielding a convex optimization problem

$$\min_{\mathbf{Z} \succeq 0} \quad \left\| \mathbf{M} \circ \left(\mathbf{T} - \tilde{\mathbf{A}}\right) \right\|_{1,1} + \alpha \mathrm{Tr}\{\mathbf{Y}\} \tag{4.29a}$$

$$\text{s.t.} \quad \mathbf{Z} = \Omega(\mathbf{x}, \mathbf{X}) \tag{4.29b}$$

$$\mathbf{T} = \phi(\mathbf{x}) \tag{4.29c}$$

$$\mathbf{Y} = \Upsilon(\mathbf{X}), \tag{4.29d}$$

where $\mathbf{M}$ is a binary mask matrix with one only on entry (i, j) if noisy TDOA $\tilde{a}_{ij}$ is available. In (4.29a), $\|\cdot\|_{1,1}$ denotes the $\ell_{1,1}$ norm of a matrix (i.e., $\|\mathbf{X}\|_{1,1} = \sum_{i,j} |X_{ij}|$). Since the number of outliers is generally unknown and smaller than that of TDOAs, we adopt the $\ell_{1,1}$ norm to enforce the data fidelity.

4.2.2.2 ADMM for TDOA estimation

Eq. (4.29) represents an SDP problem, whose optimal solution can be obtained by second-order-based interior-point methods [7, 8], which adopt Newton's method to minimize a log-barrier penalty function in each iteration and, thus, are stable and accurate at the expense of scalability. In this subsection, we shall develop an alternating direction method of multipliers (ADMM) [41–43], which involves several subproblems in each iteration. Closed-form solutions to these subproblems can be obtained simply by matrix multiplication and EVD. The augmented Lagrangian function of (4.29) is given by

$$\begin{aligned} L_\rho(\mathbf{T}, \mathbf{Y}, \mathbf{x}, \mathbf{X}, \mathbf{Z}; \Gamma, \Pi, \Theta) = & \left\|\mathbf{M} \circ \left(\mathbf{T} - \tilde{\mathbf{A}}\right)\right\|_{1,1} + \alpha \mathrm{Tr}\{\mathbf{Y}\} \\ & + \mathrm{Tr}\left\{\Pi^T \cdot [\mathbf{Z} - \Omega(\mathbf{x}, \mathbf{X})]\right\} \\ & + \mathrm{Tr}\left\{\Gamma^T \cdot [\mathbf{T} - \phi(\mathbf{x})]\right\} \\ & + \mathrm{Tr}\left\{\Theta^T \cdot [\mathbf{Y} - \Upsilon(\mathbf{X})]\right\} \\ & + \frac{\rho}{2}\|\mathbf{Z} - \Omega(\mathbf{x}, \mathbf{X})\|_F^2 \\ & + \frac{\rho}{2}\|\mathbf{T} - \phi(\mathbf{x})\|_F^2 + \frac{\rho}{2}\|\mathbf{Y} - \Upsilon(\mathbf{X})\|_F^2 . \end{aligned} \tag{4.30}$$

Then, the ADMM needs to alternately solve the following subproblems to achieve new estimates $\mathbf{T}^{(k+1)}, \mathbf{Y}^{(k+1)}, \mathbf{x}^{(k+1)}, \mathbf{X}^{(k+1)}$, and $\mathbf{Z}^{(k+1)}$:

$$\mathbf{T}^{(k+1)} = \arg\min_{\mathbf{T}} \left\{ \left\|\mathbf{M} \circ \left(\mathbf{T} - \tilde{\mathbf{A}}\right)\right\|_{1,1} + \frac{\rho^{(k)}}{2} \left\|\mathbf{T} - \hat{\mathbf{T}}^{(k)}\right\|_F^2 \right\} \tag{4.31a}$$

$$\mathbf{Y}^{(k+1)} = \arg\min_{\mathbf{Y}} \left\{ \mathrm{Tr}\{\mathbf{Y}\} + \frac{\rho^{(k)}}{2\alpha} \left\|\mathbf{Y} - \hat{\mathbf{Y}}^{(k)}\right\|_F^2 \right\} \tag{4.31b}$$

$$\mathbf{x}^{(k+1)} = \arg\min_{\mathbf{x}} \left\{ \left\|\mathbf{T}^{(k+1)} - \phi(\mathbf{x}) + \frac{\Gamma^{(k)}}{\rho^{(k)}}\right\|_F^2 \right. \tag{4.31c}$$

$$\left. + 2\left\|\mathbf{Z}^{(k)}_{2:n+1,1} - \mathbf{x} + \frac{\Pi^{(k)}_{2:n+1,1}}{\rho^{(k)}}\right\|_2^2 \right\} \tag{4.31d}$$

$$\mathbf{X}^{(k+1)} = \arg\min_{\mathbf{X}} \left\{ \left\|\mathbf{G}^{(k)} - \mathbf{X}\right\|_F^2 + \left\|\mathbf{H}^{(k)} - \Upsilon(\mathbf{X})\right\|_F^2 \right\} \tag{4.31e}$$

$$\mathbf{Z}^{(k+1)} = \arg\min_{\mathbf{Z} \succeq 0} \left\{ \left\|\mathbf{Z} - \hat{\mathbf{Z}}^{(k)}\right\|_F^2 \right\}, \tag{4.31f}$$

where

$$\hat{\mathbf{T}}^{(k)} = \phi(\mathbf{x}^{(k)}) - \frac{1}{\rho^{(k)}}\Gamma^{(k)} \tag{4.32}$$

$$\hat{\mathbf{Y}}^{(k)} = \Upsilon(\mathbf{X}^{(k)}) - \frac{1}{\rho^{(k)}}\Theta^{(k)} \tag{4.33}$$

$$\mathbf{G}^{(k)} = \mathbf{Z}^{(k)}_{2:n+1,2:n+1} + \frac{1}{\rho^{(k)}}\Pi^{(k)}_{2:n+1,2:n+1} \tag{4.34}$$

$$\mathbf{H}^{(k)} = \mathbf{Y}^{(k+1)} + \frac{1}{\rho^{(k)}}\Theta^{(k)} \tag{4.35}$$

$$\hat{\mathbf{Z}}^{(k)} = \Omega(\mathbf{x}^{(k+1)}, \mathbf{X}^{(k+1)}) - \frac{1}{\rho^{(k)}}\Pi^{(k)}. \tag{4.36}$$

Furthermore, at the end of each iteration, $\Gamma^{(k)}$, $\Theta^{(k)}$, $\Pi^{(k)}$, and $\rho^{(k)}$ are updated by

$$\Gamma^{(k+1)} = \Gamma^{(k)} + \rho^{(k)}\left[\mathbf{T}^{(k+1)} - \phi(\mathbf{x}^{(k+1)})\right] \tag{4.37a}$$

$$\Theta^{(k+1)} = \Theta^{(k)} + \rho^{(k)}\left[\mathbf{Y}^{(k+1)} - \Upsilon(\mathbf{X}^{(k+1)})\right] \tag{4.37b}$$

$$\Pi^{(k+1)} = \Pi^{(k)} + \rho^{(k)}\left[\mathbf{Z}^{(k+1)} - \Omega(\mathbf{x}^{(k+1)}, \mathbf{X}^{(k+1)})\right] \tag{4.37c}$$

$$\rho^{(k+1)} = \eta\rho^{(k)}. \tag{4.37d}$$

The optimization of (4.31a) can be implemented on elements of $\mathbf{T}$. If $M_{ij} = 0$, the optimal solution to (4.31a) is $T_{ij}^{(k+1)} = \hat{T}_{ij}^{(k)}$. Otherwise, (4.31a) becomes a soft-thresholding problem $\left|T_{ij} - \tilde{a}_{ij}\right| + \frac{\rho^{(k)}}{2}\left|T_{ij} - \hat{T}_{ij}^{(k)}\right|^2$ [44, 45], whose solution is given by $T_{ij}^{(k+1)} = \tilde{a}_{ij} - \text{soft}\left(\tilde{a}_{ij} - \hat{T}_{ij}^{(k)}, \frac{1}{\rho^{(k)}}\right)$ where $\text{soft}(u, \lambda) = \text{sign}(u) \cdot \max\{|u| - \lambda, 0\}$. Thus, we have

$$T_{ij}^{(k+1)} = \begin{cases} \hat{T}_{ij}^{(k)}, & \text{if } M_{ij} = 0 \\ \tilde{a}_{ij} - \text{soft}\left(\tilde{a}_{ij} - \hat{T}_{ij}^{(k)}, \frac{1}{\rho^{(k)}}\right), & \text{if } M_{ij} \neq 0. \end{cases} \tag{4.38}$$

The optimization of (4.31b) can be implemented on diagonal and off-diagonal elements of $\mathbf{Y}$, separately. For off-diagonal elements, the solution is simply $\mathbf{Y}^{(k+1)} = \hat{\mathbf{Y}}^{(k)}$. To achieve the optimal solution of diagonal elements, one has to minimize $\mathbf{1}^T\mathbf{y} + \frac{\rho^{(k)}}{2\alpha}\left\|\mathbf{y} - \text{diag}\left(\hat{\mathbf{Y}}^{(k)}\right)\right\|_2^2$. Therefore, the solution to (4.31b) can be simply given by

$$\mathbf{Y}^{(k+1)} = \hat{\mathbf{Y}}^{(k)} - \frac{\alpha}{\rho^{(k)}}\mathbf{I}. \tag{4.39}$$

Problem (4.31f) can be considered as a variant of singular value thresholding problem (4.2) [11]. Its solution is computed by

$$\mathbf{Z}^{(k+1)} = \mathcal{T}_+ \left(\hat{\mathbf{Z}}^{(k)}\right), \tag{4.40}$$

where, given a symmetric matrix $\mathbf{S}$, $\mathcal{T}_+(\mathbf{S})$ is an operator used to replace all the negative eigenvalues of $\mathbf{S}$ by zeros in the EVD of $\mathbf{S}$.

The computation of the optimal solution to (4.31e) is more complicated. Let the EVD of $\mathbf{E}$ be $\mathbf{E} = \mathbf{P}\begin{bmatrix} 1 & \mathbf{0}^T \\ \mathbf{0} & \mathbf{0} \end{bmatrix}\mathbf{P}^T$. Then, multiplying $\mathbf{P}$ and $\mathbf{P}^T$ on both sides of $\mathbf{Y}$ yields

$$\mathbf{P}^T\mathbf{Y}\mathbf{P} = \begin{bmatrix} \mathrm{Tr}\{\mathbf{Q}_{22}\} & \mathbf{0}^T \\ \mathbf{0} & \mathbf{Q}_{22} \end{bmatrix}, \tag{4.41}$$

where

$$\mathbf{Q} = \mathbf{P}^T\mathbf{X}\mathbf{P} = \begin{bmatrix} Q_{11} & \mathbf{Q}_{12}^T \\ \mathbf{Q}_{12} & \mathbf{Q}_{22} \end{bmatrix}. \tag{4.42}$$

Using $\mathbf{Q}$, subproblem (4.31e) can be reformulated as

$$\mathbf{Q}^{(k+1)} = \arg\min_{\mathbf{Q}} \left\{ \left\|\hat{\mathbf{G}}^{(k)} - \mathbf{Q}\right\|_F^2 + \left\|\hat{\mathbf{H}}^{(k)} - \begin{bmatrix} \mathrm{Tr}\{\mathbf{Q}_{22}\} & \mathbf{0}^T \\ \mathbf{0} & \mathbf{Q}_{22} \end{bmatrix}\right\|_F^2 \right\}, \tag{4.43}$$

where

$$\hat{\mathbf{G}}^{(k)} = \mathbf{P}^T\mathbf{G}^{(k)}\mathbf{P} = \begin{bmatrix} \hat{G}_{11}^{(k)} & \hat{\mathbf{G}}_{12}^{(k)T} \\ \hat{\mathbf{G}}_{12}^{(k)} & \hat{\mathbf{G}}_{22}^{(k)} \end{bmatrix}$$

$$\hat{\mathbf{H}}^{(k)} = \mathbf{P}^T\mathbf{H}^{(k)}\mathbf{P} = \begin{bmatrix} \hat{H}_{11}^{(k)} & \hat{\mathbf{H}}_{12}^{(k)T} \\ \hat{\mathbf{H}}_{12}^{(k)} & \hat{\mathbf{H}}_{22}^{(k)} \end{bmatrix}.$$

Since diagonal blocks Q_{11} and $\mathbf{Q}_{12}$ only exist in the first term of (4.43), the optimization of (4.43) can be implemented on Q_{11}, $\mathbf{Q}_{12}$, and $\mathbf{Q}_{22}$, separately. Obviously,

$$Q_{11}^{(k+1)} = \hat{G}_{11}^{(k)} \tag{4.44}$$

$$\mathbf{Q}_{12}^{(k+1)} = \hat{\mathbf{G}}_{12}^{(k)}. \tag{4.45}$$

Taking (4.44) and (4.45) into (4.43) yields a simplified problem

$$\mathbf{Q}_{22}^{(k+1)} = \arg\min_{\mathbf{Q}_{22}} \left\{ \left\|\hat{\mathbf{G}}_{22}^{(k)} - \mathbf{Q}_{22}\right\|_F^2 + \left\|\hat{\mathbf{H}}_{22}^{(k)} - \mathbf{Q}_{22}\right\|_F^2 + \left(\hat{H}_{11}^{(k)} - \mathrm{Tr}\{\mathbf{Q}_{22}\}\right)^2 \right\}. \tag{4.46}$$

Note that, for an off-diagonal element of Q_{ij} in $\mathbf{Q}_{22}$, (4.46) is equivalent to

$$Q_{ij}^{(k+1)} = \arg\min_{Q_{ij}} \left\{ \left(\hat{G}_{ij}^{(k)} - Q_{ij}\right)^2 + \left(\hat{H}_{ij}^{(k)} - Q_{ij}\right)^2 \right\}, \tag{4.47}$$

whose solution is given by

$$Q_{ij}^{(k+1)} = 0.5\left(\hat{G}_{ij} + \hat{H}_{ij}\right), \quad i \neq j, \quad i, j > 1. \tag{4.48}$$

For diagonal elements in $\mathbf{Q}_{22}$, (4.46) is reduced to

$$\begin{aligned} \operatorname{diag}\left\{\mathbf{Q}_{22}^{(k+1)}\right\} = \arg\min_{\mathbf{q}} \Bigg\{ & \left\| 0.5 \cdot \operatorname{diag}\left\{\hat{\mathbf{G}}_{22}^{(k)} + \hat{\mathbf{H}}_{22}^{(k)}\right\} - \mathbf{q} \right\|_2^2 \\ & + \left(\hat{H}_{11}^{(k)} - \mathbf{1}^T\mathbf{q}\right)^2 \Bigg\}, \end{aligned} \tag{4.49}$$

whose solution is finally obtained by

$$\operatorname{diag}\left\{\mathbf{Q}_{22}^{(k+1)}\right\} = [2\mathbf{I} + n\mathbf{E}]^{-1} \cdot \left(\operatorname{diag}\left\{\hat{\mathbf{G}}_{22}^{(k)} + \hat{\mathbf{H}}_{22}^{(k)}\right\} + \hat{H}_{11}^{(k)} \cdot \mathbf{1}\right). \tag{4.50}$$

Once $\mathbf{Q}^{(k+1)}$ is computed, the optimal solution $\mathbf{X}^{(k+1)}$ to (4.31e) can be recovered by $\mathbf{X}^{(k+1)} = \mathbf{P}\mathbf{Q}^{(k+1)}\mathbf{P}^T$.

4.2.2.3 Experimental results

We present two sets of experiments using synthetic data to evaluate the performance of the ADMM discussed before. Suppose that 15 sensors and one source are randomly distributed within a unit square. TDOA measurements are corrupted by additive Gaussian noise $n_{ij} \sim \mathcal{N}\left(0, \sigma_{ij}^2\right)$ and sparse outliers $s_{ij} \sim \mathcal{N}\left(0, \varsigma_{ij}^2\right)$. In our experiments, the regularization parameter α used in (4.29) is always set to 10^{-5}. The mean square error MSE $= \frac{1}{n}\sqrt{\sum_{i,j}\left(a_{ij} - \hat{a}_{ij}\right)^2}$, where a_{ij} and $\hat{a}_{ij}$ represent, respectively, the real and estimated TDOA values between sensors i and j, is utilized in our evaluation.

In the first set of experiments, we assume that there are no missing TDOA measurements. Five TDOA measurements are randomly chosen as outliers. Standard deviations σ_{ij} and ς_{ij} vary within $\left[10^{-6}, 10^{-4}\right]$ and $\left[10^{-2}, 1\right]$, respectively. The variation of average MSE is depicted in Fig. 4.3. Estimation results of the algorithm developed in [40] (represented by DeN+MC) are also

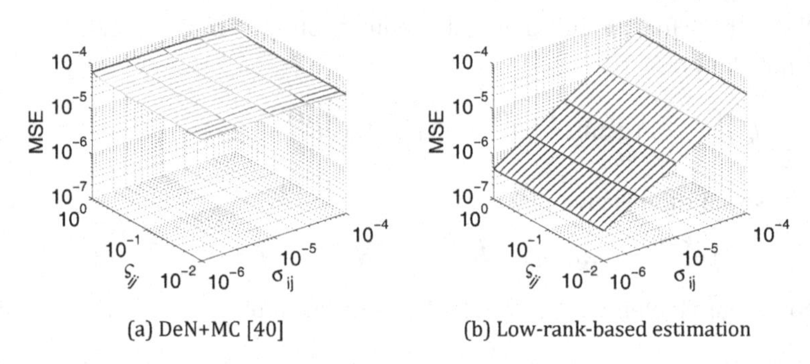

(a) DeN+MC [40] (b) Low-rank-based estimation

Figure 4.3 Average MSE with respect to standard deviations σ_{ij} and ς_{ij} of additive Gaussian noise and outliers.

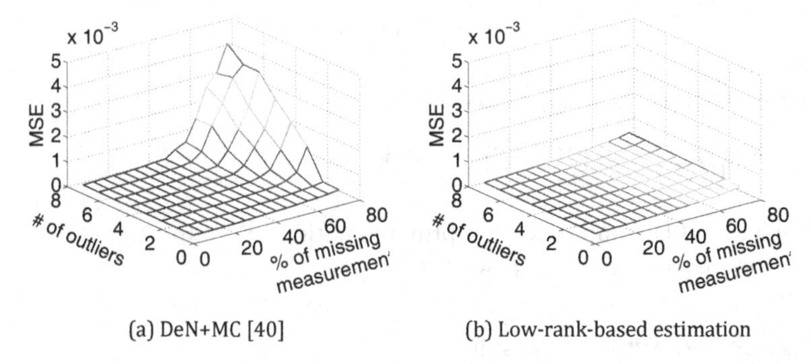

(a) DeN+MC [40] (b) Low-rank-based estimation

Figure 4.4 Average MSE with respect to various numbers of outliers and rates of missing TODA measurements.

shown in Fig. 4.3. Note that the exact number of outliers is required by the DeN+MC. Obviously, the low-rank-based approach outperforms the DeN+MC in this set of experiments. An interesting observation is that the performance of the low-rank-based approach is more vulnerable to additive noise than outliers, while the DeN+MC is more sensitive to the variation of outlier levels.

In the second set of experiments, σ_{ij} and ς_{ij} are set equal to 10^{-3} and 10^{-1}, respectively. But the number of outliers varies from 0 to 7, while the rate of missing data in the TDOA matrix increases from 0 to 70%. The variation of average MSE is depicted in Fig. 4.4. It can be observed that both approaches achieve similar estimation accuracy when the number of outliers and the rate of missing data are both low. However, with the increase of these two parameters, the low-rank-based TDOA estimation approach demonstrates more robust performance than the DeN+MC.

4.2.3 Sparse Common Spatial Pattern (CSP) for Feature Extraction in Brain–Computer Interface (BCI)

BCIs aim to help impaired people to communicate with environment using solely their brain signals [46, 47]. In a BCI system, brain activities are generally measured by electroencephalograph (EEG) signals, due to their low cost and high time resolution [48]. Recently, the research of motor imagery (MI) has attracted much attention. CSP has been considered as an effective tool to extract features for the classification of MI tasks [48, 49].

4.2.3.1 Traditional CSP

The traditional CSP is designed to capture the projection directions which maximize the power ratio of two classes of spatially filtered EEG signals. For a binary classification task, the traditional CSP problem is formulated as

$$\max_{\mathbf{w}} \frac{\mathbf{w}^T \mathbf{C}_1 \mathbf{w}}{\mathbf{w}^T \mathbf{C}_2 \mathbf{w}}, \tag{4.51}$$

where $\mathbf{w} \in \mathbb{R}^N$ represents the coefficient vector of a spatial filter, N is the number of recording channels of EEG signals, and $\mathbf{C}_i (i = 1, 2)$ denotes the covariance matrix for one class. Problem (4.51) has an optimal solution obtained by the generalized EVD

$$\mathbf{C}_1 \mathbf{w}_{opt} = \lambda_{\max} (\mathbf{C}_1 + \mathbf{C}_2) \mathbf{w}_{opt}, \tag{4.52}$$

where $\lambda_{\max}$ denotes the maximal generalized eigenvalue of $\mathbf{C}_1$ and $\mathbf{C}_2$, and the optimal solution $\mathbf{w}_{opt}$ to (4.51) is essentially the eigen vector corresponding to $\lambda_{\max}$. Similarly, to extract features of EEG signals of the second class, one can design another spatial filter by maximizing $\frac{\mathbf{w}^T \mathbf{C}_2 \mathbf{w}}{\mathbf{w}^T \mathbf{C}_1 \mathbf{w}}$ also through the generalized EVD.

4.2.3.2 Sparse CSP

Much research has revealed that one can achieve similar recognition results using fewer electrodes or channels [51–56]. In practice, EEG signals are vulnerable to measurement noise and artifacts, which adversely affect the recognition accuracy. Thus, one needs to find a way to remove highly noisy or irrelevant channels.

To this end, we can formulate the sparse CSP problem as

$$\min_{\mathbf{w}} \quad \|\mathbf{w}\|_0 \tag{4.53a}$$

$$\text{s.t.} \quad \frac{\mathbf{w}^T \mathbf{C}_1 \mathbf{w}}{\mathbf{w}^T \mathbf{C}_2 \mathbf{w}} \geq \tau, \tag{4.53b}$$

where τ is a predefined threshold used to control the final recognition accuracy. Similar to (4.16), we further introduce $\mathbf{W} = \mathbf{w}\mathbf{w}^{\mathrm{T}}$. Then, (4.53) can be equivalently cast as

$$\min_{\mathbf{w},\mathbf{W}} \quad \|\mathbf{W}\|_{0,1} \tag{4.54a}$$

$$\text{s.t.} \quad \frac{\mathrm{Tr}(\mathbf{C}_1\mathbf{W})}{\mathrm{Tr}(\mathbf{C}_2\mathbf{W})} \geq \tau, \tag{4.54b}$$

$$\mathbf{W} = \mathbf{w}\mathbf{w}^{\mathrm{T}}, \tag{4.54c}$$

where

$$\|\mathbf{W}\|_{0,1} = \sum_i \mathcal{I}\left(\left\|\mathbf{W}_{i,:}\right\|_1\right) \tag{4.55}$$

$$\mathcal{I}(x) = \begin{cases} 1, & x \neq 0, \\ 0, & x = 0. \end{cases} \tag{4.56}$$

Due to the existence of the combinatorial objective function (4.54a) and the second constraint (4.54c), (4.54) is still a nonconvex optimization problem. Using the convex relaxation technique, (4.54) is reformulated as

$$\min_{\mathbf{W}\succeq 0} \quad \|\mathbf{W}\|_{1,1} \tag{4.57a}$$

$$\text{s.t.} \quad \frac{\mathrm{Tr}(\mathbf{C}_1\mathbf{W})}{\mathrm{Tr}(\mathbf{C}_2\mathbf{W})} \geq \tau, \tag{4.57b}$$

$$\mathrm{Tr}(\mathbf{W}) = 1, \tag{4.57c}$$

which now becomes an SDP problem. Note that the constraint $\mathrm{Tr}(\mathbf{W}) = 1$ is incorporated to remove the ambiguity caused by the scaling of $\mathbf{W}$. Let $\mathbf{W}_{opt}$ be the optimal solution to (4.57). Then, the spatial filter $\mathbf{w}^1$ used to extract features of EEG signals of the first class can be achieved by the eigen vector of $\mathbf{W}_{opt}$ corresponding to the largest eigenvalue $\lambda_{\max}\left(\mathbf{W}_{opt}\right)$. Similarly, $\mathbf{w}^2$ can be obtained by solving (4.57) with the second constraint replaced by $\frac{\text{Tr}(\mathbf{C}_2\mathbf{W})}{\text{Tr}(\mathbf{C}_1\mathbf{W})} \geq \tau$. Because of the introduction of the $\ell_{1,1}$ norm in the objective function of (4.57), channels corresponding to zero elements in both spatial filters can be discarded.

4.2.3.3 Real data experiment

The data used in our experiment are taken from dataset V of BCI Competition III. EEG signals consist of 32 channels located at standard positions of the International 10-20 system. The sampling rate is 512 Hz. The data include 3 mental imagery tasks (i.e., left hand, right hand, and word associated with labels 2, 3, and 7, respectively) from 3 subjects. In our study, we only use the data with labels 2 and 7 from subject one. One hundred sampling points during the transition stage between different imagery tasks are deleted. The remaining data are divided into 447 segments with 512 points per segment. Finally, EEG signals are filtered by a bandpass filter with the passband from 9 to 35 Hz.

To find a reasonable value for τ in (4.53b) and also (4.57b), we employ the traditional CSP to compute $\tau_{\text{CSP}} = \frac{\mathbf{w}_{opt}^T\mathbf{C}_1\mathbf{w}_{opt}}{\mathbf{w}_{opt}^T\mathbf{C}_2\mathbf{w}_{opt}}$ or $\tau_{\text{CSP}} = \frac{\mathbf{w}_{opt}^T\mathbf{C}_2\mathbf{w}_{opt}}{\mathbf{w}_{opt}^T\mathbf{C}_1\mathbf{w}_{opt}}$. Then, τ is set to $\tau = \rho\tau_{\text{CSP}}$ where ρ varies from 0.6 to 1. Spatial filters obtained by the traditional CSP and the sparse CSP are both employed to compute features. Let $\mathbf{X}_i \in \mathbb{R}^{N\times M}$ be a matrix representing observed EEG signals of the ith data segment, where M is the number of samples. Then, feature extraction is implemented by

$$\mathbf{f}_i = \log\left(\text{var}\left(\mathbf{W}^T\mathbf{X}_i\right)\right), \tag{4.58}$$

where $\mathbf{W} = [\mathbf{w}^1\ \mathbf{w}^2]$. Finally, support vector machine (SVM) with Gaussian kernel function is adopted in the classification of our experiment. In the whole experiment, k-fold cross validation with $k = 10$ is employed to enhance the generalization capability of CSP-based feature extraction approaches.

Besides the tradition CSP, the other three sparse CSP approaches (namely, CSPv [56], ssCSP [52], and rCSP [53]) are also used in our experiments for performance comparison. In the CSPv, channels corresponding to the maximal coefficients of spatial pattern vectors are chosen for feature extraction. The ssCSP approach aims to obtain spatial filters by solving the following problem

$$\min_{\{\mathbf{w}_i\}} \quad (1-r)\left(\sum_{i=1}^{m} \mathbf{w}_i \mathbf{C}_2 \mathbf{w}_i^T + \sum_{i=m+1}^{2m} \mathbf{w}_i \mathbf{C}_1 \mathbf{w}_i^T\right) + r\sum_{i=1}^{2m} \frac{\|\mathbf{w}_i\|_1}{\|\mathbf{w}_i\|_2} \tag{4.59a}$$

$$\text{s.t.} \quad \mathbf{w}_i(\mathbf{C}_1 + \mathbf{C}_2)\mathbf{w}_i^T = \delta_{ij}, \; i, j = 1, 2, \ldots, 2m, \tag{4.59b}$$

where $2m$ is the number of spatial filters of two classes, and δ_{ij} is defined by

$$\delta_{ij} = \begin{cases} 1, & i = j, \\ 0, & i \neq j. \end{cases} \tag{4.60}$$

A simplified version of (4.59) is formulated and tackled in the rCSP approach

$$\min_{\mathbf{w}} \quad \mathbf{w}\mathbf{C}_i\mathbf{w}^T + r\frac{\|\mathbf{w}\|_1}{\|\mathbf{w}\|_2} \tag{4.61a}$$

$$\text{s.t.} \quad \sum_{i=1}^{2} \mathbf{w}\mathbf{C}_i\mathbf{w}^T = 1. \tag{4.61b}$$

Both approaches are highly nonconvex and have to resort to constrained nonlinear optimization techniques. In our experiments, different regularization parameter r is used to adjust the sparsity of spatial filters.

Figure 4.5 illustrates the variation of error rate of classification and the number of effective channels with respect to regularization parameter ρ. It can be observed that, with the increase of ρ, less channels are required by the subsequent classification with the expense of classification accuracy. But, even using only a half of channels, solving (4.57) can still lead to classification accuracy very close to that of the traditional CSP. Figure 4.6 shows the variation of classification accuracy with respect to the number of effective channels. Obviously, the SDP model (4.57) demonstrates more robust classification performance than the other sparse CSP approaches. It can significantly reduce the number of effective channels with a slight decrease of classification accuracy.

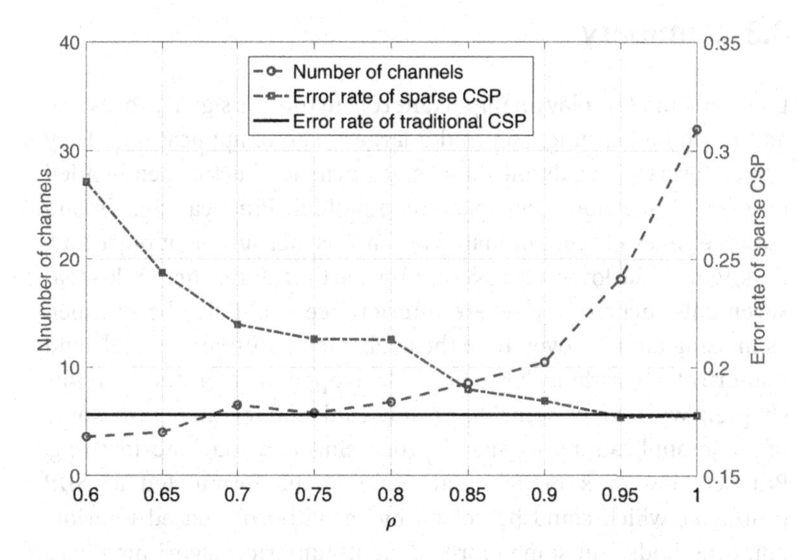

Figure 4.5 Variation of error rate of classification and the number of effective channels with respect to parameter ρ.

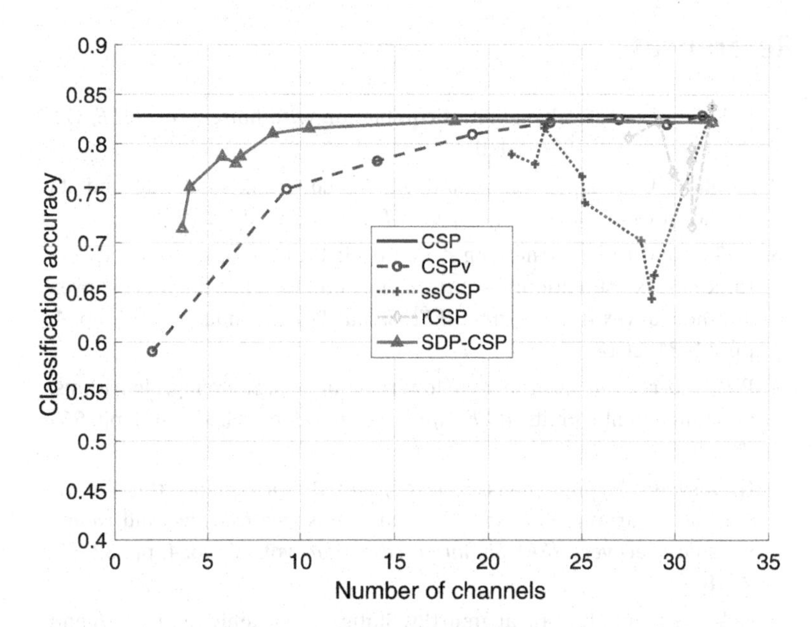

Figure 4.6 Variation of classification accuracy with respect to the number of effective channels.

4.3 Summary

Low-rank models play an important role in today's signal processing and machine learning, and find a large range of applications. They aim to describe signals and data using intrinsic structures embedded in a low-dimensional subspace or manifold. Practical signals and data are generally contaminated by various measurement noise and, thus, violate the low-rank assumption. Furthermore, unreliable measurements incurred by severe interference could also be dropped as missing data. To overcome these obstacles, different models and numerical algorithms have been developed in literature. In this chapter, we review some basic low-rank models and also some of their applications in signal processing and machine learning. Practical low-rank models can generally be formulated as SDP problems, which could be solved by second-order-based interior-point methods. But some more efficient numerical algorithms (e.g., ADMM) are more suitable when dealing with large-scale problems.

References

1. E. J. Candes and Y. Plan, Matrix completion with noise, *Proc. IEEE*, vol. 98, no. 6, pp. 925–936, 2010.
2. X. Zhou, C. Yang, H. Zhao, and W. Yu, Low-rank modeling and its applications in image analysis, *arXiv Preprint arXiv1401.3409*.
3. Y. Chen and Y. Chi, Harnessing structures in big data via guaranteed low-rank matrix estimation: Recent theory and fast algorithms via convex and nonconvex optimization, *IEEE Signal Process. Mag.*, vol. 35, no. 4, pp. 14–31, 2018.
4. P. Pal, Correlation awareness in low-rank models: Sampling, algorithms, and fundamental limits, *IEEE Signal Process. Mag.*, vol. 35, no. 4, pp. 56–71, 2018.
5. N. Vaswani, T. Bouwmans, S. Javed, and P. Narayanamurthy, Robust subspace learning: Robust PCA, robust subspace tracking, and robust subspace recovery, *IEEE Signal Process. Mag.*, vol. 35, no. 4, pp. 32–55, 2018.
6. N. Vaswani and P. Narayanamurthy, Static and synamic robust PCA and matrix completion: A review, *Proc. IEEE*, vol. 106, no. 8, pp. 1359–1379, 2018.

7. S. Boyd and L. Vandenberghe, *Convex Optimization. Cambridge*, U.K.: Cambridge Univ. Press, 2004.
8. L. Vandenberghe and S. Boyd, Semidefinite programming, *SIAM Rev.*, vol. 38, no. 1, pp. 49–95, 1996.
9. A. Lemon, A. M.-C. So, and Y. Ye, Low-rank semidefinite programming: Theory and applications, *Found. Trends Optim.*, vol. 2, no. 1–2, pp. 1–156, 2016.
10. M. Journee, F. Bach, P.-A. Absil, and R. Sepulchre, Low-rank optimization on the cone of positive semidefinite matrices, *SIAM J. Optim.*, vol. 20, no. 5, pp. 2327–2351, 2010.
11. J.-F. Cai, E. J. Candès, and Z. Shen, A singular value thresholding algorithm for matrix completion, *SIAM J. Optim.*, vol. 20, no. 4, pp. 1956–1982, 2010.
12. Z. Lin, M. Chen, L. Wu, and Y. Ma, The augmented Lagrange multiplier method for exact recovery of corrupted low-rank matrices, *UIUC Tech. Rep., UILU-ENG-09-2215*, 2009.
13. R. M. Freund, P. Grigas, and R. Mazumder, An extended Frank-Wolfe method with in-face directions, and its application to low-rank matrix completion, *SIAM J. Optim.*, vol. 27, no. 1, pp. 319–346, 2017.
14. P. Jain, P. Netrapalli, and S. Sanghavi, Low-rank matrix completion using alternating minimization, in *Proc. 45th Annu. ACM Symp. Theory of Computing*, 2013, pp. 665–674.
15. S. Burer and R. D. C. Monteiro, Local minima and convergence in low-rank semidefinite programming, *Math. Program.*, vol. 103, no. 3, pp. 427–444, 2005.
16. P. Jain and P. Netrapalli, Fast exact matrix completion with finite samples, in *Proc. Conf. Learning Theory*, 2015, pp. 1007–1034.
17. S. Burer and R. D. C. Monteiro, A nonlinear programming algorithm for solving semidefinite programs via low-rank factorization, *Math. Program.*, vol. 95, no. 2, pp. 329–357, 2003.
18. V. Cappellini, A. G. Constantinides, and P. Emiliani, *Digital Filters and their Applications.* London: Academic Press, 1978.
19. A. Antoniou, *Digital Filters: Analysis, Design, and Applications.* New York: McGraw-Hill, 2006.
20. E. C. Levy, Complex curve fitting, *IRE Trans. Autom. Control*, vol. AC-4, pp. 37–43, May 1959.
21. C. K. Sanathanan and J. Koerner, Transfer function synthesis as a ratio of two complex polynomials, *IEEE Trans. Autom. Control*, vol. AC-8, pp. 56–58, Jan. 1963.

22. W.-S. Lu, Design of stable IIR digital filters with equiripple passbands and peak-constrained least-squares stopbands, *IEEE Trans. Circuits Syst. II, Analog Digit. Signal Process.*, vol. 46, no. 11, pp. 1421–1426, Nov. 1999.
23. C. Tseng and S. Lee, Minimax design of stable IIR digital filter with prescribed magnitude and phase responses, *IEEE Trans. Circuits Syst. I, Fundam. Theory Appl.*, vol. 49, no. 4, pp. 547–551, Apr. 2002.
24. A. Jiang and H. K. Kwan, IIR digital filter design with new stability constraint based on argument principle, *IEEE Trans. Circuits Syst. I, Reg. Papers*, vol. 56, no. 3, pp. 583–593, Mar. 2009.
25. X. Lai and Z. Lin, Minimax design of IIR digital filters using a sequential constrained least-squares method, *IEEE Trans. Signal Process.*, vol. 58, no. 7, pp. 3901–3906, Jul. 2010.
26. X. Lai and Z. Lin, Minimax phase error design of IIR digital filters with prescribed magnitude and phase responses, *IEEE Trans. Signal Process.*, vol. 60, no. 2, pp. 980–986, Feb. 2012.
27. A. Jiang, H. K. Kwan, Y. Zhu, N. Xu, and X. Liu, Efficient WLS Design of IIR Digital Filters Using Partial Second-Order Factorization, *IEEE Trans. Circuits Syst. II Express Briefs*, vol. 63, no. 7, pp. 703–707, 2016.
28. W.-S. Lu and T. Hinamoto, Optimal design of IIR digital filters with robust stability using conic-quadratic-programming updates, *IEEE Trans. Signal Process.*, vol. 51, pp. 1581–1592, Jun. 2003.
29. M. C. Lang, Least-squares design of IIR filters with prescribed magnitude and phase responses and a pole radius constraint, *IEEE Trans. Signal Process.*, vol. 48, pp. 3109–3121, Nov. 2000.
30. M. Fazel, H. Hindi, and S. Boyd, Rank minimization and applications in system theory, in *Proc. Amer. Control Conf., Boston*, MA, Jul. 2004, pp. 3273–3278.
31. J. F. Sturm, Using SeDuMi 1.02, a MATLAB toolbox for optimization over symmetric cones, *Optim. Methods Softw.*, vol. 11/12, pp. 625–653, 1999.
32. K. C. Ho, X. Lu, and L. Kovavisaruch, Source localization using TDOA and FDOA measurements in the presence of receiver location errors: Analysis and solution, *IEEE Trans. Signal Process.*, vol. 55, no. 2, pp. 684–696, 2007.
33. W. K. Ma, B. N. Vo, S. S. Singh, and A. Baddeley, Tracking an unknown time-varying number of speakers using TDOA measurements: A random finite set approach, *IEEE Trans. Signal Process.*, vol. 54, no. 9, pp. 3291–3303, 2006.

34. C. H. Knapp and G. C. Carter, The generalized correlation method for estimation of time delay, *IEEE Trans. Acoust., Speech, Signal Process.*, vol. 24, no. 4, pp. 320–327, 1976.
35. W. Hahn and S. Tretter, Optimum processing for delay-vector estimation in passive signal arrays, *IEEE Trans. Inf. Theory*, vol. 19, no. 5, pp. 608–614, 1973.
36. R. Schmidt, Least squares range difference location, *IEEE Trans. Aerosp. Electron. Syst.*, vol. 32, no. 1, pp. 234–242, 1996.
37. J. S. Picard and A. J. Weiss, Time difference localization in the presence of outliers, *Signal Process.*, vol. 92, no. 10, pp. 2432–2443, 2012.
38. A. Canclini, P. Bestagini, F. Antonacci, M. Compagnoni, A. Sarti, and S. Tubaro, A robust and low-complexity source localization algorithm for asynchronous distributed microphone networks, *IEEE/ACM Trans. Audio, Speech, Lang. Process.*, vol. 23, no. 10, pp. 1563–1575, 2015.
39. M. Compagnoni, A. Pini, A. Canclini, P. Bestagini, F. Antonacci, S. Tubaro, and A. Sarti, A geometrical-statistical approach to outlier removal for TDOA measurements, *IEEE Trans. Signal Process.*, vol. 65, no. 15, pp. 3960–3975, 2017.
40. J. Velasco, D. Pizarro, J. Macias-Guarasa, and A. Asaei, TDOA matrices: Algebraic properties and their application to robust denoising with missing data, *IEEE Trans. Signal Process.*, vol. 64, no. 20, pp. 5242–5254, 2016.
41. Z. Wen, D. Goldfarb, and W. Yin, Alternating direction augmented Lagrangian methods for semidefinite programming, *Math. Program. Comput.*, vol. 2, no. 3-4, pp. 203–230, 2010.
42. T. Goldstein, B. O'Donoghue, S. Setzer, and R. Baraniuk, Fast alternating direction optimization methods, *SIAM J. Imaging Sci.*, vol. 7, no. 3, pp. 1588–1623, 2014.
43. S. Boyd, N. Parikh, E. Chu, B. Peleato, and J. Eckstein, Distributed optimization and statistical learning via the alternating direction method of multipliers, *Found. Trends Mach. Learn.*, vol. 3, no. 1, pp. 1–122, 2010.
44. D. L. Donoho, De-noising by soft-thresholding, *IEEE Trans. Inf. theory*, vol. 41, no. 3, pp. 613–627, 1995.
45. M. Elad, *Sparse and Redundant Representations: From Theory to Applications in Signal and Image Processing*. New York, USA: Springer, 2010.

46. B. J. Lance, S. E. Kerick, A. J. Ries, K. S. Oie, and K. McDowell, Brain-computer interface technologies in the coming decades, *Proc. IEEE*, vol. 100, pp. 1585–1599, 2012.
47. D. P. Subha, P. K. Joseph, R. Acharya U, and C. M. Lim, EEG signal analysis: A survey, *J. Med. Syst.*, vol. 34, no. 2, pp. 195–212, 2010.
48. M. Arvaneh, C. Guan, K. K. Ang, and C. Quek, Optimizing the channel selection and classification accuracy in EEG-based BCI, *IEEE Trans. Biomed. Eng.*, vol. 58, no. 6, pp. 1865–1873, 2011.
49. H. Wang and W. Zheng, Local temporal common spatial patterns for robust single-trial EEG classification, *IEEE Trans. Neural Syst. Rehabil. Eng.*, vol. 16, pp. 131–139, 2008.
50. B. Blankertz, R. Tomioka, S. Lemm, M. Kawanabe, and K.-R. Müller, Optimizing spatial filters for robust EEG single-trial analysis, *IEEE Signal Process. Mag.*, vol. 25, no. 1, pp. 41–56, 2008.
51. X. Yong, R. K. Ward, and G. E. Birch, Sparse spatial filter optimization for EEG channel reduction in brain-computer interface, in *Proc. IEEE Int. Conf. Acoustics, Speech Signal Process.*, 2008, pp. 417–420.
52. M. Arvaneh, C. Guan, K. K. Ang, and C. Quek, Optimizing the channel selection and classification accuracy in EEG-based BCI, *IEEE Trans. Biomed. Eng.*, vol. 58, no. 6, pp. 1865–1873, 2011.
53. M. Arvaneh, C. Guan, K. K. Ang, and C. Quek, Optimizing EEG channel selection by regularized spatial filtering and multi band signal decomposition, in *Proc. Int. Assoc. Sci. Technol. Develop. Int. Conf. Biomed. Eng.*, Innsbruck, 2010, pp. 86–90.
54. J. Farquhar, N. J. Hill, T. N. Lal, and B. Schölkopf, Regularised CSP for sensor selection in BCI, in *Proc. 3rd Int. BCI Workshop Training Course 2006*, Graz, Austria, pp. 14–15.
55. H. P. Lu, K. N. Plataniotis and A. N. Venetsanopoulos, Regularized common spatial patterns with generic learning for EEG signal classification, in *Proc. IEEE Int. Conf.*, Minneapolis, Minnesota, USA, Sep. 2009, pp. 6599–6602.
56. Y. Wang, X. Gao, and S. Gao, Common spatial pattern method for channel selection in motor imagery based brain-computer interface, in *Proc. 27th Annu. Int. Conf. IEEE Eng. Med. Biol. Soc.*, pp. 5392–5395, 2005.

Chapter 5

Introduction to Face Recognition and Recent Work

Tianrui Liu,[a,b] Wan-Chi Siu,[a] Cigdem Turan,[a] Shun-Cheung Lai,[a] Kin-Man Lam,[a] and Tania Stathaki[b]

[a] *Center for Multimedia Signal Processing,*
Department of Electronic and Information Engineering,
The Hong Kong Polytechnic University, Hong Kong
[b] *Communication and Signal Processing Group,*
Department of Electrical and Electronic Engineering,
Imperial College London, United Kingdom
enwcsiu@polyu.edu.hk

Perhaps, face is one of the most useful biometrics for person recognition. Conventionally, three steps are required for face recognition, namely face detection, face alignment and face recognition. In this chapter, we just briefly mention the first two steps and concentrate on some core technologies for face recognition. We will start the chapter with a thorough introduction to using Eigenface or Principal Component Analysis (PCA) for face recognition, since it is a classic approach, and many modern approaches take it as a reference for development. Subsequently, we briefly describe Linear Discriminant

Learning Approaches in Signal Processing
Edited by Wan-Chi Siu, Lap-Pui Chau, Liang Wang, and Tieniu Tan

ISBN 978-981-4800-50-1 (Hardcover), 978-0-429-06114-1 (eBook)
www.panstanford.com

Analysis (LDA) and other modern approaches, including subspace learning methods and deep learning neural networks for face recognition. We expect that the beginning part of this chapter will be useful for beginners, whilst the rest of the chapter could form a good reference for early researchers in the area.

5.1 Introduction

Face is one of the most commonly used biometrics by human. With the development of big data and the increasing computational power of computers, face recognition methods dealing with real-world variations, like poses, expressions, lighting, etc., have achieved an improving performance.

There are mainly three parts to the success of a face recognition system: face detection, face alignment and face recognition. Face detection is to find and localize faces in an image. A face recognition system classifies an input face image into one of the person identities within a database. The detected face images may have different scales and orientations. To enhance the robustness of the final face recognition, a face alignment algorithm performs alignment to the detected face images so that the key points on the aligned face images are of the same locations. Finally, face recognition algorithm classifies the face image into one of the identities in a database.

Face recognition has been a popular research topic since the introduction of the eigenface method. Different algorithms based on subspace analysis, handcrafted visual features, sparse representation, etc. have been proposed to address different issues for face recognition, due to the fact that face recognition is a very challenging task. Human faces have similar structures, but the appearance of a face may have large variations when it is under different poses, illumination conditions, facial expressions, and occlusion. In other words, a good face-recognition algorithm has to be able to overcome large intra-personal variations and small inter-personal differences. Therefore, to achieve high accuracy, a face-recognition system must have a high learning capacity.

5.2 Face Detection and Face Alignment

5.2.1 Face Detection

The task of face detection is to find and localize faces in an image. It seems an easy task for a human, however, it is not trivial for a computer. The Viola Jones face detector [1] is one of the most famous face detection methods. There are five main factors that lead to the success of the Viola Jones face detector:

(1) The Haar-like features are used for effective feature representation. They can be computed efficiently in one pass through the image with the integral image representation.
(2) AdaBoost (short for Adaptive Boosting) is employed to select the most representative features and construct a strong classifier as a linear combination of the weighted weak classifiers.
(3) A cascade classifier structure is used to successively combine the strong classifiers, which is able to quickly reject non-face objects.
(4) The multi-scale sliding window detection scheme allows the detector to scan over different locations on different scales of a test image.

The Viola Jones Face detector has been imbedded into the OpenCV library [1] as a common and standard face detector.

5.2.2 Face Alignment

Once the face region has been localized from an image, it is sometimes necessary to align the detected face region before performing face recognition. The alignment step aims to make the key points on the face region, namely eyes, nose, and mouth to have the same relative locations.

Given the coordinates of the landmarks of a face, alignment and normalization can be achieved by simply using a transformation translating matrix. The functionality of the transformation matrix is to translate, rotate and scale the source keypoints towards a target position by multiplying the source pixels with a pre-defined matrix, as given in Fig. 5.1.

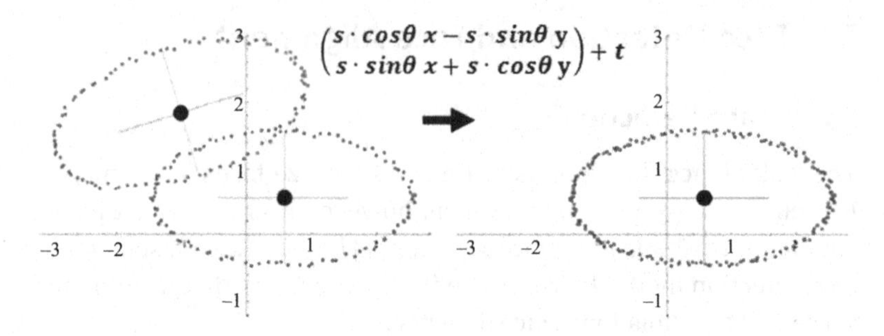

Figure 5.1 Illustration of image alignment using translating matrix.

Promising approaches for face alignment include Active Shape Models (ASMs) [2] and Active Appearance Models (AAMs) [3]. AAMs are generative models of shape and appearance typically learned by applying Principal Component Analysis to both shape and texture. The alignment problem is formulated as a minimization problem which minimizing the error between the model and a given face image with respect to the model parameters. These model parameters control the shape and appearance variation of faces.

5.3 Face Recognition

5.3.1 The Eigenface Approach

As we mentioned in the introduction, a face recognition system classifies an input face image into one of the person identities within a database. The most direct approach is template matching which compares the distances between the input face image and every face image from the database and classifies the input face image to the person identity which is with the smallest distance. There are mainly two problems with the template matching method. The first problem is that the face images are usually of high dimension. The computational complexity for distance measures can be very high. The second problem is that template matching can be easily affected by variations in the face images, such as the variation caused by different lighting condition.

The eigenface method can achieve more robust face recognition. The idea is to project the input face image onto a new basis set where these bases can capture the most significant and possibly distinguishable feature components for face recognition. Those bases that represent irrelevant feature components to the face recognition (for example, lighting condition) can be discarded. An eigenface is a vector from this basis set. This enables us to perform robust feature matching using low-dimensional features.

5.3.1.1 Main idea behind eigenfaces

Eigenfaces can be generated from a training database which contains face images of different persons using a mathematical tool called Principal Component Analysis (PCA) [4]. PCA can be viewed as a coordinate transformation to a new vector space in the directions of maximal variability. The transformation is defined in such a way that the first principal component has the largest possible variance, and each succeeding component is orthogonal to the preceding components. The resulting vectors, to be as referred the principal components, form an orthogonal basis set.

A simple example on PCA projection of two-dimensional (2D) data is illustrated in Fig. 5.2. PCA seeks the first axis (labeled as — in Fig. 5.2) which accounts for as much of the variability in the data as possible. The second axis (labeled as—in Fig. 5.2) is geometrically orthogonal to the first axis. The first principal component (denoted as 1st PC in Fig. 5.2) is along the first axis, the second principal component (denoted as 2nd PC in Fig. 5.2) is along the second axis.

Eigenface is such an approach that searches for the principal representation of face images. It is a learning-based method which requires a training process so that inference can be performed. In the training phase, the algorithm is fed with face image samples to learn the eigenface dictionary. Inference can be achieved by comparing how the query images are represented by the eigenfaces.

In this section, we introduce the inference phase by assuming that the eigenfaces have already been constructed. The procedure for eigenfaces construction will be presented in Section 5.3.2.

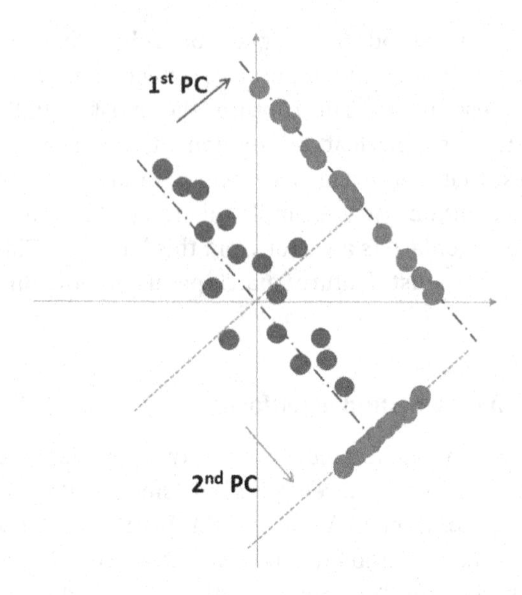

Figure 5.2 Illustration of PCA projection on the maximum-variation directions.

Let us assume that we have already obtained K eigenfaces from the training database and a vector representing the mean face. During the face recognition process, the input query face image will first be subtracted from the mean face, and then projected onto each eigenface resulting in K weight coefficients. The weights specify to what degree the feature represented by an eigenface is presented in this input face image. The input face image can be reconstructed almost perfectly as a weighted combination of the eigenfaces with the obtained weights.

An example is shown in Fig. 5.3. The input face image equals the linear combination of the mean face and the weighted eigenfaces. The weight of the k-th eigenface is the contribution of that eigenface to the input face image. These K eigenfaces capture most energy of the face image and discard the disturbing information. The distance measure between the input query image and the face images in the database can be obtained by comparing the Euclidean distance using these obtained K weight coefficients. This gives us a more computational efficient and robust method.

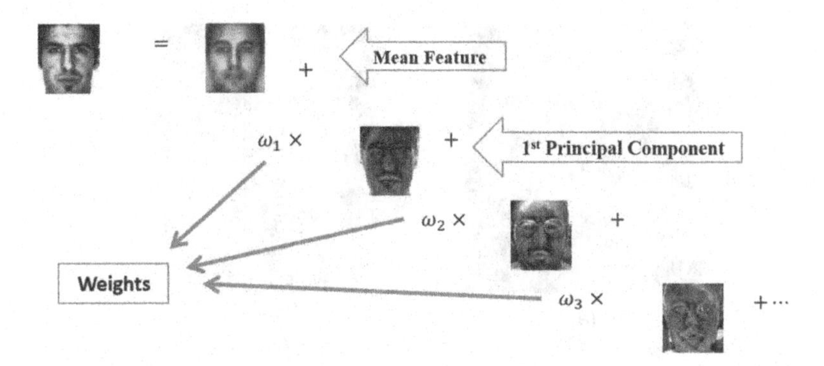

Figure 5.3 A face image can be represented as a "mean face" together with a linear combination of the principal components. We refer these principal components as eigenfaces in the face recognition problem. The weight of each eigenface is obtained by projecting the test face region onto the eigenface space.

We can say that each face image can be made up of "proportions" of all the "eigen-features" or eigenfaces, i.e.,

- Suppose $\mathbf{\Gamma}$ is an (PxQ)x1 vector, corresponding to an PxQ face image **I**.
- The idea is to represent $\mathbf{\Gamma}$ ($\varphi = \mathbf{\Gamma}$-mean face) into a low-dimensional space.
- $\hat{\varphi} - \textbf{mean} = \omega_1\mathbf{u}_1 + \omega_2\mathbf{u}_2 + \cdots \omega_M\mathbf{u}_M$, for $M < P \times Q$).

$\hat{\varphi}$ is the original face image, $\{\omega_i$ for $i = 1, 2, \ldots,$ M) are the weights and $\{\mathbf{u}_i$ for $i = 1, 2, \ldots$ M) are eigenfaces. If we use all the eigenfaces extracted from the original image, we can reconstruct the original image from the eigenfaces exactly. But if we use only a part of the eigenfaces, the reconstructed image is an approximation of the original image. However, one can ensure that the losses due to omitting some of the eigenfaces can be minimized. This can be achieved by choosing only the most important features (eigenfaces).

5.3.2 Face Recognition via Discrimination Function

A query image is projected onto the eigen-space of the eigenfaces. The projected coefficients can then be obtained and can be used to determine whether the acquired face image has the same identity as

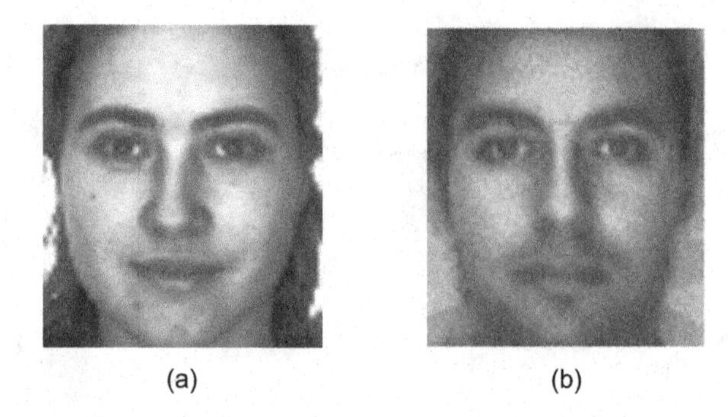

Figure 5.4 (a) An example face image. (b) Example mean face image.

any image in the face database. A certain discrimination function is acquired to determine whether the recognition is triggered. The face recognition process can be summarized as two steps: projection and recognition via distance computation.

Projection

Given the eigenfaces, the weight corresponding to each eigenface can be obtained by projecting the input face Γ onto the "eigenspace." The projection is realized by calculating the dot product of the demeaned input image with the eigenfaces dictionary, i.e.,

$$\omega_{\mathrm{k}} = \mathbf{u}_{\mathrm{k}}^{\mathrm{T}} \cdot (\mathbf{\Gamma} - \mathbf{\Psi}), \tag{5.1}$$

where Γ is an input image for questioning (see Fig. 5.4a) and $\mathbf{\Psi}$ is the mean face image (see Fig. 5.4b), k is the index of the eigenfaces, $\mathbf{u}_{\mathrm{k}}$, in the dictionary. The mean face image $\mathbf{\Psi}$ is the average intensity of all training images which have been computed during the eigenfaces dictionary construction phase (see Section 3.2 for details).

These weights represent the degree to which the face in question is similar to "typical" faces represented by the eigenfaces. Therefore, using these weights, one can determine:

(1) Whether this is a face image or not: In case the weights of a testing image differ too much from the weights of face images, the image probably is not a face. Euclidian distances between

the weights are calculated to determine whether the input image is a face image or not.

(2) Similar faces (images) possess similar features (eigenfaces) with similar degrees (weights). If we consider the face of a person, images with small Euclidean distances are likely to belong to the same person.

Recognition

Face recognition can be realized by comparing the Euclidean distance between the test image and its reconstruction image. The smaller the Euclidean distance, the higher possibility the testing potential face region refers to the true face will have. Hence, the Euclidean distance is used to compare the distance of the projected image to each of the g known face classes. Or, Euclidean distance measure is used as a distance measure metric for searching the most similar vector in the training set to the vector of the testing image. The Euclidean distance ε_k for the k-th class for $k = 1, \ldots, g$ can be written as

$$\varepsilon_k = d(\mathbf{y}, \mathbf{y}_k) = \|\mathbf{y} - \mathbf{y}_k\| = \sqrt{(\mathbf{y} - \mathbf{y}_k)^T(\mathbf{y} - \mathbf{y}_k)}, \qquad (5.2)$$

where $d(.)$ denotes the Euclidean distance metric, $\mathbf{y}_k$ is the mean of the projected training images for the k-th class.

The testing image is classified as the unknown class if $\varepsilon_i > \theta_i, \forall i = 1, \ldots, g$. Otherwise, the questioning image x_j is allocated to the j-th class if $\varepsilon_k = \|\mathbf{y} - \mathbf{y}_k\| < \theta_{\mathrm{j}}$, where θ_j is the pre-defined thresholds for the j-th class.

Flow diagram of testing phase of the eigenfaces method

Figure 5.5 summarizes the testing phase of the eigenfaces recognition method. An input image is first demeaned and projected onto the eigenspace. The obtained weights are a good representation for the input image. If the weights have the smallest distance to the weights of an identity in the database, then the input image belongs to that identity. Otherwise, if all the distances are large, this input image is not a face image.

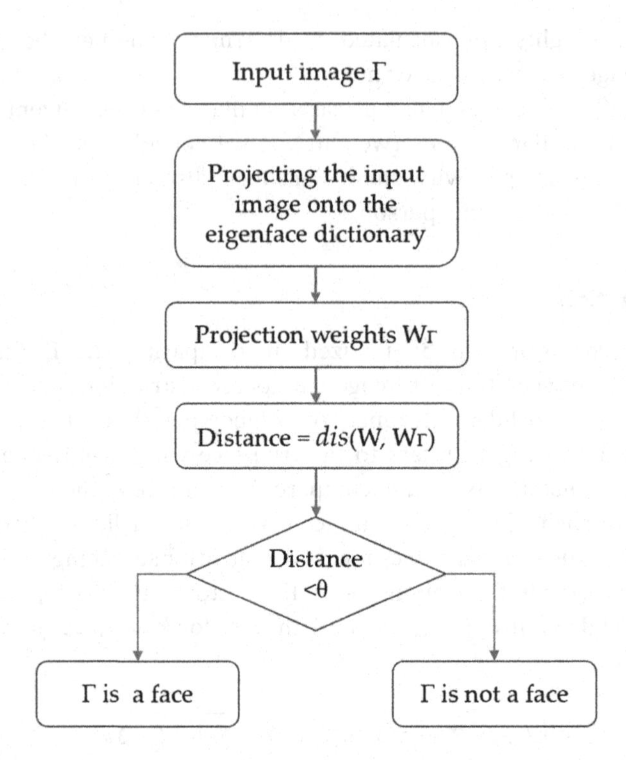

Figure 5.5 Flow diagram of the inference phase of the eigenfaces method.

5.3.3 Principal Component Analysis for Eigenfaces Dictionary Construction

Principal Component Analysis (PCA) is a statistical approach that can reduce the dimensionality of data with factor analysis. The main idea of using PCA for face recognition is to factorize a face image into principal components of a feature space. It transforms face images into a set of orthogonal vectors called principal components. The principal components are guaranteed to be orthogonal to each other so that information redundancies can be avoided. The eigenvectors $\mathbf{v}$ of the covariance matrix are the so-called principal components. By seeking a smaller number of n orthonormal vectors $\mathbf{v}$, the distribution of the data in the original high dimensional image can be described with a small number of the principal components and thus dimensional reduction can be realized.

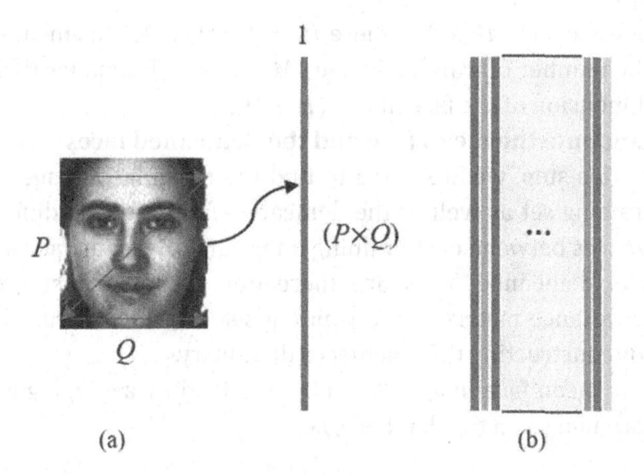

Figure 5.6 Illustration of (a) image vectorization and (b) training matrix construction.

In this section, the scheme for constructing an eigenface dictionary using PCA will be presented. A detailed (and more theoretical) description of PCA can be found in [4, 5]. The implementation procedure of PCA is as follows:

Procedures

(i) Prepare training data

In this step, we need to construct a training matrix containing pixel intensities of all the training images. All the training images should be of the same size, or have been resized into the same size. As illustrated in Fig. 5.6a, any 2D image of size $P \times Q$ in the training image set should be vectorized into a one-dimensional column vector of length $(P \times Q) \times 1$. The intensity value of the $(p, q) \forall p \in P, q \in Q$ pixel on the image is saved as the $\mathrm{Q} \times p + q$-th element of the column vector $\mathbf{\Gamma}$. For a training set containing M training images, we can construct a matrix $\mathbf{S}$ where each column is a vectorized training face image from the database, i.e.,

$$\mathbf{S} = \{\mathbf{\Gamma}_1, \mathbf{\Gamma}_2, \cdots \mathbf{\Gamma}_\mathrm{M}\} \tag{5.3}$$

The column number of $\mathbf{S}$ is equal to the number of images in the training database. As given in Fig. 5.6b, $\mathbf{S}$ is a "tall" matrix of dimension $H \times M$, where $H = P \times Q \times 1$. This means that the number of training images M is normally smaller than the dimension of the face image ($P \times Q$).

(ii) Compute the mean face and the demeaned faces

In this step, we are going to find the mean face image of the training set as well as the demeaned faces, i.e., the difference images between each training image and the mean face image. The demeaned faces are thereafter used to construct the covariance matrix of the training set, which is a crucial step for constructing the eigenfaces dictionary.

The mean face image $\mathbf{\Psi}$ can be obtained by averaging all the face images in the database, i.e.,

$$\psi = \frac{1}{M}\sum_{i=1}^{M}\mathbf{\Gamma}_i \tag{5.4}$$

The demeaned face image φ_i is the difference between each training image vector $\mathbf{\Gamma}_i$ and the mean image vector $\mathbf{\Psi}$, i.e.,

$$\varphi_i = \mathbf{\Gamma}_i - \psi \tag{5.5}$$

(iii) Calculate the covariance matrix C.

$$\mathbf{C} = \frac{1}{M}\sum_{i=1}^{M}\varphi_i\varphi_i^T = \mathbf{AA}^T, \tag{5.6}$$

where $A = \{\varphi_1, \varphi_2, \ldots, \varphi_M\}$ is a matrix of size $H \times M$, for $H = P \times Q \times 1$. The covariance matrix $\mathbf{C}$ is a square matrix of size $H \times H$.

Covariance matrix is a matrix whose element at the i, j position is the covariance between the i-th and j-th elements of a vector. Since the covariance of the i-th random variable with itself is simply that random variable's variance, each element on the principal diagonal of the covariance matrix is the variance of one of the random variables.

(iv) Compute the eigenvectors and eigenvalues of the covariance matrix C.

In this step, the eigenvectors (eigenfaces) and eigenvalues of covariance matrix $\mathbf{C}$ are calculated. In linear algebra, eigenvectors and eigenvalues have the relationship $\mathbf{Cv} = \lambda\mathbf{v}$ for non-zero vector(s) $\mathbf{v}$. That is, the linear transformation $\mathbf{C}$

of a non-zero vector $\mathbf{v}$ is the original vector $\mathbf{v}$ multiplied by a scalar λ. The scalar λ is known as an eigenvalue of the matrix $\mathbf{C}$. The vector for a particular value of λ is referred to as an eigenvector of $\mathbf{C}$.

Find Eigenvalues and Eigenvectors

We briefly introduce the algorithm to determine the eigenvectors and eigenvalues in this section. According to the definition of eigenvectors and eigenvalues,

$$\mathbf{C} \cdot \mathbf{v} = \lambda \cdot \mathbf{v}, \tag{5.7}$$

where $\mathbf{C}$ is an H-by-H matrix, $\mathbf{v}$ is a H-by-1 vector and λ is a scalar. By multiplying an identity matrix, $\mathbf{I}$ of size H-by-H to the right-hand side of the equation, we have

$$\begin{aligned} &\mathbf{C} \cdot \mathbf{v} = \lambda \cdot \mathbf{I} \cdot \mathbf{v} \\ &\mathbf{C} \cdot \mathbf{v} - \lambda \cdot \mathbf{I} \cdot \mathbf{v} = 0 \\ &(\mathbf{C} - \lambda \cdot \mathbf{I}) \cdot \mathbf{v} = 0 \end{aligned} \tag{5.8}$$

If $\mathbf{v}$ is a non-zero vector, then the determinant of $(\mathbf{C} - \lambda\mathbf{I})$, denoted as $\det(\mathbf{C} - \lambda\mathbf{I})$ or $|\mathbf{C} - \lambda\mathbf{I}|$ needs to be zero, i.e.,

$$|\mathbf{C} - \lambda\mathbf{I}| = 0 \tag{5.9}$$

To solve the eigenvalue problem is to solve this determinant function of matrix $\mathbf{C}$. For each eigenvalue λ there will be an eigenvector $\mathbf{v}$ for which the eigenvalue equation is true. In what follows, we provide a simple example of solving the eigenvalues and eigenvectors of a 2-by-2 matrix.

Example: Find Eigenvalues and Eigenvectors of a 2-by-2 Matrix

Given a covariance matrix

$$\mathbf{C} = \begin{bmatrix} 0 & 1 \\ -2 & 3 \end{bmatrix} \tag{5.10}$$

The characteristic equation is

$$|\mathbf{C} - \lambda \cdot \mathbf{I}| = \left| \begin{bmatrix} 0 & 1 \\ -2 & 3 \end{bmatrix} - \begin{bmatrix} \lambda & 0 \\ 0 & \lambda \end{bmatrix} \right| = 0 \tag{5.11}$$

$$\left| \begin{bmatrix} -\lambda & 1 \\ -2 & 3-\lambda \end{bmatrix} \right| = \lambda^2 - 3\lambda + 2 = 0 \tag{5.12}$$

The two eigenvalues can be obtained as $\lambda_1 = 1$, $\lambda_2 = 2$.

For each eigenvalue, we can compute the corresponding eigenvector. For$\lambda_1 = 1$, we denote its eigenvector as $\mathbf{v}_1$, then according to the definition of eigenvector and eigenvalue, we have

$$\mathbf{C} \cdot \mathbf{v}_1 = \lambda_1 \cdot \mathbf{v}_1 \tag{5.13}$$

Hence,

$$(\mathbf{C} - \lambda_1) \cdot \mathbf{v}_1 = \cdot 0 \tag{5.14}$$

$$\begin{bmatrix} -\lambda_1 & 1 \\ -2 & 3 - \lambda_1 \end{bmatrix} \cdot v_1 = 0 \tag{5.15}$$

Substitute $\lambda_1 = 1$ into the above equation, we have

$$\begin{bmatrix} -1 & 1 \\ -2 & 2 \end{bmatrix} \cdot \mathbf{v}_1 = 0 \tag{5.16}$$

Let us represent the two-dimensional vector v_1 as $[v_{11}, v_{12}]'$, then

$$\begin{bmatrix} -1 & 1 \\ -2 & 2 \end{bmatrix} \cdot \begin{bmatrix} v_{11} \\ v_{12} \end{bmatrix} = 0 \tag{5.17}$$

This matrix representation can be transformed into two equations, i.e.,

$$-v_{11} + v_{12} = 0$$

$$-2v_{11} + 2v_{12} = 0 \tag{5.18}$$

By solving the two equations we get,

$$v_{11} = v_{12} \tag{5.19}$$

$\mathbf{v}_1 = \begin{bmatrix} v_{11} \\ v_{12} \end{bmatrix} = \alpha_1 \begin{bmatrix} 1 \\ 1 \end{bmatrix}$, where α_1 is an arbitrary constant.

Generally, we normalize the eigenvectors so that they are unit vectors, i.e., of length 1. Thus,

$$\mathbf{v}_1 = \begin{bmatrix} v_{11} \\ v_{12} \end{bmatrix} = \frac{1}{\sqrt{2}} \begin{bmatrix} 1 \\ 1 \end{bmatrix} \tag{5.20}$$

Go through the same procedures for the second eigenvalue $\lambda_2 = 2$, i.e.,

$$\begin{bmatrix} -\lambda_2 & 1 \\ -2 & 3 - \lambda_2 \end{bmatrix} \cdot v_2 = 0, \tag{5.21}$$

we have

$$2v_{11} = v_{12}$$

$$\mathbf{v}_2 = \begin{bmatrix} v_{21} \\ v_{22} \end{bmatrix} = \frac{1}{\sqrt{5}} \begin{bmatrix} 1 \\ 2 \end{bmatrix} \tag{5.22}$$

In order to clarify the procedure, Fig. 5.7 summarizes the training phase of the eigenfaces method.

Find eigenvector and eigenvalue using MATLAB

The eigenvector and eigenvalue problem can be solved in MATLAB simply by applying the function eig( ).

Improved solution for Eigenfaces

In practice, the covariance matrix $\boldsymbol{C}$ could be of very large dimension. For a face image of size $P \times Q$ with $P = Q = 256$, the covariance matrix $\mathbf{C}$ has a size of $65{,}536 \times 65{,}536$ and we need to compute 65,536 eigenfaces. Computationally, this is not very efficient as most of those eigenfaces with small eigenvalues are not useful for their. cognition task.

```
> > C = [0, 1; -2, 3]
C = 0   1
   -2   3

> > [v,d]=eig(C)

v = -0.7071 -0.4472
    -0.7071 -0.8944

d = 1   0
    0   2
```

Calculation of eigenvectors and eigenvalues using Matlab.

Alternatively, we can construct matrix $\mathbf{L} = \mathbf{A}^{\mathrm{T}}\mathbf{A}$, (note: not $\mathbf{AA}^{T}$, with the size equal $(M \times H) \times (H \times M) = M \times M)$ and solve the eigenvectors and eigenvalues of $\mathbf{L}$. The matrix $\mathbf{L}$ is of size M-by-M, where $M^2 \ll (P \times Q)^2$ usually. As we can prove in the followings, the eigenvalues of matrix $\mathbf{C} = \mathbf{AA}^{\mathrm{T}}$ and matrix $\mathbf{L} = \mathbf{A}^{\mathrm{T}}\mathbf{A}$ are equivalent, and the eigenvectors $\mathbf{v}$ of covariance $\mathbf{C}$ are related to the eigenvectors $\mathbf{u}$ of covariance $\mathbf{C}$ in terms of $\mathbf{u} = \mathbf{Av}$. Hence, we can find the eigenfaces by evaluating the eigenvalues and eigenvectors of $\mathbf{L}$. This approach greatly reduces the computation.

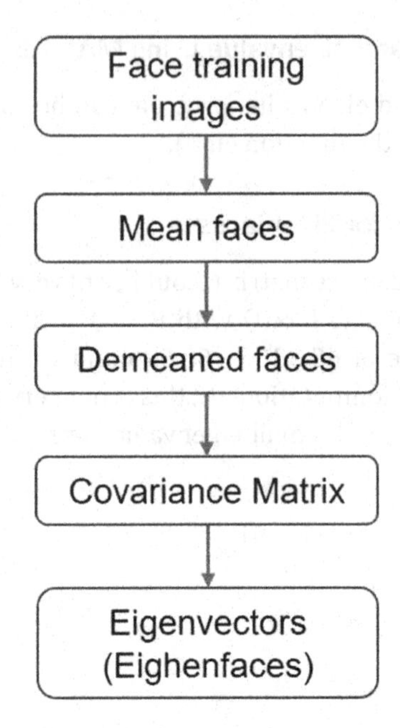

Figure 5.7 Flow diagram of the training phase of the eigenfaces method.

Step (iv) of the eigenface method can be replaced by an

Improved Step (iv) approach:

- Find the M eigenvectors **v** of **L** so that, $\mathbf{A^T A v_i} = \lambda_i \mathbf{v_i}$.
- Obtain the eigenvector of $\mathbf{C} = \mathbf{AA^T}$ by multiplying **A** on both sides of $\mathbf{A^T A v_i} = \lambda_i \mathbf{v_i}$

Proof:

- The eigenvalues of matrices **C** and **L** are equivalent
- The eigenvectors **v** of the covariance matrix **C** are related to the eigenvectors **u** of the covariance matrix **C** by $\mathbf{v} = \mathbf{A}^T \mathbf{u}$

Let us assume that we have found the eigenvalues and eigenvectors of the matrix $\mathrm{L} = \mathbf{A^T A}$ as λ_i and $\mathbf{v}_i$. Thus,

$$\mathbf{L} \cdot \mathbf{v} = \lambda \cdot \mathbf{v} \tag{5.23}$$

$$\mathbf{A^T A} \cdot \mathbf{v} = \lambda \cdot \mathbf{v} \tag{5.24}$$

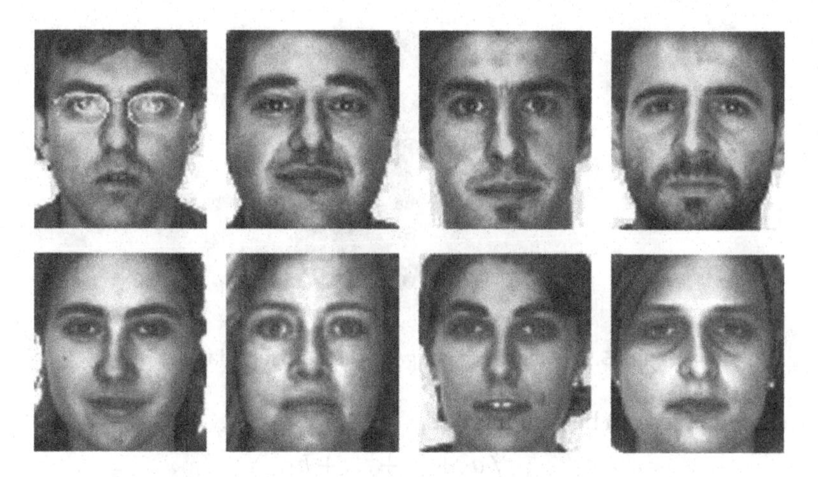

Figure 5.8 Training images of the mini face dataset.

Multiplying matrix **A** on both sides of (5.24)

$$\mathbf{A} \cdot \mathbf{A}^{\mathrm{T}}\mathbf{A} \cdot \mathbf{v} = \lambda \cdot \mathbf{A} \cdot \mathbf{v} \tag{5.25}$$

Therefore,

$$\mathbf{A}\mathbf{A}^{\mathrm{T}}\mathbf{A}\mathbf{v} = \lambda(\mathbf{A}\mathbf{v}) \tag{5.26}$$

Let us define vector $\mathbf{u} = \mathbf{A}\mathbf{v}$. According to the definition of eigenvector and eigenvalue, $\mathbf{A}\mathbf{v}$ is the eigenvector of $\mathbf{A}\mathbf{A}^{\mathrm{T}}$, that is $\mathbf{u} = \mathbf{A}\mathbf{v}$ is eigenvector of covariance $\mathbf{C}$.

A MiniExample: Eigenfaces face recognition

In order to let the readers better understand the whole procedure of the eigenfaces method. We use a mini face dataset with 8 training images and 3 testing images to demonstrate the eigenfaces recognition method. The training set which contains face images of 8 different identities is shown in Fig. 5.8. All training and testing images are of size 64×72.

The training phase

Let us denote the vector representation of the eight face images as $\mathbf{a}, \mathbf{b}, \ldots,$ and $\mathbf{h}$, respectively. Hence, the mean face vector $\mathbf{m}$ of these

Figure 5.9 Mean face image of the mini face dataset.

eight images can be computed as

$$\mathbf{m} = \frac{1}{M}\begin{pmatrix} a_1 + b_1 + \ldots + h_1 \\ a_2 + b_2 + \ldots + h_2 \\ \vdots \quad \vdots \quad \vdots \quad \vdots \\ a_H + b_H + \ldots + h_H \end{pmatrix}. \tag{5.27}$$

In this case, $M = 8$ and $\mathrm{H} = 64 \times 72 = 4608$. In order to visualize the mean face image, we need to reshape the vector **m** of size 4608 $\times$ 1 into a matrix of size 64 $\times$ 72. The obtained mean face image is shown in Fig. 5.9.

The demeaned face images, denoted as $\mathbf{a}_{\mathrm{dm}}, \mathbf{b}_{\mathrm{dm}}, \ldots,$ and $\mathbf{h}_{\mathrm{dm}}$ are computed by subtracting the mean image vector **m** from the training image vectors ***a***, ***b***, . . . , and ***h***, respectively:

$$\mathbf{a}_{dm} = \frac{1}{M}\begin{pmatrix} a_1 - m_1 \\ a_2 - m_2 \\ \vdots \quad \vdots \\ a_H - m_H \end{pmatrix}, \quad \mathbf{b}_{dm} = \frac{1}{M}\begin{pmatrix} b_1 - m_1 \\ b_2 - m_2 \\ \vdots \quad \vdots \\ b_H - m_H \end{pmatrix}, \ldots,$$

$$\mathbf{h}_{dm} = \frac{1}{M}\begin{pmatrix} h_1 - m_1 \\ h_2 - m_2 \\ \vdots \quad \vdots \\ h_H - m_H \end{pmatrix} \tag{5.28}$$

The demeaned face images are shown in Fig. 5.10.

To construct the covariance matrix, we can first rearrange the demeaned face vectors into matrix A as,

$$\mathbf{A} = [\mathbf{a}_{dm}, \mathbf{b}_{dm}, \mathbf{c}_{dm}, \mathbf{d}_{dm}, \mathbf{e}_{dm}, \mathbf{f}_{dm}, \mathbf{g}_{dm}, \mathbf{h}_{dm}] \tag{5.29}$$

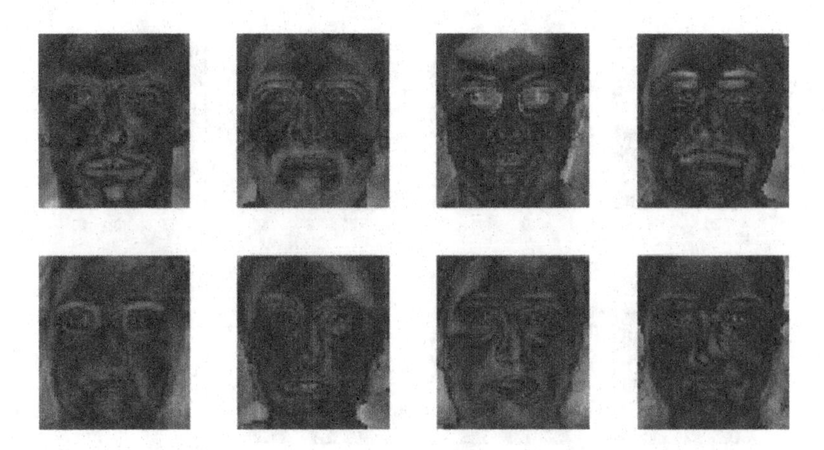

Figure 5.10 Demeaned faces of the mini face dataset.

We have to find the covariance matrix $\mathbf{C} = \mathbf{AA}^T$. As matrix $\mathbf{A}$ is of size 4608×8, matrix $\mathbf{C}$ should be a square matrix of size 4608×4608.

If we apply the MATLAB function eig( ) to resolve the eigenvectors and eigenvalues of the matrix $\mathbf{C}$ as follows:

```
for i = 1: NumOfSamples
    CovFace1 = DemeanFace(:,i)*(DemeanFace(:,i)');
end
CovFace1 = CovFace1/NumOfSamples;
tic;
[EigenVector, EigenValue] = eig(CovFace1);
toc;
Elapsed time is 11.432939 seconds.
```

It takes around 10 seconds to solve the eigenvectors and eigenvalues of a matrix $\mathbf{C}$ which is of size 4608×4608. The eigenfaces, which are reshaped eigenvector, are shown in Fig. 5.11.

Actually, we can use the improved solution for finding the eigenfaces by constructing the matrix $\mathbf{L} = \mathbf{A}^T\mathbf{A}$, and then solve the eigenvectors and eigenvalues of the matrix $\mathbf{L}$. The matrix $\mathbf{L}$ is a square matrix of size 8×8.

```
>> CovFace2 = DemeanFace'*DemeanFace;
   tic;
   [EigenVector, EigenValue] = eig(CovFace2);
```

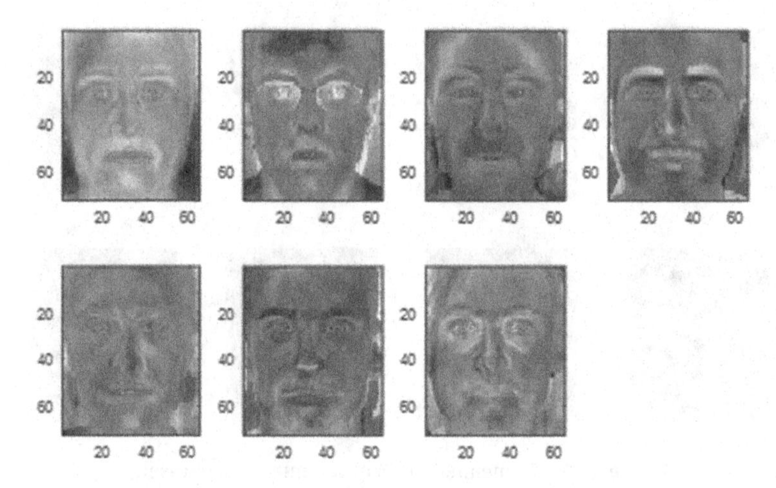

Figure 5.11 Eigenfaces of the mini face dataset.

```
    EigenFaces = DemeanFace*EigenVector;
    toc;
Elapsed time is 0.021004 seconds.
```

Time for solving the eigenvectors and eigenvalues of a matrix **L** is about 0.02 seconds only. Compared to the way of finding eigenfaces using $\mathbf{C} = \mathbf{A}\mathbf{A}^{\mathrm{T}}$, solving the eigenface problem using $\mathbf{A}^{\mathrm{T}}\mathbf{A}$ is about 500 times faster.

The testing phase

Now we have the eigenfaces dictionary of our mini face dataset. We can use three testing images, as shown in Fig. 5.12, to evaluate the constructed eigenfaces dictionary.

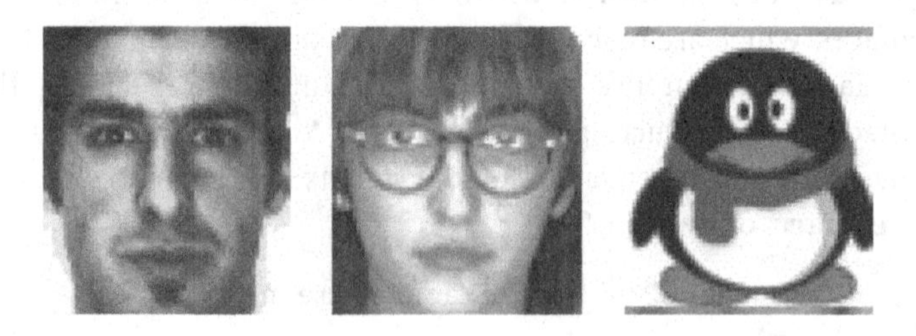

Figure 5.12 Testing images of the mini face dataset.

The first thing we need to do is to compute the weight corresponding to each eigenface. According to (5.1), the weights are obtained by computing the dot product between Γ and the respective eigenfaces, i.e., $\omega_k = \mathbf{u}_\mathbf{k}^\mathrm{T} \cdot (\Gamma - \boldsymbol{\Psi})$, where $k \leq M$ is the eigenface number in the eigenface dictionary.

The three testing images represent three different cases. The first testing image is a face image, which belongs to a person whose face image has been included in the training set; the second testing image is a face image of an unknown person; and the third testing image is a non-face image.

We computed the Euclidean distances between the weights ω of a questioning image and the corresponding weights of each of the training images. Then, we recorded three training images which have the smallest Euclidean distances.

The results are given in Fig. 5.13. As we can see, the first testing image, i.e., Test1, has the smallest distance from one of the training images, belonging to the same person.

Threshold setting is very important in this approach. Suppose we set the threshold at 5×10^6. After calculation, the minimum distance between Test 3 and the images in the training set is 7.6×10^6, which is larger than the threshold 5×10^6. Hence, the Test 3 image should be recognized as a non-face image. Both the test images Test 1 and Test 2 are then recognized as face images as the minimum distance values between images Test 1 and Test 2 and the training images are respectively 7×10^5 and 4.4×10^6, which are smaller than the threshold 5×10^6. The decision for image Test 2 is wrong. Furthermore, if the threshold is set to 5×10^5, then image Test 1 will be wrongly classified as a known face, whilst image Test 2 will be correctly classified as an unknown face.

5.3.4 Limitation of the PCA for Face Detection

PCA is suitable for data compression and data reconstruction. However, it is explicitly defined for classification problems (i.e., in cases that data come with labels). The effectiveness of the eigenface method is limited by the assumptions behind. PCA can be interpreted as a coordinate transformation to a vector space which is defined in the directions of maximal variability. When PCA

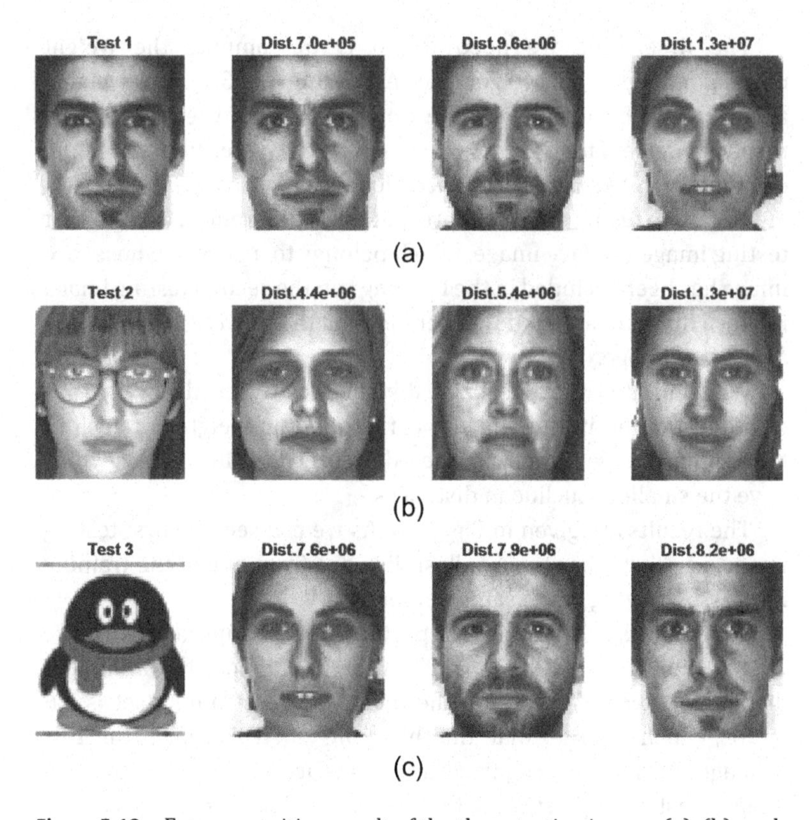

Figure 5.13 Face recognition result of the three testing images (a), (b), and (c), respectively.

is used to find the projection that maximizes the total variation, it selects the most expressive features. The PCA-based face recognition methods do not differentiate the inter-class variation and the intra-class variation. In practical face recognition problems, the variations between the images of the same face due to illumination and pose variations can sometimes be even larger than the variations between different face identities. Thus, the projections using PCA may not be accurate.

This issue can be improved by using Linear Discriminant Analysis (LDA). While PCA finds the projection that preserves maximum variation, LDA finds the projection that maximizes scatter between classes and minimizes scatter within classes, and preserves the discrimination. That is, LDA-based methods differentiate the inter-class variation and the intra-class variation. In such a way, we can separate the sources of variation that are useful for discriminating between face images and those that are not. LDA aims to find the directions that are optimal for classifying individuals.

PCA and LDA-based face recognition methods [5, 6] are conventional methods that have wide extensions [7, 8]. There are some recent works that are able to improve the PCA-based face recognition in terms of computational time [9, 10], and recognition accuracy [11, 12].

5.4 Subspace Learning Methods for Face Methods

Linear methods, like PCA and LDA, may fail to find the underlying nonlinear structure of the data under consideration, and they may lose some discriminant information of the manifolds during the linear projection. To overcome this problem, some nonlinear dimensionality reduction techniques have been proposed. In general, the nonlinear dimensionality reduction techniques are divided into two categories: kernel-based and manifold-learning-based approaches.

Kernel-based methods, as well as the linear methods mentioned above, only employ the global structure while ignoring the local geometry of the data. However, manifold-learning-based methods can explore the intrinsic geometry of the data. Popular nonlinear manifold-learning methods include ISOMAP [13], Locally Linear Embedding (LLE) [14], and Laplacian eigenmaps [15], which can be considered as special cases of the general framework for dimensionality reduction, named "graph embedding," proposed by Yan et al. [16]. Although these methods can represent the local structure of the data, they suffer from the out-of-sample problem. Locality Preserving Projection (LPP) [17] was proposed as a linear approximation of the nonlinear Laplacian eigenmaps [15] to overcome the out-of-sample problem. The manifold-learning methods presented so far are unsupervised methods, i.e., they do not consider the class information. Several supervised-based methods [18–20] have been proposed, which utilize the discriminant structure of the manifolds. With the Marginal Fisher Analysis (MFA) [16], which uses the Fisher criterion and constructs two adjacency graphs to represent the within-class and the between-class geometry of the data, several other methods have been proposed with similar ideas, such as Locality-Preserved Maximum Information Projection (LPMIP) [21], Constrained Maximum Variance Mapping (CMVM) [22], and Locality Sensitive Discriminant Analysis (LSDA) [23]. In the rest of this section, we will describe some of the selected manifold learning methods in detail.

5.4.1 Locality Preserving Projections (LPP)

The most fundamental graph-based subspace-learning method is Locality Preserving Projection (LPP) [17], which uses an intrinsic graph to represent the locality information of a data set, i.e., the neighborhood information. The idea behind LPP is that if the data points $\mathbf{x}_i$ and $\mathbf{x}_j$ are close to each other in the feature space, then they should also be close to each other in the transformed manifold subspace.

Given m face images $\{\mathbf{x}_1, \mathbf{x}_2, \ldots, \mathbf{x}_m\} \in \mathbb{R}^D$ which have a zero mean, LPP aims to find a projection vectors $\mathbf{w}$ that maps the training face images or data point to a new set of points $\{y_1, y_2, \ldots, y_m\}$, where $y_i = \mathbf{w}^{\mathrm{T}}\mathbf{x}_i$. After the transformation, the data points $\mathbf{x}_i$ and $\mathbf{x}_j$, which are close to each other, will have their projections in the manifold space $\mathbf{y}_i$ and $\mathbf{y}_j$ close to each other. This goal can be achieved by minimizing the following objective function:

$$\sum_{ij} \frac{1}{2} \left(y_i - y_j\right)^2 s_{ij}, \tag{5.30}$$

where s_{ij} represents the similarity between the training face samples $\mathbf{x}_i$ and $\mathbf{x}_j$. If s_{ij} is non-zero, y_i and y_j must be close to each other in the new subspace, in order to minimize (5.30). Taking the data points in the feature space as nodes of a graph, an edge between nodes i and j has a weight of s_{ij}, which is not zero, if they are close to each other. The similarity matrix $\mathbf{S} = [s_{ij}]$ for LPP can be defined as follows:

$$s_{ij} = \begin{cases} 1, & \text{if } \mathbf{x}_i \in N\left(\mathbf{x}_j, k\right) \text{ or } \mathbf{x}_j \in N\left(\mathbf{x}_i, k\right), \\ 0, & \text{otherwise,} \end{cases} \tag{5.31}$$

where $N(\mathbf{x}_i, k)$ represents the set of k nearest neighbors of $\mathbf{x}_i$. One shortfall of the above formulation for s_{ij} is that it is an unsupervised-learning method, i.e., not using any class-label information. Thinking that the label information can help to find a better separation between different class manifolds, Supervised Locality Preserving Projections (SLPP) was introduced in [19]. Denote $l(\mathbf{x}_i)$ as the corresponding class label of the data point $\mathbf{x}_i$. SLPP uses either one of the following formulations:

$$s_{ij} = \begin{cases} 1, & \text{if } l(\mathbf{x}_i) = l(\mathbf{x}_j), \\ 0, & \text{otherwise,} \end{cases} \tag{5.32}$$

or

$$s_{ij} = \begin{cases} 1, & \text{if } \left(\mathbf{x}_i \in N\left(\mathbf{x}_j, k\right) \text{ or } \mathbf{x}_j \in N(\mathbf{x}_i, k)\right) \text{ and } l(\mathbf{x}_i) = l(\mathbf{x}_j), \\ 0, & \text{otherwise,} \end{cases} \tag{5.33}$$

The heat kernel, i.e., $s_{ij} = \exp\left(-\left\|\mathbf{x}_i - \mathbf{x}_j\right\|^2 / t\right)$ is often used to determine the weights of the edges in the similarity graph, considering the fact that the distance between two neighboring points can also provide useful information about the manifold.

After constructing the similarity matrix with the weights, the minimization problem defined in (5.30) can be solved by using the spectral graph theory. Defining the Laplacian matrix $\mathbf{L} = \mathbf{D} - \mathbf{S}$, where $\mathbf{D}$ is the diagonal matrix whose entries are the column sum of $\mathbf{S}$, i.e., $d_{ii} = \sum_j s_{ij}$, the objective function is reduced to

$$\begin{aligned} \min \frac{1}{2} \sum_{ij} (y_i - y_j)^2 s_{ij} &= \min \frac{1}{2} \sum_{ij} \left(\mathbf{w}^{\mathrm{T}}\mathbf{x}_i - \mathbf{w}^{\mathrm{T}}\mathbf{x}_j\right)^2 s_{ij} \\ &= \min \frac{1}{2} \sum_{ij} \left(\mathbf{w}^{\mathrm{T}}\mathbf{x}_i - \mathbf{w}^{\mathrm{T}}\mathbf{x}_j\right) s_{ij} \left(\mathbf{x}_i^{\mathrm{T}}\mathbf{w} - \mathbf{x}_j^{\mathrm{T}}\mathbf{w}\right) \\ &= \min \sum_{ij} \left(\mathbf{w}^{\mathrm{T}}\mathbf{x}_i s_{ij} \mathbf{x}_i^{\mathrm{T}}\mathbf{w} - \mathbf{w}^{\mathrm{T}}\mathbf{x}_i s_{ij} \mathbf{x}_j^{\mathrm{T}}\mathbf{w}\right) \\ &= \min \mathbf{w}^{\mathrm{T}} \sum_i \left(\mathbf{x}_i d_i \mathbf{x}_i^{\mathrm{T}}\right) \mathbf{w} - \mathbf{w}^{\mathrm{T}} \sum_{ij} \left(\mathbf{x}_i s_{ij} \mathbf{x}_j^{\mathrm{T}}\right) \mathbf{w} \end{aligned} \tag{5.34}$$

Let $\mathbf{X} = [\mathbf{x}_1, \mathbf{x}_2, \ldots, \mathbf{x}_m]$, which is a matrix whose column are the face samples. Consider the following two terms:

$$\mathbf{XDX}^{\mathrm{T}} = \begin{bmatrix} \mathbf{x}_1 \cdots \mathbf{x}_m \end{bmatrix} \begin{bmatrix} d_1 & 0 & \cdots & 0 \\ 0 & d_2 & & \vdots \\ \vdots & & \ddots & 0 \\ 0 & \cdots & 0 & d_m \end{bmatrix} \begin{bmatrix} \mathbf{x}_1^{\mathrm{T}} \\ \vdots \\ \mathbf{x}_{\mathbf{m}}^{\mathrm{T}} \end{bmatrix} = \sum_i \mathbf{x}_i d_i \mathbf{x}_i^{\mathrm{T}}$$

$$\mathbf{XSX}^{\mathrm{T}} = \begin{bmatrix} \mathbf{x}_1 \cdots \mathbf{x}_m \end{bmatrix} \begin{bmatrix} s_{11} & s_{12} & \cdots & s_{1m} \\ s_{21} & s_{22} & & \vdots \\ \vdots & & \ddots & s_{m-1,m} \\ s_{m1} & \cdots & s_{m,m-1} & s_{m,m} \end{bmatrix} \begin{bmatrix} \mathbf{x}_1^{\mathrm{T}} \\ \vdots \\ \mathbf{x}_{\mathbf{m}}^{\mathrm{T}} \end{bmatrix} = \sum_i \mathbf{x}_i s_{ij} \mathbf{x}_j^{\mathrm{T}}$$

Therefore, (5.34) can be written as follows:

$$\begin{aligned} \min \frac{1}{2} \sum_{ij} \left(y_i - y_j\right)^2 s_{ij} &= \min \left(\mathbf{w}^{\mathrm{T}}\mathbf{XDX}^{\mathrm{T}}\mathbf{w} - \mathbf{w}^{\mathrm{T}}\mathbf{XSX}^{\mathrm{T}}\mathbf{w}\right) \\ &= \min \left(\mathbf{w}^{\mathrm{T}}\mathbf{X}\left(\mathbf{D} - \mathbf{S}\right)\mathbf{X}^{\mathrm{T}}\mathbf{w}\right) \\ &= \min \mathbf{w}^{\mathrm{T}}\mathbf{XLX}^{\mathrm{T}}\mathbf{w} \end{aligned} \tag{5.35}$$

Note that d_i is a measure of the importance of y_i. A constraint on the projection vector $\mathbf{w}$ is $\mathbf{y}^{\mathrm{T}}\mathbf{Dy} = 1$, which is equivalent to $\mathbf{w}^{\mathrm{T}}\mathbf{XDX}^{\mathrm{T}}\mathbf{w} = 1$. The minimization problem reduces to

$$\underset{\substack{w \\ \mathbf{w}^T \mathbf{XDS}^T \mathbf{w}=1}}{\arg\min}\ \mathbf{w}^{\mathrm{T}}\mathbf{XLX}^{\mathrm{T}}\mathbf{w} \tag{5.36}$$

Note that $\frac{\partial}{\partial \mathbf{x}}\mathbf{x}^{\mathrm{T}}\mathbf{Ax} = \left(\mathbf{A}+\mathbf{A}^{\mathrm{T}}\right)\mathbf{x}$, where $\mathbf{A}$ is a matrix. (5.36) can be solved by using Lagrange multiplier, as follows:

$$\frac{\partial}{\partial \mathbf{w}}\left(\mathbf{w}^{\mathrm{T}}\mathbf{XLX}^{\mathrm{T}}\mathbf{w} - \lambda\left(\mathbf{w}^{\mathrm{T}}\mathbf{XDX}^{\mathrm{T}}\mathbf{w} - 1\right)\right) = \mathbf{0}$$

Therefore, $\mathbf{XLX}^{\mathrm{T}}\mathbf{w} = \lambda\mathbf{XDX}^{\mathrm{T}}\mathbf{w}$, or $\left(\mathbf{XDX}^{\mathrm{T}}\right)^{-1}\mathbf{XLX}^{\mathrm{T}}\mathbf{w} = \lambda\mathbf{w}$. In other words, the optimal projection vectors are the eigenvectors of the matrix $\left(\mathbf{XDX}^{\mathrm{T}}\right)^{-1}\mathbf{XLX}^{\mathrm{T}}$. When this solution is substituted into (5.35), it becomes λ Consequently, the optimal projection matrix $\mathbf{W} = \begin{bmatrix}\mathbf{w}_1 \cdots \mathbf{w}_d\end{bmatrix}$ can be obtained by choosing the eigenvectors corresponding to the d $(d \ll D)$ smallest eigenvalues computed by solving the standard eigenvalue decomposition.

Example: Assume that there are five face images, denoted as $\mathbf{x}_1$, $\mathbf{x}_2$, $\mathbf{x}_3$, $\mathbf{x}_4$, and $\mathbf{x}_5$, and $l(\mathbf{x}_2) = l(\mathbf{x}_3)$, $l(\mathbf{x}_1) = l(\mathbf{x}_4) = l(\mathbf{x}_5)$. Then, $\mathbf{S}$ is a 5×5 matrix. If (5.32) is used, then for example, $s_{23} = s_{32} = 1$, and $s_{21} = s_{12} = 0$. The similarity matrix $\mathbf{S}$ is as follows:

$$\mathbf{S} = \begin{bmatrix} 1 & 0 & 0 & & 1 \\ 0 & 1 & 1 & 0 & 0 \\ 0 & 1 & 1 & 0 & 0 \\ 1 & 0 & 0 & 1 & 1 \\ 1 & 0 & 0 & 1 & 1 \end{bmatrix}$$

The diagonal matrix $\mathbf{D} = \begin{bmatrix} 3 & 0 & 0 & 0 & 0 \\ 0 & 2 & 0 & 0 & 0 \\ 0 & 0 & 2 & 0 & 0 \\ 0 & 0 & 0 & 3 & 0 \\ 0 & 0 & 0 & 0 & 3 \end{bmatrix}$

and the Laplacian matrix $\mathbf{L} = \mathbf{D} - \mathbf{S} = \begin{bmatrix} 2 & 0 & 0 & -1 & -1 \\ 0 & 1 & -1 & 0 & 0 \\ 0 & -1 & 1 & 0 & 0 \\ -1 & 0 & 0 & 2 & -1 \\ -1 & 0 & 0 & -1 & 2 \end{bmatrix}$

5.4.2 Marginal Fisher Analysis (MFA)

Marginal Fisher Analysis (MFA) [16], which is based on graph embedding as LPP, uses two graphs, the intrinsic and penalty graphs, to characterize the intra-class compactness and the interclass separability, respectively. In MFA, the intrinsic graph s_{ij}^w, i.e., the within-class graph, is constructed using the neighborhood and class information as follows:

$$s_{ij}^w = \begin{cases} 1, & \text{if } \mathbf{x}_i \in N\left(\mathbf{x}_j, k_1^+\right) \text{ or } \mathbf{x}_j \in N\left(\mathbf{x}_i, k_1^+\right), \\ 0, & \text{otherwise,} \end{cases} \tag{5.37}$$

Similarly, the penalty graph s_{ij}^b, i.e., the between-class graph, is constructed as follows:

$$s_{ij}^b = \begin{cases} 1, & \text{if } \mathbf{x}_i \in N\left(\mathbf{x}_j, k_2^-\right) \text{ or } \mathbf{x}_j \in N\left(\mathbf{x}_i, k_2^-\right), \\ 0, & \text{otherwise.} \end{cases} \tag{5.38}$$

MFA employs the Fisher criterion, which aims to maximize the ratio between the scattering of the between-class matrix $\mathbf{L}_b$ and that of the within-class Laplacian matrix $\mathbf{L}_w$, i.e.,

$$\frac{\mathbf{W}^{\mathrm{T}}\mathbf{X}\mathbf{L}_w\mathbf{X}^{\mathrm{T}}\mathbf{W}}{\mathbf{W}^{\mathrm{T}}\mathbf{X}\mathbf{L}_b\mathbf{X}^{\mathrm{T}}\mathbf{W}} \tag{5.39}$$

5.4.3 Soft Locality Preserving Projections (SLPM)

Although the application of the Fisher criterion shows its robustness, it involves taking the inverse of a high-dimensional matrix to solve a generalized eigenvalue problem. To tackle this problem, the Soft Locality Preserving Map (SLPM) [46] defines the objective function as the difference between the intrinsic and the penalty-graph matrices.

SLPM, similar to other manifold-learning algorithms, constructs two graph-matrices, the between-class matrix $\mathbf{W}_b$ and the within-class matrix $\mathbf{W}_w$, to characterize the discriminative information, based on the locality and class-label information. Given m training face images or data points $\{\mathbf{x}_1, \mathbf{x}_2, \ldots, \mathbf{x}_m\} \in \mathbb{R}^D$ and their corresponding class labels $\{l(\mathbf{x}_1), l(\mathbf{x}_2), \ldots, l(\mathbf{x}_m)\}$, we denote $N_w(\mathbf{x}_i, k_w) = \{\mathbf{x}_i^{w_1}, \mathbf{x}_i^{w_2}, \ldots, \mathbf{x}_i^{w_{k_w}}\}$ as the set of k_w nearest neighbors with the same class label as $\mathbf{x}_i$, i.e., $l(\mathbf{x}_i) = l\left(\mathbf{x}_i^{w_1}\right) = l\left(\mathbf{x}_i^{w_2}\right) = \ldots = l\left(\mathbf{x}_i^{w_{k_w}}\right)$, and $N_b(\mathbf{x}_i, k_b) = \left\{\mathbf{x}_i^{b_1}, \mathbf{x}_i^{b_2}, \ldots, \mathbf{x}_i^{b_{k_b}}\right\}$ as the set of its k_b nearest neighbors with different class labels from $\mathbf{x}_i$, i.e., $l(\mathbf{x}_i) \neq l\left(\mathbf{x}_i^{w_j}\right)$, where $j = 1, 2, \ldots, k_b$. Figure 5.14 illustrates the construction of these two similarity graphs. Then, the inter-class weight matrix $\mathbf{W}_b = [w_{ij}^b]$ and the intra-class weight matrix $\mathbf{W}_w = [w_{ij}^w]$ can be computed as follows:

$$w_{ij}^b = \begin{cases} \exp\left(\frac{-\|\mathbf{x}_i - \mathbf{x}_j\|^2}{t}\right), & \mathbf{x}_j \in N_b(\mathbf{x}_i, k_b), \\ 0, & \text{otherwise.} \end{cases} \tag{5.40}$$

$$w_{ij}^w = \begin{cases} \exp\left(\frac{-\|\mathbf{x}_i - \mathbf{x}_j\|^2}{t}\right), & \mathbf{x}_j \in N_w(\mathbf{x}_i, k_w), \\ 0, & \text{otherwise.} \end{cases} \tag{5.41}$$

SLPM considers the problem of creating a subspace, such that the data points from different classes, i.e., represented as edges in the between-class matrix $\mathbf{W}_b$, stay as distant as possible, while data

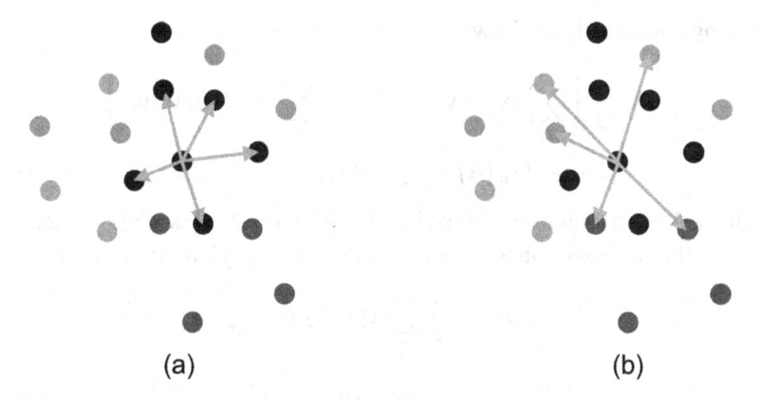

Figure 5.14 Construction of the similarity graphs of SLPM, where circles of the same color represent face images of the same class: (a) the between class, and (b) the within class, with $k_b = k_w = 5$.

points from the same class, i.e., represented as edges in the within-class matrix $\mathbf{W}_w$, stay close to each other. To achieve this, two objective functions are defined as follows:

$$\max \frac{1}{2} \sum_{ij} \left(y_i - y_j\right)^2 w_{ij}^b \tag{5.42}$$

and

$$\min \frac{1}{2} \sum_{ij} \left(y_i - y_j\right)^2 w_{ij}^w. \tag{5.43}$$

(5.40) ensures that the samples from different classes will stay as far as possible from each other, while (5.41) is to make samples from the same class stay close to each other after the projection. However, as shown in [24] and [25], small variations in the manifold subspace can lead to overfitting in training. To overcome this problem, a parameter β is added to control the intra-class spread. Note that, the SDM in [25] uses the within-class scatter matrix $\mathbf{S}_W$—as defined for LDA—to control the intra-class spread.

SLPM adopts the graph-embedding method, which uses the locality information about each class, in addition to the class information. Hence, the two objective functions (5.40) and (5.41) can be combined as follows:

$$\max \frac{1}{2}\left(\sum_{ij}(\mathbf{y}_i-\mathbf{y}_j)^2 w_{ij}^b-\beta\sum_{ij}(\mathbf{y}_i-\mathbf{y}_j)^2 w_{ij}^w\right)$$
$$=\max\left(J_b(\mathbf{A})-\beta J_w(\mathbf{A})\right), \tag{5.44}$$

where $\mathbf{A}$ is a projection matrix, i.e., $\mathbf{Y}=\mathbf{A}^{\mathrm{T}}\mathbf{X}$ and $\mathbf{X}=[\mathbf{x}_1,\mathbf{x}_2,\ldots,\mathbf{x}_m]$. Then, the between-class objective function J_b (A) can be reduced to

$$J_b(\mathbf{A})=\frac{1}{2}\sum_{ij}(\mathbf{y}_i-\mathbf{y}_j)^2 w_{ij}^b$$
$$=\mathbf{A}^{\mathrm{T}}\mathbf{X}\mathbf{L}_b\mathbf{X}^{\mathrm{T}}\mathbf{A} \tag{5.45}$$

where $\mathbf{L}_b=\mathbf{D_b}-\mathbf{W_b}$ is the Laplacian matrix of $\mathbf{W}_b$ and $\mathbf{D_b}=\left[d_{b_{ii}}\right]$ with $d_{b_{ii}}=\sum_j w_{ij}^b$, is a diagonal matrix. Similarly, the within-class objective function $J_w(\mathbf{A})$ can be written as

$$J_w(\mathbf{A})=\frac{1}{2}\sum_{ij}(\mathbf{y}_i-\mathbf{y}_j)^2 w_{ij}^w$$
$$=\mathbf{A}^{\mathrm{T}}\mathbf{X}\mathbf{L}_w\mathbf{X}^{\mathrm{T}}\mathbf{A}, \tag{5.46}$$

where $\mathbf{L}_w=\mathbf{D}_w-\mathbf{W}_w$ and $d_{w_{ii}}=\sum_j w_{ij}^w$. If J_w and J_b are substituted to (5.42), the objective function becomes as follows:

$$\max J_T(\mathbf{A})=\max\left(J_b(\mathbf{A})-\beta J_w(\mathbf{A})\right)$$
$$=\max\left(\mathbf{A}^{\mathrm{T}}\mathbf{X}\mathbf{L}_b\mathbf{X}^{\mathrm{T}}\mathbf{A}-\beta\mathbf{A}^{\mathrm{T}}\mathbf{X}\mathbf{L}_w\mathbf{X}^{\mathrm{T}}\mathbf{A}\right)$$
$$=\max\mathbf{A}^{\mathrm{T}}\mathbf{X}\left(\mathbf{L}_b-\beta\mathbf{L}_w\right)\mathbf{X}^{\mathrm{T}}\mathbf{A} \tag{5.47}$$

which is subject to the constraint $\mathbf{A}^{\mathrm{T}}\mathbf{A}-\mathbf{I}=0$ so as to guarantee orthogonality. By using Lagrange multiplier, we obtain

$$\mathrm{L}(\mathbf{A})=\mathbf{A}^{\mathrm{T}}\mathbf{X}\left(\mathbf{L}_b-\beta\mathbf{L}_w\right)\mathbf{X}^{\mathrm{T}}\mathbf{A}-\lambda\left(\mathbf{A}^{\mathrm{T}}\mathbf{A}-\mathbf{I}\right). \tag{5.48}$$

By computing the partial derivative of L(**A**), the optimal projection matrix **A** can be obtained, as follows:

$$\frac{\partial\mathrm{L}(\mathbf{A})}{\partial\mathbf{A}}=\mathbf{X}\left(\mathbf{L}_b-\beta\mathbf{L}_w\right)\mathbf{X}^T\mathbf{A}-\lambda\mathbf{A}=0, \tag{5.49}$$

i.e., $\mathbf{X}\left(\mathbf{L}_b-\beta\mathbf{L}_w\right)\mathbf{X}^T\mathbf{A}=\lambda\mathbf{A}$. The projection matrix **A** can be obtained by computing the eigenvectors of $\mathbf{X}\left(\mathbf{L}_b-\beta\mathbf{L}_w\right)\mathbf{X}^T$. The columns of **A** are the d leading eigenvectors, where d is the dimension of the subspace. LDA, LPP, MFA, and other manifold-learning algorithms, whose objective functions have a similar structure, lead to a generalized eigenvalue problem. Such methods suffer from the matrix-singularity problem, because the solution involves computing the inverse of a singular matrix. The proposed objective function is designed in such a way as to overcome this singularity problem. However, in this algorithm, PCA is still applied to data, so as to reduce its dimensionality and to reduce noise. The overall flow of the subspace learning methods can be listed as below:

1. Obtain features from face images such as intensity information or Gabor features: $\mathbf{X}_{fea}$.
2. Learn the projection matrix $\mathbf{W}_{pca}$ via PCA.
3. Construct the within-class graph matrix $\mathbf{W}_w$ and the between-class similarity matrices $\mathbf{W}_b$.
4. Calculate the Laplacian matrices $\mathbf{L}_w$ and $\mathbf{L}_b$.
5. Solve the eigenvalue decomposition of $\mathbf{X}\left(\mathbf{L}_b - \beta\mathbf{L}_w\right)\mathbf{X}^T$, where $\mathbf{X}$ is the facial features in eigenspace.
6. Choose the eigenvectors corresponding to the d largest eigenvalues, $\mathbf{W}_{mL}$.
7. $\mathbf{Y}_{desc} = \mathbf{W}_{mL}^{\mathrm{T}}\mathbf{W}_{pca}^{\mathrm{T}}\mathbf{X}_{fea}$.
8. Learn the nearest neighbor classifier.

These advanced subspace analysis methods have been widely applied to face recognition and facial image analysis. Interested readers may refer to [26–33].

5.5 Face Recognition by Deep Learning

Because of the use of deep neural networks, performances of face-recognition algorithms have been significantly improved over last few years, and have surpassed that of humans. A deep neural network (DNN) contains a cascade of a large number of convolutional layers with non-linear activation functions. This kind of deep model is called convolutional neural network (CNN). In this section, the basic structure and operations of CNN will be presented and explained. Then, the training the network for face recognition, with the use of different loss functions.

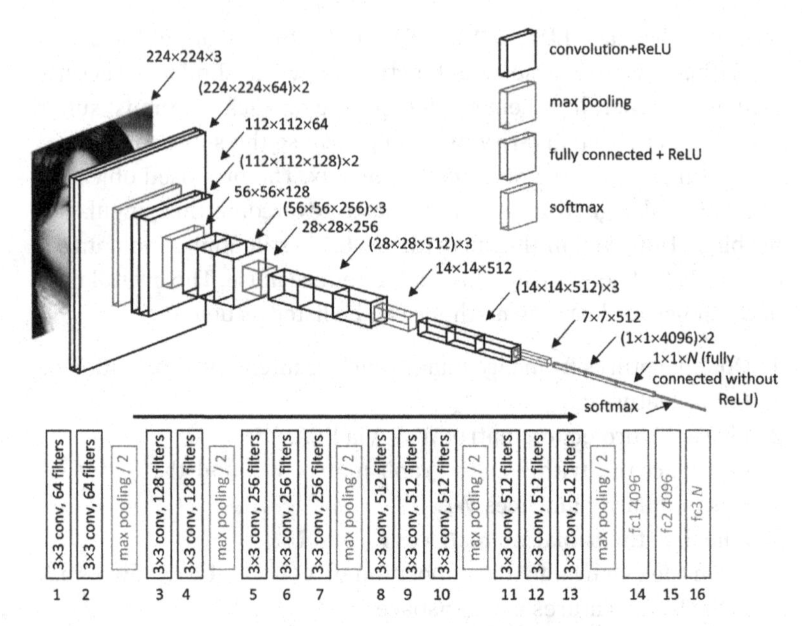

Figure 5.15 The network architecture for VGG-Face, which consists of 16 layers of convolution (conv) + activation, convolution + activation + max pooling (pool), and fully-connected (fc) network. The number before and after "conv" refer to the size of the filter and the number of the filters used, respectively. N is the number of identities to be recognized. In VGG-Face, $N = 2{,}622$.

5.5.1 Architecture of Deep Neural Network

Many deep models have been proposed for improving their discrimination and generalization capability for face recognition. Nevertheless, the basic elements used are the convolutional layers, non-linear activation, and pooling. A commonly used model for face recognition is the VGG-Face [34], whose implementation is based on the VGG-Very-Deep-16 CNN architecture.

Figure 5.15 shows the structure of this network. The input to the network has the size of 224×224, with three channels for the three color components. Low-level visual features are extracted from the input color image by using 64 filters, each with the size of 3×3. In order to keep the size of the filter output the same as the input, zero padding is adopted to increase the input size to 226×226.

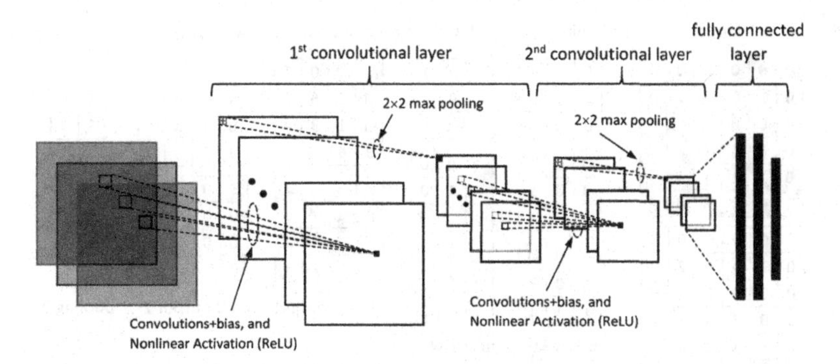

Figure 5.16 A simplified CNN, with two convolutional layers and maximum pooling performed after each layer.

Therefore, 64 feature maps/output channels, with the size of 224 $\times$ 224, are generated by 64 filters. Furthermore, each of the output values is transformed by a non-linear activation function, namely the rectified linear unit (ReLU). Assuming that y is the output from a filter, ReLU performs the following operation:

$$h(y) = \max(0, y) \tag{5.50}$$

where $h(\cdot)$ is the ReLU function. In other words, the outputs after ReLU will always be positive, and its gradient will not be vanished, even if y has a large value. The use of ReLU, or a similar function, is important for training up the deep networks using the back-propagation algorithm.

The use of the filters to extract local features from the input by convolution and performing non-linear transformation form a convolutional layer. Those convolutional layers near the input extract primitive low-level features. Those later layers extract more abstract, semantic features. In other words, the architecture extracts representations of the input corresponding to different levels of abstraction. From Fig. 5.15, it can be seen that the network contains 16 convolutional layers. In order to reduce computational complexity and make the features robust to translational variations, pooling is performed after a certain number of convolutional layers. Usually, 2 $\times$ 2 maximum pooling, with a stride of 2, is adopted. With this operation, the horizontal and vertical dimensions of the layer are both reduced by half. Figure 5.16 illustrates a simplified

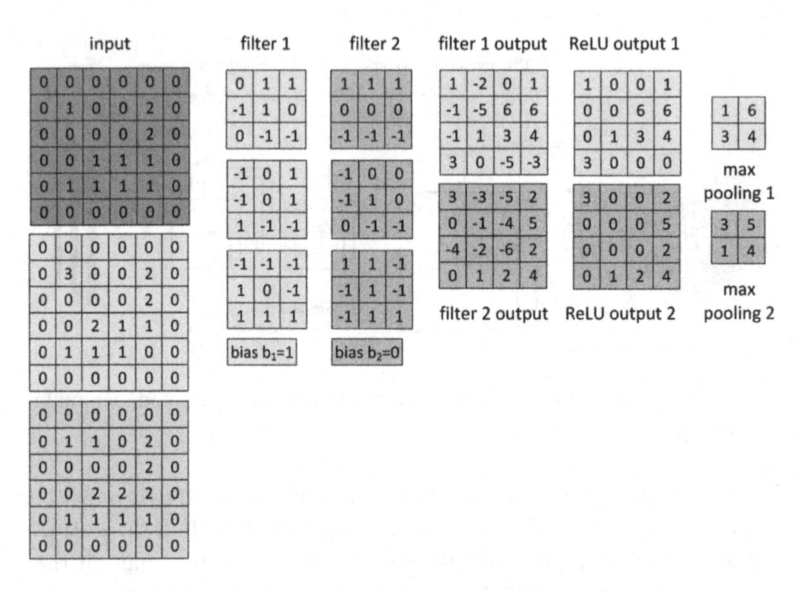

Figure 5.17 Using two filters to extract features from a three-layer input, with ReLU and maximum pooling performed. The input is zero padded, so its size for each layer is increased from 4×4 to 6×6. After filtering, the size of each output reduces to 4×4.

CNN, whose input is a color image. Therefore, the input has three channels. A number of filters are used to extract features from the input. To illustrate how to calculate outputs using convolution, Fig. 5.17 shows an example, where the input is of size 6×6, with 3 channels. Although the filters are convolved with the input to extract features to form a feature map, the correlation operation is usually performed. Assume that a filter is convolved with an image. If the filter is flipped vertically and horizontally, the correlation of this flipped filter with the image is equivalent to performing convolution on the image.

To generate one feature map, a 3×3 filter is needed to extract features from each channel. Therefore, three 3×3 filters, with a bias value, are used to produce one feature map. The coefficients of the filters, as well as the bias, are learned based on the training samples. The number of parameters involved for generating one feature map is $3 \times 3 \times 3 + 1 = 28$. If the filter size is larger and more convolutional layers are used, the number of parameters involved will be larger. In a DNN, the number of parameters involved is usually tremendous. To avoid the overfitting problem, a very big training set is indispensable.

The last three layers in the deep model shown in Fig. 5.16 are fully connected layers. The first two fully connected layers are associated with the ReLU operation, while the last uses the softmax loss function for classification. The number of neurons in the last layer is equal to the number of classes to be distinguished. The softmax loss can encourage the deep features learned for different classes separable. However, this is insufficient for face recognition as we need the features to be discriminative as well. This can enhance the generalization capability of the features. Different kinds of discriminative loss function will be discussed in the next section.

The deep architecture described previously is a typical CNN architecture. Similar structures, such as AlexNet [35], GoogleNet [36], ResNet [37], etc., are used as a baseline model for face recognition. There are also other novel deep architectures designed for face recognition.

5.5.2 Loss Functions

A loss function is used to evaluate how accurate a network can classify training samples. The softmax loss function is usually employed. This loss function considers the scores or the neuron outputs in the output layer are unnormalized log probabilities of the classes. If z_i and p_i' represent the score and the unnormalized log probabilities, respectively, of the i-th output neuron, then we have

$$z_i = \log p_i' \quad \text{or} \quad p_i' = e^{z_i}. \tag{5.51}$$

The normalized probability, $p_i = \frac{\exp(z_i)}{\sum_{j=1}^{N} \exp(z_j)}$, where N is the number of classes. The softmax loss $\mathcal{L}_{SM}$ is then defined as follows:

$$\mathcal{L}_{SM} = -\sum_{t=1}^{M} \log \left(\frac{\exp(z_i)}{\sum_{j=1}^{N} \exp(z_j)} \right) \tag{5.52}$$

where M is the number of training samples, and $z_i = \boldsymbol{W}_i^T \boldsymbol{x}_t + b_i$. Here $\boldsymbol{x}_t \in \mathbb{R}^d$ denotes the deep feature, of dimensionality d, of the

t-th training sample, $\boldsymbol{W}_i \in \mathbb{R}^d$ is the i-th column of the weight matrix $\boldsymbol{W} \in \mathbb{R}^{d\times N}$ in the last fully connected layer, and $\mathrm{b_i}$ is the bias term for the i-th class.

Under the supervision of this loss function, it has been found that the learned deep features are separable, but are not sufficiently discriminative. The reason for this is that the features still exhibit significant intra-class variations. Consequently, the deep networks trained with this loss function have limited accuracy. More discriminative loss functions are necessary to learn features with large margin, which have better generalization capability.

Contrastive Loss

Both the contrastive loss [38, 40] and triplet loss [39, 40] are Euclidean-distance-based losses. With these loss functions, the features learned by a CNN, such as the output at fc2 in Fig. 5.17, will have their intra-variance compressed, while the inter-variance is enlarged. In other words, the features belonging to the same class will become more compact, while features belonging to different classes become farther apart. For a pair of data points $(\boldsymbol{x}_i, \boldsymbol{x}_j)$, whose corresponding features produced by the CNN are denoted as $f(\boldsymbol{x}_i)$ and $f(\boldsymbol{x}_j)$, respectively. If $\boldsymbol{x}_i$ and $\boldsymbol{x}_j$ are semantically similar or of the same class, they form a positive pair, otherwise a negative pair is formed. Let $\mathcal{S} = \{(i, j)\}$ be the index set of positive pairs, and $\mathcal{D} = \{(i, j)\}$ be the index set of negative pairs. Then, the contrastive loss function $\mathcal{L}_{CT}$ is given as follows:

$$\mathcal{L}_{CT} = \sum_{(i,j)\in S} h\left(\left\|f(\boldsymbol{x}_i) - f(\boldsymbol{x}_j)\right\|_2 - \tau_1\right) + \sum_{(i,j)\in D} h\left(\tau_2 - \left\|f(\boldsymbol{x}_i) - f(\boldsymbol{x}_j)\right\|_2\right) \tag{5.53}$$

where $h(x) = \max(0, x)$ is the hinge loss function. To minimize the loss, x should be negative. τ_1 and τ_2 are two positive thresholds used to control the margins of the positive and negative pairs, respectively, and $\tau_1 < \tau_2$. With the loss function, the distance $\left\|f(\boldsymbol{x}_i) - f(\boldsymbol{x}_j)\right\|_2$ between positive pairs will be smaller than τ_1, while the distance between negative pairs will be larger than τ_2. However, it is difficult to choose the optimal values for τ_1 and τ_2.

Assume that five samples $(\mathbf{x}_i, y_i)$, where $i = 1, 2, \ldots, 5$ and y_i is the class label of $\mathbf{x}_i$, are selected to form a mini-batch for optimization. The five samples are $(\mathbf{x}_1, 1)$, $(\mathbf{x}_2, 1)$, $(\mathbf{x}_3, 2)$, $(\mathbf{x}_4, 3)$ and $(\mathbf{x}_5, 2)$, so they belong to three different classes, 1, 2, and 3. The positive pairs produced include $(\mathbf{x}_1, \mathbf{x}_2)$ and $(\mathbf{x}_3, \mathbf{x}_5)$, while negative pairs are $(\mathbf{x}_1, \mathbf{x}_3)$, $(\mathbf{x}_1, \mathbf{x}_4)$, $(\mathbf{x}_1, \mathbf{x}_5)$, $(\mathbf{x}_2, \mathbf{x}_3)$, $(\mathbf{x}_2, \mathbf{x}_4)$, $(\mathbf{x}_2, \mathbf{x}_5)$, $(\mathbf{x}_3, \mathbf{x}_4)$ and $(\mathbf{x}_4, \mathbf{x}_5)$. Therefore, the number of pairs produced is more than the number of samples used for training.

Triplet Loss

For the triplet loss [39, 40], three samples are considered in computing the loss function. With a training sample $\mathbf{x}$, a positive sample $\mathbf{x}^+$ and a negative sample $\mathbf{x}^-$ are selected to form a triplet $(\mathbf{x}, \mathbf{x}^+, \mathbf{x}^-)$, and $\mathbf{x}$ is the anchor sample. Therefore, $\mathbf{x}$ and $\mathbf{x}^+$ are of the same class, while $\mathbf{x}$ and $\mathbf{x}^-$ belong to different classes. The triplet loss function is defined as follows:

$$\mathcal{L}_{TP} = \sum_i h\left(\tau + \left\| f(\mathbf{x}_i) - f\left(\mathbf{x}_i^+\right)\right\|_2 - \left\| f(\mathbf{x}_i) - f\left(\mathbf{x}_i^-\right)\right\|_2\right), \tag{5.54}$$

where $\tau > 0$ is a margin. If $\left\| f(\mathbf{x}_i) - f\left(\mathbf{x}_i^-\right)\right\|_2 - \left\| f(\mathbf{x}_i) - f\left(\mathbf{x}_i^+\right)\right\|_2$ is larger than τ for the anchor sample $\mathbf{x}_i$, then its corresponding loss is zero. Therefore, the negative samples of each anchor sample will be pushed away, while the positive samples will be pulled close to the anchor sample.

Consider the mini-batch $(\mathbf{x}_1, 1)$, $(\mathbf{x}_2, 1)$, $(\mathbf{x}_3, 2)$, $(\mathbf{x}_4, 3)$ and $(\mathbf{x}_5, 2)$ once again. The possible triplets to be generated are $(\mathbf{x}_1, \mathbf{x}_2, \mathbf{x}_3)$, $(\mathbf{x}_1, \mathbf{x}_2, \mathbf{x}_4)$, $(\mathbf{x}_1, \mathbf{x}_2, \mathbf{x}_5)$, $(\mathbf{x}_2, \mathbf{x}_1, \mathbf{x}_3)$, $(\mathbf{x}_2, \mathbf{x}_1, \mathbf{x}_4)$, $(\mathbf{x}_2, \mathbf{x}_1, \mathbf{x}_5)$, $(\mathbf{x}_3, \mathbf{x}_5, \mathbf{x}_1)$, $(\mathbf{x}_3, \mathbf{x}_5, \mathbf{x}_2)$, $(\mathbf{x}_3, \mathbf{x}_5, \mathbf{x}_4)$, $(\mathbf{x}_5, \mathbf{x}_3, \mathbf{x}_1)$, $(\mathbf{x}_5, \mathbf{x}_3, \mathbf{x}_2)$, and $(\mathbf{x}_5, \mathbf{x}_3, \mathbf{x}_4)$.

Center Loss

Another effective loss function to enhance the discriminative power of the learned deep features is the center loss [41], which learns the center for the deep features of each class and penalize each deep feature according to its distance to the corresponding center. It has been proven that CNNs can be trained easily with the center loss function. This is because the contrastive loss and the triplet loss will have the number of training pairs or triplets increased dramatically when the number of training samples increases.

In the course of training, the center for each class and the distances between deep features and their corresponding centers are simultaneously optimized, under the joint supervision of the softmax loss and center loss. The center loss $\mathcal{L}_C$ is formulated as follows:

$$\mathcal{L}_C = \frac{1}{2}\sum_{i=1}^{M} \| f(\mathbf{x}_i) - \mathbf{c}_{y_i} \|_2^2 \tag{5.55}$$

where y_i is the class label of $\mathbf{x}_i$, and $\mathbf{c}_{y_i}$ is the center or the mean of the deep features belonging to class y_i. This formulation can characterize the intra-variance effectively. To jointly supervise the CNNs with the softmax loss and the center loss, the overall loss function used, based on (5.52) and (5.55), is

$$\mathcal{L} = \mathcal{L}_{SM} + \lambda \mathcal{L}_C \tag{5.56}$$

where λ is used to balance the two loss functions.

Angular-margin-based Loss

The angular margin losses [42–45] aim to learn the deep features that are angularly separable. In face verification, it has been demonstrated that cosine similarity is more suitable for measuring the similarities of two deep features, when the softmax loss function is used during training. This observation reveals that the angular distance is related to intra-class variance, while the feature's magnitude is related to inter-class variance. The idea of angular-margin-based losses can be simply explained by using Fig. 5.18. The angular margin losses have a stricter decision boundary to encourage the learned features lying within the target regions of each class. As shown in Fig. 5.18, the dotted line is the softmax boundary. With the use of the angular-margin loss, the boundary for each of the classes is much tighter.

The softmax loss function, as shown in (5.52), encourages $\frac{\exp(y_i)}{\sum_{j=1}^{N}\exp(y_j)} = 1$, which is equivalent to maximize y_i such that $y_i > y_j$,

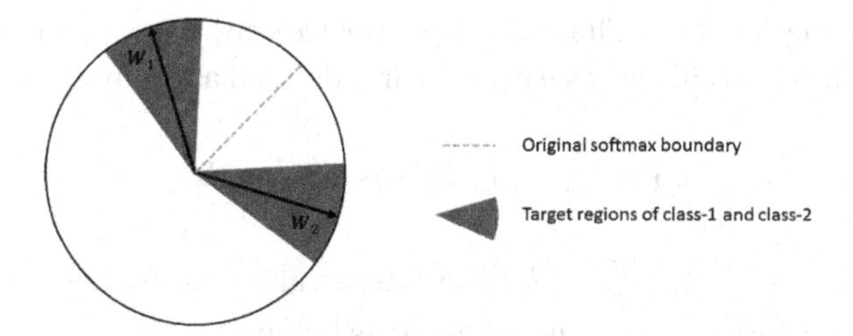

Figure 5.18 Interpretation of the original softmax boundary and the target regions of angular margin losses. Assume there are only two classes.

$\forall i \neq j$. It can be rewritten as follows:

$$\begin{aligned} & y_i > y_j \\ \Rightarrow\ & \boldsymbol{W}_i^T \mathbf{x}_t > \boldsymbol{W}_j^T \mathbf{x}_t \\ \Rightarrow\ & \|\boldsymbol{W}_i\| \, \|\mathbf{x}_t\| \cos\theta_{i,t} > \left\|\boldsymbol{W}_j\right\| \|\mathbf{x}_t\| \cos\theta_{j,t} \\ \Rightarrow\ & \|\boldsymbol{W}_i\| \cos\theta_{i,t} > \left\|\boldsymbol{W}_j\right\| \cos\theta_{j,t}, \end{aligned} \tag{5.57}$$

where $\mathbf{x}_t$ is the deep feature of a test sample, and $\theta_{i,t}$ is the angle between $\boldsymbol{W}_i$ and $\mathbf{x}_t$. The bias terms are removed for better interpretation, i.e., $b_i = b_j = 0$. From the above equation, we can observe that the softmax loss is independent of the feature magnitude, $\|\mathbf{x}_t\|$, as y_i and y_j share the same $\mathbf{x}_t$. The decision boundary of the original softmax loss is not strict enough, and it depends on both the weight magnitudes and angular distances. Angular softmax (A-softmax) loss [43], the loss used for face recognition is called SphereFace, which was first proposed to deal with this issue. It modifies the original softmax such that (5.57) is rewritten as follows:

$$\begin{aligned} & \|\boldsymbol{W}_i\| \cos\theta_{i,t} > \left\|\boldsymbol{W}_j\right\| \cos\theta_{j,t} \\ \Rightarrow\ & \cos m\theta_{i,t} > \cos\theta_{j,t}, \end{aligned} \tag{5.58}$$

where $\|\boldsymbol{W}_i\| = \|\boldsymbol{W}_k\| = 1$, and $m \geq 2$. Note that A-softmax loss normalizes the weights such that the loss function depends on the angular distance only. Also, it uses a stricter angular margin, $m\theta_{i,t}$ to make sure the largest intra-class angular distance is still smaller than the smallest inter-class angular distance. Therefore, the learned features are more discriminative. However, it is infeasible to implement this function, because cosine is not a monotonically decreasing function. Therefore, SphereFace [43] uses a piece-wise function, φ, to replace $\cos m\theta_{i,t}$, which is defined as follows:

$$\varphi(m, \theta_{i,t}) = (-1)^k \cos m\theta_{i,t} - 2k, \tag{5.59}$$

where $\theta_{i,t} \in \left[\frac{k\pi}{m}, \frac{(k+1)\pi}{m}\right]$, k is an integer and $k \in [0, m-1]$. The A-softmax loss $\mathcal{L}_{A-SM}$ is then defined as follows:

$$\mathcal{L}_{A-SM} = -\sum_{t=1}^{M} \log\left(\frac{\exp(\|\mathbf{x}_t\| \varphi(m, \theta_{i,t}))}{\exp(\|\mathbf{x}_t\| \varphi(m, \theta_{i,t})) + \sum_{j, j \neq i} \exp(\|\mathbf{x}_t\| \cos\theta_{j,t})}\right) \tag{5.60}$$

where $\mathbf{x}_t$ belongs to the ith class.

However, directly using the A-softmax loss will lead to divergence, as this loss function is too strict for the network to find the optimal set of parameters. In the original implementation of A-softmax, it uses an annealing strategy to train the network. To be specific, $\|\mathbf{x}_t\| \varphi(m, \theta_{i,t})$ in (5.60) is changed to $f = \frac{\lambda\|\mathbf{x}_t\| \cos\theta_{i,t} + \|\mathbf{x}_t\|\varphi(m, \theta_{i,t})}{1+\lambda}$. The hyperparameter λ is initialized to be a very large value, and it decreases gradually during training. Therefore, the loss function is equivalent to the original softmax loss at the beginning. Besides, ArcFace [44] and CosFace [45] were proposed to introduce the angular margin loss by replacing $\cos\theta_{i,t}$ with $\cos(\theta_{i,t} + m)$ and $\cos\theta_{i,t} - m$, respectively. Although they share a similar idea to the A-softmax loss, they are easier to implement.

5.6 Conclusions

In this chapter, we have provided a detailed account of face recognition using the eigenface/PCA approach, including the basic formulation, illustrative examples, and experimental results. However, these are just basic concepts for face recognition. We subsequently introduce a few modern approaches, such as subspace learning methods and deep neural networks for face recognition. These are very promising methods, which are extremely useful for practical practitioners and fruitful directions for future research.

References

1. G. Bradski and A. Kaehler (2000). OpenCV. *Dr. Dobb's Journal of Software Tools*, https://docs.opencv.org/3.4.1/d7/d8b/tutorial_py_face_detection.html.
2. T. F. Cootes, C. J. Taylor, D. H. Cooper, and J. Graham, Active shape models-their training and application, *Comput. Vis. Image Understand.*, 61(1), pp. 38–59, 1995.
3. T. F. Cootes, G. J. Edwards, and C. J. Taylor, Active appearance models, *IEEE Trans. Pattern Anal. Mach. Intell.*, 23(6), pp. 681–685, 2001.
4. Dunteman, G. H., *Principal Components Analysis*, No. 69. Sage, 1989.
5. M. Turk and A. Pentland, Face recognition using eigenfaces, in *Proceedings of the IEEE Conference on Computer Vision and Pattern Recognition*, pp. 586–591, 1991.
6. P. Belhumeur, J. Hespanha, and D. Kriegman, Eigenfaces vs. Fisherfaces: Recognition using class specific linear projection, *IEEE Trans. Pattern Anal. Machine Intell.*, 19(7), 1997.
7. M. H. Yang, Face recognition using kernel eigenfaces, *Proc. Int. Conf. Image Process*, 1, pp. 37–40, 2000.
8. X. Xie and K.-M. Lam, Gabor-based kernel PCA with doubly nonlinear mapping for face recognition with a single face image, *IEEE Trans. Image Process.*, 15(9), pp. 2481–2492, Sept. 2006.
9. R. Cendrillon and B. Lovell, Real-time face recognition using eigenfaces, *Visual Commun. Image Process.*, pp. 269–276, 2000.
10. Y. Woo, C. Yi and Y. Yi, Fast PCA-based face recognition on GPUs, in *2013 IEEE International Conference on Acoustics, Speech and Signal Processing*, Vancouver, BC, 2013, pp. 2659–2663.
11. W. Deng, J. Hu, J. Lu, and J. Guo, Transform-invariant PCA: A unified approach to fully automatic face alignment, representation, and recognition, *IEEE Trans. Pattern Anal. Machine Intell.*, 36(6), pp. 1275–1284, June 2014.
12. G. Ghinea, R. Kannan, and S. Kannaiyan, Gradient-orientation-based PCA subspace for novel face recognition, *IEEE Access*, 2, pp. 914–920, 2014.
13. J. B. Tenenbaum, V. De Silva, and J. C. Langford, A global geometric framework for nonlinear dimensionality reduction, *Science*, 290, pp. 2319–2323, 2000.
14. S. T. Roweis and L. K. Saul, Nonlinear dimensionality reduction by locally linear embedding, *Science*, 290, pp. 2323–2326, 2000.

15. M. Belkin and P. Niyogi, Laplacian eigenmaps and spectral techniques for embedding and clustering, in *Neural Information Processing Systems*, pp. 585–591, 2001.
16. S. Yan, D. Xu, B. Zhang, H.-J. Zhang, Q. Yang, and S. Lin, Graph embedding and extensions: A general framework for dimensionality reduction, *IEEE Trans. Pattern Anal. Machine Intell.*, 29, pp. 40–51, 2007.
17. X. Niyogi, Locality preserving projections, in *Neural Information Processing Systems*, pp. 153, 2004.
18. C. Shan, S. Gong, and P. W. McOwan, Appearance manifold of facial expression, in *International Workshop on Human-Computer Interaction*, Springer, pp. 221–230, 2005.
19. W. K. Wong and H. Zhao, Supervised optimal locality preserving projection, *Pattern Recognit.*, 45, pp. 186–197, 2012.
20. W.-L. Chao, J.-J. Ding, and J.-Z. Liu, Facial expression recognition based on improved local binary pattern and class-regularized locality preserving projection, *Signal Process.*, 117, pp. 1–10, 2015.
21. H. Wang, S. Chen, Z. Hu, and W. Zheng, Locality-preserved maximum information projection, *IEEE Trans. Neural Netw.*, 19, pp. 571–585, 2008.
22. B. Li, D.-S. Huang, C. Wang, and K.-H. Liu, Feature extraction using constrained maximum variance mapping, *Pattern Recognit.*, 41, pp. 3287–3294, 2008.
23. D. Cai, X. He, K. Zhou, J. Han, and H. Bao, Locality sensitive discriminant analysis, in *IJCAI*, pp. 708–713, 2007.
24. H.-W. Kung, Y.-H. Tu, and C.-T. Hsu, Dual subspace nonnegative graph embedding for identity-independent expression recognition, *IEEE Trans. Info. Forensics Secur.*, 10, pp. 626–639, 2015.
25. R. Liu and D. F. Gillies, Overfitting in linear feature extraction for classification of high-dimensional image data, *Pattern Recognit.*, 53, pp. 73–86, 2016.
26. C. Turan and K.-M. Lam, Histogram-based local descriptors for facial expression recognition (FER): A comprehensive study, *J. Visual Commun. Image Representation*, 55, pp. 331–341, 2018.
27. M. Jian and K.-M. Lam, Simultaneous hallucination and recognition of low-resolution faces based on singular value decomposition, *IEEE Trans. Circuits Syst. Video Technol.*, 25(11), pp. 1761–1772, 2015.
28. K.-H. Pong and K.-M. Lam, Multi-resolution feature fusion for face recognition, *Pattern Recognit.*, 47(2), pp. 556–567, 2014.

29. W.-P. Choi, S.-H. Tse, K.-W. Wong, and K.-M. Lam, Simplified Gabor wavelets for human face recognition, *Pattern Recognit.*, 41(3), pp. 1186–1199, 2008.
30. C. Turan, K. M. Lam, and X. He, Soft locality preserving map (SLPM) for facial expression recognition, *arXiv preprint*, arXiv:1801.03754, 2018.
31. K.-W. Wong, K.-M. Lam and W.-C. Siu, A robust scheme for live detection of human face in color images, *Signal Process. Image Commun.*, 18(2), pp. 103–114, February 2003, The Netherlands.
32. K.-W. Wong, K.-M. Lam and W.-C. Siu, An efficient algorithm for human face detection and facial feature extraction under different conditions, *Pattern Recognit.*, 34, pp. 1993–2004, 2001.
33. K.-W. Wong, K.-M. Lam, and W.-C. Siu, An efficient low bit-rate video coding algorithm focusing on moving regions, 11(10), *IEEE Trans. Circuits Systems Video Technol.*, pp. 1128–1134, 2001.
34. M. Parkhi, A. Vedaldi, and A. Zisserman, Deep face recognition, in *British Machine Vision Conference*, 2015.
35. A. Krizhevsky, I. Sutskever, and G. E. Hinton, ImageNet classification with deep convolutional neural networks, in *Proceedings of Neural Information Processing Systems (NIPS)*, pp. 1106–1114, 2012.
36. C. Szegedy, W. Liu, Y. Jia, P. Sermanet, S. Reed, D. Anguelov, D. Erhan, V. Vanhoucke, A. Rabinovich, et al., Going deeper with convolutions, in *IEEE Conference on Computer Vision and Pattern Recognition*, 2015.
37. K. He, X. Zhang, S. Ren, and J. Sun, Deep residual learning for image recognition, in *Conference on Computer Vision and Pattern Recognition*, pp. 770–778, 2016.
38. Y. Sun, Y. Chen, X. Wang, and X. Tang, Deep learning face representation by joint identification-verification, in *Neural Information Processing Systems*, pp. 1988–1996, 2014.
39. F. Schroff, D. Kalenichenko, and J. Philbin, FaceNet: A unified embedding for face recognition and clustering, in *IEEE Conference on Computer Vision and Pattern Recognition*, 2015.
40. J. Lu, J. Hu, and J. Zhou, Deep metric learning for visual understanding: An overview of recent advances, *IEEE Signal Process. Mag.*, pp. 76–84, November 2017.
41. Y. Wen, K. Zhang, Z. Li, and Y. Qiao, A discriminative feature learning approach for deep face recognition, in *Proceedings of the European Conference on Computer Vision*, 2016.

42. W. Liu, Y. Wen, Z. Yu, and M. Yang, Large-margin softmax loss for convolutional neural networks, in *International Conference on Machine Learning*, pp. 507–516, 2016.
43. W. Liu, Y. Wen, Z. Yu, M. Li, B. Raj, and L. Song, Sphereface: Deep hypersphere embedding for face recognition, in *Conference on Computer Vision and Pattern Recognition*, vol. 1, 2017.
44. J. Deng, J. Guo, and S. Zafeiriou, ArcFace: Additive angular margin loss for deep face recognition, *arXiv preprint*, arXiv:1801.07698, 2018.
45. H. Wang et al., CosFace: Large margin cosine loss for deep face recognition, in *IEEE Conference on Computer Vision and Pattern Recognition*, 2018.
46. C. Turan, K. M. Lam, and X. He, Soft locality preserving map (SLPM) for facial expression recognition, *arXiv preprint*, arXiv:1801.03754 2018.

Part II

Filter Design and Multirate Signal Processing

Chapter 6

The Ensemble Kalman Filter

Pedro A. M. Fonini and Paulo S. R. Diniz
COPPE/Poli, Universidade Federal do Rio de Janeiro (UFRJ), Rio de Janeiro, Brazil
diniz@smt.ufrj.br

Kalman Filtering is the signal processing tool of choice when an application requires knowledge of the internal, non-observable, state variables of a complex dynamical system. Given the observed output of the system and its mathematical model, the Kalman Filter estimates adaptively, optimally, or near-optimally depending on the linearity of the system, the internal system state. While one of the most important factors for the success of the Kalman Filter is the adaptive estimation of the state-vector covariance matrix, this estimation becomes computationally infeasible once the dimensionality of the state-space becomes too large. The Ensemble Kalman Filter is designed to circumvent this restriction using a sample of estimations of the state variables, whose sample covariance is a low-rank approximation to the underlying covariance matrix. In this chapter, we describe the non-linear Ensemble Kalman Filter and include some illustrative examples exploiting its features.

Learning Approaches in Signal Processing
Edited by Wan-Chi Siu, Lap-Pui Chau, Liang Wang, and Tieniu Tan

ISBN 978-981-4800-50-1 (Hardcover), 978-0-429-06114-1 (eBook)
www.panstanford.com

6.1 Introduction

Mathematical models based on dynamical systems are used in virtually all branches of the natural sciences: They describe the evolution of the size of populations in an ecosystem, chemical reactions in a controlled experiment, the altitude a rigid body in a free fall through the atmosphere, and the voltage difference across a particular circuit element during the operation of an electronic device. As computers become more powerful each year, and our technological needs become more ambitious, the systems used to model these phenomena have become more complex, and so the mathematical tools used to work with these models also need to be updated.

When a signal processing application involves keeping track of non-observable state variables of such models, for example because it needs to predict the output of the model ahead of time, the mathematical tool used is the Kalman Filter [1, 2]. The Kalman Filter,[1] or KF, reads information about the input of the system, and combines it with knowledge about the internal workings of the model to output an estimate of the internal state of the system, along with a covariance matrix for this estimate.

The linear Kalman Filter guarantees that this estimate is optimal, while other variations based on it, like the extended Kalman Filter and the ensemble Kalman Filter, do not. The extended Kalman Filter is used when the dynamical system under consideration is not linear. The methodology is to approximate the system, in each iteration, using a Taylor series, and then apply the standard KF.

The second variation, the Ensemble Kalman Filter (EnKF), is used to avoid expensive computations with huge covariance matrices. The estimation of the covariance matrix for the state-vector mentioned above is one of the most important attributes of the Kalman Filter, and therefore cannot be dropped even though this estimation might sometimes be too computationally expensive. More specifically,

[1]It should be noted that the KF is used for *discrete* dynamical systems. When the system under consideration is continuous in time (for instance, given by a differential equation), it should be discretized before the Kalman Filter can be implemented in a digital machine.

when we need to tackle high-dimensional state-spaces, i.e., when our model complexity requires too many state variables, the problem of estimating this matrix becomes more and more infeasible [3], to the point it cannot be afforded every iteration of the Kalman Filter.

Examples of applications which require such high-dimensional state-spaces are weather forecasting, and control of oil wells. These systems are modeled considering that the value of the relevant physical quantities at each point in space is a state variable. In a weather forecasting application, for instance, the state variables could be temperature, humidity, and atmospheric pressure in a grid of points on the surface of the Earth. It is clear now how big the number of state variables can be. It is not difficult, considering the weather forecasting example above, to imagine a 1000×1000 grid, in order to cover the area of interest and at the same time achieve enough geographical resolution to be used as the discretization of a differential equation. This system would have 3 000 000 state variables.

While the KF and the extended KF maintain an estimate of the vector of state variables, and update this estimate at each iteration, the EnKF maintains an *ensemble* of estimates, which are all updated at each iteration. This ensemble serves a two-fold purpose: While its sample mean can be used as a final estimate of the state vector, its sample covariance is a low-rank approximation of the true state-vector covariance matrix.

6.1.1 Notation and Conventions

In the context of the Kalman Filter, we consider a linear system given by the following formulas:

$$\mathbf{x}(k) = \mathbf{F}\mathbf{x}(k-1) + \mathbf{w}(k)$$
$$\mathbf{y}(k) = \mathbf{H}\mathbf{x}(k) + \mathbf{v}(k)$$

Here, $\mathbf{x}(k)$ is the state-vector at the time instant k. $\mathbf{F}$ and $\mathbf{H}$ are the state-update matrix and the observation matrix, respectively. The observed output of the system is $\mathbf{y}(k)$.

The noise signals $\mathbf{w}(k)$ and $\mathbf{v}(k)$ are added directly to their targets, and may have arbitrary covariance matrices $\mathbf{Q}$ and $\mathbf{R}$. Some authors consider the noise vectors to have an identity covariance matrix, and pre-multiply the vectors by transformation matrices, before adding them to their targets, so that the covariance matrix will be as desired. Both approaches are interchangeable. We choose not to pre-multiply the noise vector by any matrices, in attempt to make the descriptions of the algorithms as clean as possible.

Also, with the same point in mind, we will write $\mathbf{Q}$ instead of $\mathbf{Q}(k)$, and similarly $\mathbf{R}$ instead of $\mathbf{R}(k)$, even though the noise processes might be non-stationary. At last, we write $\mathbf{F}$ instead of $\mathbf{F}(k)$ and $\mathbf{H}$ instead of $\mathbf{H}(k)$, even though the systems might be time-variant.

In the context of the Extended Kalman Filter, we consider non-linear systems:

$$\begin{aligned}\mathbf{x}(k) &= \mathbf{f}(\mathbf{x}(k-1), \mathbf{w}(k))\\ \mathbf{y}(k) &= \mathbf{h}(\mathbf{x}(k), \mathbf{v}(k))\end{aligned}$$

The linear update matrix $\mathbf{F}$ has been replaced by the more general update function $\mathbf{f}$. Also, to be as general as possible, we have replaced the additive noise by source of randomness in the input to the non-linear update function. However, in this presentation we will consider only deterministic non-linear systems in the presence of additive noise:

$$\begin{aligned}\mathbf{x}(k) &= \mathbf{f}(\mathbf{x}(k-1)) + \mathbf{w}(k)\\ \mathbf{y}(k) &= \mathbf{h}(\mathbf{x}(k)) + \mathbf{v}(k)\end{aligned} \tag{6.1}$$

Of course, all of the comments above, written with the update equation in mind, are valid for the observation equation, with $\mathbf{H}$ having been replaced by $\mathbf{h}$.

6.2 The Extended Kalman Filter

The Kalman Filter makes two predictions at each iteration:

(1) Given an unbiased estimate $\widehat{\mathbf{x}}(k-1 \mid k-1)$ of the previous state vector $\mathbf{x}(k-1)$, the KF makes an initial estimate $\widehat{\mathbf{x}}(k \mid k-1)$ of the next state vector, $\mathbf{x}(k)$. This step is called the *forecast* step, and $\widehat{\mathbf{x}}(k \mid k-1)$ is called the *a priori estimate* of the system state.

(2) Once the output $\mathbf{y}(k) = \mathbf{h}(\mathbf{x}(k)) + \mathbf{v}(k)$ is known, the KF calculates a final estimate for the state vector, $\widehat{\mathbf{x}}(k \mid k)$, called the *a posteriori estimate*. This second step is the *data assimilation* step.

In this section, we will describe how the extended KF realizes these two steps. The linear KF is a special case of the extended KF, namely when $\mathbf{f}(\mathbf{x}) = \mathbf{F}\mathbf{x}$ and $\mathbf{h}(\mathbf{x}) = \mathbf{H}\mathbf{x}$.

The forecast step is simple: The estimate $\widehat{\mathbf{x}}(k \mid k-1)$ is defined as the expected value of the state vector:

$$\begin{aligned}\widehat{\mathbf{x}}(k \mid k-1) &= \mathsf{E}[\mathbf{x}(k)] \\ &= \mathsf{E}[\mathbf{f}(\mathbf{x}(k-1)) + \mathbf{w}(k)] \\ &= \mathsf{E}[\mathbf{f}(\mathbf{x}(k-1))] + \mathsf{E}[\mathbf{w}(k)]\,. \end{aligned} \tag{6.2}$$

The expectations above are tacitly conditioned to the information that we have available until now. This information is comprised of the outputs of the system up to the instant $k-1$, that is, $\mathbf{y}(k-1), \mathbf{y}(k-2), \mathbf{y}(k-3)$, and so on. The only assumption we will make about the noise process $\mathbf{w}(k)$ is that it is zero-mean: Any non-zero mean that the noise presents can be incorporated into $\mathbf{f}$. Hence, the second expected value above, $\mathsf{E}[\mathbf{w}(k)]$, can be dropped.

In order to calculate $\mathsf{E}[\mathbf{f}(\mathbf{x}(k-1))]$, we use the first-order Taylor expansion for $\mathbf{f}$:

$$\begin{aligned}\mathbf{f}(\mathbf{x}(k-1)) \approx\ & \mathbf{f}\left(\widehat{\mathbf{x}}(k-1 \mid k-1)\right) \\ & + \mathbf{J}_{\mathbf{f}}(k) \cdot \left(\mathbf{x}(k-1) - \widehat{\mathbf{x}}(k-1 \mid k-1)\right). \end{aligned} \tag{6.3}$$

Here, $\mathbf{J}_{\mathbf{f}}(k) = \mathbf{f}'\left(\widehat{\mathbf{x}}(k-1 \mid k-1)\right)$ is the Jacobian matrix of $\mathbf{f}$ evaluated at $\widehat{\mathbf{x}}(k-1 \mid k-1)$. In the linear KF, we just plug in $\mathbf{J}_{\mathbf{f}}(k) = \mathbf{F}$. Taking the expected value in both sides of the above equation:

$$\mathsf{E}[\mathbf{f}(\mathbf{x}(k-1))] \approx \mathbf{f}\left(\widehat{\mathbf{x}}(k-1 \mid k-1)\right),$$

since the estimate $\widehat{\mathbf{x}}(k-1 \mid k-1)$ is unbiased. Going back to equation (6.2), we arrive at the forecast equation for the extended Kalman Filter:

$$\widehat{\mathbf{x}}(k \mid k-1) = \mathbf{f}\left(\widehat{\mathbf{x}}(k-1 \mid k-1)\right). \tag{6.4}$$

It is possible to implement higher-order Kalman Filters by adding more terms to the first-order Taylor expansion above. As an example, we show how to calculate the second-order update equation for the Kalman Filter. The second-order Taylor expansion for $\mathbf{f}$ is

$$\mathbf{f}(\mathbf{x}(k-1)) \approx \mathbf{f}(\widehat{\mathbf{x}}(k-1 \mid k-1)) + \mathbf{J_f}(k) \cdot \boldsymbol{\varepsilon}(k-1) + \mathbf{s} \tag{6.5}$$

where

$$\boldsymbol{\varepsilon}(k-1) = \mathbf{x}(k-1) - \widehat{\mathbf{x}}(k-1 \mid k-1). \tag{6.6}$$

and $\mathbf{s}$ is the second-order correction vector. This vector's entries are given by

$$s_i = \frac{1}{2}\boldsymbol{\varepsilon}(k-1)^T \mathbf{H}_\mathbf{f}^{(i)}(k)\, \boldsymbol{\varepsilon}(k-1) \tag{6.7}$$

where $\mathbf{H}_\mathbf{f}^{(i)}(k) = f_i''(\widehat{\mathbf{x}}(k-1 \mid k-1))$ is the Hessian matrix of the i-th coordinate of $\mathbf{f}$, evaluated at $\widehat{\mathbf{x}}(k-1 \mid k-1)$. In the linear case, this is zero, of course. The only thing we are missing in the above equation is the expected value of $\mathbf{s}$. We will show the calculations with the time-index omitted. Using the spectral decompositions $\mathbf{H}_\mathbf{f}^{(i)} = \mathbf{U}^{(i)}\boldsymbol{\Lambda}^{(i)}\mathbf{U}^{(i)T}$, where $\boldsymbol{\Lambda}^{(i)}$ are diagonal and $\mathbf{U}^{(i)}$ are orthogonal, we can write

$$\begin{aligned} 2s_i &= \boldsymbol{\varepsilon}^T \mathbf{H}_\mathbf{f}^{(i)} \boldsymbol{\varepsilon} = \left(\boldsymbol{\varepsilon}^T \mathbf{U}^{(i)}\right) \boldsymbol{\Lambda}^{(i)} \left(\mathbf{U}^{(i)T} \boldsymbol{\varepsilon}\right) \\ &= \sum_j \lambda_j^{(i)} \left(\mathbf{u}_j^{(i)T} \boldsymbol{\varepsilon}\right)^2 = \sum_j \lambda_j^{(i)} \mathbf{u}_j^{(i)T} \boldsymbol{\varepsilon}\boldsymbol{\varepsilon}^T \mathbf{u}_j^{(i)} \end{aligned} \tag{6.8}$$

and therefore

$$\mathsf{E}[s_i] = \frac{1}{2}\mathsf{E}\left[\boldsymbol{\varepsilon}^T \mathbf{H}_\mathbf{f}^{(i)} \boldsymbol{\varepsilon}\right] = \frac{1}{2}\sum_j \lambda_j^{(i)} \mathbf{u}_j^{(i)T} \mathbf{P} \mathbf{u}_j^{(i)} \tag{6.9}$$

where $\lambda_j^{(i)}$ are the eigenvalues of $\mathbf{H}_\mathbf{f}^{(i)}$, $\mathbf{u}_j^{(i)}$ are the corresponding normalized eigenvectors, and $\mathbf{P} = \mathsf{E}\left[\boldsymbol{\varepsilon}\boldsymbol{\varepsilon}^T\right]$ is the covariance matrix of the estimation error at the instant $k-1$. Finally, we can write the full, component-wise, update equation for the second-order extended Kalman Filter:

$$\begin{aligned} \widehat{x}_i(k \mid k-1) = {} & f_i(\widehat{\mathbf{x}}(k-1 \mid k-1)) \\ & + \frac{1}{2}\sum_j \lambda_j^{(i)} \mathbf{u}_j^{(i)T} \mathbf{P}(k-1 \mid k-1)\, \mathbf{u}_j^{(i)}, \end{aligned} \tag{6.10}$$

where the notation $\mathbf{P}(k-1 \mid k-1)$ indicates that this is the covariance matrix of the error of the *a posteriori* estimate. This matrix should have been calculated in the previous iteration.

It goes without saying that calculating a Hessian matrix for each state variable, along with their spectral decompositions, at each iteration is hardly feasible in most online applications—specially the large-matrix applications which are the focus of this chapter. From now on, we will consider exclusively the first-order Kalman Filter.

As mentioned above, the covariance matrix of $\widehat{\mathbf{x}}(k-1 \mid k-1)$ has been calculated in the previous iteration:

$$\text{Var}\left[\widehat{\mathbf{x}}(k-1 \mid k-1)\right] = \mathbf{P}(k-1 \mid k-1)$$

We continue the forecast step of the extended KF by updating this matrix to account for the fact that we have not yet received any information about this iteration. This updated matrix will be the covariance of the state $\mathbf{x}(k)$. The mean of this state vector—given the information that we have up to now—is by definition $\widehat{\mathbf{x}}(k \mid k-1)$.

We recall the Taylor expansion used above:

$$\mathbf{x}(k) = \mathbf{f}\left(\widehat{\mathbf{x}}(k-1 \mid k-1)\right) + \mathbf{J}_{\mathbf{f}}(k) \cdot \boldsymbol{\varepsilon}(k-1) + \mathbf{w}(k) \tag{6.11}$$

Since the transformation $\mathbf{f}$ and the previous *a posteriori* estimate $\widehat{\mathbf{x}}(k-1 \mid k-1)$ are known, the first term above is deterministic. It is, as already noted, the mean of $\mathbf{x}(k)$, and this can also be seen by noting that the other two terms in the right-hand side of (6.11) have zero mean. Putting these observations together with the fact that the state-update noise $\mathbf{w}(k)$ is independent of all other variables, we can calculate the new *a priori* state-vector covariance matrix:

$$\begin{aligned}
\mathbf{P}(k \mid k-1) &= \text{Var}\left[\mathbf{x}(k)\right] \\
&= \text{Var}\left[\mathbf{J}_{\mathbf{f}}(k) \cdot \boldsymbol{\varepsilon}(k-1) + \mathbf{w}(k)\right] \\
&= \text{Var}\left[\mathbf{J}_{\mathbf{f}}(k) \cdot \boldsymbol{\varepsilon}(k-1)\right] + \text{Var}\left[\mathbf{w}(k)\right] \\
&= \mathbf{J}_{\mathbf{f}}(k)\,\text{Var}\left[\boldsymbol{\varepsilon}(k-1)\right]\mathbf{J}_{\mathbf{f}}(k)^T + \text{Var}\left[\mathbf{w}(k)\right] \\
&= \mathbf{J}_{\mathbf{f}}(k)\,\mathbf{P}(k-1 \mid k-1)\,\mathbf{J}_{\mathbf{f}}(k)^T + \mathbf{Q}
\end{aligned}$$

We now calculate the output forecast:

$$\widehat{\mathbf{y}}(k \mid k-1) = \mathsf{E}[\mathbf{y}(k)]$$

Again, the expected value is tacitly conditioned to the information available up to the time instant $k-1$. Using now a Taylor expansion for $\mathbf{h}$ around $\widehat{\mathbf{x}}(k \mid k-1)$:

$$\widehat{\mathbf{y}}(k \mid k-1) = \mathbf{h}\left(\widehat{\mathbf{x}}(k \mid k-1)\right).$$

The last calculation needed in the forecast step of the algorithm is the covariance matrix of the output. The calculations are analogous to those of $\mathbf{P}(k \mid k-1)$ shown above, and the result is

$$\mathbf{S}(k) = \mathbf{J_h}(k)\,\mathbf{P}(k \mid k-1)\,\mathbf{J_h}(k)^T + \mathbf{R} \tag{6.12}$$

The Jacobian matrix $\mathbf{J_h}(k)$ of $\mathbf{h}$ is evaluated at $\widehat{\mathbf{x}}(k \mid k-1)$.

The purpose of this first step, the *forecast*, was to use our knowledge of the dynamical system model to compute an *a priori* estimate of the next state-vector. This estimate is blind in the sense that we do not have access to absolutely any information about the value of the update noise $\mathbf{w}(k)$ at the time we make the prediction. If we were to run the Kalman filter using only forecast steps as the one described above, this lack of information about the update noise would accumulate and, sooner or later, our state vector estimate would be completely unrelated to the true system state. We would have no way to get information about the true system state, because what we have described is an open-loop system.

This missing information about the update noise will be gained, implicitly, when we measure the value of the system output $\mathbf{y}(k)$. The goal of second step of the Kalman iteration, named *data assimilation*, is now clear: To use this information to close the loop left open in the forecast step. Rephrasing, what we want is to obtain a new unbiased estimate $\widehat{\mathbf{x}}(k \mid k)$ of $\mathbf{x}(k)$, optimal in the sense of having the least possible variance given the information encoded in $\mathbf{y}(k)$. This new $\widehat{\mathbf{x}}(k \mid k)$ will be called the *a posteriori* estimate.

Considering a linear estimator, the zero-bias restriction implies the following structure for the estimation formula (after applying the usual assumption that the non-linear functions can be substituted by their first-order Taylor expansions):

$$\widehat{\mathbf{x}}(k \mid k) = \widehat{\mathbf{x}}(k \mid k-1) + \mathbf{K}(k) \cdot \left(\mathbf{y}(k) - \widehat{\mathbf{y}}(k \mid k-1)\right)$$

The matrix $\mathbf{K}(k)$ is called the *Kalman gain*, and we are free to choose it. As already mentioned, we will optimize it to yield the least possible variance in the state vector estimate. Since the state vector has more than one coordinate, what will actually be minimized is the *total* final (*a posteriori*) uncertainty, represented by the trace of the covariance matrix $\mathbf{P}(k \mid k)$ (that is, the sum of the variances of each state variable). The reader should note that this is a flexible point of the algorithm: If the application requires, for example, that some specific state variables be estimated more accurately than others, we could use a weight vector to ponder this optimization step.

The covariance matrix $\mathbf{P}(k \mid k)$ is by definition the covariance of $\mathbf{x}(k)$. It is instructive to calculate the formula for the *a posteriori* estimation error:

$$
\begin{aligned}
\mathbf{x}(k) - \widehat{\mathbf{x}}(k \mid k) &= \mathbf{x}(k) - \left(\widehat{\mathbf{x}}(k \mid k-1) + \mathbf{K}(k) \cdot [\mathbf{y}(k) - \widehat{\mathbf{y}}(k \mid k-1)]\right) \\
&= \mathbf{x}(k) - \left(\widehat{\mathbf{x}}(k \mid k-1) + \mathbf{K}(k) \cdot [\mathbf{h}(\mathbf{x}(k)) + \mathbf{v}(k) - \mathbf{h}(\widehat{\mathbf{x}}(k \mid k-1))]\right) \\
&= \left(\mathbf{x}(k) - \widehat{\mathbf{x}}(k \mid k-1)\right) - \mathbf{K}(k) \cdot [\mathbf{h}(\mathbf{x}(k)) - \mathbf{h}(\widehat{\mathbf{x}}(k \mid k-1))] - \mathbf{K}(k)\,\mathbf{v}(k) \\
&= \left(\mathbf{x}(k) - \widehat{\mathbf{x}}(k \mid k-1)\right) - \mathbf{K}(k)\,\mathbf{J}_{\mathbf{h}}(k) \cdot [\mathbf{x}(k) - \widehat{\mathbf{x}}(k \mid k-1)] - \mathbf{K}(k)\,\mathbf{v}(k) \\
&= \left(\mathbf{I} - \mathbf{K}(k)\,\mathbf{J}_{\mathbf{h}}(k)\right)\left(\mathbf{x}(k) - \widehat{\mathbf{x}}(k \mid k-1)\right) - \mathbf{K}(k)\,\mathbf{v}(k).
\end{aligned}
$$

We have expressed the *a posteriori* estimation error as a combination of the *a priori* error $\mathbf{x}(k) - \widehat{\mathbf{x}}(k \mid k-1)$ and the measurement noise $\mathbf{v}(k)$. The Kalman gain shapes both of the terms in this combination. Qualitatively, if the Kalman gain is too low, the *a posteriori* error will be close to the *a priori* error, which is clearly non-optimal; if, on the other hand, it is too high, the noise contribution will be amplified.

As mentioned, we want to choose the Kalman gain $\mathbf{K}(k)$ which minimizes the trace of $\mathbf{P}(k \mid k)$. This covariance matrix can be calculated using the combination formula above, together with equation (6.12):

$$
\begin{aligned}
\mathbf{P}(k \mid k) &= \left(\mathbf{I} - \mathbf{K}(k)\,\mathbf{J}_{\mathbf{h}}(k)\right)\mathbf{P}(k \mid k-1)\left(\mathbf{I} - \mathbf{K}(k)\,\mathbf{J}_{\mathbf{h}}(k)\right)^T \\
&\quad + \mathbf{K}(k)\,\mathbf{R}\mathbf{K}(k)^T \\
&= \mathbf{P}(k \mid k-1) - \mathbf{P}(k \mid k-1)\,\mathbf{J}_{\mathbf{h}}(k)^T\,\mathbf{K}(k)^T \\
&\quad - \mathbf{K}(k)\,\mathbf{J}_{\mathbf{h}}(k)\,\mathbf{P}(k \mid k-1) + \mathbf{K}(k)\,\mathbf{S}(k)\,\mathbf{K}(k)^T
\end{aligned}
$$

To find the Kalman gain $\mathbf{K}(k)$ which minimizes the trace of this matrix, we differentiate the trace of the above expression:

$$
\frac{d}{d\mathbf{K}(k)}\{\operatorname{tr}\mathbf{P}(k \mid k)\} = -2\mathbf{P}(k \mid k-1)\,\mathbf{J}_{\mathbf{h}}(k)^T + 2\mathbf{K}(k)\,\mathbf{S}(k)
$$

Since this derivative must be zero, the optimal Kalman gain is

$$
\mathbf{K}(k) = \mathbf{P}(k \mid k-1)\,\mathbf{J}_{\mathbf{h}}(k)^T\,\mathbf{S}(k)^{-1}. \tag{6.13}
$$

The matrix $\mathbf{M}(k) = \mathbf{P}(k \mid k-1)\,\mathbf{J}_{\mathbf{h}}(k)^T$ is called the cross-covariance between the state and the output predictions.

It is instructive to notice [4] that, according to equation (6.13), $\mathbf{K}(k)$ will push high gains to the energy of the update vector in the directions emphasized by $\mathbf{P}(k \mid k-1)$. These are exactly the directions in which the variance $\mathbf{P}$, which represents the known inaccuracies in the prediction of the state vector, is high. In other words, the Kalman gain makes for a data assimilation step update which is more energetic in the directions in which "the algorithm is unsure" about the value of the state vector $\mathbf{x}$, and gives more importance to the measurement error $\mathbf{y}(k) - \widehat{\mathbf{y}}(k \mid k-1)$. Likewise, because of the $\mathbf{S}(k)^{-1}$ term in equation (6.13), the directions in the output space along which the variance of the measurement error is high have a proportionately low impact in the data assimilation step update.

We finish the Kalman iteration by noting that, when $\mathbf{K}(k)$ is the optimal Kalman gain, the formula for the *a posteriori* state covariance reduces to

$$\mathbf{P}(k \mid k) = \mathbf{O}(k)\,\mathbf{P}(k \mid k-1)$$

where we have defined the *observation matrix*:

$$\mathbf{O}(k) = \mathbf{I} - \mathbf{K}(k)\,\mathbf{J_h}(k)$$

6.3 Ensemble Kalman Filter

As well motivated in the introduction to this chapter, the ensemble Kalman Filter (EnKF), introduced in [5], avoids the direct manipulation of the state covariance matrix $\mathbf{P}$. To that end, instead of keeping a single state vector estimate, the algorithm keeps an ensemble of distinct estimates of the same vector. The actual, effective estimate is the ensemble mean.

6.3.1 The Algorithm

We represent the estimate ensemble by a matrix of ensemble members:

$$\widehat{\mathbf{X}}(k \mid k) = \left[\widehat{\mathbf{x}}^{(0)}(k \mid k) \,\middle|\, \cdots \,\middle|\, \widehat{\mathbf{x}}^{(E-1)}(k \mid k)\right]$$

The ensemble mean can be calculated as

$$\overline{\mathbf{x}}(k \mid k) = \frac{1}{E}\widehat{\mathbf{X}}(k \mid k)\,\mathbf{1},$$

where $\mathbf{1}$ is the column-vector of all ones, and E is the ensemble size. We also define the matrix of *anomalies*:

$$\delta\widehat{\mathbf{X}}(k \mid k) = \widehat{\mathbf{X}}(k \mid k) - \overline{\mathbf{x}}(k \mid k)\,\mathbf{1}^T.$$

The columns of the matrix of anomalies are the differences between each ensemble member $\widehat{\mathbf{x}}^{(i)}(k \mid k)$ and the ensemble mean $\overline{\mathbf{x}}(k \mid k)$. The ensemble $\widehat{\mathbf{X}}(k \mid k)$ can be thought of as its mean plus a cloud of anomalies—naturally, the mean of the anomalies is zero.

The main purpose of the matrix of anomalies is to be able to estimate the state covariance matrix:

$$\overline{\mathbf{P}}(k \mid k) = \frac{1}{E-1}\delta\widehat{\mathbf{X}}(k \mid k)\,\delta\widehat{\mathbf{X}}(k \mid k)^T. \tag{6.14}$$

It is important to note, though, that this matrix concerns only the theoretical development of the ensemble Kalman filter; we have already made clear that it will not be directly manipulated during the execution of the algorithm.

Another important observation about equation (6.14) is that it reveals one of the drawbacks of the EnKF when compared to the traditional KF: Since the ensemble size is less than the number of state variables (otherwise there would be no computational gains in using the EnKF), the estimate $\overline{\mathbf{P}}(k \mid k)$ is a low-rank approximation to the true state covariance matrix. In other words, since the number of ensemble vectors is *less* than the dimensionality of the space in which they are embedded, the information that they give about the second-order statistics of the subjacent random variable is incomplete: The cloud of ensemble members does not really cover all dimensions around its center, because it is embedded in a low-dimensional hyper-plane. This lack of information can lead the ensemble Kalman Filter to underestimate the state vector covariance matrix, building up *overconfidence*, until it becomes unstable.

In the forecast step, the EnKF updates, independently, each ensemble member:

$$\widehat{\mathbf{X}}(k \mid k-1) = \mathbf{f}\left(\widehat{\mathbf{X}}(k-1 \mid k-1)\right) + \mathbf{W}(k).$$

In this equation, $\mathbf{f}(\widehat{\mathbf{X}}(k-1 \mid k-1))$ is the matrix in which the i-th column is the result of applying the function $\mathbf{f}$ to the i-th column of $\widehat{\mathbf{X}}(k-1 \mid k-1)$. Also, the columns of the matrix $\mathbf{W}(k)$ are E artificial realizations of the noise $\mathbf{w}(k)$; this is meant to update our uncertainty of the state vector, and can be done in more sophisticated manners by considering, for instance, covariance inflation [3] and the time auto-correlation of the noise [6].

The output forecast happens in a similar manner:

$$\widehat{\mathbf{Y}}(k \mid k-1) = \mathbf{h}\left(\widehat{\mathbf{X}}(k \mid k-1)\right) + \mathbf{V}(k).$$

Given $\widehat{\mathbf{Y}}(k \mid k-1)$, we can also compute the anomalies observed in the output:

$$\delta\widehat{\mathbf{Y}}(k \mid k-1) = \widehat{\mathbf{Y}}(k \mid k-1)\left(\mathbf{I} - \frac{1}{E}\mathbf{1}\mathbf{1}^T\right).$$

For the data assimilation step, the first thing we need is the Kalman gain $\mathbf{K}(k)$. According to equation (6.13), we need the matrices of output covariance, $\mathbf{S}(k)$, and of cross-covariance between state and output $\mathbf{M}(k)$. These matrices can be estimated from the ensemble anomalies calculated above:

$$\overline{\mathbf{M}}(k) = \frac{1}{E-1}\delta\widehat{\mathbf{X}}(k \mid k-1)\,\delta\widehat{\mathbf{Y}}(k \mid k-1)^T$$

$$\overline{\mathbf{S}}(k) = \frac{1}{E-1}\delta\widehat{\mathbf{Y}}(k \mid k-1)\,\delta\widehat{\mathbf{Y}}(k \mid k-1)^T$$

This suggests the following formula for the Kalman gain:

$$\mathbf{K}(k)^T \approx \left(\delta\widehat{\mathbf{Y}}\delta\widehat{\mathbf{Y}}^T\right)^{-1}\delta\widehat{\mathbf{Y}}\delta\widehat{\mathbf{X}}^T. \tag{6.15}$$

In the above formula, we have omitted the parenthesis with the time-instants in $\delta\widehat{\mathbf{X}}(k \mid k-1)$ and $\delta\widehat{\mathbf{Y}}(k \mid k-1)$.

The right-hand side of equation (6.15) can be seen as the solution to the least-squares system:

$$\delta\widehat{\mathbf{Y}}(k \mid k-1)^T\,\mathbf{K}(k)^T = \delta\widehat{\mathbf{X}}(k \mid k-1)^T. \tag{6.16}$$

Since the ensemble size can sometimes be larger than the number of system outputs, this system can be under-determined. In this case, the matrix $\delta\widehat{\mathbf{Y}}(k \mid k-1)\,\delta\widehat{\mathbf{Y}}(k \mid k-1)^T$ is singular. For that reason, we use the following, more general, equation for the Kalman gain:

$$\mathbf{K}(k)^T \approx \delta\widehat{\mathbf{Y}}(k \mid k-1)^{\dagger}\,\delta\widehat{\mathbf{X}}(k \mid k-1)^T, \tag{6.17}$$

where $\delta\widehat{\mathbf{Y}}(k \mid k-1)^{\dagger}$ is the Moore–Penrose inverse of $\delta\widehat{\mathbf{Y}}(k \mid k-1)$.

Finally, having calculated the Kalman gain, we perform the *a posteriori* estimation of the state vector ensemble:

$$\widehat{\mathbf{X}}(k \mid k) = \widehat{\mathbf{X}}(k \mid k-1) + \mathbf{K}(k)\left(\mathbf{y}(k)\,\mathbf{1}^T - \widehat{\mathbf{Y}}(k \mid k-1)\right).$$

6.3.2 Complexity Analysis

It is not difficult to analyze both algorithms, with and without the ensemble, and understand how the EnKF reduces the amount of computation necessary. Let us call N and M the dimensionalities of the input and output spaces, respectively (in other words, N and M are the lengths of $\mathbf{x}$ and $\mathbf{y}$). Also, as above, let E be the ensemble size.

Four assumptions will ease our task. We assume that: multiplications and additions are equally expensive; the quantities N, M, E obey the inequalities $N > M$ and $N > E$; the evaluations of the nonlinear functions $\mathbf{f}$ and $\mathbf{h}$ and their respective Jacobian matrices have negligible computational costs when compared to the cost of the rest of the Kalman iteration; and, finally, we assume that the noise covariance matrices do not change with time[2].

The computational complexity of the extended Kalman filter is dominated by the update equation for the covariance matrix $\mathbf{P}$ [3], which boils down to $O(N^3)$ operations each loop. Most other steps have complexities of third-order polynomials in N and M (e.g., $O(N^2M + NM^2)$). Since we are considering $N > M$, the asymptotic order of one extended Kalman iteration is $O(N^3)$

The most complicated part of the EnKF is the computation of the Kalman gain, which involves computing the Moore–Penrose inverse of an $M \times E$ matrix. This is usually done via an a singular value

[2]This last requirement is because the task of generating artificial noise samples $\mathbf{W}$ given the covariance $\mathbf{Q}$ requires computing a Cholesky factorization of $\mathbf{Q}$, which should be done beforehand.

decomposition (SVD) [7], which has a complexity of $O(ME^2)$ if $M < E$, or $O(M^2E)$ if $M > E$. Because the numbers are small, thanks to the fact that the ensemble size is usually much smaller than N, this is not even close to $O(N^3)$. In fact, we called this the most *complicated* step, avoiding the word complex, because the actual computational complexity of the EnKF loop is dominated by the generation of the artificial noise matrix $\mathbf{W}$, which is of the order $O(N^2E)$.

Comparing the asymptotic complexities of both methods, we conclude that the operation count gain of moving from the classical extended Kalman filter to the Ensemble Kalman Filter is of the order $O(E\,N^2/N^3) = O(E/N)$. As will be shown in the computational examples, this ratio can be made very low when N grows large.

As a final remark, we mention that the assumptions mentioned above are not very restrictive. The lack of importance of the time of evaluation for the non-linear functions is perhaps the most dubious. If the application requires that it be taken into account, the extended KF iteration evaluates $\mathbf{f}$, $\mathbf{h}$, $\mathbf{J_f}$, and $\mathbf{J_h}$ once each, while the EnKF iteration evaluates $\mathbf{f}$ and $\mathbf{h}$ E times each, but makes no use of the Jacobians. As for the time invariance of the noise covariance, if it does change with time, a Cholesky factorization in each EnKF iteration would have a dramatic impact in the total complexity of the iteration. It can, however, be avoided if the change is slow enough, so that the update to the factorization can be done sparsely.

6.4 Computational Examples

We will now inspect the application of the Kalman filter varieties described above to two non-linear systems described by equations of the form given by (6.1).

6.4.1 System I: Particle around the Origin

The first studied system can be described by the following functions:

$$\mathbf{f}(\mathbf{x}) = A(\|\mathbf{x}\|)\,\mathbf{R}(\theta(\|\mathbf{x}\|))\,\mathbf{x}$$

$$\mathbf{h}(\mathbf{x}) = \mathbf{h}\left(\begin{bmatrix} x_0 \\ x_1 \end{bmatrix}\right) = \begin{bmatrix} x_1/\|\mathbf{x}\| \\ x_0 \end{bmatrix}$$

where we are using the following definitions for convenience:

$$
\begin{aligned}
A(\|\mathbf{x}\|) &= 1 + \lambda \left(\frac{1}{\|\mathbf{x}\|} - 1 \right) \\
\mathbf{R}(\theta(\|\mathbf{x}\|)) &= \begin{bmatrix} \cos\theta(\|\mathbf{x}\|) & -\sin\theta(\|\mathbf{x}\|) \\ \sin\theta(\|\mathbf{x}\|) & \cos\theta(\|\mathbf{x}\|) \end{bmatrix} \\
\theta(\|\mathbf{x}\|) &= \theta_0 \|\mathbf{x}\| \exp\left\{ \frac{1 - \|\mathbf{x}\|^2}{2} \right\} \\
\theta_0 &= 2\pi/20 \\
\|\mathbf{x}\| &= \sqrt{x_0^2 + x_1^2} \\
\lambda &= 1/100
\end{aligned}
$$

This system describes the dynamics of a particle moving around the origin in the plane $\mathbb{R}^2$. Its angular velocity, given by $\theta(\|\mathbf{x}\|)$ in radians per iteration, depends on its distance to the origin: This velocity is maximal (equal to θ_0) when the distance is unitary, and it is zero when the distance to the origin is 0 or infinite. Also, the term $A(\|\mathbf{x}\|)$ pulls the particle towards the unit circle, thus making the system converge (when there is no noise) to the cyclo-stationary state in which the particle just runs around the unit circle. The presence of noise pushes the particle away from the unit circle, preventing the system from staying in a stationary state for too long.

6.4.1.1 Calculating the derivatives

In order to apply the extended Kalman filter to this system, we need the expressions for the derivatives of **f** and **h**. We can start by rewriting the expression for **f** without the parenthesis with function arguments, and defining some intermediate variables:

$$
\begin{aligned}
\mathbf{f} &= A\mathbf{R}\mathbf{x} \\
r &= \|\mathbf{x}\| \\
E &= \exp\frac{1 - r^2}{2} \\
\theta &= \theta_0 r E \\
c &= \cos\theta \\
s &= \sin\theta
\end{aligned}
$$

With these definitions, differentiating $\mathbf{f} = A\mathbf{Rx}$ is just a matter of applying the chain rule repeatedly. We first calculate the intermediate derivatives:

$$\frac{d\theta}{dr} = \theta_0 \left[1 - r^2\right] E$$

$$\frac{dA}{dr} = -\lambda \frac{1}{r^2}$$

$$\frac{dr}{dx_i} = \frac{x_i}{r}$$

We are now in position to calculate the elements of the matrix of derivatives of $\mathbf{f}$:

$$\begin{aligned}
\frac{df_0}{dx_0} &= \frac{d}{dx_0}\{A \cdot (x_0 c - x_1 s)\} \\
&= A \cdot \frac{d}{dx_0}\{x_0 c - x_1 s\} + \frac{dA}{dx_0} \cdot (x_0 c - x_1 s) \\
&= A \cdot \left(c + (-x_0 s - x_1 c)\frac{d\theta}{dr}\frac{dr}{dx_0}\right) + \frac{dA}{dr}\frac{dr}{dx_0} \cdot (x_0 c - x_1 s)
\end{aligned}$$

The rest of the calculations are similar, and the results can be summarized in the following formulas:

$$\begin{aligned}
\frac{df_0}{dx_0} &= A \cdot \left(c + (-x_0 s - x_1 c)\frac{d\theta}{dr}\frac{dr}{dx_0}\right) + \frac{dA}{dr}\frac{dr}{dx_0} \cdot (x_0 c - x_1 s) \\
\frac{df_0}{dx_1} &= A \cdot \left(-s + (-x_0 s - x_1 c)\frac{d\theta}{dr}\frac{dr}{dx_1}\right) + \frac{dA}{dr}\frac{dr}{dx_1} \cdot (x_0 c - x_1 s) \\
\frac{df_1}{dx_0} &= A \cdot \left(s + (\ x_0 c - x_1 s)\frac{d\theta}{dr}\frac{dr}{dx_0}\right) + \frac{dA}{dr}\frac{dr}{dx_0} \cdot (x_0 s + x_1 c) \\
\frac{df_1}{dx_1} &= A \cdot \left(c + (\ x_0 c - x_1 s)\frac{d\theta}{dr}\frac{dr}{dx_1}\right) + \frac{dA}{dr}\frac{dr}{dx_1} \cdot (x_0 s + x_1 c)
\end{aligned}$$

Since the formulas for $\mathbf{h}$ are simpler:

$$h_0 = x_1/r, \quad h_1 = x_0$$

its derivatives are also much more trivial to calculate:

$$\begin{aligned}
\frac{dh_0}{dx_0} &= x_1 \cdot \left(-\frac{1}{r^2}\frac{dr}{dx_0}\right) = -\frac{x_1}{r^2}\frac{x_0}{r} = \frac{-x_0 x_1}{r^3} \\
\frac{dh_0}{dx_1} &= x_1 \cdot \left(-\frac{1}{r^2}\frac{dr}{dx_1}\right) + \frac{1}{r} = \frac{1}{r}\left(1 - \frac{x_1^2}{x_0^2 + x_1^2}\right) = \frac{x_0^2}{r^3} \\
\frac{dh_1}{dx_0} &= 1 \\
\frac{dh_1}{dx_1} &= 0
\end{aligned}$$

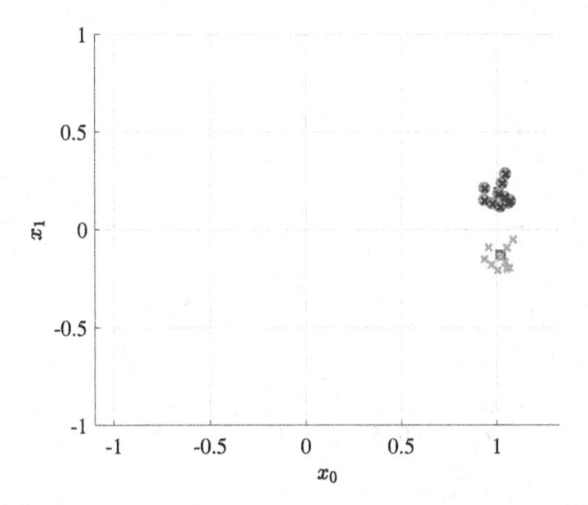

Figure 6.1 An illustration of the dynamics of the particle-around-the-origin system. We show an arbitrary point, a cloud of points around it, and also the images of both the point and the cloud under the non-linear transformation **f**. Figures 6.2 and 6.3 are zooms into this figure, and better explore this issue.

Figures 6.1, 6.2, and 6.3 illustrate the meaning of the first-order approximation given by the derivative of a function **f**. Figures 6.2 and 6.3 are zoomed views into the first. They show that, locally, the behavior of a differentiable function is linear, and therefore can be

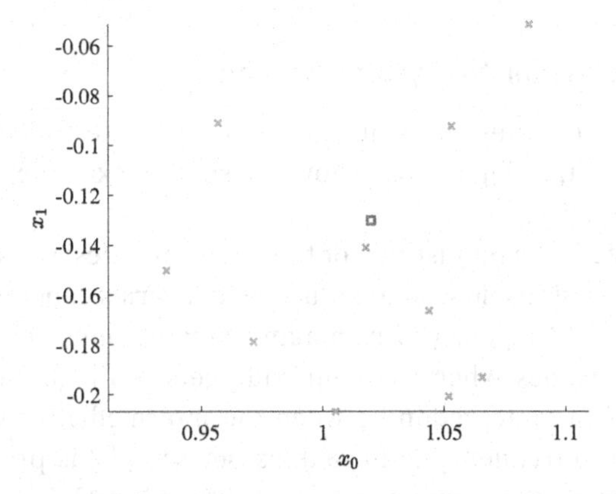

Figure 6.2 This is the lower cloud of points in Fig. 6.1. In the center, the blue, square-shaped point $\mathbf{x}_0$ represents a particular state in which the system could be found. The cloud of yellow, $\times$-shaped points around it, $\mathbf{x}_0 + \boldsymbol{\delta}_i$, represent small deviations from this state.

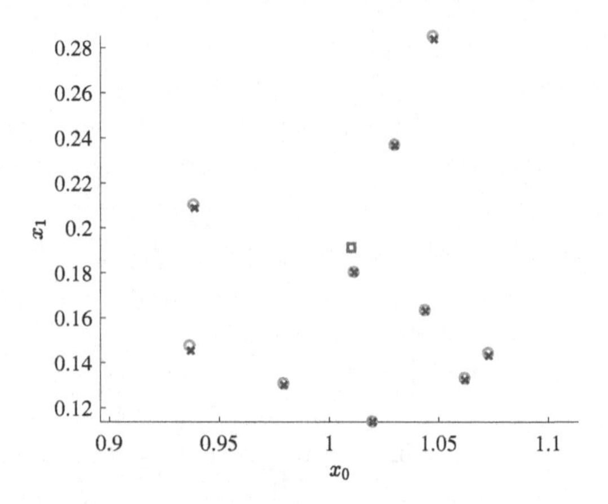

Figure 6.3 This is the upper cloud of points in Fig. 6.1. The red, square-shaped point in the center was calculated by applying the function **f** to the central point in Fig. 6.2; that is, its position is given by $\mathbf{f}(\mathbf{x}_0)$. The purple, $\times$-shaped points around it were also calculated by repeatedly applying **f** to each of the $\times$-shaped points in Fig. 6.2; their positions are therefore $\mathbf{f}(\mathbf{x}_0 + \boldsymbol{\delta}_i)$. The second cloud of points, green and circle-shaped, were calculated by evaluating **f** only once, at $\mathbf{x}_0$, and then using this result to evaluate the estimates $\mathbf{f}(\mathbf{x}_0 + \boldsymbol{\delta}_i) \approx \mathbf{f}(\mathbf{x}_0) + \mathbf{J}_\mathbf{f}(\mathbf{x}_0)\boldsymbol{\delta}_i$.

approximated by a linear transformation (which is the derivative, or the Jacobian).

6.4.1.2 Examining the system dynamics

Figure 6.4 illustrates the behavior of the particle trajectory in a noiseless setting. Figure 6.5 shows a similar example, with the presence of noise.

When we focus on just one of the state variables, we see that it behaves like a sinusoid with a frequency that varies depending on its amplitude. The frequency is maximum when the sinusoid amplitude is 1, and vanishes when the amplitude gets too high, or too low (because of the interaction between the intermediate variables r, E, and θ: The frequency θ, in radians per sample, is proportional to rE, which is maximum when the amplitude r is 1). In Fig. 6.6, for example, the wavelength is noticeably longer when the amplitude is 2, and, also, the wavelength is smallest around the sample 130, when the amplitude is 1.

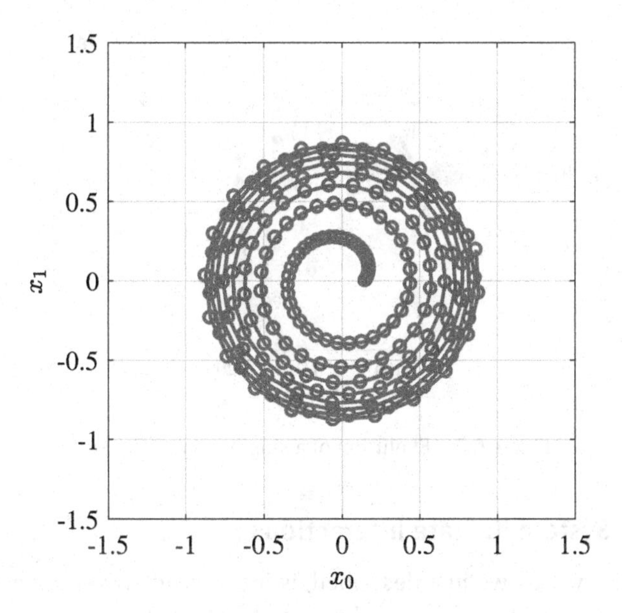

Figure 6.4 State trajectory in a noise-free case.

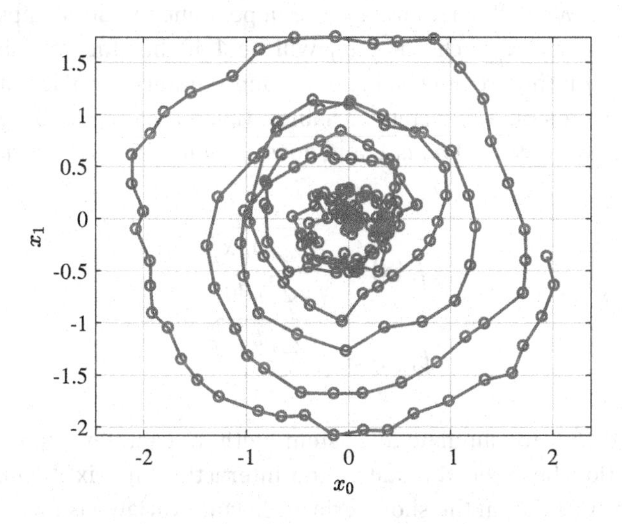

Figure 6.5 State trajectory in a noisy case.

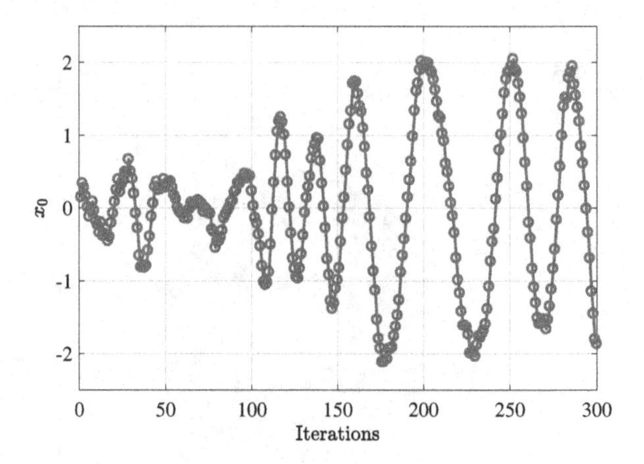

Figure 6.6 Evolution of a single state variable.

6.4.2 System II: State Interactions

System I, which we just described, is intrinsically two-dimensional (a two-dimensional dynamical system has two state variables). We will now analyze a multidimensional system in which each state variable depends on the previous value of a few of the other state variables. We will summarize these dependency relationships in a sparse matrix **A**: Each entry a_{ij} will be 1 if the state variable x_i depends on the previous value of x_j, and 0 otherwise. Each output also depends on a few state variables: Given a matrix **B**, output y_i depends on state x_j if b_{ij} is 1, and does not if it is zero. The formulas are

$$f_i(\mathbf{x}) = \sqrt{\frac{\sum_j a_{ij}x_j^2}{\sum_j a_{ij}}}$$

$$h_i(\mathbf{x}) = \sqrt{\frac{\sum_j b_{ij}x_j^2}{\sum_j b_{ij}}}$$

In order to simulate a system with a complex, non-linear interaction between the states, the interaction matrix **A** must be chosen such that in the short term, each state variable is influenced and influences just a few others, but in the long run all variables depend on all others.

To illustrate, we will analyze a 10-dimensional system with four outputs and interactions given by

$$\mathbf{A} = \begin{bmatrix} 1 & & & 1 & & 1 & 1 & & & \\ & & & 1 & 1 & & & & & \\ 1 & & 1 & 1 & & & & & & \\ & & 1 & & & & 1 & & & \\ & & & & 1 & & 1 & & 1 & \\ & 1 & & & & & & & 1 & \\ 1 & & & & & 1 & & 1 & & \\ & & & & & 1 & 1 & 1 & & 1 \\ & & 1 & 1 & & 1 & & & & 1 \\ 1 & & 1 & 1 & & & & 1 & & \end{bmatrix}$$

$$\mathbf{B} = \begin{bmatrix} & & & & & 1 & 1 & & & 1 \\ 1 & & & & & & 1 & & 1 & \\ & & & 1 & & & & & 1 & \\ & 1 & & 1 & & 1 & 1 & & & \end{bmatrix} \tag{6.18}$$

Since each coordinate f_i is an RMS mean of a few others, in the absence of noise all states will converge to a single value between the least and the greatest between the starting values of the state variables. This can be seen in Fig. 6.7. When noise is added, the concavity of the functions $x_i \mapsto \sqrt{x_i^2 + C}$ and the fact that this constant C accumulates the noises of the previous iterations make each state variable grow unbounded. Figure 6.8 shows this dynamic behavior.

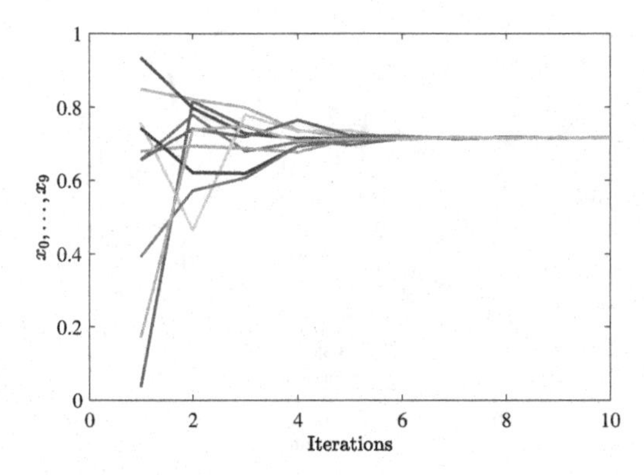

Figure 6.7 Convergence of the state variables in a noiseless setting.

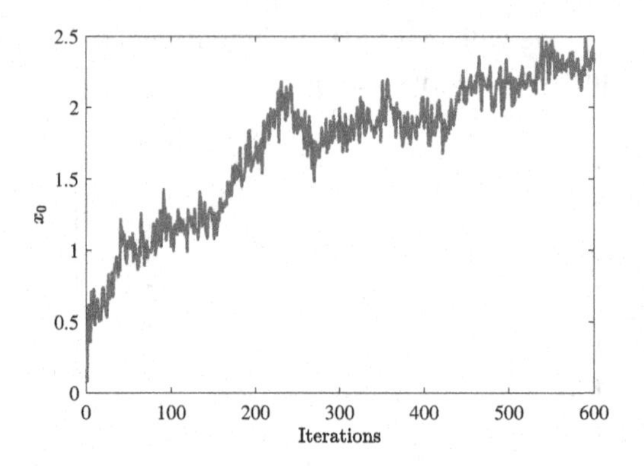

Figure 6.8 Time series of the x_0 state variable in a noisy environment.

6.4.3 Results

In Fig. 6.9, we can see the extended KF and the EnKF tracking the state x_1 of System I. We are using an ensemble of size 6 for the EnKF. The state x_0 is easier to estimate, since it is present explicitly in the output h_1. Figure 6.10 shows an initial sign of instability in the EnKF.

If we repeat the experiment using a smaller ensemble size, the filter becomes explicitly unstable. Figure 6.11 shows an example with only 4 ensemble points. It should be noted that 4 is still above the dimensionality of the state space, showing that the computational efficiency of the ensemble Kalman Filter is only achieved at high dimensionalities.

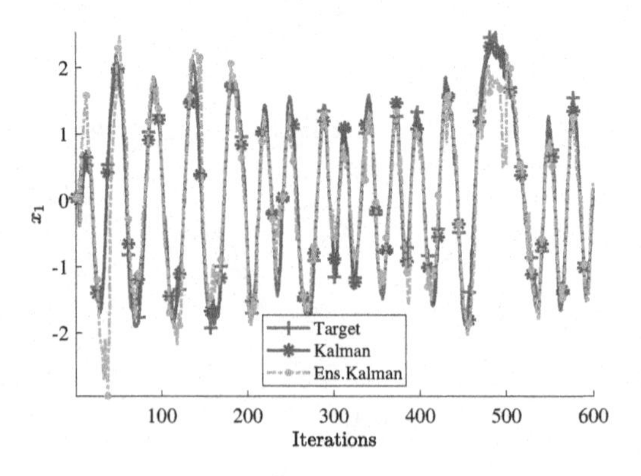

Figure 6.9 Kalman filters for System I.

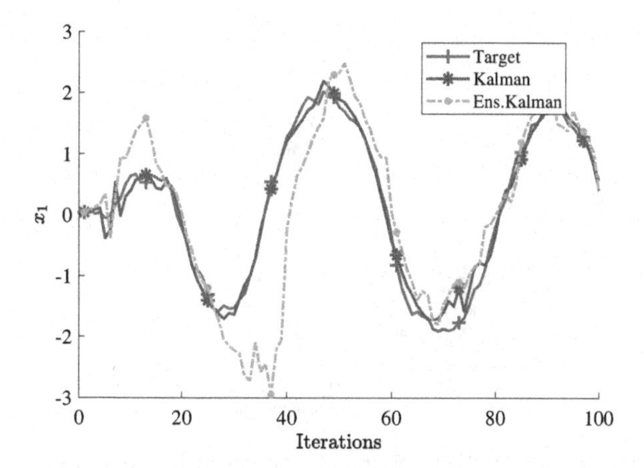

Figure 6.10 Zoom into Fig. 6.9.

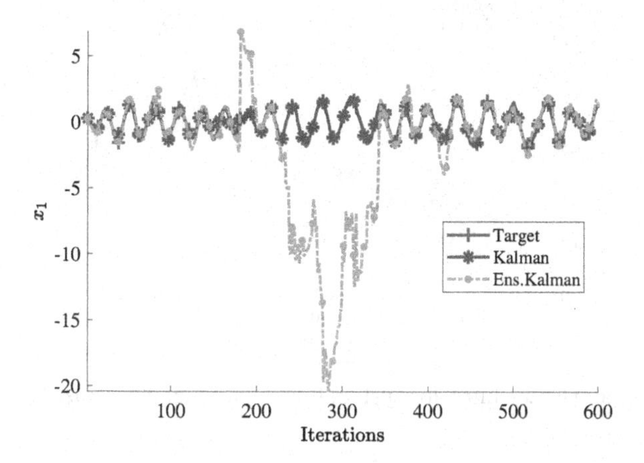

Figure 6.11 Instability in the EnKF.

We finish this exposition by analyzing System II. Figure 6.12 shows the behavior of the ensemble estimation of x_0 when the ensemble size is 7. The filter quickly grows unstable. When the ensemble size is increased, no such instability is observed (Fig. 6.13). Prolonged instabilities were observed in only 3 out of 50 simulations for an ensemble size of 8 (not counting eventual spikes, such as in Fig. 6.14). Using an ensemble size of 9, zero instabilities were observed.

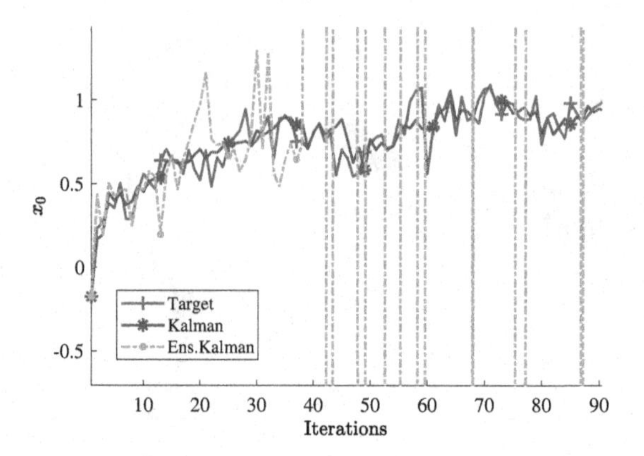

Figure 6.12 EnKF for System II with an ensemble of size 7.

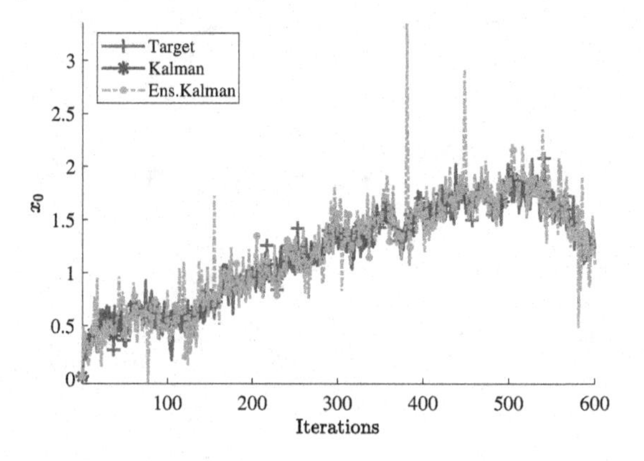

Figure 6.13 EnKF for System II with an ensemble of size 8.

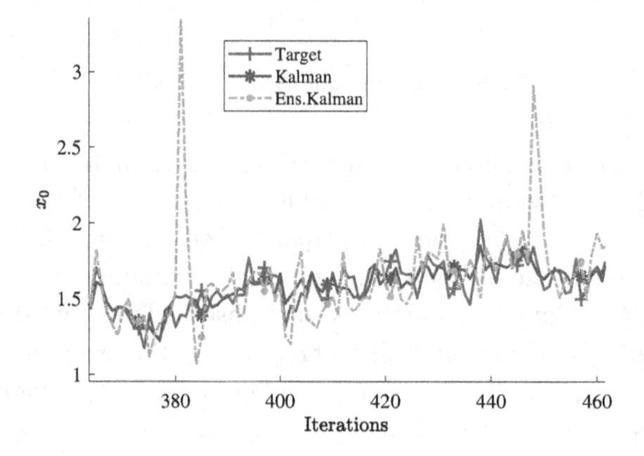

Figure 6.14 Zoom into Fig. 6.13.

As mentioned in Section 6.3.2, in order to benefit from the reduction in computational complexity when substituting the classical KF by the Ensemble KF, the ratio E/N should be small.[3] Although N is usually fixed in a given application, E is a design variable. When E is chosen too small, the ensemble will not correctly represent the state prediction covariance, and the filter becomes unstable; if E is chosen too high, the ratio E/N will also be high, and then there will be no reason to use the Ensemble Kalman Filter at all.

In Systems I and II above, the minimum ratio E/N necessary to prevent instabilities was found to be, through experimentation, 3 and 0.9, respectively, showing that the EnKF exhibits no compelling advantages over the classical KF when N is low.

The dynamics of System II, however, depend only on the matrices $\mathbf{A}$ and $\mathbf{B}$ (see equation (6.18)). They are sized $N \times N$ and $M \times N$, respectively, and we could construct a system with arbitrary N and M by generating random matrices $\mathbf{A}$ and $\mathbf{B}$. We used the following random matrix generation procedure. We start by fixing $M = 2$ and $N = 100$. Then, we generate a random binary matrix $\mathbf{A}$ by choosing, for each row $n = 1, \ldots, 100$, a random integer q_n between 2 and 10. Finally, for each row n we choose q_n of the N elements to be 1; the rest are filled with zeroes. The exact same procedure is used for generating a random $\mathbf{B}$ matrix. With this setting, we can use as few as $E = 15$ ensemble members, which means $E/N = 0.15$. If we repeat for $N = 1000$, $E = 20$ (and therefore $E/N = 0.02$) showed no instabilities during 50 simulations with 600 samples each, indicating that when $N \to \infty$, then computational gain of using the Ensemble Kalman Filter grows.

6.5 Conclusions

This chapter discusses the Ensemble Kalman Filter, an extension of the classical Kalman filtering estimation to the practical case where

[3]We recall that N and M are the dimensionalities of the input and output spaces, respectively, and E is the ensemble size, as defined in Section 6.3.2.

the number of parameters is large, rendering the classical approach to the tracking of the state covariance matrix computationally infeasible. The strategy consists of substituting the state-vector prediction by an ensemble of predictions which can itself be used to approximate the state covariance, eliminating the need for the direct (and expensive) covariance matrix estimation used in the classical case.

This adaptation to large state-space dimensionalities proves useful in natural sciences. In particular, it benefits reservoir modeling and tracking during production [8]. The EnKF can be used, for example, during history-matching, the process of updating the prediction of future daily production rates to match the most recent rates observed. Another area which benefits from the extension is the discipline of weather forecasting [9].

References

1. R. E. Kálmán, A new approach to linear filtering and prediction problems, *Journal of Basic Engineering*, no. 82, pp. 35–45, 1960.
2. R. E. Kálmán, A new approach to linear filtering and prediction problems, *SIAM Journal of Control*, vol. 1, pp. 152–192, 1963.
3. M. Roth, G. Hendeby, C. Fritsche, and F. Gustafsson, The ensemble Kalman filter: A signal processing perspective, *EURASIP Journal on Advances in Signal Processing*, vol. 2017, no. 1, pp. 56, August 2017.
4. Y. Bar-Shalom, X. R. Li, and T. Kirubarajan, *Estimation with Applications to Tracking and Navigation: Theory Algorithms and Software*, John Wiley & Sons, 2001.
5. G. Evensen, Sequential data assimilation with a nonlinear quasi-geostrophic model using Monte Carlo methods to forecast error statistics, *Journal of Geophysical Research: Oceans*, vol. 99, no. C5, pp. 10143–10162, May 1994.
6. G. Evensen, The ensemble Kalman filter: Theoretical formulation and practical implementation, *IEEE Control Systems Magazine*, no. 29, pp. 83–104, 2009.
7. L. N. Trefethen and D. Bau, III, *Numerical Linear Algebra*, Society for Industrial and Applied Mathematics, Philadelphia, PA, USA, 1997.

8. X.-H. Wen and W. H Chen, Real-time reservoir model updating using ensemble Kalman filter with confirming option, *SPE Journal*, no. 11, pp. 431–442, 2006.
9. G. Burgers, P. J. van Leeuwen, and G. Evensen, Analysis scheme in the ensemble Kalman filter, *Monthly Weather Review*, no. 126, pp. 1719–1724, 1998.

Chapter 7

Teaching Programming and Debugging Techniques for Multirate Signal Processing

Fred Harris[a] **and Chris Dick**[b]

[a]*University Department, University of California San Diego, 9500 Gilman Drive, Jacobs Hall, Room 3803, La Jolla, California 92093-0407, USA*
[b]*Xilinx Corp., 2100 Logic Drive, San Jose, California 95124, USA*
fjharris@eng.ucsd.edu, chris.dick@xilinx.com

Much of the material we learn and teach in our DSP classes deal with linear time invariant (LTI) systems. Linearity and time invariance are important properties that led to insight and comprehension through capable analysis and synthesis tools. We learn to appreciate the importance of impulse response, convolution, transfer functions, frequency response, and the Nyquist sampling criterion. When we introduce multirate signal processing, we introduce the process of resampling and we find ourselves in unfamiliar territory. The quote "Toto, I've a feeling we're not in Kansas anymore!" comes to mind. Dorothy was right! We are now in the land of linear time varying (LTV) systems. In this land, we embed one or more sample rate changes in the signal flow path, which result in systems with multiple impulse responses. We become uncomfortably aware

Learning Approaches in Signal Processing
Edited by Wan-Chi Siu, Lap-Pui Chau, Liang Wang, and Tieniu Tan

ISBN 978-981-4800-50-1 (Hardcover), 978-0-429-06114-1 (eBook)
www.panstanford.com

that the system does not have a transfer function and that our analysis and synthesis tools don't work here. We must develop new perspectives while teaching new ways of analyzing and synthesizing LTV systems. We must give our students new approaches and tools to test and probe implementations to verify proper operation as well as to debug and repair faulty realizations of multirate systems.

7.1 Introduction

When we teach digital signal processing we assign homework projects requiring our students to program the algorithms we are teaching. To help them better understand the algorithms and finite arithmetic effects we require the students to write code that mimics the processing hardware. We start with simple programming tasks such as a second order bi-quadratic low pass filter or a cascade of such filters to synthesize a higher order filter. The code must include declarations of input and output data arrays, an array of feed-back and feed-forward coefficients, and an array of internal state registers. It contains a segment that forms one or more input signals to be processed by the algorithm and then the actual algorithm. The algorithm will include an input for-loop that presents successive input signal samples to the process, performs a sum of products between coefficients and register contents of feed-back and of feed-forward paths, inter-stage scaling, output sample scaling, and update of register contents. We instruct the students to apply specific input signals to their algorithm to exercise and demonstrate their performance. The input signals may include an impulse, a step, one or more fixed frequency tones, linear FM sweeps over specific spectral spans, noise, and noisy sinewaves. We ask the student to present time responses and spectra of the various input and output signals as well as for specific internal state signal sequences. We may ask the student to repeat the assignment with quantized coefficients, quantized arithmetic, saturation arithmetic, or with quantized arithmetic coupled to a tough specification such as a large ratio of sample rate to bandwidth.

Figure 7.1 shows an example of an MP3 filter problem; design a small system that accepts data samples at 96 kHz, filters the series

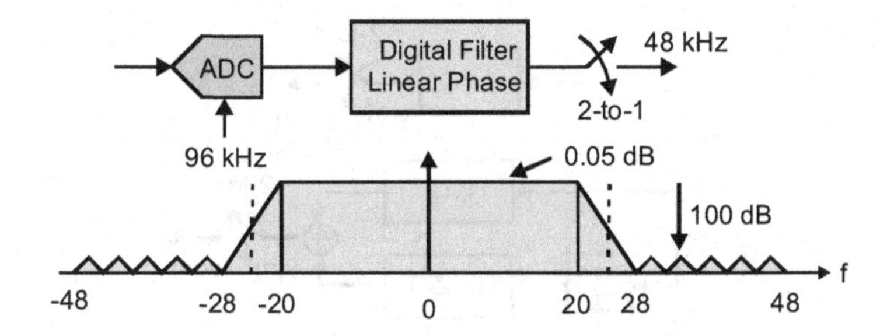

Figure 7.1 Filter design problem with filter specifications.

with a linear phase half band filter, with shown specifications, and reduces the sample rate to 48 kHz. We require the filter process that minimizes the computational workload.

The obvious solution is a half band FIR filter. The Remez algorithm will design a 57-tap filter meeting this specification or using P.P. Vaidyanathan's half band trick we can design a 77-tap true half-band filter with even indexed tap weights equal to zero. Using the even symmetry of the tap weights, the 57-tap filter requires 28 multiplies and the 77-tap filter requires 19 multiplies respectively per output sample. Knowing that we will discard alternate output samples at the filter's output we can avoid forming those outputs by down-sampling prior to the filtering operation. This of course invites aliasing but when performed correctly the aliasing is canceled at the filter output. Correctly means, we form a two-path polyphase filter as shown in (7.1).

$$\begin{aligned} H(Z) = \sum_{n=0}^{N-1} h(n)\, Z^{-n} &= \sum_{n=0}^{N/2-1} h(2n)\, Z^{-2n} + \sum_{n=0}^{N/2-1} h(2n+1)\, Z^{-(2n+1)} \\ &= \sum_{n=0}^{N/2-1} h(2n)\, Z^{-2n} + Z^{-1} \sum_{n=0}^{N/2-1} h(2n+1) Z^{-2n} \\ &= H_0(Z^2) + Z^{-1} H_1(Z^2) \end{aligned} \tag{7.1}$$

We then form the 2-path filter, and invoking the noble identity, pull the 2-to-1 down sampler through the filter paths to reverse the order of filtering and resampling. This partition and reordering of filter and resampler is shown in Figs. 7.2a,b. The block diagram of Fig. 7.2a has a transfer function because it operates at the input rate while the

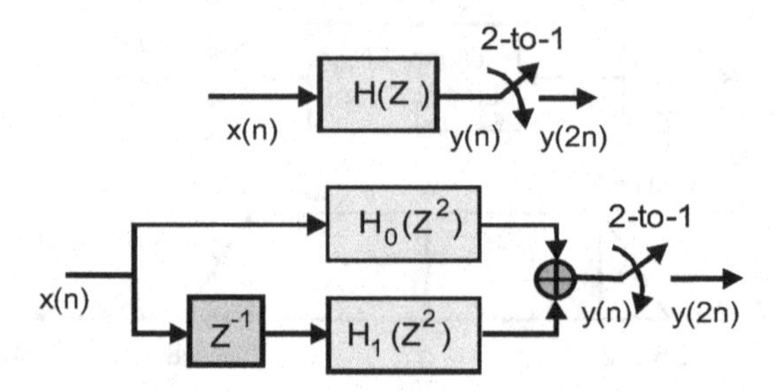

Figure 7.2a Half band filter followed by 2-to-1 down sample and two-path polyphase partition at input sample rate.

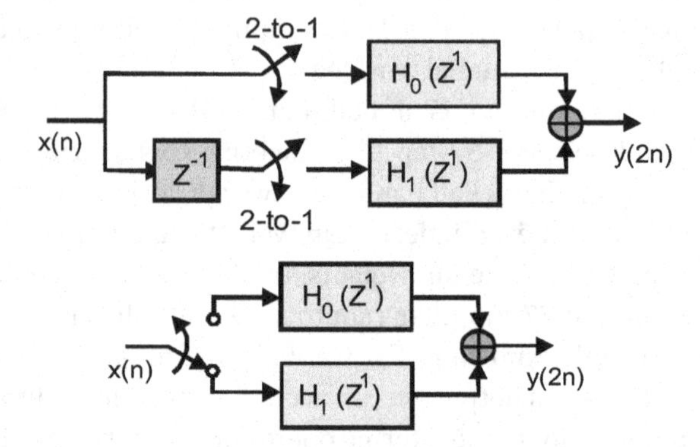

Figure 7.2b Two-path polyphase filter at output sample rate, and input commutator driven two-path filter.

block diagram of Fig. 7.2b does not have a transfer function because of the embedded 2-to-1 resampler. The question we pose here is how does the student probe the filter to verify its proper operation, and how does the student avoid a common programming error that would cause improper operation?

The most common programming error is the student delivering successive samples first to the upper path and then to lower path when in fact the first sample should be delivered to the lower path and the next sample to the upper path. It isn't a coin toss. One option is correct and the other is not. What would guide a student to choose the incorrect option? Well, look at the last line of (7.1) and the upper subplot of Fig. 7.2b. The student properly interprets the Z^{-1} as a delay which means the later, so the path with the delay is offered the later sample and the one without the delay is offered the earlier sample. An easy mistake to make! The proper interpretation is that at time n an input sample arrives and is stored in the delay line and at time $n+1$ when the next input arrives, the two switches close and the current input is delivered to the top path while the previous input is pulled from the delay and is deliver to the lower path. See the *correct* and the *incorrect* script files in Tables 7.1a and 7.1b.

What could a student have done to detect this error? Not surprisingly, the impulse and step responses give no hint as to the correctness of the input commutator operation. We found the best probe signal to be a low frequency in-band sinusoid and a high frequency out-of-band sinusoid to be the best probe signal. Figure 7.3a shows the input and output time series for a low frequency in band sinusoid and for a high frequency out-of-band

Table 7.1a Correct script for 2-path 2-to-1 resampling half band filter

```
h1=[remez(56,[0 20 28 48]/48,[1 1 0 0],[1 150]) 0];
g1=reshape(h1,2,29);

reg=zeros(2,29);
x1=[1 zeros(1,60)];
m=1;
for n=1:2:60
     reg(2,:)=[x1(n)    reg(2,1:28)]; % 1-st sample
     reg(1,:)=[x1(n+1)  reg(1,1:28)]; % 2-nd sample
     y1(m)=reg(1,:)*g1(1,:)'+reg(2,:)*g1(2,:)';
     m=m+1;
end
```

Table 7.1b Incorrect script for 2-path 2-to-1 resampling half band filter

```
h1=[remez(56,[0 20 28 48]/48,[1 1 0 0],[1 150]) 0];
g1=reshape(h1,2,29);

reg=zeros(2,29);
x1=[1 zeros(1,60)];
m=1;
for n=1:2:60
     reg(1,:)=[x1(n)    reg(1,1:28)]; % 1-st sample
     reg(2,:)=[x1(n+1)  reg(2,1:28)]; % 2-nd sample
     y1(m)=reg(1,:)*g1(1,:)'+reg(2,:)*g1(2,:)';
     m=m+1;
end
```

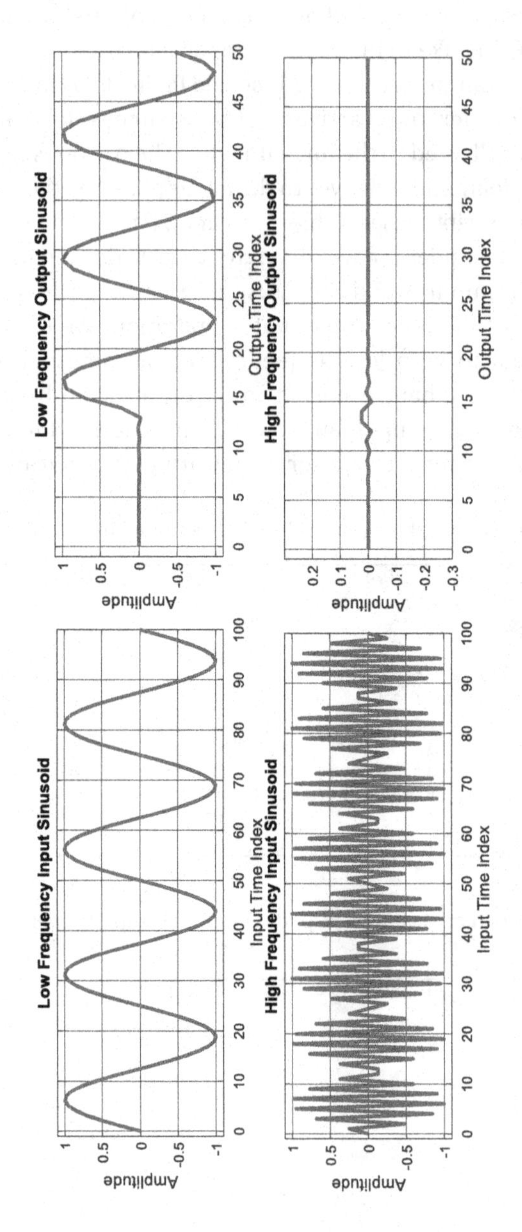

Figure 7.3a Low frequency in-band and high frequency out-of-band sinusoids at input and output of correctly operating two-path polyphase low-pass filter.

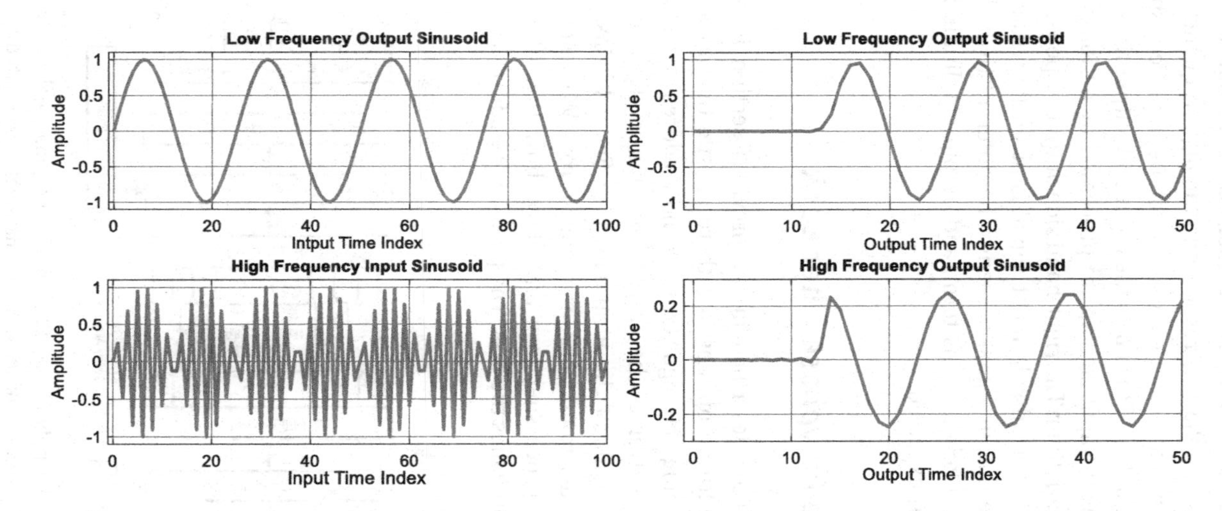

Figure 7.3b Low frequency in-band and high frequency out-of-band sinusoids at input and output of incorrectly operating two-path polyphase low-pass filter.

sinusoid of a properly operating two-path filter. The right-hand side output responses are the correct responses to the left-hand side input signals. We see that the response to the high frequency input signal is a low-level transient copy of the filter's impulse response. On the other hand, Fig. 7.3b shows the input and output time series for the same pair if input signals are processed by an improperly operating two-path filter. The right-hand side output response to the high frequency left-hand side input signal is absurdly wrong. The frequency of the output signal is seen to be different than that of the input signal, likely an alias term, and output amplitude is not near zero level as expected.

7.2 Cascade Polyphase Filter Bank

The multirate filter described in the previous section is one of the simplest multirate options and there is little that we can do wrong when coding the algorithm. We now address an example of the opposite extreme, the cascade non-maximally decimated polyphase filter bank, of the form shown in Fig. 7.4. This design offers multiple ways to make programming errors. We first present and examine a processing task and show how this filter bank efficiently accomplishes the task. The problem we address is a filter

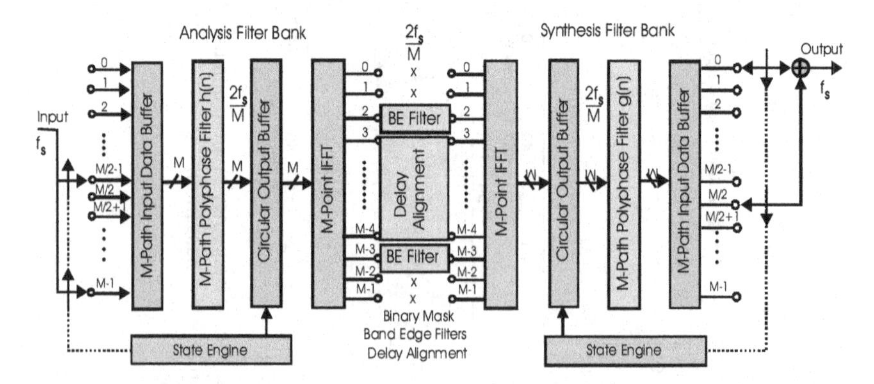

Figure 7.4 Cascade polyphase filter bank, non-maximally decimated analysis filter bank, binary mask, and non-maximally decimated synthesis filter bank.

in the 5G-NR (New Radio), specifically the one that matches the specifications shown in Fig. 7.5.

The Remez algorithm designed a 549-tap FIR filter that satisfied the shown specifications. We are tasked to implement this filter with a smaller number of arithmetic operations. The block diagram of Fig. 7.4 is one of many such reduced computational workload solutions we designed. We describe how it works and then examine its algorithmic pitfalls. The input stage of this diagram is a 60-path polyphase filter that partitioned a 479-tap Nyquist filter formed by a Kaiser windowed sinc filter. An appended "0" formed a 480-tap filter reshaped to a 60-path filter with 8 coefficients per path. The output stage is a 60-path polyphase filter that partitioned a 419-tap modified Remez filter. Here too, an appended "0" formed a 420-tap filter reshaped to a 60-path polyphase filter with 7-coefficients per path. Their spectra are shown in Fig. 7.6.

The analysis polyphase filter contains 60 paths which will form channel filters with 6-dB bandwidths and channels spacings of fs/60 or 2.048 MHz. We operate the filter in an M/2-to-1 or 30-to-1 down sampling mode to obtain an output sample rate of 4.096 MHz. We deliver data in 30 sample input vectors to the input filter which forms a 60-sample output vector in response to the newest 30 sample input vector. The 60-point output vector from the 60-path filter is delivered to a 60-point circular buffer which performs a 30-point circular shift on alternate input vectors. This operation is required to translate to baseband, the odd indexed frequency bins, which alias to the half sample rate due to the M/2 down sampling. The 60-point IFFT unaliases the 60-aliased bands. A subset of 49 channels spanning the desired output bandwidth are tagged to be assembled by the output synthesizer. The two end bands are processed by a pair of 21-tap band edge filters to reduce their excess bandwidths to the specified bandwidth. These spectra are shown in Fig. 7.8. There is a transport delay of 10 samples in these filters. The other 47 channels in the tagged frequency span are moved through a 10-tap delay line to be aligned and merged with the delayed outputs of the two band edge filters. This merged vector is processed by the synthesizer's 60-point IFFT then by its circular buffer, and finally by its 60-path output polyphase filter which mergers output samples

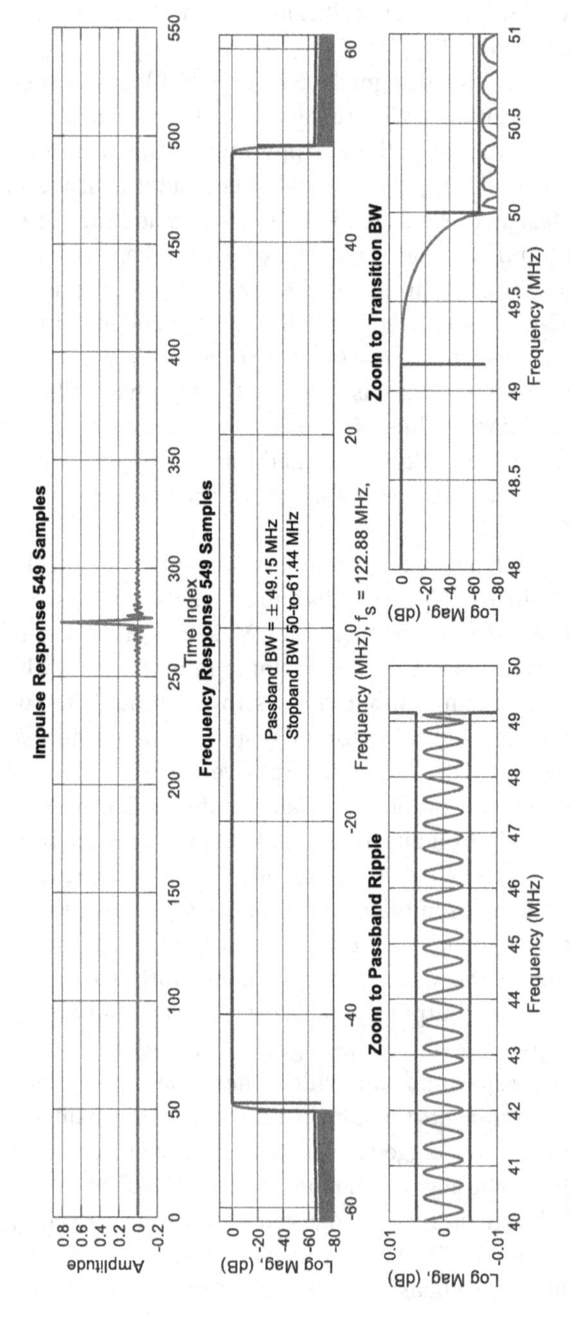

Figure 7.5 5G-new radio impulse response, passband and stopband frequency filter specifications, and zoom to specific filter attributes.

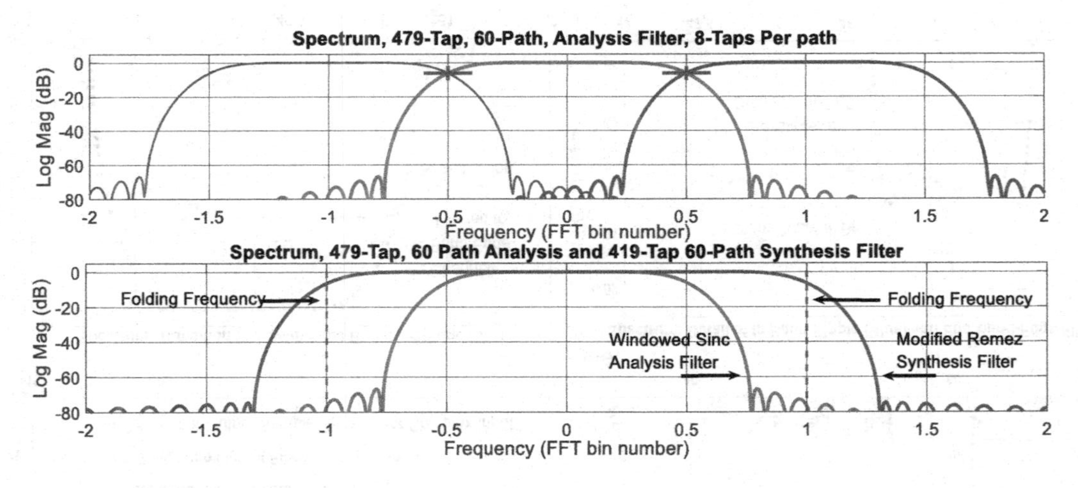

Figure 7.6 Spectra of adjacent channel analysis bands and analysis and synthesis low pass filters.

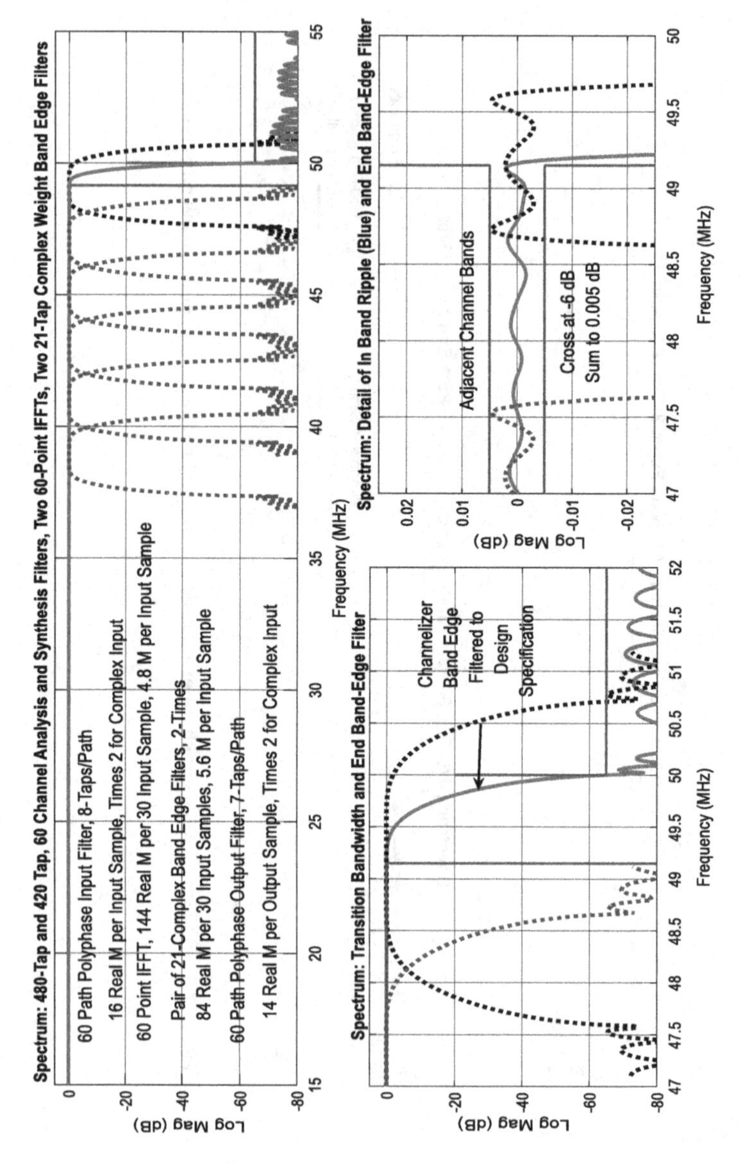

Figure 7.7 Spectra of synthesized filter from non-maximally decimated filter banks with internal band edge filters.

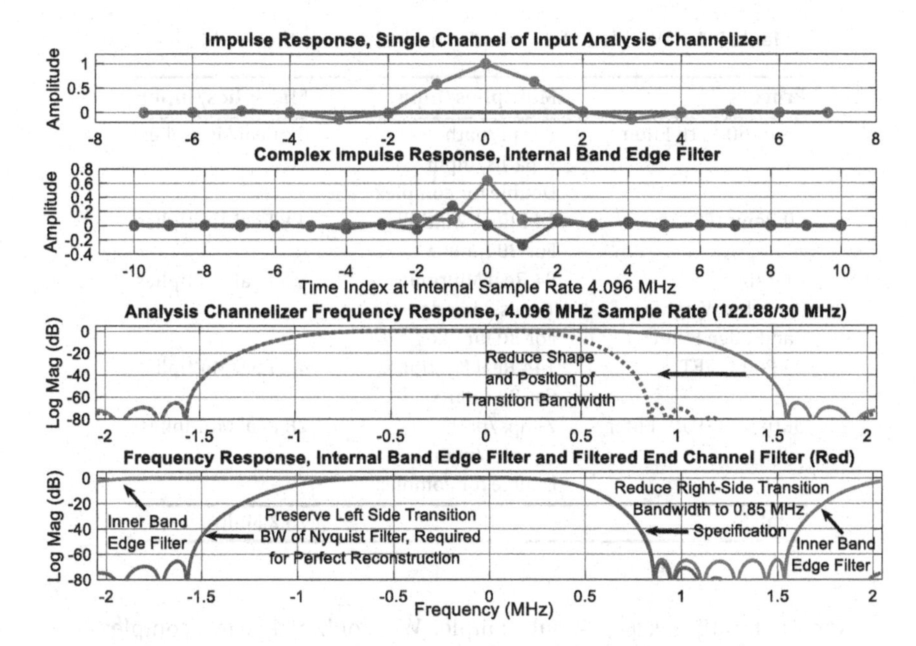

Figure 7.8 Spectra of end channel, band edge filter and reduced BW channel.

separated by 30 samples, a process which is the dual of the input 30 sample offset.

Details of the synthesized frequency response formed by the cascade analysis and synthesis filter banks of Fig. 7.4 is shown in Fig. 7.7. The top subplot shows the frequency response of the composite system and is overlaid by the spectra of the synthesized narrowband analysis filters in the composite output. The lower left-hand subplot shows the detail of the transition bandwidth and its target spectral mask along with the spectra of the band edge filter that was filtered to the design transition band by the internal band edge filter. The lower right-hand subplot shows the summation of the two most right channel bandwidths that cross 6 dB below the displayed region but still sum to the specified ripple levels of 0.005 dB. Note that on this scale, −6 dB is a mile away!

We now compare the workload for the two implementations. The score card is itemized in Table 7.2. The input channelizer has 8 coefficients per path but operates 2-paths per input sample

Table 7.2 Workload for cascade channelizer synthesized filter

Process	Multiplies/Input	Multiplies/Input
Input 60-Path Filter	8-taps/path 2-paths/input Double for complex	32 Real Multiplies
60 Point IFFT	144 Real Multiplies For 30 Inputs	4.8 Real Multiplies
21-Tap Complex Weights Band Edge Filter	84 Real Multiplies Times 2 bands For 30 inputs	5.6 Real Multiplies
60 Point IFFT	144 Real Multiplies For 30 Inputs	4.8 Real Multiplies
Output 60-Path Filter	7-taps/path 2-paths/input Double for complex	28 Real Multiplies
Total		75.2 Multiplies

for 16 multiplies per input sample. We double this for complex input samples. The 60-point IFFT is performed by a Good-Thomas, prime factor algorithm which requires only 144 real multiplies per transform. The 144 multiplies are distributed over 30 input samples which is equivalent to 4.8 multiplies per input sample. We perform both an input and output IFFT so we count this work twice. The inner 21-tap band edge filter with complex weights and complex channel samples has a workload of 84 real multiplies per filter. We double this to account for the band edge filters on each side of the spectrum. Finally, we have the output IFFT and the output 60 path filter with 7-coefficients per path. Table 7.2 shows the workload for the cascade channelizers to be 75.2 multiplies per input sample. Contrary to our intuition, the two IFFTs and the pair of inner band edge filters account for only 20% of the workload. The filter we are synthesizing is a pair of 549-tap FIR filters. Making use of the even symmetry of the target filter weights, we are replacing a pair of filters requiring 549 real multiplies with a process that requires 75.2 real multiplies. The workload ratio is 75.2/549 or 13.7% representing a savings of 86.3%.

This workload reduction is purchased by a more sophisticated processing scheme which almost cries out that all these processing steps can't possibly be less expensive than the original tapped delay line FIR filter; but it is! What we exchange for the reduced workload is a longer delay through the filter chain, a result we expected because we pass the input signal through three filters.

7.3 Debugging Cascade Polyphase Filter Script

We now have synthesized a target filter with a cascade polyphase analysis and synthesis filter. The cascade filter is connected by a programmable binary mask, a pair of band-edge FIR filters, and delay array. Now the question is how do we verify that all the pieces are working correctly and performing the task they were designed to accomplish. Even though the filter bank does not have a transfer function, it can still be probed by standard test signals. The two standards probes are an impulse and a sinusoidal tone burst. The probe signal starts the process, but we still must select test points in the processing chain to observe and analyze. The first data collection point we examine is the set of time series at the output of the analysis filter's IFFT. To collect the probe response signal, we divert the successive output vectors from the IFFT into a 60×16 array for an impulse response test or into a 60×160 array for a sinusoidal tone test. When we apply a probe signal to the analysis filter we must remember that data slides through the polyphase filter in stride steps of 30 samples and even though each path contains only 8 weights, the unit amplitude impulse sees successive interleaved weights in two different paths, path n and path $n+30$ for a total of 16 samples. The shortest length probe signal must be 30 times 16 or 480, which of course is the length of the prototype FIR filter designed for the analysis channelizer. The impulse response and frequency response of the first 3 of the 60 filter responses formed by the analysis filter bank are shown in Fig. 7.9. Figures 7.10a,b show the tone burst time response and windowed spectrum of a bin centered tone and of bin center offset tone. The bin centered tone entered the filter as a sinusoid but left the filter as a DC term. Aliasing at its finest! Recall that our channel bin centers are separated by 2.048 MHz and the channel sample rate is 4.096 MHz. Note the low-level tone images in bin 2, to the right side of bin 1. Frequency zero of

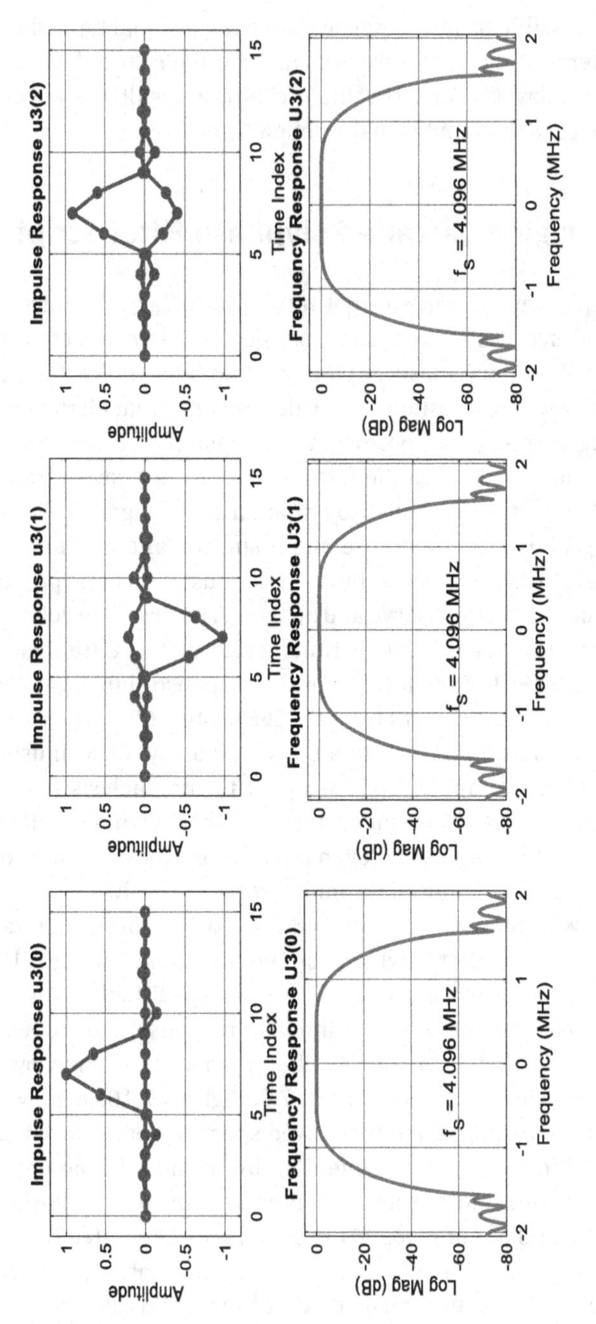

Figure 7.9 Impulse response and spectrum of three baseband channels.

bin 1 is the same as frequency –2.048 of bin 2 The tone to the right of 0 in bin 1 is the same tone to the right of –2.048 in bin 2. The most common script errors deal with the MATLAB *fftshift* command which shifts the DC bin between index (1) and index (31) for ease of accessing positive and negative bin center frequencies. Is the DC bin at the end of the array at *fx(1)* or at its center *fx(31)*? This tone probe helps resolve bin location confusion.

The next probe is the impulse response and pulse tone response of the cascade analysis and synthesis filter chains. This probe tests the validity of the perfect reconstruction. We conduct this probe with all output channel ports of the analysis filter presented to the input channel ports of the synthesis filter. In this test, we probe with a single impulse and expect to construct a single delayed impulse at the output.

Figure 7.11 shows the impulse response of the cascade along with a zoom to the low-level time domain echo artifacts approximately 4 orders of magnitude below the primary unit amplitude impulse. The primary response, at position 431 matches the synthesis filter length (419) added to the 12-sample offset of the input probe pulse. The echo artifact terms are separated from the primary pulse position in multiples of 60 samples. The bottom subplot shows the 0 dB level spectrum of the unit impulse along with the 0.003 dB ripple level responsible for its time domain echoes as well as the ripple levels of the analysis channel bank filters translated to their center frequencies. One problem here is that the system has multiple impulse responses, one for each of the possible 30 positions of the input vector. The student should be asked to examine the time and frequency artifact responses generated by the impulse response probe for the input impulse at more than one initial position. We now insert the binary mask, the band-edge filters and the bulk delay line to form the desired bandwidth with specified transition bandwidth. And repeat the impulse response test. We did this and intentionally introduced an incorrect delay in the delay block so that the delayed samples are not properly time aligned with the filtered edge samples as they are passed on to the synthesis filter block. When we examined the spectral response as shown in Fig. 7.12, the assembled synthesized ripple at first appeared correct until we zoomed into the crossover frequency span. We start seeing

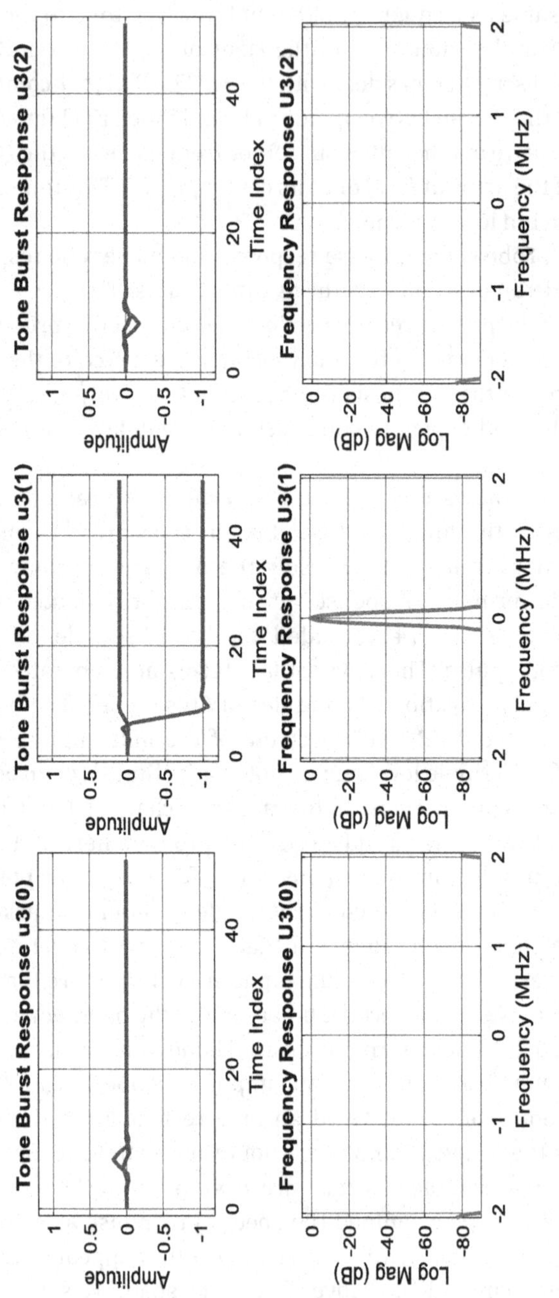

Figure 7.10a Bin centered tone burst response and spectrum of three baseband channels.

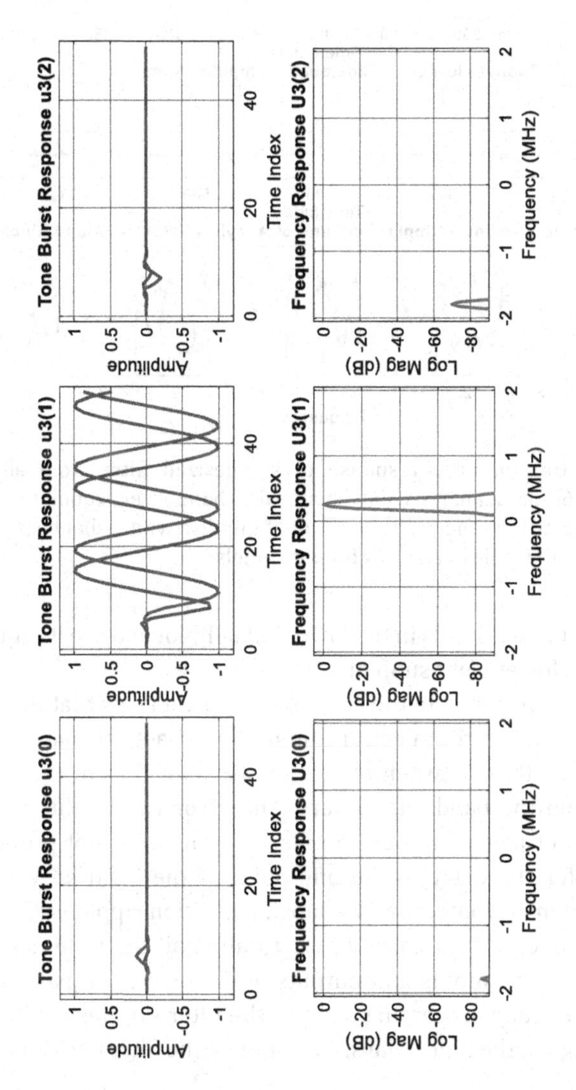

Figure 7.10b Bin offset tone burst response and spectrum of three baseband channels.

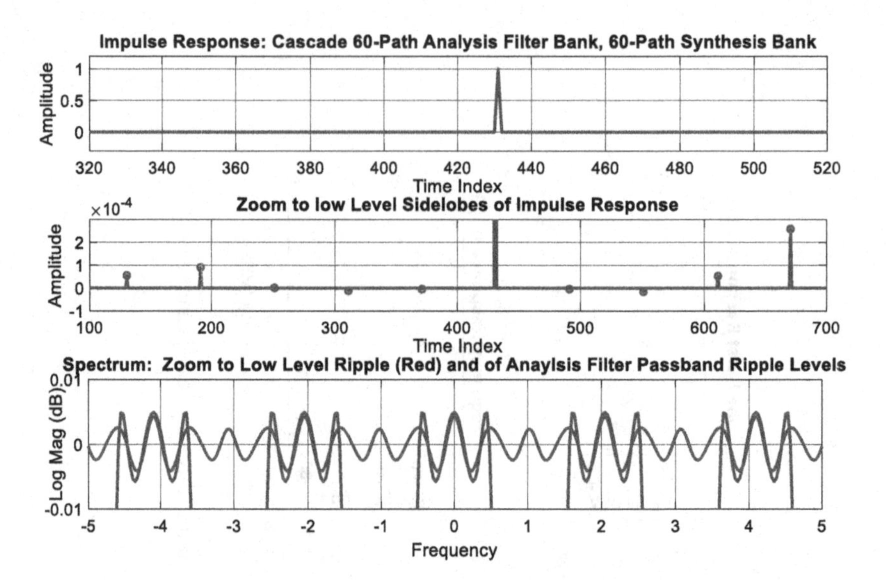

Figure 7.11 Unit impulse response of synthesized filter from all 60 channels of 60-path analysis and synthesis channelizer, zoom to time domain echo artifacts, and 0 dB spectrum of impulse with in-band spectral ripple along with ripple of analysis filter channels.

the imperfect ripple levels in the lower left subplot and see it acquire clearly in the lower right subplot.

The transition ripple in the crossover ripple tells us that there is a phase mismatch in the spectral summation. Phase mismatch here means a time offset between the terms we aligned from the delay line block and the band edge filters. The error can be due to the edge filters not having an odd number of weights or not properly accounting for the delay to the mid-point of the band edge filter impulse response. That letter W-shaped transition ripple error was seen many times as we learned how to align all the time domain components at the inputs and outputs of the IFFT as signal vector traversed the many processing tasks in the filter synthesis process. When we aligned the time delays, we obtained the filter performance shown in Fig. 7.7.

Other possible phase mismatch error sources is the selection of an even number instead of an odd number of coefficients in the prototype low-pass Nyquist filter presented to the polyphase M-path

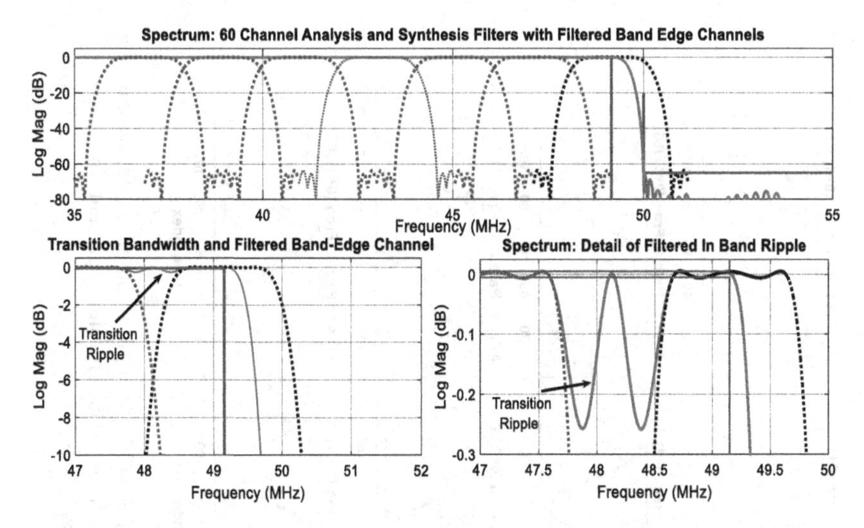

Figure 7.12 Spectrum of synthesized filter from 49 channels of 60-path analysis and synthesis channelizer with filtered band edge channels and mismatched delay with the non-filtered channel components.

partition. Yet another source of phase mismatch is the zero extension from an odd number to an even number of coefficients in the polyphase partition. Has the single zero extension been appended to the beginning or the end of the coefficient sequence prior to the polyphase partition.

While on the topic of phase match we are reminded that we are discussing a polyphase filter bank that derives its unusual capabilities through intentional aliasing due to the M/2 down sampling at its input ports. We have commented several times that due to the down sampling the filter bank doesn't have a transfer function. One response to this is to perform the mapping to the M-path filter and examine the transfer function of the separate paths prior to invoking the Noble identity and down sampling the path. An alternate approach is to probe the polyphase M/2-1 down sampled response with a single sinusoid and examine the phase profiles of the aliased tone. This works out to be a nice diagnostic tool that often guides us past code errors. We have found a sequence of single tones near the center of any of the Nyquist zones are good probe signals. We suggest the small offset from the center, because tones at the

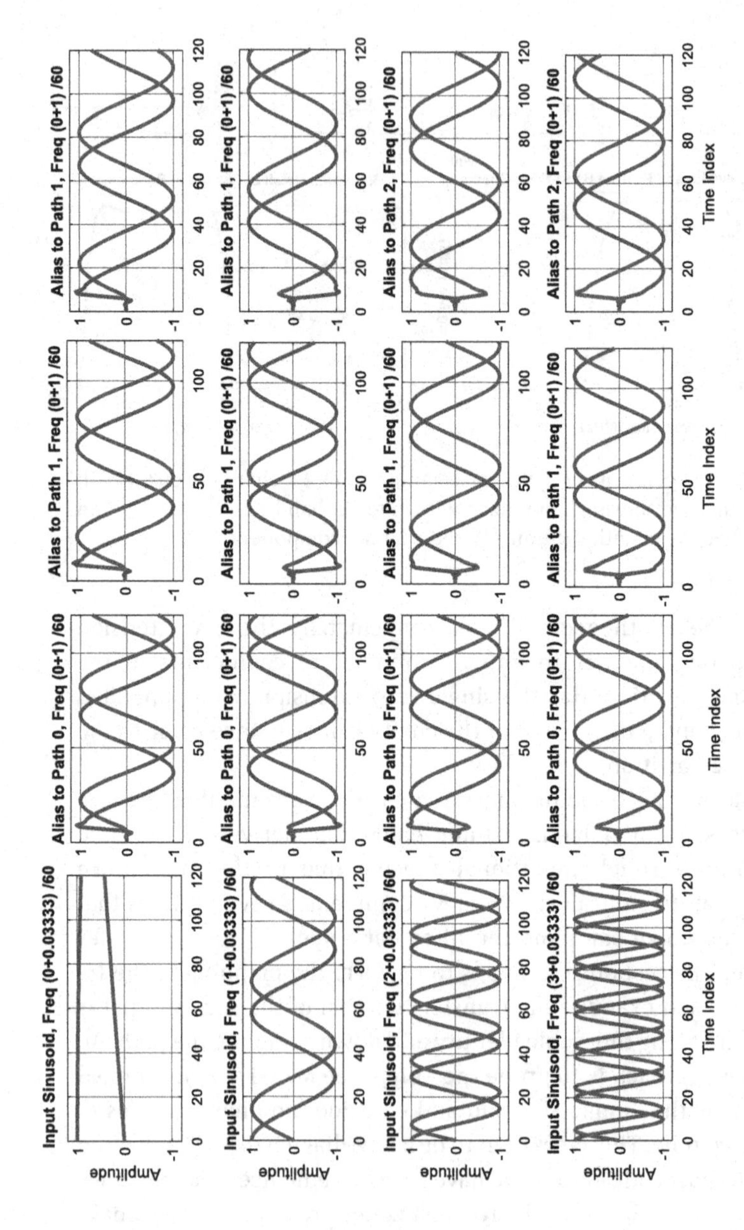

Figure 7.13 Time series for four successive input sinusoids at frequencies $(k + 1/30)/60$ input to 60-path filter operating at 30-to-1 down sample and aliased time responses to first three paths of analysis filter bank. Left column presents input sinusoid. Three right columns present aliased time sinusoids with successive phase shifts.

center, this is k fs/M, alias to zero or the half sample rate frequency in each of the M path. Both frequencies are fine, but the time signal figures are a bit boring. The small offset leads to complex sinusoids which offer more interesting time signal figures.

Figure 7.13 shows the input and output time series for 4 probe sinusoids. The input sinusoids are offset from the center frequencies k $(2\pi/60)$, for $k = 0, 1, 2$, and 3 with offsets of $1/30$ $(2\pi/60)$. The center frequencies are the channelizer center frequencies and the offset matches down sampling index which aliases the offset to 1 cycle per 60 output time samples. The left column of this figure presents 120 samples of each of the input signals with frequencies near 0, 1, 2, and 3 cycles per 60 samples. We clearly see the aliasing of the input sequences in the time series observed in the outputs of the first 3 paths of the 60 paths. The aliased time signals are seen to have the same frequency of 1 cycle per 60 output samples for the four different input signals. This is the reason we don't have a transfer function here: the same output frequency for different input frequencies. While the output frequencies are the same for the different input frequencies, the phase progression of output sinusoids in successive paths differs. We can see the phase changes in successive output sequences by noting the sample amplitude of the sine and cosine at their right boundary, time index 120. An easier way to see these phase increments is to take the 120-point FFT of each sequence, being sure to omit the starting transient. Each FFT has one spectral component with its phase angle matching the phase of the input time series. We plot all 60 of the transforms, I against Q, on the same figure to see the phase progression across the 60 paths. Remember we are only examining the polyphase partition of the analysis filter, prior to its IFFT. Figure 7.14 presents Four FFT phasor plots for the four probe sinusoids presented in Fig. 7.13. The upper left sub-plot shows that all 60 paths have the same phase corresponding to zero frequency. The IFFT in the polyphase filter bank would recognize that this is the DC channel and would collect the samples from the polyphase partition and map them to the DC bin. The phasor, in the first quadrant, is the main lobe of the FFT while the spill into the third quadrant are the FFT side-lobes. The upper right sub-plot shows that the 60 paths have phases separated by multiples of $2\pi/60$, corresponding to 1 cycle in 60 samples. The

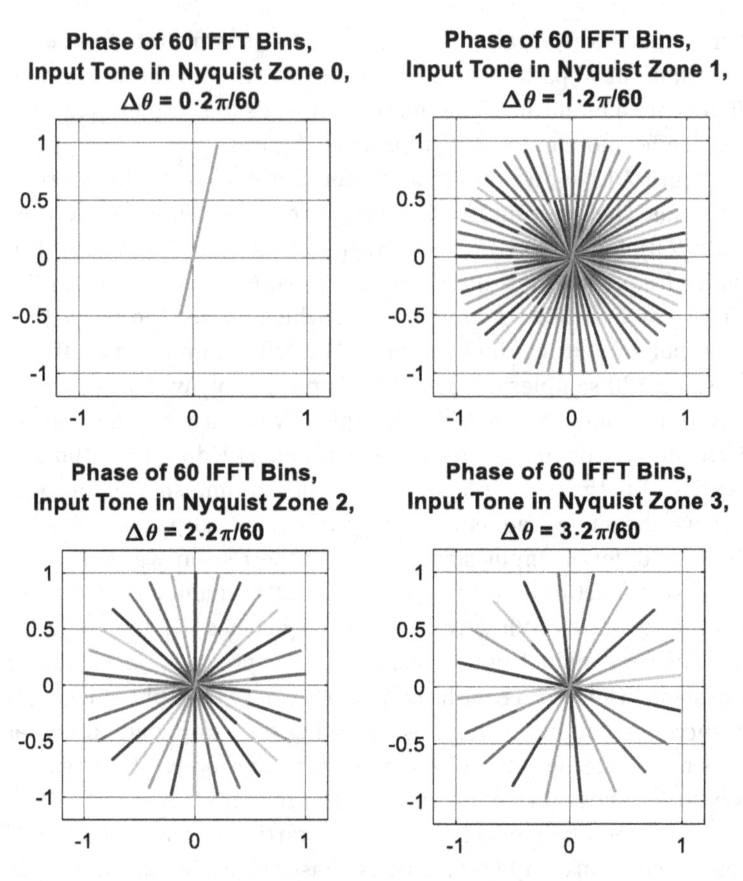

Figure 7.14 Polar plots of complex ffts for all 60 aliased time series output by 60-path polyphase filter for four input sinusoids slightly offset from center of successive Nyquist zones. In upper left, all 60 phasors have same phase angle or differ by $0(2\pi/60)$. In upper right, 60 successive phasors differ by $1(2\pi/60)$. In lower left, 60 successive phasors differ by $2(2\pi/60)$. In lower right, 60 successive phasors differ by $3(2\pi/60)$.

IFFT in the polyphase filter bank would recognize that this is the 1-cycle per interval channel and would collect the samples from the polyphase partition and map them to the index $+1$ bin. The lower left sub-plot shows that the 60 paths have phases separated by multiples of $2(2\pi/60)$, corresponding to 2 cycles in 60 samples. The IFFT in the polyphase filter bank would recognize that this is the 2-cycle per interval channel and would collect the samples from the polyphase partition and map them to the index +2 bin. The lower right sub-plot shows that the 60 paths have phases separated by multiples of $3(2\pi/60)$, corresponding to 3 cycles in 60 samples. The IFFT in the polyphase filter bank would recognize that this is the 3-cycle per interval channel and would collect the samples from the polyphase partition and map them to the index +3 bin.

What we have demonstrated is that while tones offset from different Nyquist zones alias to the same frequency sequence in each path of the polyphase filter, the phase progression across the 60 paths of the polyphase partition identifies from which Nyquist zone the alias came and can be extracted from the multiple aliases by the phase aligning properties of the IFFT. One of the values of the property just illustrated is it gives students another perspective of how an analysis channelizer operates. It also offers a diagnostic tool to probe the operation of a polyphase filter in a channelizer. If a set of probes of the polyphase partition do not exhibit the property illustrated, then there is problem with the partition.

7.4 Comments and Conclusions

We have presented the structure of a simple multirate and then of a sophisticated multirate signal processing routine. In this description, we commented on the sample rate changes that converted the systems from linear time invariant to linear time varying. We noted that the time varying systems offer interesting implementation options that attract us to their use. A problem with time varying systems is they don't have transfer functions and despite that we can probe them with standard probes of impulses and tone bursts. Sometimes the impulse response probe is useful and occasionally it is not and the same can be said of the tone burst test. We have tried to demonstrate how probing different parts of the sophisticated system can lead to insight and detection of simple conceptual errors. Space limitations keep us from presenting all the ways we can probe the time varying system. Earlier we did mention an important one. We find it useful to examine our models prior to the resampling operation known as the *Noble Identity* which swaps the order of the filtering and resampling tasks following the initial mapping to the M-path partition. Each path of the partition still has a transfer function prior to the resampling operation and we garner useful insight by examining the set of spectral phase profiles prior to their aliasing. We normally employ this option with low-order polyphase filters: 6 or 12 paths, for instance.

References

1. F. J. Harris, C. Dick, X. Chen, and E. Venosa, Wideband 160-channel polyphase filter bank cable TV Channelizer, *Signal Processing, IET*, vol. 5, no. 3, pp. 325–332, June 2011.
2. F. J. Harris, E. Venosa, X. Chen, and B. D. Rao, Polyphase analysis filter bank down-converts unequal channel bandwidths with arbitrary center frequencies, *Analog Integrated Circuits and Signal Processing*, vol. 71, no. 3, pp. 481–494, 2012.
3. T. Karp and N. J. Fliege, Modified DFT filter banks with perfect reconstruction, *IEEE Transactions on Circuits and Systems. II, Analog and Digital Signal Processing*, vol. 46, no. 11, pp. 1404–1414, December 1999.
4. X. Chen, F. J. Harris, E. Venosa, and B. D. Rao, Non-maximally decimated analysis/synthesis filter banks: Applications in wideband digital filtering, *Signal Processing, IEEE Transactions on*, vol. 62, no. 4, pp. 852, 867, February 15, 2014.
5. F. J. Harris, *Multirate Signal Processing for Communication Systems*, Prentice Hall, Upper Saddle River, New Jersey 07458, 2004.
6. P. P. Vaidyanathan, Theory and design of M-channel maximally decimated quadrature mirror filters with arbitrary M, having the perfect-reconstruction property, *Acoustics, Speech and Signal Processing, IEEE Transactions on*, vol. 35, no. 4, pp. 476, 492, April 1987.
7. M. Vetterli and D. Le Gall, Perfect reconstruction FIR filter banks: Some properties and factorizations, *Acoustics, Speech and Signal Processing, IEEE Transactions on*, vol. 37, no. 7, pp. 1057, 1071, July 1989.
8. V. K. Jain and R. E. Crochiere, A novel approach to the design of analysis/synthesis filter banks, *Acoustics, Speech, and Signal Processing, IEEE International Conference on ICASSP '83.*, vol. 8, no. 3, pp. 228, 231, April 1983.
9. F. Harris, C. Dick, X. Chen, and E. Venosa, M-path channelizer with arbitrary center frequency assignments, WPMC 2010, October 11–14, 2010, Recife, Brazil.

10. F. Harris, B. Behrokh, X. Chen, and E. Venosa, Multi-resolution PR NMDFBs for programmable variable bandwidth filter in wideband digital transceivers, *Int'l Conf. on DSP-2014*, Hong Kong, August 20–23, 2014.
11. F. Harris, C. Dick, X. Chen, and E. Venosa, Interleaving different bandwidth narrowband channels in perfect reconstruction cascade polyphase filter banks for efficient flexible variable bandwidth filters in wideband digital transceivers, *DSP-2015 Conference*, Singapore, July 21–24, 2015.
12. F. Harris, X. Chen, E. Venosa, and C. Dick, Cascade non-maximally decimated filter banks form efficient variable bandwidth filters for wideband digital transceivers, *DSP-2017 Conference*, London, August 23–25, 2017.

Part III

Imaging Technologies

Chapter 8

Learning Approaches for Super-Resolution Imaging

Wan-Chi Siu,[a] Zhi-Song Liu,[a] Jun-Jie Huang,[b] and Kwok-Wai Hung[c]

[a]*Center for Multimedia Signal Processing,*
Department of Electronic and Information Engineering,
The Hong Kong Polytechnic University, Hong Kong
[b]*Electrical and Electronic Engineering Department, Imperial College London*
[c]*College of Computer Science and Software Engineering, Shenzhen University*
enwcsiu@polyu.edu.hk, enwcsiu@polyu.edu.hk

Image super-resolution is a topic of great interest. It has significant applications in ultra-high-definition TV display, image resizing, face recognition, object recognition, video coding and surveillance. The objective is to construct a high-resolution image from one or several low-resolution images, while minimizing visual artifacts. Classical approaches have come to a quality limit because of the constraint on manual filter design and limited design structures. Learning approaches allow super-resolution algorithm design to be adaptive to training data and automatically form thousands of filters/adapters for the best super-resolution. In this chapter, we will introduce (i) some conventional learning approaches, (ii) random forests, and (iii) Convolutional Neural Network (CNN) for effective image super-resolution.

Learning Approaches in Signal Processing
Edited by Wan-Chi Siu, Lap-Pui Chau, Liang Wang, and Tieniu Tan

ISBN 978-981-4800-50-1 (Hardcover), 978-0-429-06114-1 (eBook)
www.panstanford.com

8.1 Introduction

There have been some significant advancements in the processing of 8K videos [1] in the industry. For example, Pentax K-3 II DSLR Digital Camera [2] captures 4 images of the same scene by shifting the image sensor by a single pixel to improve resolution and reduce noise. One can also take multiple photos of the same scene and to use Photoshop [3] to perform auto-alignment and sharpening to a super-resolution image. The NHK [4] uses super-resolution techniques to transmit 8K videos by transmitting some side data. The study of super-resolution is not uncommon in the image research discipline, but only recently more attention has been paid to machine learning for image/video super-resolution production. Image super-resolution is the enlargement of an image to have larger size (see Fig. 8.1), but in our applications, it means a denser number of pixels per inch as shown in Fig. 8.2. Super-resolution can also refer to as a restoration problem for which the low-resolution (LR) image "were" formed by blurring the high-resolution (HR) image and down-sampling it to form LR with the presence of noise. Image interpolation can be considered as a special case of image super-resolution, for which the LR image was formed by just down-sampling without the blurring function and the noise effect.

Image super-resolution (SR) usually refers to an increase of the resolution of a single input low-resolution (LR) image by up-sampling, deblurring and denoising, while the resultant high-resolution (HR) image should preserve the characteristics of natural image, such as sharp edges and rich texture. This is a difficult and ill-posed problem, since a number of unknown pixels have to be inferred from very limited information. However, due to the demand of hi-tech applications, including image/video upsizing for mobiles, ultra-high-definition TV and large panel advertisement display, satellite imaging, medical imaging, video surveillance, human and face recognition, this topic has recently drawn the attention of many researchers. Conventional approaches starting from simple bi-cubic interpolation to reconstruction-based approaches using geometric duality, non-local similarity, etc. had been investigated thoroughly. However, machine learning by making use of k-nearest neighbors (k-NN), sparse coding, random forests and convolutional neural

Figure 8.1 (a) Targeted super-resolution video for marriage ceremony. (b) Video super-resolution in surveillance.

network (CNN) appears to win and give substantial improvement over the conventional approaches. Random forests approach is a fast method while the CNN makes use of deep learning with various neural network structures.

Super-Resolution and Interpolation

We assume that the LR (Low Resolution) image is a natural image and there is no so called HR (High Resolution) image in practical use of the super-resolution process. However, during training, we must assume a model for the relationship between the LR and HR images. Usually, the LR image is assumed to be obtained from a HR image through a blurring and down-sampling process as shown:

$$\mathbf{X} = \mathbf{DHY} + \varepsilon \tag{8.1}$$

where $\mathbf{X}$ is the LR image which could be generated by the blurring operator $\mathbf{H}$ and down-sampling operator $\mathbf{D}$ from the original HR image $\mathbf{Y}$ with additive noise ε. For image interpolation, the LR image

Figure 8.2 Diagram shows the problem to be resolved: (a) Low-resolution image/frame. (b) High-resolution image/frame to be formed from super-resolution.

$\mathbf{X}$ is obtained by directly down-sampling from the HR image $\mathbf{Y}$, i.e. $\mathbf{X} = \mathbf{DY}$. For saving space we do not discuss further its difference with super-resolution and this chapter is on image super-resolution. Note that image super-resolution algorithms may not be effective for image interpolation problem and vice versa.

Patch Processing

Patch processing is used in many parts of this chapter, and the size of which is usually $N \times N$ for $N = 5$ for most of our discussion. For the sake of convenience, a patch is usually represented by columnwise lexicographically ordered matrix notations. The size of $\mathbf{X}$ is 25 (for $5^2 = 25$) and $\mathbf{Y}$ is d $= 4n^2$ ($4 \times 5^2 = 100$). For most parts of this chapter, the low resolution is initially enlarged (unless specially specified) by bicubic interpolation, say from $N \times N$ size to $2N \times 2N = 4N^2$ size as

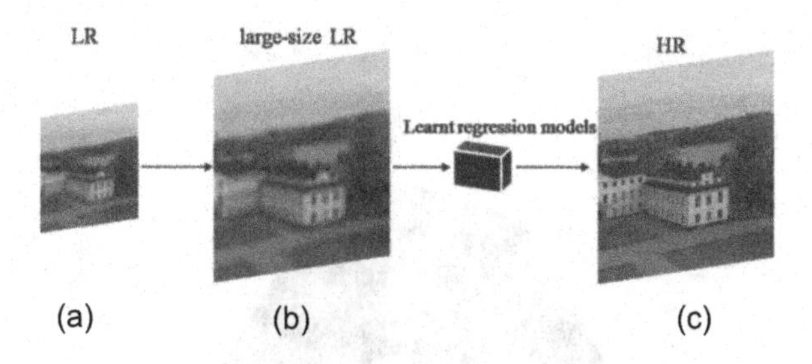

Figure 8.3 Image/Video Super-resolution process.

shown in Fig. 8.3 and for convenience we still call it LR image which has a size of $4N^2$, the same size as the ground true image for training.

8.1.1 Super-Resolution by Bicubic Interpolation

The simplest way to do image interpolation is the bicubic interpolation, which means to use a non-adaptive filter as shown in Eq. 8.2:

$$y = \sum_{i=0}^{3} \sum_{j=0}^{3} w_{ij} x_{ij} \tag{8.2}$$

where y is the pixel to be interpolated, and w_{ij}'s are the non-adaptive filter coefficients, (ij) are coordinates of its available neighbor pixels x_{ij}. Hence, the bi-cubic interpolated pixel y is obtained by a weighted average of its neighboring pixels in the LR image. The nearby pixels would have larger weights than the far away pixels as shown in Fig. 8.4. Thus, the nearby pixels contribute more to the predictions of the unknown pixel. However, the bicubic interpolation suffers from blurring and aliasing effects, as shown in Fig. 8.5.

8.1.2 New Edge Directed Interpolation

An early approach for interpolation with substantial improvement in visual quality is done by the New Edge Directed Interpolation (NEDI) [5–7] approach which makes use of geometric duality property to estimate HR covariances by LR covariances and interpolates the HR pixels using the estimated covariances. It actually makes use of the FIR Wiener filter, or equivalently the linear minimum mean squared

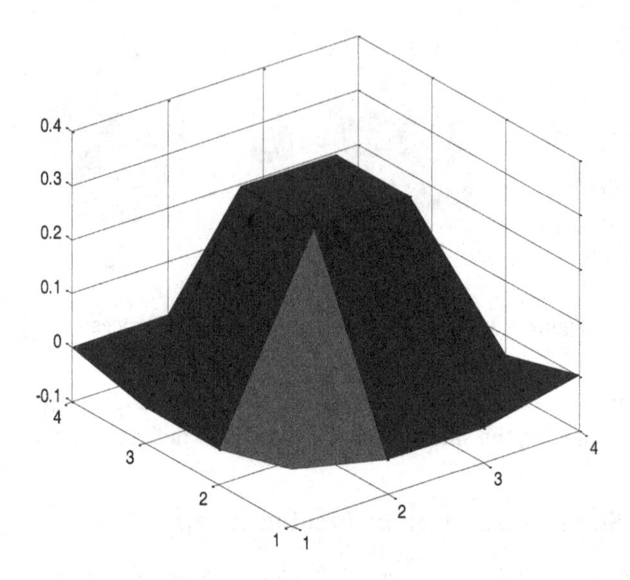

Figure 8.4 Bicubic filter coefficients.

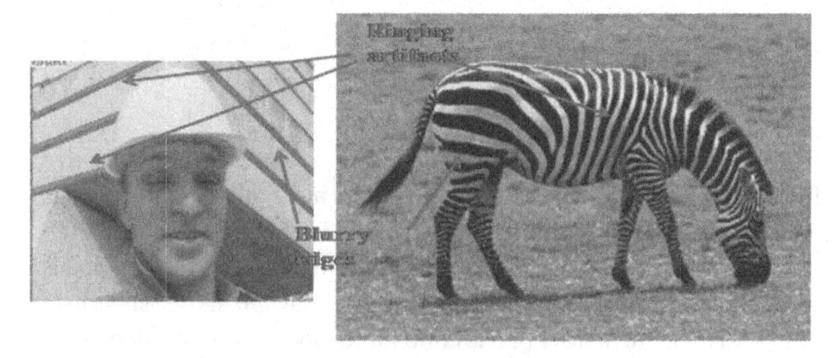

Figure 8.5 Sample results of bicubic super-resolution.

error estimator [19] for linear prediction. Let us briefly describe the formulation of NEDI as an introduction.

Model Learning

The 4-order linear estimation model is given by

$$x_i = \sum_{k=1}^{4} c_k \cdot x_{ik} + \varepsilon_i, \quad \text{for} \quad i = 1, 2, \ldots, N, \tag{8.3}$$

where ε_i is the estimation error, N is the number of data samples which is usually a large number, in the range of $30 \sim 40$ for NEDI or over 1000 for non-local-based learning approaches, c_k is the model parameters and each interpolated data point x_i has four neighboring data points, x_{ik}. Figure 8.6a shows the spatial positions of x_{ik} (4 blue circles) and x_i (big blue circle in the center). Note that in this example, x_i is only one of many sample points with known interpolating elements, for i ranges from 1 to N, where N is the number of sample points for building this model. The model parameters c_k can be found by using the least squares estimation:

$$\{c_k^*\} = \arg\min_{\{c_k\}} \sum_{i=1}^{N} \left[x_i - \sum_{k=1}^{4} c_k \cdot x_{ik} \right]^2, \tag{8.4}$$

where the matrix form of Eq. 8.4 is given by

$$\mathbf{C}^* = \arg\min_{\mathbf{C}} \|\mathbf{X} - \mathbf{X}_\mathrm{C}\mathbf{C}\|_2^2 \tag{8.5}$$

and the matrices are defined as

$$\mathbf{X} = \{x_i\}^\mathrm{T}, \mathbf{X}_\mathrm{C} = \{x_{ik}\}^\mathrm{T}, \mathbf{C} = \{c_k\}^\mathrm{T}, \text{ for } i = 1, 2, \ldots N, \ k = 1, 2, 3, 4. \tag{8.6}$$

The sizes of matrices $\mathbf{X}$, $\mathbf{X}_\mathrm{C}$ and $\mathbf{C}$ are $N \times 1$, $N \times 4$ and 4×1 respectively. The closed form solution of Eq. 8.5 is given as

$$\mathbf{C}^* = (\mathbf{X}_\mathrm{C}^\mathrm{T}\mathbf{X}_\mathrm{C})^{-1}\mathbf{X}_\mathrm{C}^\mathrm{T}\mathbf{Y}, \tag{8.7}$$

where $\mathbf{C}^*$ is called the ordinary least squares (OLS) estimator. Note that $\mathbf{C}^*$ represents the set of optimized filter coefficients of this model. However, in many cases we use $\mathbf{C}$ instead of $\mathbf{C}^*$ for the sake of simplicity.

Prediction

Equation 8.7 gives us the model parameters of a class of LR pixels. Due to "geometry duality" [5–7], which means the statistical model in Fig. 8.6a could be similar to that in Fig. 8.6b, the missing data point $\{y_i, i = 1, 2, \ldots, N\}$ (hollow red dot in Fig. 8.6b) can be interpolated by its four neighboring data points, x_{ik} (hollow blue dots in Fig. 8.6b), as follows:

$$y_i = \sum_{k=1}^{4} c_k \cdot x_{ik} + \varepsilon_i, \quad \text{for} \quad \mathrm{i} = 1, 2, \ldots, N \tag{8.8}$$

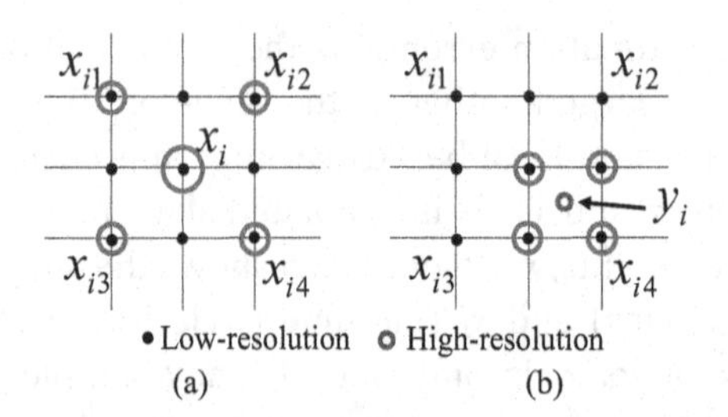

Figure 8.6 Graphical illustration of spatial position of LR and HR pixels.

NEDI uses the LR image to estimate the HR covariances for the FIR Wiener filter. There are many algorithms which are based on the idea of NEDI [5–11]. Say for example the order of the filter does not need to be 4 but it can be 6 [6] or 8 [7]. Let us clarify this filtering process with the following over-simplified example.

Example: A data sequence is given by $\{x(0), x(1), x(2), x(3), x(4), x(5), x(6), x(7), x(8), x(9)\}$. We have to find the next element in the sequence, i.e. $x(10)$ $(= y(10))$ using the FIR Wiener filter. Let also the length of the filter be 2, i.e. $M = 2$.

Model: Hence, we form the observed sample data set **Y** and their corresponding interpolating points, $\mathbf{X}_\mathrm{C}$, as shown below:

$$\begin{aligned}\mathbf{Y} &= [\mathrm{x}(2)\mathrm{x}(3)\mathrm{x}(4)\mathrm{x}(5)\mathrm{x}(6)\mathrm{x}(7)\mathrm{x}(8)\mathrm{x}(9)]^\mathrm{T} \\ &= [y(2)y(3)y(4)y(5)y(6)y(7)y(8)y(9)]^\mathrm{T}\end{aligned}$$

$$\mathbf{X}_\mathrm{C} = \begin{bmatrix} x(0)\ x(1)\ x(2)\ x(3)\ x(4)\ x(5)\ x(6)\ x(7) \\ x(1)\ x(2)\ x(3)\ x(4)\ x(5)\ x(6)\ x(7)\ x(8) \end{bmatrix}^\mathrm{T}$$

This means that $\{x(9), x(8), \ldots\ x(2)\}$ are used as sample points to look for the required statistics. Points $x(7)$ and $x(8)$ are used to interpolate $y(9)$ (i.e. $x(9)$, etc.), $x(6)$ and $x(7)$ are used to interpolate $y\{8\}, \ldots, x(0)$ and $x(1)$ are used to interpolate $y(2)$. Let us further assume that the cross-correlation matrix $\mathbf{X}_\mathbf{C}^\mathrm{T}\ \mathbf{Y} = [18\ \ 11]^\mathrm{T}$ and the auto-correlation matrix $\mathbf{X}_\mathbf{c}^\mathrm{T}\mathbf{X}_\mathbf{c} = \begin{bmatrix} 13 & 8 \\ 8 & 5 \end{bmatrix}$. The filter weights can be estimated by Eq. 8.7, as follows:

$$\mathbf{C} = \begin{bmatrix} c_1 \\ c_2 \end{bmatrix} = (\mathbf{X}_\mathrm{C}^\mathrm{T}\mathbf{X}_\mathrm{C})^{-1}\mathbf{X}_\mathrm{C}^\mathrm{T}\mathbf{Y} = \begin{bmatrix} 13 & 8 \\ 8 & 5 \end{bmatrix}^{-1} \begin{bmatrix} 18 \\ 11 \end{bmatrix} = \begin{bmatrix} 2 \\ -1 \end{bmatrix}.$$

Hence, the data point $y(10)$ can be estimated by Eq. 8.8, i.e.

$$y(10) = c_1 x(8) + c_2 x(9) = 2x(8) - x(9).$$

This is an over-simplified example. Generally, many more data points are required to form a more meaningful realization in order to satisfy the statistical requirement. Note also that this example appears to be applicable to one-dimensional case only. However, it is applicable to images super-resolution since, as we discussed in the Patch Processing part, a patch of an image is often represented in lexicographically ordered column matrix, i.e. in 1-D form.

8.1.3 Linear Regression

Depending on different choices of the regression models $f(\cdot)$, the regression function can be linear and non-linear. Many real processes are non-linear, however, they can be approximated with linear models or a combination of piecewise linear models. The advantage of linear model is that it can be solved analytically with a closed form solution.

In general, a linear model can be written as below:

$$y_j = f(\mathbf{X}, \mathbf{C}) = \sum_{i=1}^{M} c_i x_{ij}, \quad \text{for } j = 1, 2, \ldots, N, \tag{8.9}$$

where N is the number of data points and M is number of parameters, or in matrix form:

$$\mathbf{Y} = \mathbf{CX} \tag{8.10}$$

without considering the regularization term, the objective function in Eq. 8.10 minimizes the fitting error. Let us use a squared loss function for the minimization:

$$\mathbf{C}^* = \arg\min_{\mathbf{C}} ||\mathbf{Y} - \mathbf{CX}||_2^2 \tag{8.11}$$

The cost function for estimation is the squared error as follows:

$$J(\mathbf{C}) = (\mathbf{Y} - \mathbf{CX})^{\mathrm{T}}(\mathbf{Y} - \mathbf{CX}) \tag{8.12}$$

Take the derivative of the cost function:

$$\begin{aligned} \frac{\partial J(\mathbf{C})}{\partial \mathbf{C}} &= \frac{\partial}{\partial \mathbf{C}}(\mathbf{Y} - \mathbf{CX})^{\mathrm{T}}(\mathbf{Y} - \mathbf{CX}) \\ &= \frac{\partial}{\partial \mathbf{C}}[\mathbf{Y}^{\mathrm{T}}\mathbf{Y} - \mathbf{X}^{\mathrm{T}}\mathbf{C}^{\mathrm{T}}\mathbf{Y} - \mathbf{Y}^{\mathrm{T}}\mathbf{CX} + \mathbf{X}^{\mathrm{T}}\mathbf{C}^{\mathrm{T}}\mathbf{CX}] \\ &= -2\mathbf{YX}^{\mathrm{T}} + 2\mathbf{CXX}^{\mathrm{T}} \end{aligned} \tag{8.13}$$

In order to obtain the minimum error, we make the derivative of the objective function equal to zero. We have

$$2\mathbf{YX}^{\mathrm{T}} = 2\mathbf{CXX}^{\mathrm{T}} \tag{8.14}$$

If $\mathbf{XX}^{\mathrm{T}}$ is invertible:

$$\mathbf{C}^{*} = \mathbf{YX}^{\mathrm{T}}(\mathbf{XX}^{\mathrm{T}})^{-1} \tag{8.15}$$

Equation 8.15 shows the closed form solution for a linear regression problem. To calculate Eq. 8.15, it requires that the number of samples be larger than the dimension of the data ($N \gg M$), otherwise, the matrix $\mathbf{XX}^{\mathrm{T}}$ may not be full-rank to find its inversion.

Let us add a regularization term Ω ($\mathbf{C}$) in l_2 norm, the linear regression then becomes ridge regression. The 2nd term in Eq. 8.16 tries to make every element of the model parameters very small (approaching zero) and leads to more robust and stable solution.

$$\mathbf{C}^{*} = \arg\min_{\mathbf{C}} ||\mathbf{Y} - \mathbf{CX}||_2^2 + \lambda||\mathbf{C}||_2^2, \tag{8.16}$$

where λ is the regularization factor that determines the importance of the regularization term. If λ is too big, it would cause the under-fitting or for the other extreme, it would cause over-fitting. The regularized quadratic cost function:

$$J(\mathbf{C}) = (\mathbf{Y} - \mathbf{CX})^{\mathrm{T}}(\mathbf{Y} - \mathbf{CX}) + \lambda\mathbf{C}^{\mathrm{T}}\mathbf{C} \tag{8.17}$$

Similar to the derivations in Eqs. 8.12–8.15:

$$\begin{aligned} \frac{\partial J(\mathbf{C})}{\partial \mathbf{C}} &= \frac{\partial}{\partial \mathbf{C}}(\mathbf{Y} - \mathbf{CX})^{\mathrm{T}}(\mathbf{Y} - \mathbf{CX}) + \lambda\mathbf{C}^{\mathrm{T}}\mathbf{C} \\ &= \frac{\partial}{\partial \mathbf{C}}[\mathbf{Y}^{\mathrm{T}}\mathbf{Y} - \mathbf{X}^{\mathrm{T}}\mathbf{C}^{\mathrm{T}}\mathbf{Y} - \mathbf{Y}^{\mathrm{T}}\mathbf{CX} + \mathbf{X}^{\mathrm{T}}\mathbf{C}^{\mathrm{T}}\mathbf{CX} + \lambda\mathbf{C}^{\mathrm{T}}\mathbf{C}] \\ &= -2\mathbf{YX}^{\mathrm{T}} + 2\mathbf{CXX}^{\mathrm{T}} + 2\lambda\mathbf{IC} \end{aligned} \tag{8.18}$$

Let us make the derivative equal zero again, we have

$$2\mathbf{YX}^{\mathrm{T}} = 2\mathbf{CXX}^{\mathrm{T}} + 2\lambda\mathbf{IC} \tag{8.19}$$

From the learning perspective, l_2 norm can prevent overfitting and increase generalization ability. From optimization perspective, l_2 norm can help to improve the condition number for matrix inversion problem (make the ill-condition problem become well-condition).

The regularization term also makes $(\mathbf{XX}^{\mathrm{T}} + \lambda\mathbf{I})$ invertible.

$$\mathbf{C}^* = \mathbf{YX}^{\mathrm{T}}(\mathbf{XX}^{\mathrm{T}} + \lambda\mathbf{I})^{-1} \tag{8.20}$$

8.1.4 Classification and Regression

The major idea of Regression is to find filter coefficients, w_{ij} in Eq. 8.2 or **C** in Eqs. 8.10 or 8.15, to interpolate a point or a patch in Eq. 8.10 (let us generally refer the discussion to a patch). The filter coefficients of a model are obtained by making use of a set of data (**X** and **Y**), to construct **C** in Eq. 8.14. How can we find the set of data? For NEDI and other conventional approaches, they use a predetermined set according to a geometric model or make use of some adaptive testing techniques. Different sets of filter coefficients can be obtained for different patch features (hence, feature vectors). These sets of filter coefficients (with respect to their features) can be obtained offline via training, which can then form a dictionary (library). A set of filter coefficients becomes a class in the dictionary, and classes are obtained via training with a large amount of data. This initiates the research on efficient classification for super-resolution. Hence, the super-resolution (interpolation) of images can be done in two steps: Classification and then Regression. Some features in the form of a feature vectors of a patch are used to access the pre-trained library to obtain the class of the patch; hence, filter coefficients, **C**, for the interpolating filter are obtained. For example, the classification can be done via random forests with simple split functions and unsupervised learning. This will form many precise classes, and samples of the same class have very similar properties. Regression models are built during training. With some simple tests (equivalent to feature extraction), the class (hence, **C** of the regression model) of a patch can quickly be determined with the pre-built library **C**. The super-resolution can then be done using Eq. 8.10. In the coming three sections, we are going to discuss super-resolution with conventional learning approaches, the random forests and the convolutional neural network (CNN). In particular, the random forests approach makes explicitly use of the concepts of "Classification and then Regression," but the CNN usually mixes the classification and regression in the network structure.

8.2 Super-Resolution with Conventional Learning Approaches

Classic image super-resolution (including interpolation) approaches model the image characteristics using explicit priors, which are constructed by analyzing the image statistics. For example, the bicubic interpolation uses two-dimensional polynomial to model the stationary image signal. However, the image priors are usually good at modeling global features with limited localization ability [15]. Bicubic interpolation is not good at modeling abrupt signal changes at local regions. In recent years, approaches using adaptive learning from input images and external images are widely available. In this chapter, we entitle them as conventional learning [5–12, 14, 15].

Conventional learning approaches suggest that super-resolution learns from training patches, using the self image or external database to exploit the explicit relationship between the low-resolution and high-resolution pixels. Conventional learning approaches can be classified as online learning [5–11, 14, 15] and offline learning [12]. Online learning approaches learn from recently observed information to estimate the relationship of low-resolution and high-resolution pixels for the super-resolution estimation using a per-defined cost function. Offline learning approaches pre-learn the relationship of low-resolution and high-resolution pixels, and search for the relevant relationship/model during the online estimation to accomplish the super-resolution process [12].

8.2.1 Super-Resolution Using Online Learning from Self-Image and External Database

Online learning approaches often make use of the minimization of mean squared error as the optimization criterion to solve for model parameters for high-resolution estimation. As we discussed in Section 1 of this chapter, one of the widely used models is Auto-Regressive (AR) model [5–7, 9, 11], which is based on the

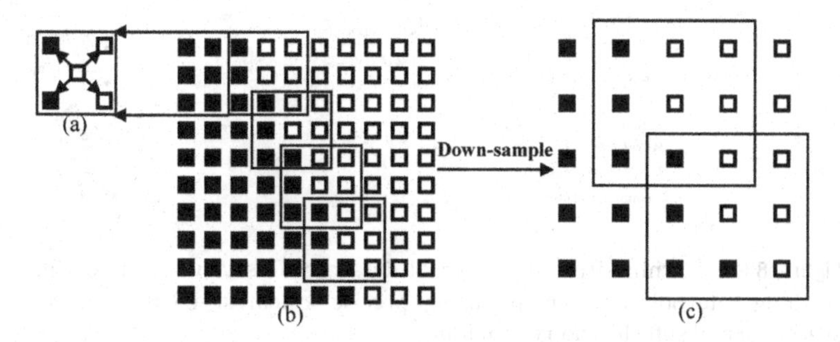

Figure 8.7 Learning from self image: illustration of "Geometric Duality": (a) a HR geometric structure, (b) repeating in the HR image and (c) repeating in the down-sampled LR image.

statistically stationary information within a local region. There are many variations proposed using the AR model, such as NEDI variations [5–7], soft-decision interpolation [11] and *k*-NN MMSE estimation [24, 25, 30–33], which exploit the self-image and the external training patches as the sampling points to formulate the minimal mean squared estimation for estimating the high-resolution pixels. Figure 8.7 shows a sample pattern of a local region for seeing the usefulness of formulating the auto-regressive modeling using the observed signals. The edge patterns shown in Fig. 8.7a is repeated in the HR image (Fig. 8.7b) and also the LR image (Fig. 8.7c), which indicates that the linear relationship between HR and LR signal can be reconstructed by using auto-regressive model.

The bottleneck of online learning from the self-image is resolved by exploiting more external information from the external database. Although the local information is very relevant to estimate the auto-regressive model parameters, there are some cases, say abrupt changes of signals, which are very complicated. For these cases, if there are abundant information from external database for referencing, the model parameters can be accurately estimated. Hence, the conventional learning-based methods using AR model try to search for relevant information in terms of LR-HR image pairs for combing local and non-local information, in order to estimate better model parameters. The representative methods for searching

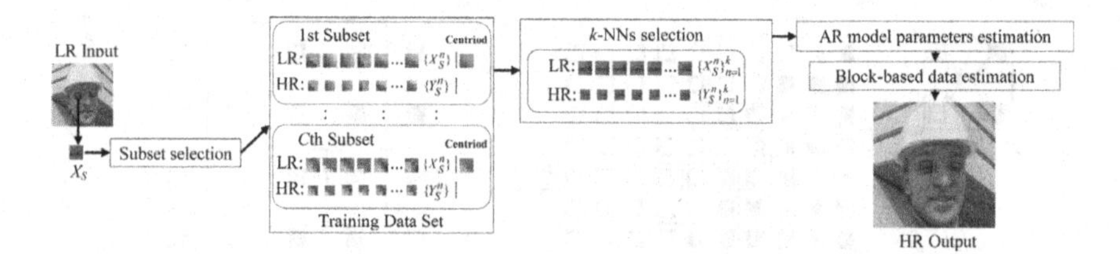

Figure 8.8 Learning from the external database: Overview of the online learning interpolation scheme using precise *k*-nearest LR-HR patches searching and soft-decision estimation.

the external training pairs is the *k*-NN MMSE methods [24, 25, 30–33], which try to find adaptive *k*-nearest neighboring LR-HR image pairs from a pre-built dictionary. The *k*-NN pairs can be grouped to formulate the minimum mean squared error estimation for estimating the high-resolution pixels, which resembles the auto-regressive model. Figure 8.8 illustrates the basic algorithm flow of the external training patch-based approach for conventional online learning: by searching an external training image set, we first extract the corresponding LR-HR patches, then group data into same classes based on the similarity of their feature patterns, and each group can find the patch which is the most representative one as the centroid. During the super-resolution, we extract the LR patch from the LR image and compare it with different centroid patches and use k-NNs to search for the nearest neighbor LR-HR patch and finally, the reconstructed HR patch is calculated by the AR model.

In the following section, let us improve the original AR model to Soft-decision Adaptive Interpolation (SAI) to achieve image super-resolution with high quality.

8.2.2 Super-Resolution via Auto-Regressive (AR) Model

Physically, we can explain the "geometry duality" as that the similar geometric characteristics repeat themselves across the image. These repeated patterns also suggest that the parameter **C** remain constant or near constant in a small region (except at different regions or sudden changes at boundaries). Using AR process, we can perform super-resolution as to look for missing pixels by using the available

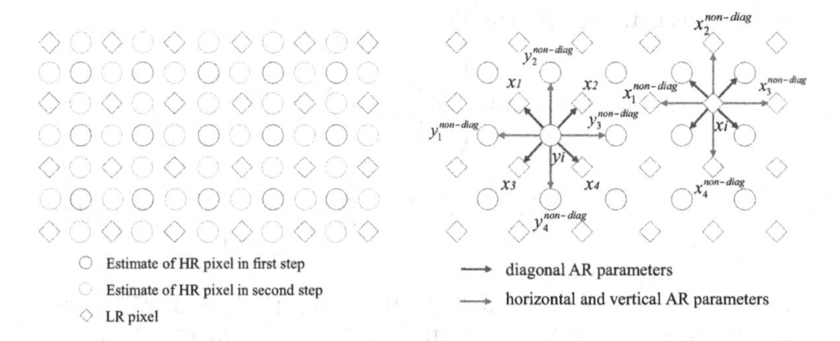

Figure 8.9 Soft-decision interpolation approach.

LR pixels, but it also suffers from the lack of feedback mechanism to reduce the estimation error.

8.2.3 Soft-Decision Interpolation (SAI) Using 2D AR Model

Let us visualize the interpolation problem with a local window as shown in Fig. 8.9. The observed LR pixels x in the LR image (blue diamond in Fig. 8.9) are used to predict HR pixels y in the HR image (circles in Fig. 8.9). Before investigating the advantages of SAI algorithm, we firstly introduce how the SAI works.

Based on the location of missing HR pixels, the SAI algorithm predicts in two steps: (1) predict center HR pixel (red circles in Fig. 8.9, which has 4 diagonal LR pixels) in the first step, (2) predict adjacent HR pixels (green circles in Fig. 8.9 which has two non-diagonal LR pixels) in the second step. In this chapter, we just concentrate on step 1, and step 2 should be similarly done as in Ref. [33]. To calculate the missing pixels in the first step, we use 4 nearest diagonal and/or non-diagonal LR pixels to form a AR model as follows:

$$y_i = \sum_{1 \leq t \leq 4} c_t^{diag} x_t + \varepsilon_i \tag{8.21}$$

where c_t^{diag} is the diagonal AR parameter and ε_i is the random noise. For SAI, the whole local region (window) D is considered. We need to calculate N missing HR pixels $\mathbf{Y} = \{y_i, i = 1, 2, \ldots, N\}$ by the least squares estimation as follows:

$$\{y_i^*, i \in D\} = \underset{y_i, i\in D}{\arg\min} \left\{ \sum_{i\in D} \left\| y_i - \sum_{1\leq t\leq 4} c_t^{diag} x_{it}^{diag} \right\|^2 + \sum_{i\in D} \left\| x_i - \sum_{1\leq t\leq 4} c_t^{diag} y_{it}^{diag} \right\|^2 \right\} \quad (8.22)$$

where $y*_i$ is the i-th predicted HR pixel, y_{it}^{diag} and x_{it}^{diag} are respectively the neighborhood HR and LR pixels in the diagonal directions. The reason that we use the least squares estimation is because we can form a close-form solution similar to linear regression model (please refer to Section 8.1). Equation 8.22 is obviously different from the conventional AR model in Eq. 8.11, which is the second term on the right hand side of the equation. The second term is the error for predicting the known LR pixels from the predicted HR pixels. For conventional AR model, each missing pixel is estimated by its neighborhood LR pixels so that it is independent from missing pixels. However, as we discussed in Section 8.21, pixels in the local window generally follow the stationary process which includes not only the available LR pixels, but also the missing HR pixels (like what we have shown in Fig. 8.7, both HR and LR patch share the same spatial pattern). Assume that we super-resolve a LR image by 4 times, which means using 1/16 LR pixels to predict the rest 15/16 HR pixels. Without including the estimated missing HR pixels, we actually waste many possible pieces of statistical information for evaluation. We entitle Eq. 8.22 as soft-decision interpolation (SAI) because it resembles the block decoding of error correction codes. Within the local window D, the missing HR pixels are better jointly estimated to achieve the minimum least squares error. Furthermore, the reader may find that the second term uses the same set of AR parameters as the first term. This equation requires the estimated missing HR pixels to feed back the known LR pixels with the same AR model. This feedback mechanism gives a soft-decision estimation to prevent the violation caused by miscalculation of missing HR pixels. Meanwhile, not only the diagonal pixels can be used to form the AR model, the non-diagonal (horizontal & vertical) pixels can also be incorporated into the SAI function as follows:

$$\{y_i^*, i \in W\} = \arg\min_{y_i, i\in D} \left\{ \sum_{i\in W} \left\| y_i - \sum_{1\le t\le 4} c_t^{diag} x_{it}^{diag} \right\|^2 + \sum_{i\in D} \left\| x_i - \sum_{1\le t\le 4} c_t^{diag} y_{it}^{diag} \right\|^2 + \lambda \sum_{i\in D} \left\| y_i - \sum_{1\le t\le 4} c_t^{non-diag} y_{it}^{non-diag} \right\| \right\}$$

$$subject\ to \quad \sum_{i\in D} \left\| y_i - \sum_{1\le t\le 4} c_t^{non-diag} y_{it}^{non-diag} \right\|^2 = \sum_{i\in D} \left\| x_i - \sum_{1\le t\le 4} c_t^{non-diag} x_{it}^{non-diag} \right\|^2 \quad (8.23)$$

where $c_t^{non-diag}$ contains the non-diagonal AR parameters on the horizontal and vertical direction, $y_{it}^{non-diag}/x_{it}^{non-diag}$ is the neighborhood HR/LR pixel on the non-diagonal directions and λ is a regularization factor to control the importance of horizontal and vertical correlations. The second part of Eq. 8.23 implies a restricted condition that we assume that both HR and LR pixels can obtain the same estimation error when sharing the horizontal and vertical AR model. The reason is that all the horizontal and vertical neighborhood pixels $y_{it}^{non-diag}$ are unknown for estimation so we use the regularization factor to control the contribution from those pixels.

To calculate AR parameters $\{c_t^{diag}, c_t^{non-diag}, 1\le t \le 4\}$, we can use the available LR pixels to follow the same "geometric characteristics" for model estimation. Figure 8.10 shows a zoom in situation on a local region. The horizontal and vertical parameters can be easily calculated by using the available LR pixels. As for the diagonal parameters, SAI use the idea from NEDI [5] (introduced in Section 8.1) that the nearest diagonal LR pixels share the same distribution as the missing diagonal HR pixels. The accuracy replies strongly on the correlation between pixels in different scales as long as the local window is filled with simple edge or texture pattern.

After introducing the SAI algorithm, there is another point we have not touched yet. The assumption of stationary process is only

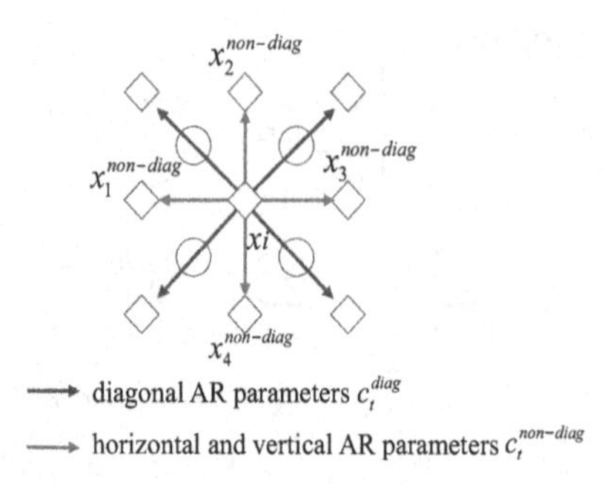

Figure 8.10 AR model estimation.

an ideal situation when we perform super-resolution. Most of the time, the local region of an image contains some outliers or noise that can jeopardize the advantage of the AR model estimation. To solve this problem, in next section we will discuss a robust SAI algorithm for image super-resolution.

8.2.4 Weighted Least Squares to Improve the Robustness

SAI makes use of the least squares estimation (for both parameter and data estimation steps), which is equivalent to the Maximum likelihood and Maximum a Posterior estimation using the white Gaussian noise assumption in the image model. It is well known that the least squares estimation is not robust to outliers. During the last decades, the majority of solutions suggested using a robust norm, such as l_1 norm used for super-resolution and adaptive norm as Huber function, instead of using l_2 norm which results in the use of least squares estimation. Weighted least squares estimation is another more intuitive solution, which multiplies a weighting matrix to residuals, making the minimum modification to the original solution.

For AR model, the "geometric duality" is not well satisfied in some cases. Let us refer to Fig. 8.11, which shows a HR patch extracted from the *lighthouse* image to illustrate this situation. Figure 8.11d shows a randomly sampled geometric structure in the observed LR patch. Most samples in the observed LR patches are irrelevant to the HR geometric structure of interest (Fig. 8.11a),

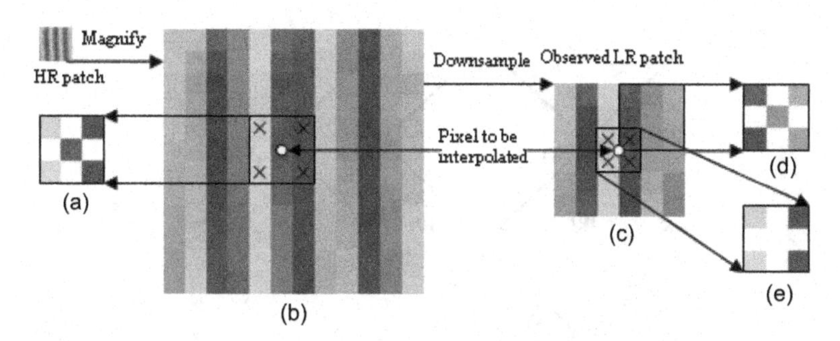

Figure 8.11 Illustration of the failure of "Geometric Duality": (a) a HR geometric structure, (b) the HR patch, (c) illustration of the observed LR patch due to a mismatch of "geometric duality," (d) a randomly sampled LR geometric structures and (e) the partial HR geometric structure without the center value.

such that the estimated parameters using the samples similar to this pattern in the observed LR patches cannot match well to the HR geometric structure for interpolating the missing pixel, which means a mismatch of "geometric duality." Answers to this problem have been addressed using weighted least squares estimation in the literature, which is called the robust soft-decision interpolation [31–33]. First, let us recall the least squares estimation on AR (Eq. 8.11) of the first step by using LR pixels in the matrix formation as

$$\mathbf{C}^{*diag} = \arg\min_{\mathbf{C}^{diag}} ||\mathbf{X} - \mathbf{C}^{diag}\mathbf{X}^{diag}||^2. \tag{8.24}$$

Note that for better understanding, we define AR model $\mathbf{C}^{diag}$ as matrix form for c_t^{diag} in Eq. 8.22, and $\mathbf{X} = \{x_i, i = 1, \ldots, N\}$ is in matrix form for training LR pixels in local region D, $\mathbf{X}^{diag} = \{(x_i^0, x_i^1, x_i^2, x_i^3), i = 1, \ldots, N\}$ is the diagonal LR pixels around the training LR pixels (we call them interpolating pixels). For the least squares estimation, we can assume the estimation error follows the multivariate Gaussian as follows:

$$\mathbf{C}^{*diag} = \arg\min_{\mathbf{C}^{diag}} ||\mathbf{W}^{1/2}(\mathbf{X} - \mathbf{C}^{diag}\mathbf{X}^{diag})||^2, \tag{8.25}$$

where $\mathbf{C}^{*diag}$ is the predicted AR model, $\mathbf{W}$ is the weighting matrix that assigns different weighting parameters for the residuals in $(\mathbf{X} - \mathbf{C}^{diag}\mathbf{X}^{diag})$. In other words, in a local window, each training LR pixel is given a weighted value based on its importance for evaluate

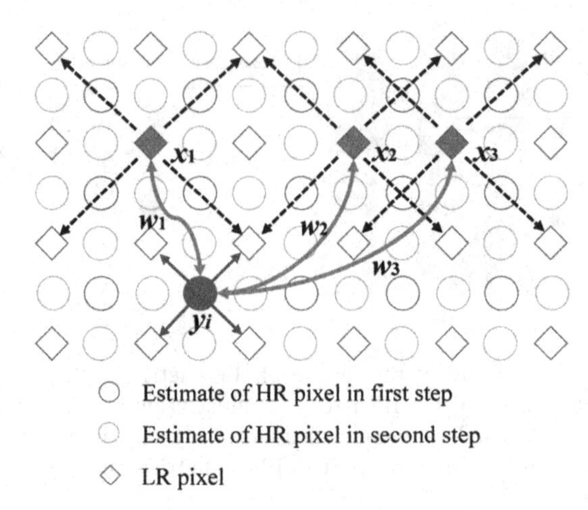

Figure 8.12 Weighting parameters for SAI algorithm.

the AR model. As shown in Fig. 8.12, to estimate the HR pixel (solid red), we take the three training LR pixels (solid blue diamond) for learning the AR parameters. According to the distance between the missing HR pixel and the three training LR pixels, we can assign different weights for estimation. The most natural and intuitive way is to use the Euclidean distance to measure the importance for each training pixel. In [33], we use the bilateral filter to calculate the distance between the HR pixel and the training LR pixels. The definition of bilateral filter is as follows:

$$W_i = e^{(-||y_c - x_i||^1/125)} \cdot e^{(-||z_c - z_i||^2/2)}, \tag{8.26}$$

where W_i is i-th element of the weighting matrix $\mathbf{W}$, y_c is the initial HR pixel calculated by bilinear interpolation. $||y_c - x_i||^1$ is the radiometric distance (pixel intensity difference) and $||z_c - z_i||^2$ is the geometric distance. The value 2 and 125 can be found empirically. Note that the bilateral filter is used because it provides a better edge selective smoothing mechanism for distance measurement. Similarly, we can also add the weighing matrix to the non-diagonal AR model on the second step as follows:

$$\mathbf{C}^{*\text{non-diag}} = \underset{\mathbf{C}^{\text{non-diag}}}{\arg\min} ||\mathbf{W}^{1/2}(\mathbf{X} - \mathbf{C}^{\text{non-diag}}\mathbf{X}^{\text{non-diag}})||^2 \tag{8.27}$$

where $\mathbf{C}^{*non-diag}$ is the non-diagonal AR model that is used for estimating HR pixel at the second step and $\mathbf{X}^{non-diag}$ is the interpolating pixels at the non-diagonal locations.

In the above introduction, we have discussed about the SAI algorithm and the weighted robust SAI algorithm for image super-resolution. However, both SAI and robust SAI algorithms are restricted to the local region to guarantee the stationary process for AR estimation. In order to explore more and more possible statistical characteristics for unknown HR pixels prediction, we will introduce how to make use of the external database for image super-resolution.

8.2.5 Super-Resolution via Learning from External Database Using *k*-NN MMSE Soft-Decision Estimation

In the last section, we have discussed the conventional image interpolation using online learning from the self image. However, the performance of online learning from self image can be significantly improved by searching from external database [14–16]. In this section, a robust and precise *k*-Nearest Neighbors (*k*-NN) searching scheme is discussed to form an accurate AR model of the local statistic [33]. We make use of both LR and HR information obtained from a large amount of training data, in order to form a coherent soft-decision estimation of both AR parameters and HR pixels. Experimental results show that the learning-based AR interpolation algorithm has a very competitive performance compared with the recent image interpolation algorithms in terms of PSNR (to measure the average fidelity of super-resolution) and SSIM (to measure the pattern similarity of super-resolution) values.

8.2.6 *k*-NN MMSE Estimation for Image Interpolation

In this section, a learning-based scheme for estimating the AR model parameters using a precise *k*-NN searching and a coherent soft-decision estimation is introduced. Before we move on, let us recall that both SAI and robust SAI algorithms consider interpolation within in a local region. There is a large variety of patterns on images. If we estimate a patch based on the collection of a group of images with similar statistical properties, this forms the idea of patch-based super-resolution. We can classify patches from external images into classes based on their feature distance. Ref. [33] gives an approach that estimates the AR model parameters by searching the k-nearest LR-HR training pairs from an external data set. Both LR and HR information from the k-nearest training pairs are used in the k-NN MMSE approach. This is essentially different from previous AR model parameters estimation approaches which rely on the information from the same LR input image or a pre-computed set of AR parameters. We will discuss it in the next section.

8.2.7 Precise Search for k-Nearest LR-HR Training Pairs

Let us denote the training data set for the LR patch $\mathbf{X}_S$ as $\{\mathbf{X}_S^n = \{x_i^n | i \in S\}\}$ and the training data set for the HR patch $\mathbf{Y}_S$ as $\{\mathbf{Y}_S^n = \{y_i^n | i \in S\}\}$, where n is the patch index. Let us also illustrate the process for interpolating the HR patch $\mathbf{Y}_S = \{y_i | i \in S\}$ in Fig. 8.12 during the online estimation. The procedure is shown as follows:

Algorithm 1: Learning-based image AR interpolation scheme
Input: LR patches $\{\mathbf{X}_S\}$, training data $\{\mathbf{X}_S^n\}$ and $\{\mathbf{Y}_S^n\}$
Output: Estimated HR patches $\{\mathbf{Y}_S\}$
(1) Initialization: For each $\mathbf{X}_S$

(a) Use normalized correlation coefficient $NCC(\mathbf{X}_S, \mathbf{X}_S^n)$ to find the k nearest neighbors, $\{\mathbf{X}_S^n\}_{n=1}^k$ and $\{\mathbf{Y}_S^n\}_{n=1}^k$, by Eq. 8.28.
(b) Use the k-NNs to find the AR model parameters, $\mathbf{C}$.
(c) Use AR model parameters to estimate the HR patch, $\mathbf{Y}_S$.

First, we use the k-means approach to roughly cluster the training LR-HR patches into several groups and save the training patches as a dictionary for online search. Second, we use the normalized correlation coefficient (NCC) to measure the similarity for patch matching. When the absolute value of the NCC equals to one, then there exists a linear relation between the two samples; on the other hand, when the value of the NCC equals zero, the two

Figure 8.13 images used in the k-NN MMSE scheme.

samples have no linear relation. Generally, the higher the absolute value of the coefficient, the stronger the linear relation between the two samples. We choose *NCC* because of its scale invariance and robustness to small noises. To search for the k-nearest LR-HR training pairs, let us define the *NCC* measure of the LR patch from $\mathbf{X}_S$ and all LR patches in the training set $\{\mathbf{X}_S^n\}$ as,

$$\mathrm{NCC}(\mathbf{X}_S, \mathbf{X}_S^n) = \sum_{i \in S} \frac{(x_i - \mathrm{Mean}(\mathbf{X}_S))(x_i^n - \mathrm{Mean}(\mathbf{X}_S^n))}{\sigma_{\mathbf{X}} \sigma_{\mathbf{X}}^n} \quad \text{for } \forall n \tag{8.28}$$

where *Mean*(.) is the mean operation, and $\sigma_{\mathbf{X}}$ and $\sigma_{\mathbf{X}}^n$ are standard derivations of patch $\mathbf{X}_S$ and patch $\mathbf{X}_S^n$ respectively. The measured *NCC* values are then sorted in a descending order, such that $NCC(\mathbf{X}_S, \mathbf{X}_S^n) \geq NCC(\mathbf{X}_S, \mathbf{X}_S^{n+1})$. Then, k-nearest LR training patches $\{\mathbf{X}_S^n\}_{n=1}^k$ which have the k-largest *NCC* values are chosen by the adaptive k criterion [14],

$$\hat{k} = \arg\min_k k \text{ s.t. } \sum\nolimits_{n=[1,k]} NCC(\mathbf{X}_S, \mathbf{X}_S^n) > T_1 \tag{8.29}$$

where T_1 is a threshold to determine the degree of similarity for k-nearest neighbors. Specifically, the largest *NCC* value, $NCC(\mathbf{X}_S, \mathbf{X}_S^1)$, is summed with the second largest *NCC* value, $NCC(\mathbf{X}_S, \mathbf{X}_S^2)$, etc., until the sum of k-largest *NCC* values is larger than T_1. The adaptive k-NN criterion in Eq. 8.29 is used to search for a large amount of k-NNs depending on the sum of costs, which can resolve the problem of outliers by increasing the number of k-NNs when there are insufficient relevant matches. Moreover, the precise and accurate searching is retained by decreasing the number of k-NNs when there are just a few very relevant matches. After finding the k-NNs training patches, we can use the AR model we introduced before to predict the HR patch and then we overlap the reconstructed HR patch together to form the complete HR image.

8.2.8 Experimental Results of Conventional Learning Methods for Interpolation

The training data set was formed by extracting the 25 images as shown in Fig. 8.13. The 25 training images cover rich image details, including edges with different orientations, various contours, and irregular structures, etc. The size of the training data set is 1,400,000, which was clustered into 32 subsets by *k*-means clustering.

During the testing, the natural images were down-sampled directly by two times and there is no compression used during the process. For color images, the chrominance components were interpolated using the bicubic interpolation. The *k*-NN MMSE algorithm was implemented using MATLAB codes, while the codes of the comparing algorithms [5, 9, 10, 15] were obtained from respective authors. Subjective evaluations in Fig. 8.14 show that the *k*-NN MMSE algorithm reconstructs edges with less aliasing effects such as jags around the leaves, and produces the smoothest textures (starfish in Fig. 8.14). Furthermore, the *Cameraman* reconstructed by the *k*-NN MMSE algorithm retains the boundaries accurately as shown in Fig. 8.15.

8.2.9 Summary of Conventional Learning for Super-Resolution and Interpolation

In this section, we have discussed the super-resolution and interpolation approaches using online learning based on auto-regressive model, which are the conventional learning tool for 2D image signals. The applications of AR model for interpolation are explained in details for two classes of methods using self image and external database as the signal sampling sources for the minimum mean squared error estimation. The key essence of conventional learning is to build an explicit and adaptive local model to relate the low-resolution and high-resolution pixels, where the local and non-local sampling window play the key role for estimating the model

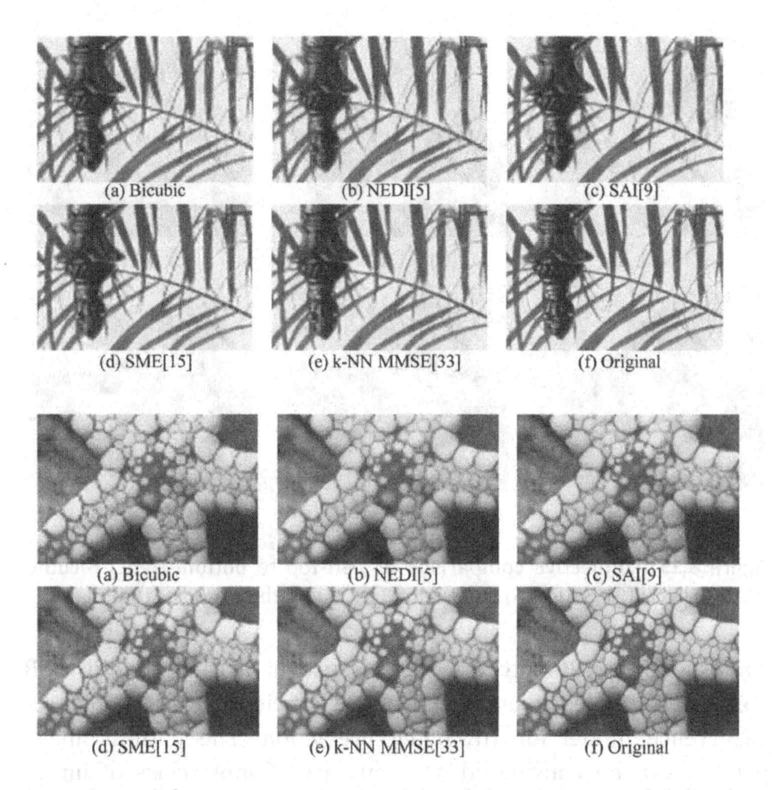

Figure 8.14 Subjective comparison of (top-left to bottom-right) bicubic, NEDI [5], SAI [9], SME [15], k-NN MMSE [33] and the original HR image.

parameters. Through MMSE estimation, it can extract the stationary property within a local and/or non-local regions for optimization. This property holds for a wide variety of natural images, such that the conventional learning approaches are still one of the top performers compared with the recent approaches for the same level of computational complexity.

8.3 Super-Resolution with Random Forests and Its Variations

The learning-based Single Image Super-resolution (SISR) methods try to learn the mapping function from the LR patches to the

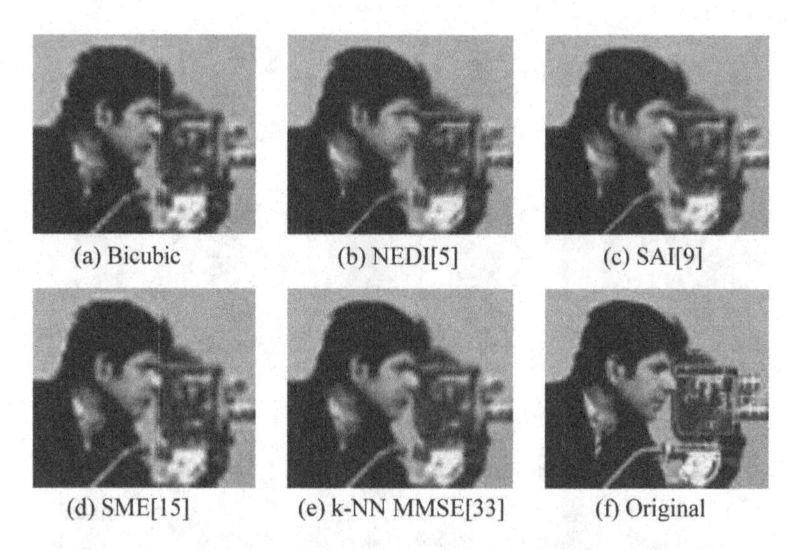

Figure 8.15 Subjective comparison of (left-top to bottom-right) Bicubic, NEDI [4], SAI [5], SME [15], k-NN MMSE [33] and the original HR image.

corresponding HR patches. The essence of learning-based SISR methods is to learn an effective and efficient LR-HR patch co-occurrence model for HR patch prediction. The natural image patches can be considered as a mixture of multi-class of image patches. We can make an assumption that the image patches in the same class form a linear subspace where the relationship between the HR image patches and the LR image patches can be modeled by a linear mapping function. Based on the above-mentioned fact and assumption, the SISR problem can be solved by a combination of classification and regression process. During the training phase, the training HR-LR patch pairs can be classified into a number of classes according to the appearances of the LR patches. The relationship between HR and LR patches is modeled as a regression problem to learn a linear mapping function for each class.

In this section, random forests and its variations will be applied for single image super-resolution based on the procedure of classification and the regression idea. The rest of this Section is organized as follows: Section 8.3.1 introduces the image super-resolution algorithm using random forests including the detailed training and testing procedures. Sections 8.3.2 and 8.3.3 further describe single image super-resolution methods with a hierarchical decision tree/random forests framework which further improve the super-resolution performance as well as the algorithm efficiency.

8.3.1 Image Super-Resolution via Random Forests

Random forests are an ensemble learning algorithm for classification and regression. Random forests can achieve high accuracy and very low computation cost. Each random forests contains multiple decorrelated decision trees. Random forests reduce the prediction error induced by estimation variance by averaging the results from multiple decorrelated decision trees. As random forests have been discussed in Chapter 1, we will not review it in detail.

In this section, we will show how to use random forests for fast image super-resolution. Let us call this method image super-resolution via random forests (SRRF). The schematic diagram for SISR with random forests is shown in Fig. 8.16. During testing, each decision tree $(T_i, i = 1, \ldots, n)$ within the random forests recursively classifies the LR patches into one of the leaf nodes using a simple decision (comparing two pixel values with a threshold) in each non-leaf node and maps the LR patches to the corresponding HR patches through the linear regression models $(\mathbf{C}^i_j, i = 1, \ldots, n;$ $j = 1, \ldots, m)$ stored in the j-th leaf nodes of i-th decision tree. The final estimated HR patch is the average of the HR patches estimated by all decision trees in the random forests. Both the classification

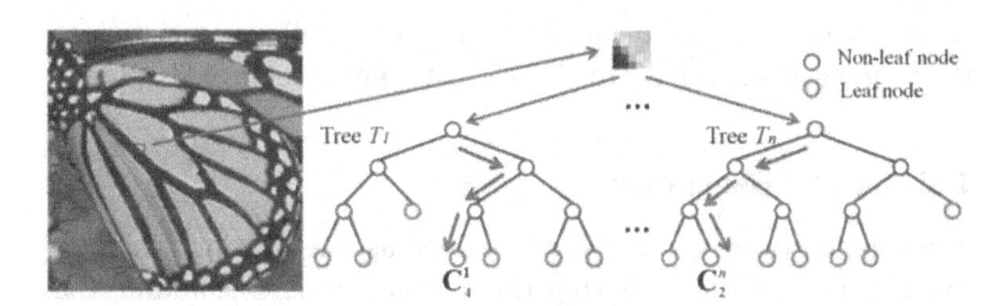

Figure 8.16 A random forest which consists of a set of decision trees. Each decision tree recursively classifies the input LR patch into left or right child node, until a leaf node is reached. By using the linear regression models (e.g. $\mathbf{C}^1_4, \ldots, \mathbf{C}^n_2$) stored in the reach leaf node, the input LR patch can be projected to the HR patch space.

process and the regression process take extremely small amount of computation. Besides, different from the K-means clustering algorithm which performs classification by minimizing l_2 distance, the proposed approach achieves classification via minimizing the reconstruction mean squared error (MSE) which relates directly to Peak Signal-to-Noise Ratio (PSNR). This leads to excellent super-resolution performance.

8.3.1.1 Training algorithm

Compared with a single decision tree, random forests achieve better generalization and higher stability by combining multiple de-correlated decision trees. Each decision tree $T_t(t = 1, 2, \ldots, n)$ in the random forests is trained using a subset of the training data to increase randomness. Besides, the randomized node optimization enables us to further increase the randomness by randomly sampling a subset of binary tests.

In the training phase, each non-leaf node in the decision tree chooses one binary test from a set of randomly selected binary tests, which can maximize the error reduction, to split the training data at this node into its two child nodes. When the error reduction is less than zero, that node is declared as a leaf node. Hence, a regression model is learned using the HR-LR patch pairs which reached this leaf node.

The focus of this subsection is not the random forests training procedure, but the key elements which enable random forests for the single image super-resolution task. For a detailed description of the training algorithm for random forests, please refer to Chapter 1.

Training Data Generation

The training time of a decision tree increases as the increasing of the training data size. During training and super-resolution, the LR image is initially up-sampled to the same size as the desired HR image using bicubic interpolation. The bicubic interpolation will introduce artifacts along strong edges, while performs well in smooth regions. To relieve the burden on training and increase the learning efficiency, the focus of training is on the image patches where edge pixels can be detected. Hence, we only collect patches containing edge pixels, while exclude patches from smooth regions in order to reduce both the training time and testing time.

The edge pixels are determined by applying the Canny edge detection (one commonly used edge extraction operator) on the bicubic interpolated image. The objective of applying the Canny edge detection on the bicubic interpolated image is not to accurately detect the edges, but to detect the possible regions where there are artifacts introduced by bicubic interpolation. If the edge magnitude of a pixel is larger than a certain threshold, this pixel is regarded as an edge pixel. The edge threshold is determined by considering both the testing time and the image super-resolution quality. With an increase in edge threshold, the image super-resolution quality would decrease, since fewer prominent edges would be processed. We select 60 as the threshold, because there is no significant drop in PSNR compared with processing the whole image, while it reduces to half of its processing time.

The extracted HR-LR patch pairs $\{\mathbf{P}_i = (\mathbf{Y}_i, \mathbf{X}_i)\}$ are used for training, where $\mathbf{Y}_i \in \mathbf{R}^d$ is the vectorized HR patch sampled from the original training image $\mathbf{T}$. $\mathbf{X}_i \in \mathbf{R}^d$ is the corresponding vectorized LR patch on $\mathbf{T_U}$ which is interpolated from the $\mathbf{T}$ down-sampled image $\mathbf{T_D}$ (using bicubic interpolation), and $\sqrt{d}$ is the patch size.

Binary Test and Split Function

The task of each non-leaf node is to classify a patch into one of its two child nodes. Since this is a binary classification problem, the classification at each non-leaf node is controlled by a binary test and a split function which will split the training data into its left or right child node according to the defined cost function. Here, the binary test adopted is parameterized by $\theta = (p, q, \tau)$ and it evaluates the difference of two pixel values in the LR patch vector $\mathbf{X} \in \mathbf{R}^d$ at position p and q with a threshold τ. This kind of binary tests has the intensity invariant property which is helpful to adapt to various intensity conditions in natural images and increase training efficiency.

The split function $h(\mathbf{X}, \theta)$ splits the HR-LR patch pair into left or right node according to the results of the binary test θ evaluated on the LR patch $\mathbf{X} \in \mathbf{R}^d$. That is a HR-LR patch pair will be classified into the left child node if the split function returns value 0, and into the right child node otherwise.

$$h(\mathbf{X}, \theta) = \begin{cases} 0, & \text{if } \mathbf{X}(p) < \mathbf{X}(q) + \tau \\ 1 & \text{otherwise} \end{cases} \tag{8.30}$$

where $1 \leq p, q \leq d, 0 \leq \tau \leq 255$ and d is the vector length of $\mathbf{X}$.

Linear Regression Model

After passing through the non-leaf nodes, the image patches arrived at a leaf node are similar to each other. We assume that a linear model can well approximate the relationship between the HR image patches and the LR image patches in this leaf node. At leaf node j, a linear regression model $\mathbf{C}_j \in \mathbf{R}^d \times \mathbf{R}^d$ is used to predict the corresponding HR patches of the given LR patches. Assume l training patch pairs $S_j = \{P_i | i = 1, \ldots, l\}$ arrived at node j. We group all the training patches into matrix forms $\mathbf{Y} \in \mathbf{R}^d \times \mathbf{R}^l$ and $\mathbf{X} \in \mathbf{R}^d \times \mathbf{R}^l$ for HR and LR training patches, respectively. The linear regression model $\mathbf{C}_j \in \mathbf{R}^d \times \mathbf{R}^d$ at node j can be obtained by minimizing the mean squared error between the ground-truth HR patches and the reconstructed HR patches:

$$\mathbf{C}_j = \arg\min_{\mathbf{C}_j} ||\mathbf{Y} - \mathbf{C}_j\mathbf{X}||_2^2 \tag{8.31}$$

Equation 8.31 has a closed form solution and can be solved by the least squares method.

$$\mathbf{C}_j = \mathbf{Y}\mathbf{X}^{\mathrm{T}}(\mathbf{X}\mathbf{X}^{\mathrm{T}})^{-1} \tag{8.32}$$

To obtain de-correlated decision trees in a random forest, the minimum member of patch pairs is usually set to two times of the feature dimension of the HR patch.

Cost Function

Image super-resolution problem is a regression problem. The cost function for evaluating the goodness of a binary test can be defined as the mean squared error between the ground-truth high-resolution patches and the estimated high-resolution patches. Let us define the fitting error $E(\mathrm{S})$ as the mean squared error between the original HR patches $\mathbf{Y}$ and the reconstructed HR patches $\mathbf{Y}^R$ of training data S.

$$E(S) = \frac{1}{|S|}||\mathbf{Y} - \mathbf{Y}^R||_2^2 \tag{8.33}$$

With this defined fitting error, the standard random forests learning algorithm (please refer to Chapter 1) can be applied to learn a random forest with the given HR-LR patch pairs.

8.3.1.2 Testing algorithm

In the image up-sampling phase, the input LR image is initially up-sampled to the same size as the desired HR image (using bicubic interpolation). Image patches containing edge pixels are extracted from the initial up-sampled image. Extracted image patches are passed into the learned random forests. In each decision tree, the input patch $\mathbf{X}$ is classified into left or right node according to the binary test stored at each non-leaf node. When the patch reaches a leaf node, the linear regression model $\mathbf{C}$ stored at the leaf node is used for the up-sampling. In order to efficiently combine all the reconstruction results generated from N decision trees, the obtained N linear regression models are firstly averaged. The final reconstructed HR patch is obtained by multiplying the averaged linear regression model with the patch $\mathbf{X}$.

$$Y^R = \left(\frac{1}{N}\sum_{i=1}^{N}\mathbf{C}_i\right)X \tag{8.34}$$

As the input patches are overlapped with their neighbors, each pixel will have multiple prediction values. The final super-resolved HR image is obtained by overlapping the reconstructed HR patches (averaging all prediction values of a pixel) to preserve the consistence between neighboring pixels.

8.3.1.3 Regression model fusion

With a learned random forests/decision tree, the image super-resolution performance can be further enhanced with a regression model fusion idea. For a leaf node ($leaf_0$) in the learned decision tree for image super-resolution, there are 3 relevant leaf nodes

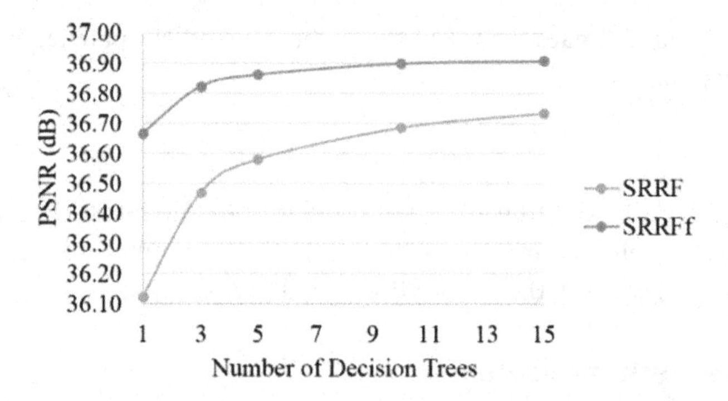

Figure 8.17 Comparison between SRRF and SRRFf with different number of decision trees on Set5 with upscaling factor 2.

$leaf_i$ ($i = 1, 2, 3$) where the reached patches are approximately the $i \times 90°$ rotated versions of that at $leaf_0$, and another 4 relevant leaf nodes $leaf_i (i = 4, \ldots, 7)$ where the reached patches are approximately the flipped versions (left flip to right) of that at $leaf_i (i = 0, \ldots, 3)$. Thus after transformation, the regression models at $leaf_i$ ($i = 1, \ldots, 7$) can have very similar coefficients as that at $leaf_0$.

Based on this observation, an input LR patch can retrieve 8 regression models from a single decision tree. By combining 8 regression models from each decision tree, we could gather 8 times more regression models for aggregation without extra training efforts. The additional testing complexity is small and comes from retrieving relevant leaf nodes and additions between regression models. We take the advantage of the fusion model idea and retrieve $8T$ regression models from random forests with T decision trees. We denote this method as SR random forests with fusion model (SRRFf).

Figure 8.17 shows the average PSNR of SRRF and SRRFf evaluated on Set5 [16] for upscaling factor of 2 with varying number of decision trees in the random forests. By combining a larger number of regression models, the quality of the HR patch estimation can be further enhanced. We can find that it is beneficial to fuse regression models. When $T = 1$, SRRFf provides over 0.5 dB improvement in PSNR compared with SRRF. SRRFf with a single decision tree has similar performance as SRRF with 8 decision trees in the random forests. When $T = 5$, the improvement is around 0.3 dB. As T increases, the gap between SRRFf and SRRF slowly drops, but remains around 0.2 dB.

8.3.2 Image Super-Resolution with Hierarchical Random Trees

In the previous section, SRRF method has been introduced. By combining the regression models from multiple de-correlated decision trees, a robust regression model can be obtained for HR patch prediction. The regression model fusion idea further improves the super-resolution performance without extra learning. To further improve the super-resolution performance and the testing efficiency, a hierarchical decision tree/random forest-based SISR method will be introduced in this section.

As shown in Fig. 8.17, the performance of SRRF can be improved by increasing the number of decision trees in the random forests. From the testing efficiency point of view, SRRF with a large number of decision trees is not the optimal choice since after saturation the improvement of the super-resolution quality is minor when compared with the increment of computational time. From both efficiency and effectiveness aspects, we cannot get further improvement (in the direction for the exploitation of fast and high-quality image SR algorithm) from random forests structure.

An extreme case of the SRRF method is to only learn a large single decision tree within the random forests. Let us call this approach as super-resolution decision tree (SRDT) method. The training algorithm for SRDT method is similar to the one introduced in Section 8.3.1 earlier. Although a large number of leaf nodes reduces ambiguity and improves the HR patch prediction accuracy, the training efficiency of SRDT method drops as the training data size increases. Using more training data to obtain a bigger SRDT may not be an effective way to get better image SR quality. Besides, the size of the learned SRDT is exponentially increasing as the depth of the decision tree ascends. This could hinder the feasibility of the super-resolution algorithm.

A better alternative is to perform super-resolution using a hierarchical decision trees (SRHDT) framework to further boost the

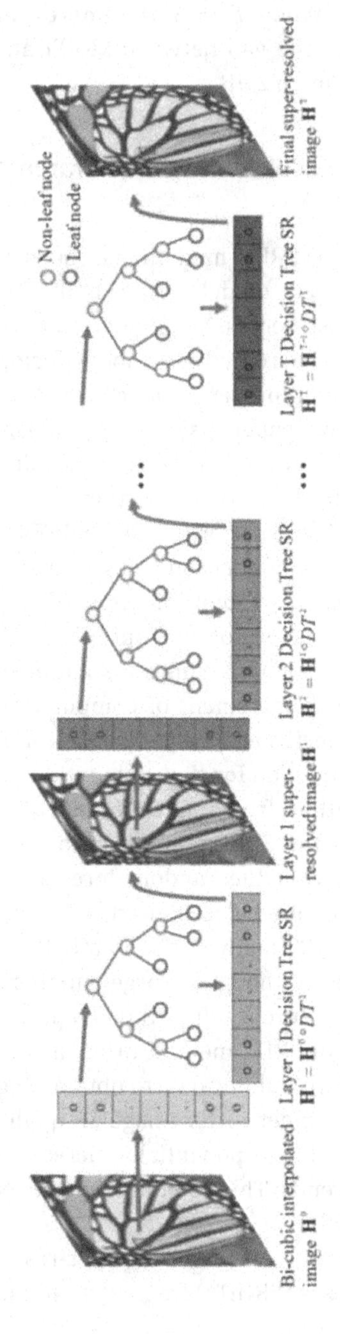

Figure 8.18 Flowchart of the image super-resolution using hierarchical decision tree method.

image SR quality of the SRDT method with short computational time. Figure 8.18 presents the flow diagram of the SRHDT method. The general idea is that each layer of the SRHDT framework is a super-resolution decision tree which pushes the estimated HR patch closer to the ground-truth HR patch. The SRHDT method progressively refines the initial bi-cubic up-sampled image using hierarchical decision trees $HD^{\mathrm{T}} = \{DT^1, \ldots, DT^{\mathrm{T}}\}$ where T is the number of layers in the hierarchical decision trees. Each refinement is performed by adding one more layer of decision tree DT^{T} which is trained with the LR-HR patch pairs from the previous layer super-resolved images and the original HR images for a set of training images. Layer i decision tree predicts the HR image $\mathbf{Y}^i$ and is denoted as

$$\mathbf{Y}^i = \mathbf{Y}^{i-1} \circ DT^i. \tag{8.35}$$

The operation '$\circ$' consists of LR patch extraction, HR patch prediction and HR image reconstruction. The learned SRDT in each layer can also follow the regression model fusion strategy as described in Section 8.3.1.3 by which the desired HR patch of each LR patch is predicted according to the fused regression model using up to 8 relevant regression models in a decision tree for higher accuracy. We denote the SRHDT method with fused regression model as SRHDTf.

The training procedure of SRDT in each layer in the SRHDT takes the LR patches as the extracted patches from the super-resolved images applied by the previous layer decision trees and the HR patches from the original HR training images, except that there is only one decision tree in the random forests. Since there is only one decision tree in each layer, the de-correlation technique should not be applied for learning SRDT. The minimum number of patch pairs for leaning a regression model should be increased. With the learned hierarchical decision trees $HD^{\mathrm{T}} = \{DT^1, \ldots, DT^{\mathrm{T}}\}$, the initially bicubic up-sampled input LR image $\mathbf{Y}^0$ will be enhanced layer by layer.

8.3.3 Hierarchical Random Forests

With similar idea, random forests can also be utilized to build a hierarchical structure for SISR. Instead of using decision trees

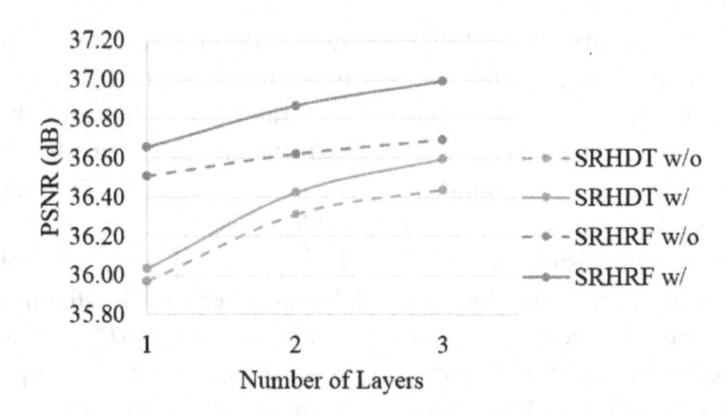

Figure 8.19 Comparison between SRHDT and SRHRF with and without model fusion using different number of layers on Set5 with upscaling factor 2. Each random forests contains 5 decision trees.

as building block, super-resolution hierarchical random forests (SRHRF) applies a random forests in each layer for the pursuit of better performance. The regression model fusion idea can also be used.

Figure 8.19 shows the average PSNR in dB of the super-resolved images using SRHDT and SRHRF with and without model fusion. SRHRF with model fusion offers around 0.3 dB improvement, while the average PSNR of SRHDT with fusion model is only 0.15 dB higher than that without model fusion. When model fusion is applied in both methods, SRHRF has around 0.4 dB higher PSNR than SRHDT.

8.4 Deep Learning Approaches for Image Super-Resolution

The recent development of Convolutional Neural Network (CNN) benefits from the advent computing machines and the availability of varying huge data sets. The traditional machine learning methods fail to learn a general model to achieve high accuracy performance on computer vision tasks. One of the problems is that most of the traditional machine learning methods require to design manually the computation structures and feature extraction techniques to fulfill the tasks. It means that the model is pre-determined by prior information of certain distribution and the training data would be clustered into groups based on their feature classification.

On the contrary, Convolutional Neural Network is a data-driven model that it randomly initializes kernels for training and learns the optimized parameters. This mechanism can train the CNN model automatically to extract multiple layers of deep features in order to discover the non-linear mapping relationship between the input and the target. However, it also brings us to another problem: to make the thousands of parameters of the CNN model converge at certain point, which requires fast computation to process a huge amount of data for training. Due to the rapid development of Graphics Processing Unit (GPU) in the recent decades which means high-speed computation is possible, more and more researchers have moved on to the study of computer vision via CNN model. Recent publications and experiments also prove the promising future of Convolutional Neural Network.

8.4.1 Revision of CNN Model

CNN model is developed from the traditional Neural Network (NN) as shown in Fig. 8.20. In the past, the NN model fails to train a deep structure to handle a huge data set, like ImageNet [42], because of the vanishing gradient and huge parameter optimization. From the study of deep learning theory, CNN is one of ways to stack a hierarchical layer to study multi-level feature, pattern

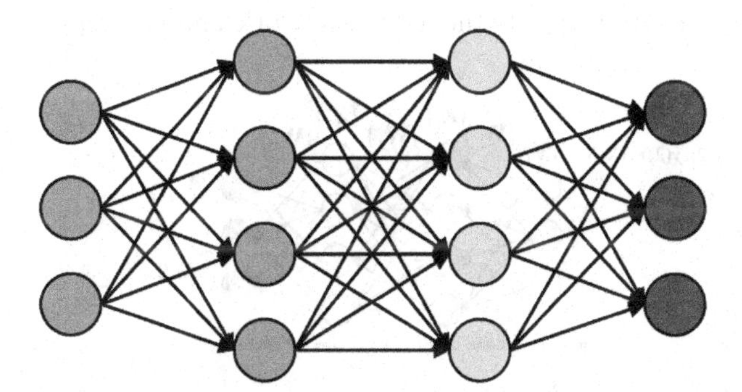

Figure 8.20 CNN model.

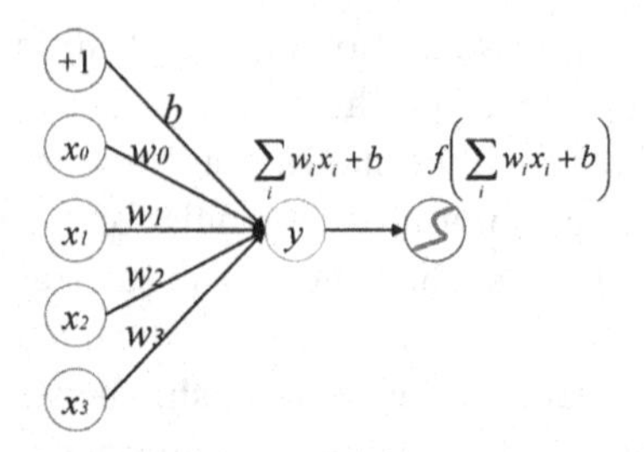

Figure 8.21 Neuron structure.

or transformation from the training data. The basic structure of CNN consists of input layer, hidden layer and output layer. Each layer consists a number of neurons, where activation and linear combination are performed. The structure is shown in Fig. 8.21. For each neuron as shown in Fig. 8.21, it has different weights $\{w_i\}$ assigned to the inputs and one additional biased value b. The activation function $f(\cdot)$ using non-linear function to map the linear combination result to obtain the output.

During the training process, as shown in Fig. 8.22, we have the loss function (function E shown in Fig. 8.22) to measure the learning accuracy and back propagate the error into the layers for parameter update, where D is the batch size representing the number of sub-sample for one iteration of backpropagation. Assuming a training data set that contains n training examples, $\{(\mathbf{X}(1), \mathrm{y}(1)), \ldots, (\mathbf{X}(\mathrm{N}), \mathrm{y}(\mathrm{N}))\}$, where $(\mathbf{X}(\mathrm{i}), \mathrm{y}(\mathrm{i}))$ is the i-th example, $\mathbf{X}(i)$ represents the feature vector (the input to the neural network, and $y(i)$ represents

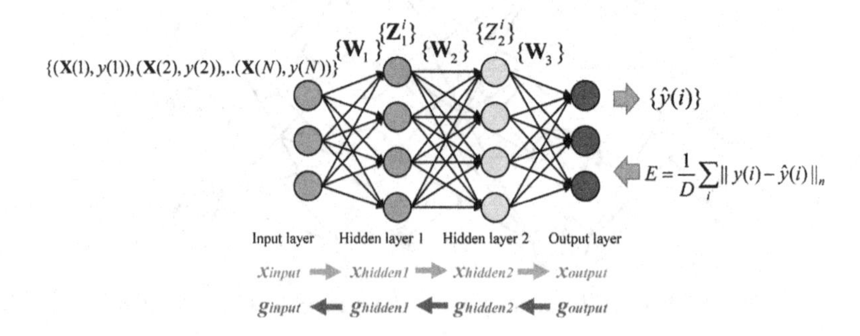

Figure 8.22 Backpropagation.

the ground-truth label, is used to train a neural net with a single output $\hat{y}(i)$ and $\mathbf{Z}_j^i$ is the j-th convolution layer output. Assuming $\hat{y}(i)$ is the predicted output of the i-th example by forward propagating the i-th feature vector ($\mathbf{X}$(i), y(i)) through the neural network. As shown in Fig. 8.22, the forward path (the blue arrow) passes the results to the output layer, and the output layer calculates the loss (using the predicted label and the ground truth label to calculate the l_n-norm distance) and passes the gradient value along the backward path (the red arrow) to update the parameters.

8.4.2 Super-Resolution via Convolutional Neural Network

The first CNN model for image super-resolution is SRCNN [20]. It is a 3-layer end-to-end CNN structure that can input one LR image and generate a SR image. The basic structure is shown in Fig. 8.23.

The SRCNN model contains three convolution layers. The filter (also refer as kernel in some literature) size at each convolution layer (presented in red block in Fig. 8.23) is described as

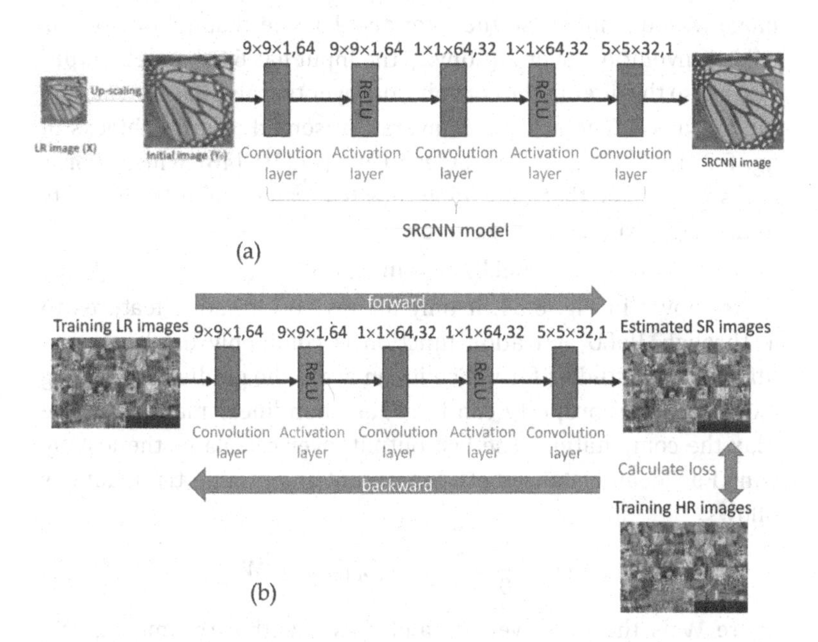

Figure 8.23 SRCNN model: (a) Testing procedure. (b) Training procedure.

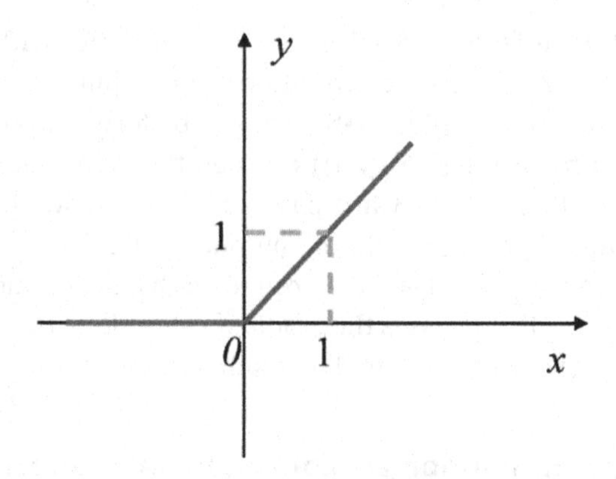

Figure 8.24 ReLU function.

$(M \times N \times C, N)$, where M, N and C is the dimension of the filters and N is the number of filters, i.e., $(9 \times 9 \times 1, 64)$ means that Convolution layer 1 (Conv1) has 64 filters with a filter size $(9 \times 9) \times 1$ channel. The channel number is equal to the number of input images. For convolution layer one, the input is the initial gray-level image with one image so the filter also has the channel number as 1. For convolution layer 2 (Conv2), the input has 64 channels (input images) so the filter in the convolution layer two also has the channel number as 64. The activation layers (presented as green blocks in Fig. 8.23) all use Rectified Linear Unit (ReLU) [46] which works as a sparse filter that only allow positive value going through. Its mathematic expression is as follows:

$$\text{ReLU}(x) = \max(x, 0) \tag{8.36}$$

As shown in Fig. 8.24, it only allows non-negative features to go through. ReLU activation function is commonly used in many different CNN studies for which it can avoid the gradient vanishing and its sparsity property can introduce non-linear mapping while relax the computation. The last output layer calculates the loss by using Euclidean distance between output and ground truth data as follows:

$$L(\mathbf{W}) = \frac{1}{D}\sum_{i=1}^{D} ||\mathbf{Y}_i - \mathbf{Z}_i^3||_2^2 + \lambda||\mathbf{W}||_2^2 \tag{8.37}$$

where $\mathbf{W}$ is the total weights and biases within the model, $\{\mathbf{Y}_i,$ $i = 1, 2, \ldots, D\}$ is the i-th ground truth HR image, $\{\mathbf{Y}^R = \mathbf{Z}_i^3,$

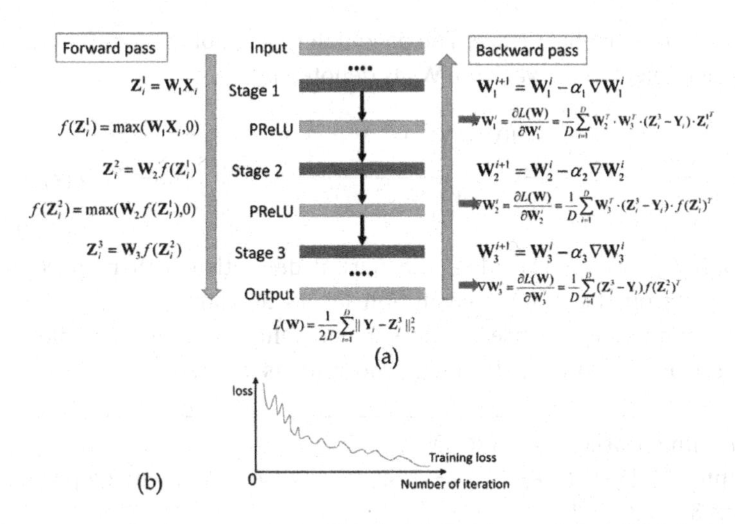

Figure 8.25 Backpropagation in SRCNN. (a) Weights updating. (b) Relationship between training loss and iteration.

$i = 1, 2, \ldots, D\}$ is the i-th predicted HR image from the output of convolution layer 3 and λ is the regularization factor. D is the batch number of training LR-HR image pairs for one iteration.

For training process, the loss in Eq. 8.37 is calculated based on a batch of LR-HR image pairs which is the concept of Stochastic Gradient Descent (SGD) to statistically minimize to the loss function. To make it simpler, SGD randomly samples a batch of data from the training set to calculate the gradient of weights $\mathbf{W}$ at each iteration and eventually find the minimal point of the loss. Finally, the weight updating for SRCNN training is shown in Fig. 8.25. Using the loss calculated in Eq. 8.37, the input data go through the forward pass as the left arrow shown Fig. 8.25a. The gradient of weights $\{\nabla \mathbf{W}_i\}$ can be calculated at i-th iteration for weights updating following the right arrow shown in Fig. 8.25a. During training process, the weights are initially assigned with random values and updated as shown in Fig. 8.25a. Let us take convolution layer 3 ($\mathbf{W}_3$) as an example to explain how the weight updates.

The loss function $L(\mathbf{W})$ can be rewritten as

$$L(\mathbf{W}) = \frac{1}{2D}\sum_{i=1}^{D} ||\mathbf{Y}_i - \mathbf{Z}_i^3||_2^2 = \frac{1}{2D}\sum_{i=1}^{D} ||\mathbf{Y}_i - \mathbf{W}_3 f(\mathbf{Z}_i^2)||_2^2, \quad (8.38)$$

where D is the batch size. The partial derivative of the loss function with respect to the variable $\mathbf{W}_3$ is denoted by

$$\frac{\partial L(\mathbf{W})}{\partial \mathbf{W}_3} = \frac{\partial \left\{ \frac{1}{2D} \sum_{i=1}^{D} ||\mathbf{W}_3 f(\mathbf{Z}_i^2) - \mathbf{Y}_i||_2^2 \right\}}{\partial \mathbf{W}_3} = \frac{1}{D} \sum_{i=1}^{D} (\mathbf{Z}_i^3 - \mathbf{Y}_i) f(\mathbf{Z}_i^2)^{\mathrm{T}} \tag{8.39}$$

Similarly, we can calculate the partial derivative with respect to weights on the other convolution layers as shown in Fig. 8.25a. To summarize, we use image super-resolution as an example to describe the complete training procedure as follows:

Training procedure of SRCNN

Input: 16 LR-HR $\{\mathbf{X}_i, \mathbf{Y}_i, i = 1, 2, \ldots, 16\}$ gray-level image pair of size $32 \times 32 \times 1$

1. Initialize weights $\{\mathbf{W}_i, i = 1, 2, 3\}$ with randomly generated values for convolution layers 1–3.
2. Set iteration $T = 500{,}000$
3. Start training

For iter $= 1{:}T$

a. Calculate the forward pass to obtain the output of convolution layer 3 as estimated SR images $\{\mathbf{Y}^R = \mathbf{Z}_i^3, i = 1, 2, \ldots, 16\}$

$$\mathbf{Z}_i^3 = \mathbf{W}_3 f(\mathbf{W}_2 f(\mathbf{W}_1 \mathbf{X}_i))$$

b. Calculate the loss as

$$L(\mathbf{W}) = \frac{1}{2 \times 16} \sum_{i=1}^{16} ||\mathbf{Y}_i - \mathbf{Z}_i^3||_2^2$$

if $L(\mathbf{W}) <$ threshold or iter $> T$
Stop the training

else
Update the weights as shown in Fig. 8.25a
end

end

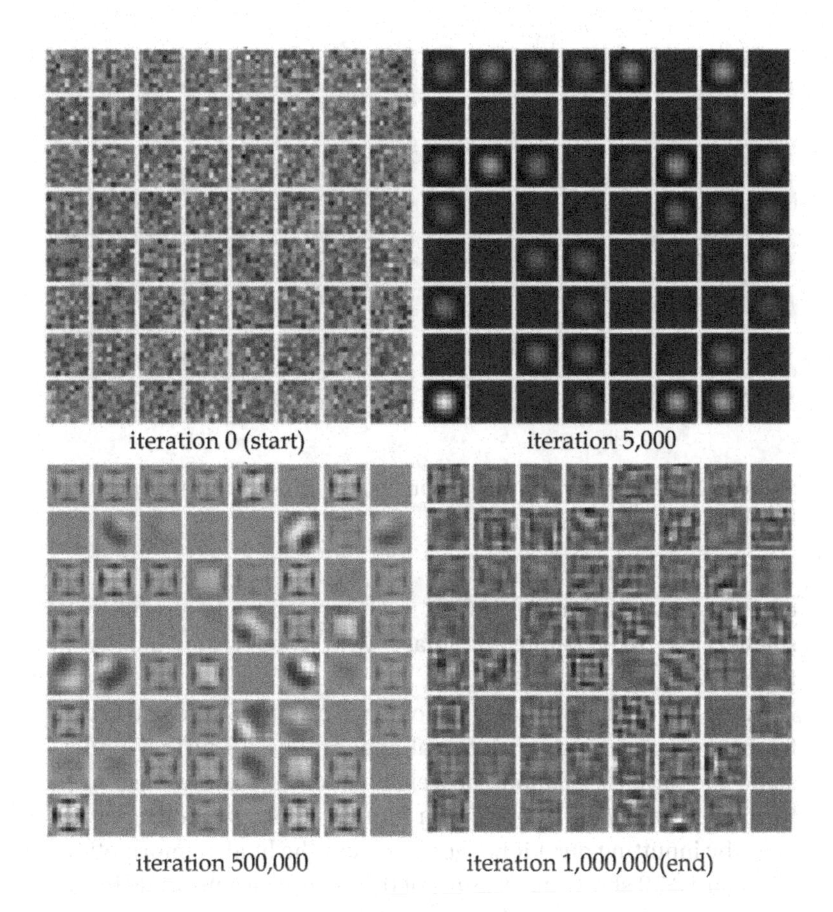

Figure 8.26 Filter updating in Convolution layer 1.

Based on the above training procedure, Fig. 8.25b shows that the more iterations, we train the model, the smaller the loss we obtain. It is easier to observe the loss reducing with the iteration increasing. However, to understand how the weights change during the training, we need to extract the weights from SRCNN model and visualize them as gray-level scale filter image. In order to demonstrate the updating process, we have extracted the 64 filters of size $9 \times 9 \times 1$ at convolution layer one in Fig. 8.26.

From Fig. 8.26, we can see that the filters in convolution layer 1 pick up the meaningful features from random noise to certain

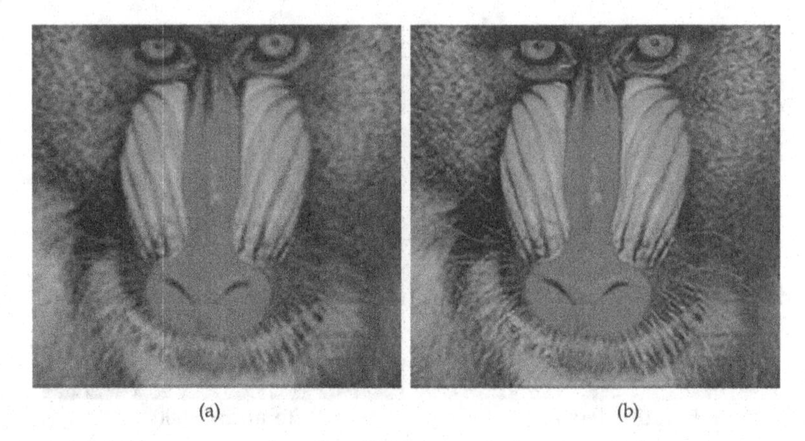

Figure 8.27 Visualization of Baboon image. (a) Bicubic interpolation. (b) SRCNN.

edge patterns. Eventually, SRCNN model converges to a result with smaller loss to generate SR image with good visual quality in Fig. 8.27.

After the training process is done, we can use SRCNN structure for image super-resolution in real applications. To further understand the physical meaning of weights at each layer, we also need to visualize the filers at each layer and their corresponding feature maps by inputting one LR image to find out the hidden information.

Figure 8.28 shows an example of the feature visualization. In this example, we use *butterfly* image as a testing image. For convolution layer 1, we have 64 filters of size 9×9. Each filter convolutes with the LR image and go through the activation function ReLU to obtain one feature map (as indicated by the black arrows, each feature map is described as ReLU($\mathbf{W}_1^i \otimes \mathbf{X}$)) where $\otimes$ represents the convolution operation and $\mathbf{W}_1^i$ is the i-th filter of the Conv1 layer. For example, the filter at the left upper corner convolutes with the LR image to obtain the left upper corner feature map on the Conv1 result. Hence, we obtain 64 feature maps with size 255×255. We can find that some filters at Conv1 layer extract low frequency components of the image while other filters extract the middle or high frequency components. This corresponds to patterns observed on varying conventional filters for visualization. This layer is what most of the

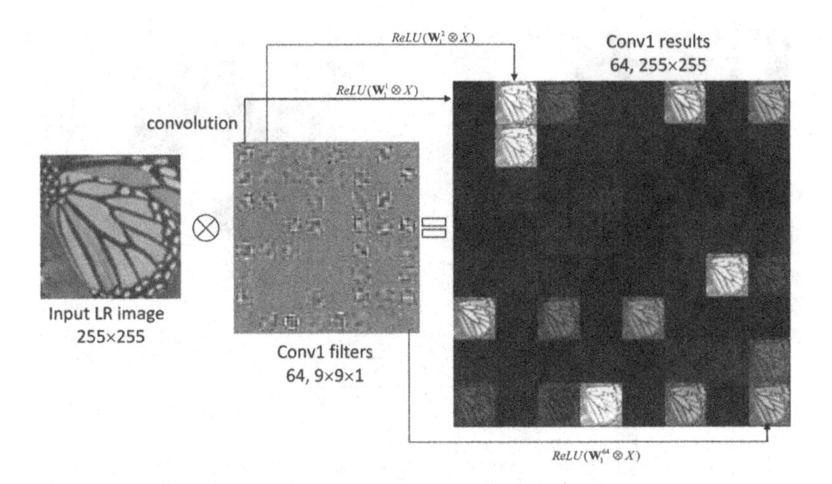

Figure 8.28 Convolution layer 1.

researchers describe as feature extraction. It decomposes the input LR image into the feature space as sampling the image by using different frequencies in conventional approaches.

Figure 8.29a shows the convolution layer 2. The input is the 64 feature maps of the output from convolution layer one; hence, the input size is $64 \times 255 \times 255$. There are 32 filters at convolution layer two with size $1 \times 1 \times 64$. Number 64 means that each filer has channel 64 corresponding to 64 input feature maps. Since the filter size is $1 \times 1 \times 64$, each Conv2 result can be considered as a weighted sum of the 64 conv1 results. Mathematically, the calculation is

$$\mathrm{ReLU}(\mathbf{W}_2^i \otimes \mathbf{Z}^1) = \sum_{c=1}^{64} \mathbf{W}_2^{i,c} \cdot \mathbf{Z}^{1,c}, \; i = 1, 2 \ldots, 32 \qquad (8.40)$$

where $\mathbf{W}_2^{i,c} \in \mathbf{R}^{1\times1}$ is the i-th Conv2 filter on channel c and $\mathbf{Z}^{1,c} \in \mathbf{R}^{255\times255}$ is the c-th feature map of the Conv1 result. Eventually, 32 feature maps are generated after convolution layer 2. This layer uses 1×1 filters to achieve data compression, that is, reducing the feature maps from 64 to 32. In this way, the complexity of computation for the next layer is reduced. From the observation of the feature maps at convolution layer 2, we can find that layer 2 generates more image-like feature maps that contain more details of the original image and the feature maps with high frequency

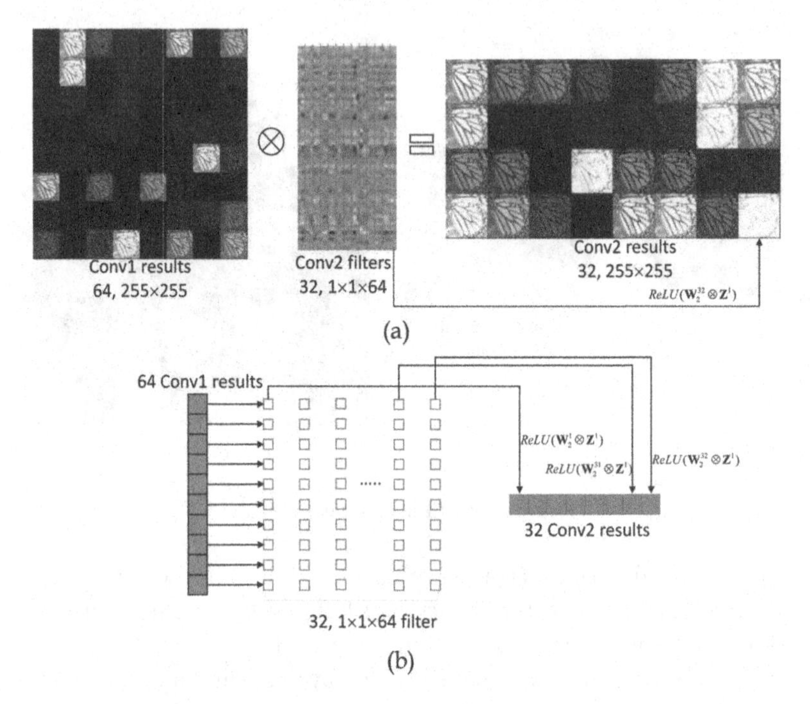

Figure 8.29 Convolution layer 2. (a) Convolution at layer 2. (b) The calculation of convolution at layer 2.

components are much reduced compared with convolution layer 1. The $1 \times 1 \times 64$ filter can be regarded as a weighting vector of length 64. It assigns a different weight for different input to compress relevant feature maps while encourages the significant ones.

As shown in Fig. 8.30, convolution layer 3 has 1 filter with size $5 \times 5 \times 32$. The input is the 32 feature maps of convolution layer 2, and it outputs one SR image as the final output. Convolution layer 3 works similar to convolution layer 2. Each channel of the filer convolutes with one Conv2 feature map and add them together to obtain the final output. The difference is that the filter size is 5×5 instead of 1×1 and there is no ReLU activation function. Mathematically, the calculation is

$$\mathbf{W}_3 \otimes \mathbf{Z}^2 = \sum_{c=1}^{32} \mathbf{W}_3^c \cdot \mathbf{Z}^{2,c} \tag{8.41}$$

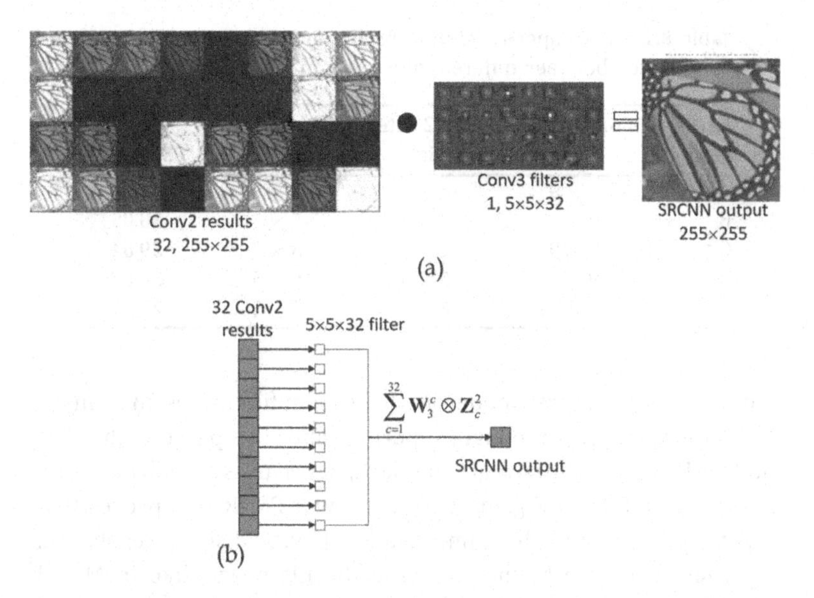

Figure 8.30 Convolution layer 3. (a) Convolution at layer 3. (b) The calculation at convolution layer 3

where $\mathbf{W}_3^c \in \mathbf{R}^{5\times5}$ is the c-th channel Conv3 filter and $\mathbf{Z}^{2,c} \in \mathbf{R}^{255\times255}$ is the c-th feature map of the Conv2 result. From the observation of Conv3 filters, we can find that filters extract different frequency components from the input feature maps and combine them together to obtain the SR image. It can be regarded as a synthesis process of further refinement of features and assigned different weights to the feature maps. The reason of using 5×5 filter instead of 1×1 is to let the SRCNN learn more complicated weighting model to extract the local information to avoid the uniform weighting process done by 1×1. To objectively measure the differences of using different filter sizes, we have done some experiments and let us use Peak Signal-to-Noise Ratio (PSNR) to calculate the quality of super-resolution. Because of the small change on PSNR can hardly be observed from the image visualization, we only record the PSNR values to illustrate the influence of different filter sizes in Table 8.1.

The test is $4 \times$ super-resolution done on Set5 [16] data set which is commonly used in many image super-resolution approaches for comparison. Model A is the original SRCNN we introduced before

Table 8.1 4 × super-resolution PSNR performance on Set5 set comparison between different filter sizes in SRCNN model

Layer/ Model	Conv1 filter size	Conv2 filter size	Conv3 filter size	PSNR(dB)
A	9 × 9	1 × 1	5 × 5	29.98
B	9 × 9	3 × 3	5 × 5	30.05
C	9 × 9	1 × 1	3 × 3	29.84
D	9 × 9	5 × 5	5 × 5	30.07
E	11 × 11	5 × 5	5 × 5	28.89

and B, C and D are three models with different filter sizes. In terms of PSNR, model D gives the highest performance compared with other models. If we compare A and C, model C, which uses a small filter size on convolution layer 3 gives a slightly lower PSNR, and proves that larger filter size can help to improve the quality of super-resolution. However, when we further increase the filter size, like in Model E which uses 11 × 11 filter for Conv1 layer, the PSNR decreases dramatically compared to model D. The question arises from this observation: How do we choose the filter size to obtain optimal results? It seems impossible to increase the filter size to pursue better performance on image super-resolution. This question can be answered in the following section.

8.4.3 Super-Resolution via Joint Back Projection and Residual Network

After introducing and analyzing SRCNN model, let us further discuss more interesting findings in CNN study. Joint Back Projection and Residual Network (BPRN) [44] for image super-resolution combines two recent CNN techniques together to design a better network for image super-resolution: residual learning and back projection. Before talking about BPRN model, we firstly need to understand the concept of residual learning and back projection.

8.4.3.1 Residual learning

The idea of residual learning comes from the ResNet [47]. ResNet is proposed for image classification by using more than hundreds

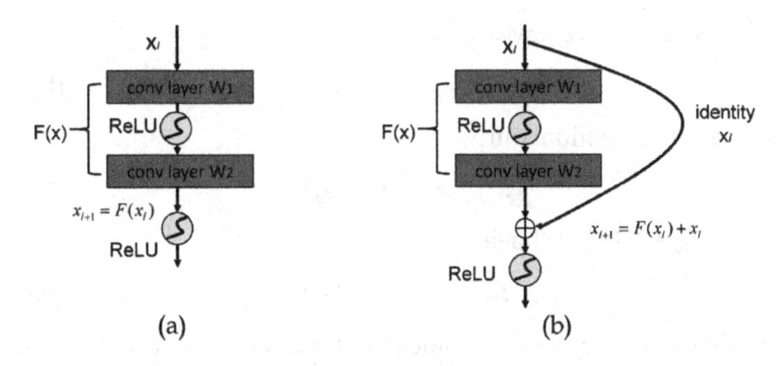

Figure 8.31 Residual block. (a) Direct mapping. (b) Residual mapping.

of convolution layers. Before ResNet, there are some other CNN models, like the VGG [48] and GoogLeNet [49], that can only design CNN networks with tens convolution layers because if a network is too deep, we will face the degradation on performance. It is actually a common problem we usually encounter. ResNet in [47] stacks the proposed residual blocks up to one hundred in order to exploit the deeper feature maps without facing the same problem and the performance is further improved.

A residual block is the basic module of the ResNet. Its basic structure is shown in Fig. 8.31. To make comparison, we also draw the direct mapping module in VGG and other non-residual block-based models.

As shown in Fig. 8.31a, the direct mapping is to learn the mapping relationship from the input to the output directly as Eq. 8.42:

$$x_{l+1} = F(x_l), \text{ where } F(x_l) = \mathbf{W}_2 \otimes f(\mathbf{W}_1 \otimes x_l) \tag{8.42}$$

where x_l is the input, f is the ReLU activation function and $\otimes$ represents the convolution operation. F is used to represent the combination of two layers of convolution with one extra activation calculation.

The direct mapping model is very straightforward, however, the residual mapping is to learn the residual information between the input and output. The difference is that residual mapping focuses on pushing the residual to zero by using shortcut to link the input to the output. By doing so, the degradation and also vanishing gradient problems can be avoided. To understand the advantage of residual mapping, we need to look into how the backpropagation works. Note that in Fig. 8.31a, we have

$$x_{l+1} = F(x_l). \tag{8.43}$$

If we want the residual output as

$$x_{l+1} = x_l + F(x_l), \tag{8.44}$$

or from Fig. 8.31b, we have

$$x_{l+1} = x_l + F(x_l). \tag{8.45}$$

Note also that the residual block is designed with 2 convolution layers and two activation layers. The residual output is obtained after these two convolution and activation because we want to make use the activation layer to learn a sparse residual model for better generalization and also feasibility in real application.

Then at $l+1$-th layer, we have

$$x_{l+2} = x_{l+1} + F(x_{l+1}). \tag{8.46}$$

Finally, at the last layer L, we add all the previous residual output together, we have the final output as

$$x_L = x_l + \sum_{i=l}^{L-1} F(x_i). \tag{8.47}$$

On the contrary, the direct mapping for last layer L, we have

$$x_L = \prod_{i=l}^{L-1} \mathbf{W}_i x_l. \tag{8.48}$$

Assuming we have the loss function as E, by using chain rule of gradient, the partial derivative of the loss function E with respect to l-th layer for the residual mapping has the following backpropagation,

$$\frac{\partial E}{\partial x_l} = \frac{\partial E}{\partial x_L} \cdot \frac{\partial x_L}{\partial x_l} = \frac{\partial E}{\partial x_L}\left(1 + \frac{\partial}{\partial x_l}\sum_{i=i}^{L-1} F(x_i)\right). \tag{8.49}$$

Similarly, the back propagation for direct mapping, we have

$$\frac{\partial E}{\partial x_l} = \prod_{i=l}^{L-1} \mathbf{W}_i \frac{\partial E}{\partial x_L} \tag{8.50}$$

From Eqs. 8.49 and 8.50, we can find that using residual learning, we can always calculate the gradient because any partial derivative

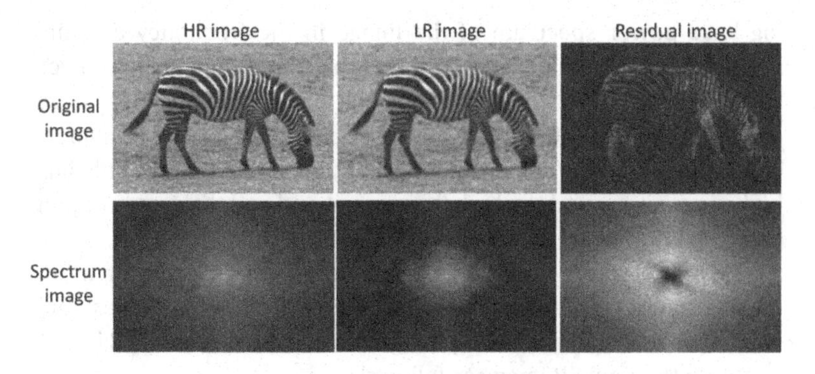

Figure 8.32 HR, LR and Residual image.

to x_L $(L > l)$ is directly back propagated to x_l with extra residual information. However, direct mapping is multiplicative product that the gradient for l-th layer is likely getting smaller or bigger. To the extreme, there could be vanishing or exploding gradient that cause the degradation problem.

From the perspective of image super-resolution, residual learning is not a strange concept. From the sparse-based approaches to random forest-based approaches, residual learning has been commonly used for image super-resolution. The basic idea is based on the observation that the LR image and the HR image share with the same low frequency information and the only difference is the missing high frequency component. In other words, compensating the residual around the edge and texture regions for LR images can resolve the blurring and ringing artifacts to generate good SR images.

In Fig. 8.32, we show the image of *Zebra*. The LR image is obtained by using simple interpolation approach to enlarge to the same size as the HR image. The residual image is generated by subtracting the LR image from the HR image. We can observe that the residual image has bigger values around the strides of the zebra and the grass, which indicates that the missing information is around the edge areas. In order to further illustrate the physical meaning of the residual image, we can use Fourier transform to convert the image data to the frequency domain and shift the zero-frequency component to the center of the array and visualize the log base power spectrum of the image in the frequency domain. From the spectrum image, we can find that the HR image has power spectrum from low to high frequency bands and the LR image only has power spectrum concentrates around center low frequency band. Obviously, the spectrum image of the residual image only has higher power around the high frequency band and the center region has very low power.

This observation is widely used in image super-resolution as an application of residual learning. The target is to learn the residual between the HR (**Y**) and LR (**X**) image and add it back to the LR image to obtain the final SR image as follows:

$$\mathbf{Y} = \mathbf{X} + f(\mathbf{X}) \tag{8.51}$$

where residual information $\mathbf{R} = f(\mathbf{X})$. Based on Eq. 8.51, there are some CNN-based residual networks for image super-resolution. Very Deep convolutional network for image Super-Resolution (VDSR) [41] is one of the first trial using residual learning for image super-resolution. The model is to input LR images to learn the residual between the LR and HR image and then add the learned residual to the LR image to obtain final LR image.

As shown in Fig. 8.33, during the training process, we force the output of the VDSR as $argmin\{\mathbf{Y}-(\mathbf{X}+\mathbf{R})\}$ so that the VDSR structure uses the LR image learns the residual image via a 20-layer convolution model instead of directly mapping to the HR image. However, this is only merely a basic implementation based on the concept of residual learning introduced in Fig. 8.32. It does not embed the residual learning into the CNN model as the ResNet in Fig. 8.31. This is the reason why VDSR can only train a 20-layer CNN model to achieve the moderate performance. Recently, image Super-Resolution via ResNet (SRResNet) [43] uses the ResNet as the backbone structure for image super-resolution to achieve a much

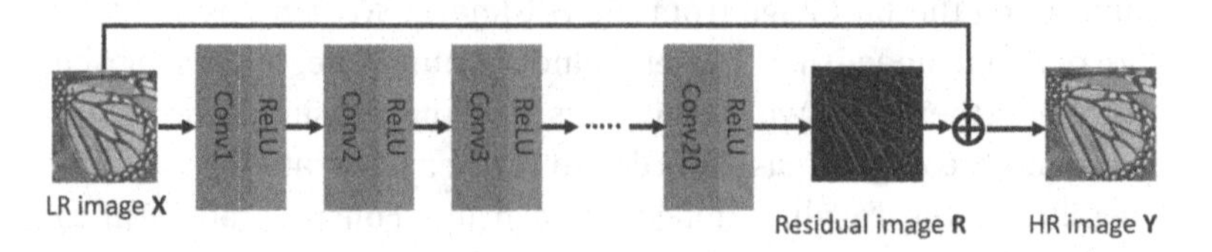

Figure 8.33 Training process of VDSR.

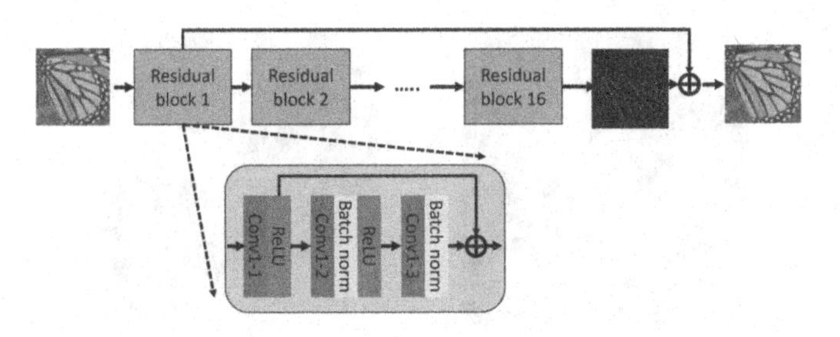

Figure 8.34 Training process of SRResNet.

better super-resolution performance. The authors manage to design a 37-layer convolution model with 16 residual blocks (similar as the Fig. 8.31b to increase the depth of the CNN model and exploit deeper features.

SRResNet is shown in Fig. 8.34. The key difference between SRResNet and VDSR is that SRResNet uses shortcuts inside each residual block to force them to learn the intermediate residual feature maps rather than just the final residual image. Note that the residual block in SRResNet is different from the residual block shown in Fig. 8.31b. At the end, there is also a global connection from residual block 1 to residual block 16 to learn the global feature maps.

To have a better understand what features are extracted at each residual block, we visualize the feature maps at different residual block by making use of the *butterfly* LR image. Figure 8.35 shows the all 64 feature maps of the input, output and residual of the residual block 7. We can see that the residual feature maps (Fig. 8.35b) learned at residual block 7 show more meaningful edge patterns around the wings of the butterfly while it is not obvious to see the differences from the input and output feature maps of residual block 7. By stacking more residual blocks, SRResNet can gradually learn the residual information to compensate the differences between the LR and HR image to generate the final SR image.

8.4.3.2 Back projection

The idea of back projection [50] was firstly proposed to resolve tomographic reconstruction problem. It is used in image super-resolution based on the idea that the down-sampled SR image

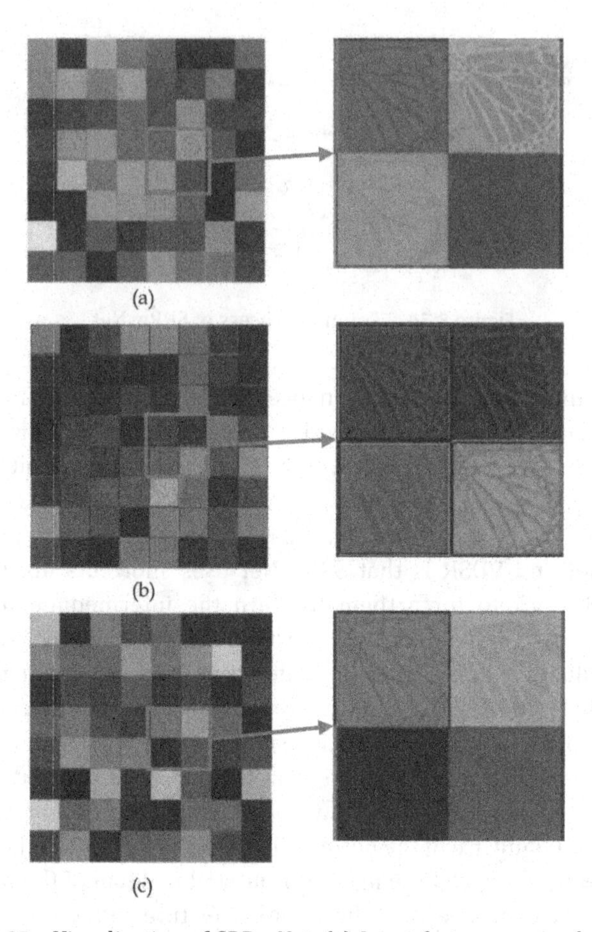

Figure 8.35 Visualization of SRResNet. (a) Input feature maps of residual block 7. (b) Residual feature maps learned at residual block 7. (c) Output feature maps of residual block 7.

should share the same distribution as the original LR image so we can ensure the SR image is close to the original HR image. To make it simpler: for back projection, assuming we generate a SR image close enough to the original HR image, then their low-resolution images should also share the same distribution. The residual between the down-sampled SR image and the original LR image can be added back to the SR image iteratively until the performance of the SR image saturates. Mathematically, we can summarize the back

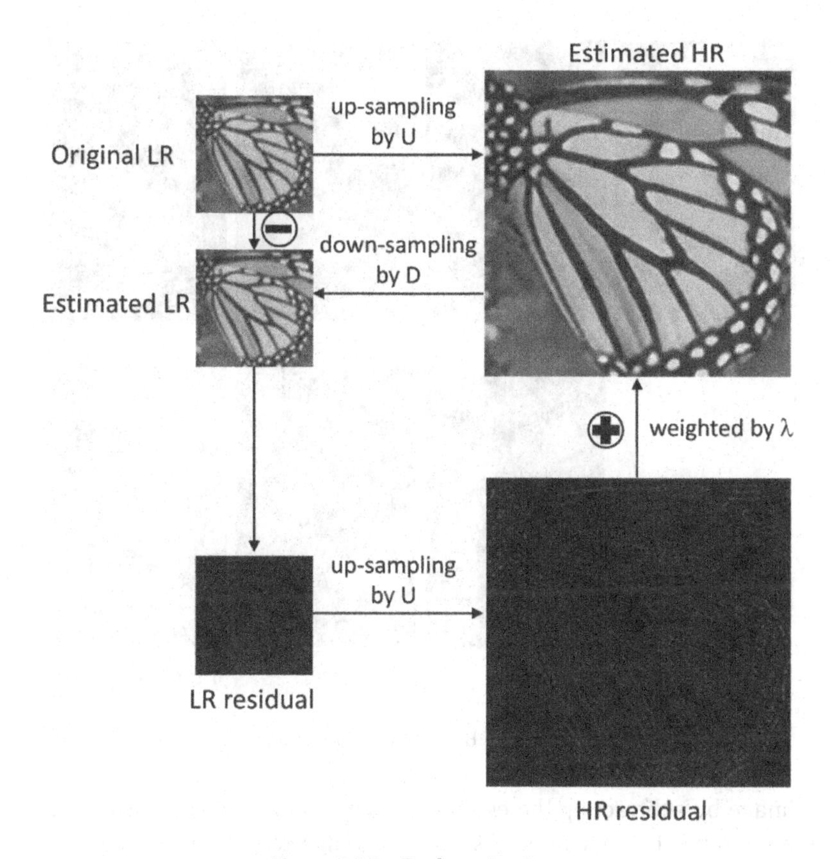

Figure 8.36 Back projection.

projection as follows:

$$\mathbf{Y}(t+1) = \mathbf{Y}(t) - \lambda \cdot \mathbf{U} \otimes (\mathbf{D} \otimes \mathbf{Y}(t) - \mathbf{X}), \tag{8.52}$$

where $\mathbf{Y}(t+1)$ is the $(t+1)$-th estimated HR image, $\mathbf{X}$ is the original low-resolution image, $\mathbf{D}$ and $\mathbf{U}$ are the down-sampling and up-sampling operators, λ is the weighting parameter and t is the number of iterations. Each iteration, λ controls the amount of residual information to be added back to the new SR image. To easy understand the back projection, Fig. 8.36 shows the complete process of back projection.

The estimated HR image is up-sampled by operator **U** and then down-sampled by operator **D** to obtain the estimated LR image. Knowing the original LR image, we can calculate the residual LR

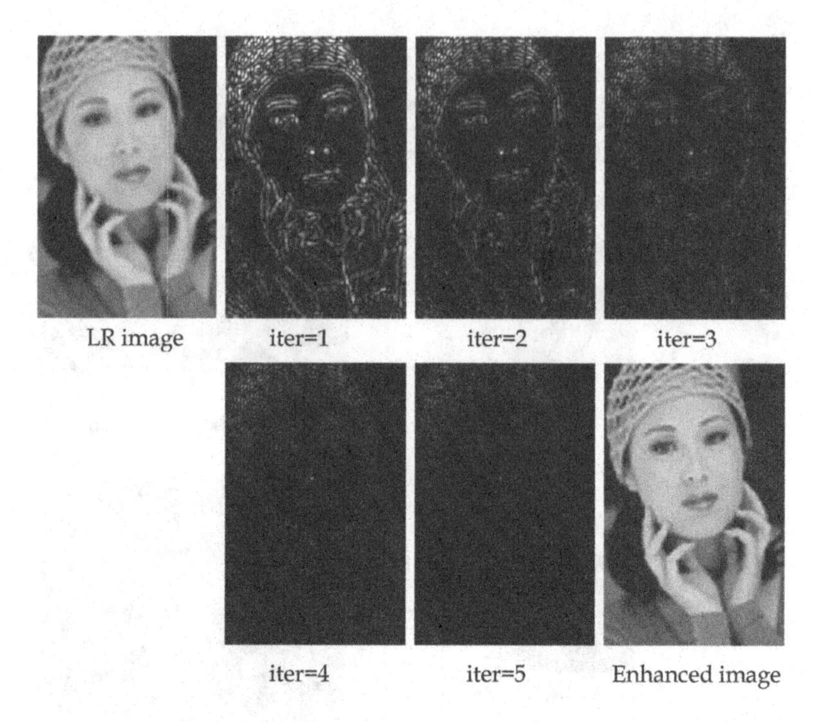

Figure 8.37 Back projection process.

image by subtracting the estimated LR image and then we can use operator **U** to up-sample residual LR image to obtain residual HR image. With a weighting factor λ, we can add the residual HR image back to the estimated HR image to complete one iteration. The idea is that the more genuine SR images we obtain, the smaller residual between LR images. This process can be iteratively repeated until it reaches some criterion (i.e., number of iteration) to stop. Back projection has been widely used in image super-resolution as a pre- or post-procedure for refinement [16, 31, 32, 34–39]. Here, let us use a simple example to show the performance of back projection.

Figure 8.37 shows a LR image was up-sampled and down-sampled by using Bicubic interpolation to simulate the back projection procedure. By setting $\lambda = 0.5$, we used 5 iterations to update the residual and finally we obtain the enhanced SR image. For better visual quality, we multiply the residual image at each iteration by 10 to enhance the differences. We can find that with

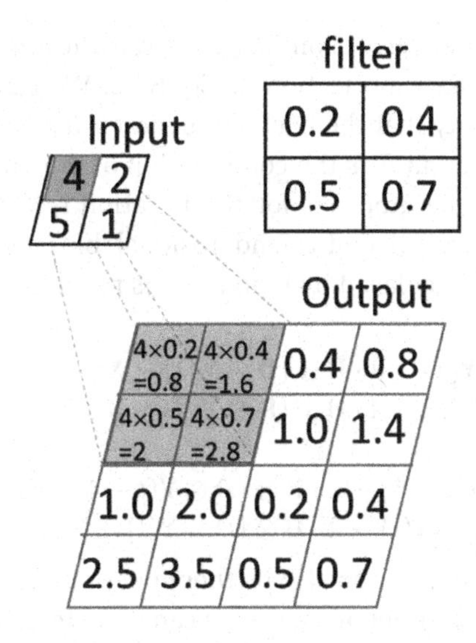

Figure 8.38 Deconvolution.

more iterations of back projection, the smaller residual information for update. Moreover, the major residual values mostly concentrate on the eyes and hat parts which resonate to the basic observation we know from Fig. 8.32 that the missing high frequency information are predicted for the better super-resolution quality.

The problem of back projection is that the up- and down-sampling operator **U** and **D** are fixed at each iteration. However, the estimated HR image is updated by the residual information iteratively so that the data distribution changes continuously. Fixed operators would quickly fail to predict the missing residual information for update and this is why researchers only use back projection as a refinement process to provide limited improvement in image quality.

8.4.3.3 Joint back projection and residual network for image super-resolution

After introducing the residual learning and back projection, we can now start to discuss a recent CNN-based image super-resolution via BPRN [44]. The basic idea is to combine the concepts of residual learning and back projection together to design a CNN model for better image super-resolution. Let us recall the residual mapping of Eq. 8.45 and back projection of Eq. 8.52. We can observe that there is a similarity that the back projection is a specific case when the mapping function F is the combination of the up-sampling and down-sampling operators. As for the iteration t, if we can expand Eq. 8.45 as a stacked end-to-end residual block where $F(x)$ in Eq. 8.45 hereby is replaced by $\mathbf{U}$ and x_{l+1} is replaced by $\mathbf{D}\otimes\mathbf{Y}(t)$, we have

$$\begin{aligned}
&\mathbf{Y}(2) = \mathbf{Y}(1) - \lambda \cdot \mathbf{U} \otimes (\mathbf{D} \otimes \mathbf{Y}(1) - \mathbf{X}) \\
&\mathbf{Y}(3) = \mathbf{Y}(2) - \lambda \cdot \mathbf{U} \otimes (\mathbf{D} \otimes \mathbf{Y}(2) - \mathbf{X}) \\
&\ldots\ldots \\
&\mathbf{Y}(t) = \mathbf{Y}(t-1) - \lambda \cdot \mathbf{U} \otimes (\mathbf{D} \otimes \mathbf{Y}(t-1) - \mathbf{X}) \\
&\mathbf{Y}(t+1) = \mathbf{Y}(t) - \lambda \cdot \mathbf{U} \otimes (\mathbf{D} \otimes \mathbf{Y}(t) - \mathbf{X})\,.
\end{aligned} \tag{8.53}$$

For back projection, we can also further generalize its concept to the inverse super-resolution problem, we can rewrite Eq. 8.53 as

$$\mathbf{X}(t+1) = \mathbf{X}(t) + \lambda \cdot \mathbf{D} \otimes (\mathbf{U} \otimes \mathbf{X}(t) - \mathbf{Y})\,, \tag{8.54}$$

for which we know the updated SR image and we want to obtain the LR image as close as possible to the original LR image. In this way, we can iteratively update the SR and LR image using the residual mapping model to generate final SR image.

From Eqs. 8.52 and 8.54, the LR and SR image are known to us and our target is to predict the up-sampling $\mathbf{U}$ and down-sampling operator $\mathbf{D}$ for real applications. We can use CNN model to simplify the $\mathbf{U}$ and $\mathbf{D}$ as Deconvolution process and Convolution process. Knowing the convolution layer in CNN, deconvolution layer is just an inverse calculation to convolution. The goal of deconvolution is to recreate the feature maps for HR image as it exists before the convolution for LR image. One simple example of deconvolution is shown as follows:

This example uses a 2×2 input convolutes with a filter of size 2×2 with convolution step 2 and output a 4×4 feature map. Each value of the input corresponds to 2×2 pixels on the feature map.

Now let us see the complete BPRN model in Fig. 8.39. The BPRN model contains multiple stages of back projection block to learn different down- and up-sampling operators (filters) to explore deeper feature for image super-resolution. It can overcome the

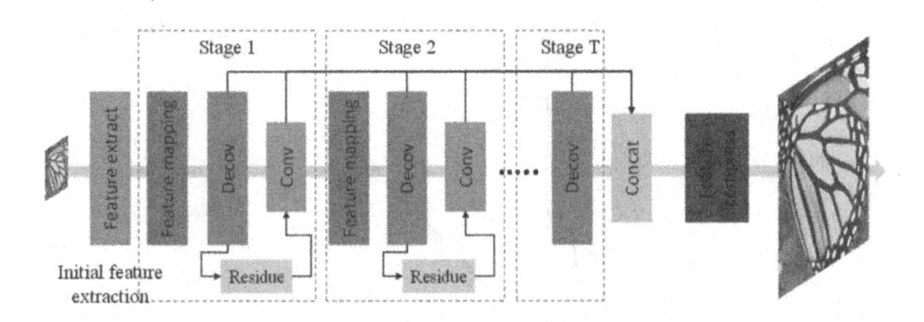

Figure 8.39 BPRN model.

disadvantage we mentioned before of the traditional back projection in Fig. 8.37. For back projection block in BPRN, each stage is made of Convolution and Deconvolution layer. There is also feature extraction layer at the beginning and Feature compression in the end. All the activation layers we use Parametric Rectified Linear Units (PReLUs) [63] which works as a sparse filter that allows positive value going through while compresses the negative value. Its mathematic expression is as follows:

$$\text{PReLU}(x) = \begin{cases} x & \text{if } x > 0 \\ \alpha x & \text{otherwise} \end{cases} \tag{8.55}$$

PReLU function is popular used in many CNN-based image super-resolution study recently for which it can avoid the gradient vanishing and its sparsity property can introduce non-linear mapping while relax the computation. As shown in Fig. 8.40, PReLU function uses a small weighting factor α to constrain the activation for negative input. Usually, $0 < \alpha < 1$ so that it does not block all the negative inputs. Meanwhile, instead of assigning a fixed value, parameter α is trained based on the training data adaptively. Now let us introduce the basic module of BPRN, which we call it back projection block, Fig. 8.41 shows the details of its structure.

From the back projection block as shown in Fig. 8.36, we can easily find the four shortcuts that represent the residual learning that skip the middle convolution layer. The four shortcuts describe two types of calculation: up-sampling and down-sampling process:

$$\begin{aligned} \mathbf{X}_{l+1}^{SR} &= \mathbf{W}_{up}\left(\mathbf{X}_l^{LR} - \mathbf{W}_{down}\left(\mathbf{W}_{up}(\mathbf{X}_l^{LR})\right)\right) + \mathbf{W}_{weight}(\mathbf{W}_{up}(\mathbf{X}_l^{LR})) \\ \mathbf{X}_{l+1}^{LR} &= \mathbf{W}_{down}\left(\mathbf{X}_{l+1}^{SR} - \mathbf{W}_{up}\left(\mathbf{W}_{down}(\mathbf{X}_{l+1}^{SR})\right)\right) + \mathbf{W}_{weight}(\mathbf{W}_{down}(\mathbf{X}_{l+1}^{SR})) \end{aligned} \tag{8.56}$$

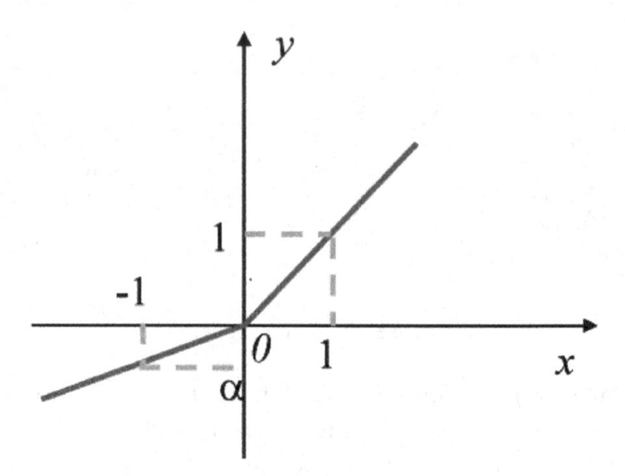

Figure 8.40 PReLU function.

The output feature maps of LR and SR is the summation of the weighted input feature maps and the updated residual information. It is different from residual block of ResNet in Fig. 8.31. The added 1×1 convolution layer works as a weighting model to assign weights to the input feature maps. In this way, the residual information can be tuned adaptively to achieve better reconstruction. It is easy to observe the training details by showing the trained filters and feature maps. Similar to SRResNet, we use the same butterfly image to test BPRN model and we have the following figures.

Figure 8.42 shows the output feature maps of 4 back projection blocks in BPRN. We can observe that at different back projection blocks, the output feature maps extract different residual information. The deeper of the back projection block, the finer details it can exploit. Note that the deconvolution layer needs a stride larger than 1 to achieve up-sampling process which would cause boundary artifact problem. The blocking artifact gradually disappears from the back projection block 1 to 5 indicates the learning process of deconvolution for residual reconstruction.

Finally, let us compare SRCNN, VDSR and BPRN on a real testing image on $4 \times$ super-resolution in Fig. 8.43. From these two image, we can clearly see the improvement using BPRN for image super-

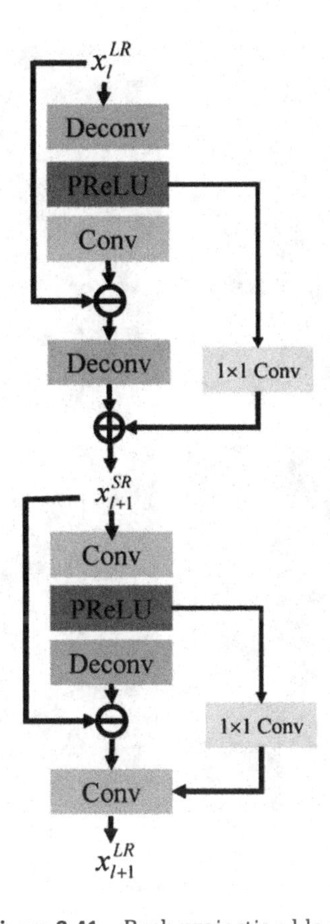

Figure 8.41 Back projection block.

resolution. The textural characteristics of the wooden bridge in the red box in Fig. 8.43a and the steel frame of the ceiling in the red box in Fig. 8.43b are reconstructed with good edgy appearance. This proves that using residual learning and back projection can significantly improve the super-resolution quality.

8.4.3.4 Conclusion on CNN-based image super-resolution

There are many other state-of-the-art CNN-based image super-resolution approaches. Generally, the deeper model we design, the better performance we can achieve. However, as introduced before, even the ResNet can only reach approximately 152 convolution

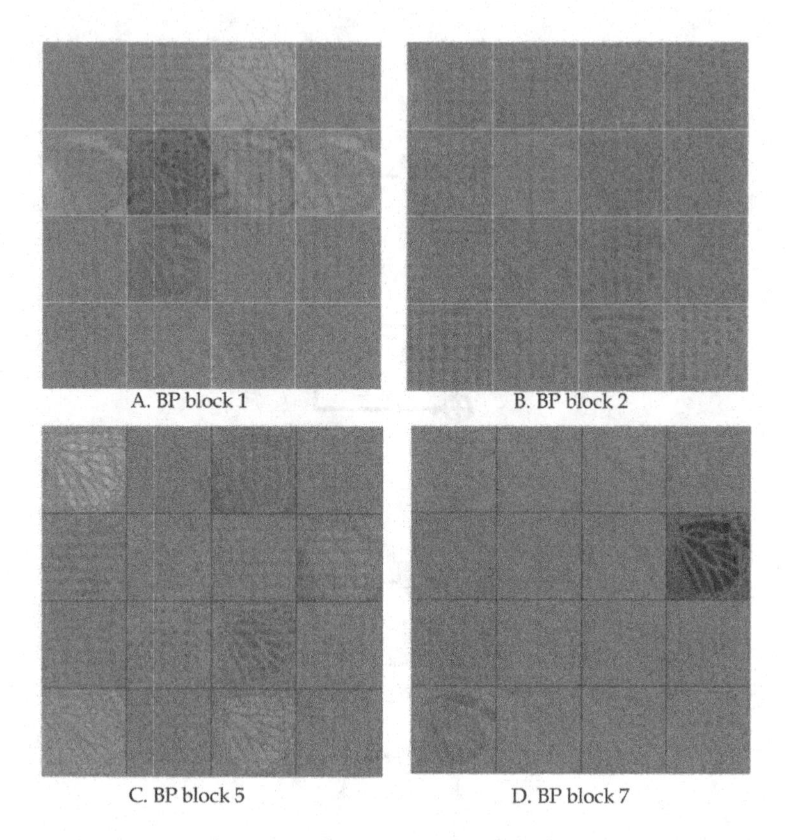

Figure 8.42 Visualization of feature maps of Back Projection (BP) blocks.

layers and the performance starts to saturate. It is meaningless to stack more and more convolution layers to pursue better performance without considering computation cost. Better structures need to be found from other research discovery to simplify the CNN model and yet provide good results. The success of combining residual learning with back projection encourages us to explore more solutions to image super-resolution from different perspectives. One of the recent promising direction is Generative Adversarial Network (GAN) for image super-resolution for its ability to learn the joint distribution of the LR and HR image to generate the estimated SR image instead of learning the feature differences for super-resolution. Besides, subjective measurement for image super-

Figure 8.43 Visual comparison among SRCNN, VDSR and BPRN.

resolution, style transformation, online pattern matching and other directions are worthy for further study.

8.5 Literature Review

The simplest way to do image interpolation is the bicubic interpolation, which means to use a non-adaptive filter. It suffers from blurring and aliasing effects. An early approach for interpolation with substantial improvement in visual quality is done by NEDI [5]. The quality has been improved by a number of methods, including our extended the NEDI filter [6, 7], DFDF (directional filtering and data fusion) [8], SAI (soft-decision adaptive interpolation) [9], localized linear regression model [10], Robust SAI using weighted least square [11] in 2012, sparse representation [12] in 2013, our proposed random forest for fast interpolation [13] in 2015.

Image Super-Resolution went through a similar path. The process changes from pixel based to patch processing [14, 15]. Substantial improvement was obtained from approaches with sparse representations [16, 17], and sparse representation together with a number of other techniques, such as PCA for input features and K-SVD for dictionary learning [18] in 2012 and using anchored neighborhood [19] in 2013. The state-of-the-art approaches are the use of the beta process joint dictionary learning in 2013, convolutional neural network (CNN) with deep learning [22, 41–45] in 2014 to 2018 and the adjusted anchored neighbor regression in 2015. These are good approaches and able to improve the super-resolution steadily along the technology road map in the literature. However, better approaches are getting more complex, with long processing time for their practical realization.

As we said, super-resolution can be applied to various areas, including video coding and transcoding [21–23]. We have done a substantial amount of work on these areas. Interested readers may refer to references [21–40, 44]. These include image/video enlargement using initially filter-based approaches [6, 7, 21–22], transform-based approaches [31–33], online learning approaches [31–33] and then learning-based approach with random trees [13, 34–40] and CNN-based deep learning [44].

8.6 Conclusion

In this chapter we start with the definition of super-resolution, and its basic realization using regression models with bicubic and Wiener filters. To obtain better quality, classification has to be performed before regression. This leads to the development of conventional learning algorithms. Random forests and Convolutional Neural Network (CNN) are two recent promising approaches for super-resolution, for which we have discussed in some details in this chapter. The random forests approach has the advantage of fast realization and hence potentially is very useful for real-time applications. CNN with deep learning is developing very fast which always produces realization results with leading quality. Can we merge these two approaches together, making advantages of both methods? An obvious extension of these works is to extend the training to the temporal domain for video super-resolution. Can we use manifold formulation to increase the frequency coverage, and use sparse representation for noise removal and quality improvement with random forests? Making further use of residual networks, back projection and good convolution layer design should also make the CNN realization to achieve better quality and faster speed. This is a very fruitful area of research.

Acknowledgment

Thanks are given to our colleagues Wai-Lam Hui and Chu-Tak Li for their assistance in the preparation of this manuscript.

References

1. More 4K UHD TV Sets Shipped in Q2'14 than in ALl of 2013, NPD DisplaySearch, Aug. 28, 2014. http://www.displaysearch.com/cps/rde/xchg/displaysearch/hs.xsl/140828_more_4k_uhd_tv_sets_shipped_q214_than_all_2013.asp.
2. http://www.imaging-resource.com/PRODS/pentax-k3-ii/pentax-k3-iiTECH2.HTM.
3. http://photoncollective.com/enhance-practical-superresolution-in-adobe-photoshop.
4. http://www.nhk.or.jp/strl/open2015/en/tenji_5.html.
5. X. Li and M. T. Orchard, New edge-directed interpolation, *IEEE Trans. Image Process.*, vol. 10, no. 10, pp. 1521–1527, Oct. 2001.

6. W.-S. Tam, C.-W. Kok, and W.-C. Siu, A modified edge directed interpolation for images, *Journal of Electronic Imaging*, vol. 19, no. 1, 013011, pp. 13011-1–20, Jan–March 2010 [One of the Top 10 Most Cited JEI Articles Published in 2010–2013].
7. C.-S. Wong and W.-C. Siu, Improved edge-directed interpolation and fast EDI for SDTV to HDTV conversion, *Proceedings of the 18th European Signal Processing Conference (EUSIPCO'2010)*, August 23–27, 2010, Aalborg Denmark, pp. 309–313.
8. L. Zhang and X. Wu, An edge guided image interpolation algorithm via directional filtering and data fusion, *IEEE Trans. Image Process.*, vol. 15, no. 8, pp. 2226–2238, Aug. 2006.
9. X. Zhang and X. Wu, Image interpolation by adaptive 2D autoregressive modeling and soft-decision estimation, *IEEE Trans. Image Process.*, vol. 17, no. 6, pp. 887–896, Jun. 2008.
10. X. Liu, D. Zhao, R. Xiong, S. Ma, W. Gao, and H. Sun, Image interpolation via regularized local linear regression, *IEEE Trans. Image Process.*, vol. 20, no. 12, pp. 3455–3469, Dec. 2011.
11. K. W. Hung and W. C. Siu, Robust soft-decision interpolation using weighted least squares, *IEEE Trans. Image Process.*, vol. 21, no. 3, pp. 1061–1069, March 2012.
12. W. Dong, L. Zhang, G. Shi, and X. Wu, Image deblurring and super-resolution by adaptive sparse domain selection and adaptive regularization, *IEEE Trans. Image Process.*, vol. 20, no. 7, pp. 1838–1857, July 2011.
13. J.-J. Huang, W.-C. Siu, and T.-R. Liu, Fast image interpolation via random forests, *IEEE Trans. Image Process.*, vol. 24, no. 10, pp. 3232–3245, October 2015.
14. H. Chang, D. T. Yeung, and Y. Xiong, Super-resolution through neighbor embedding, *Proceedings of the IEEE International Conference on Computer Vision and Pattern Recognition* (CVPR 2004), vol. 1, 2004.
15. S. Mallat and Guoshen Yu, Super-resolution with sparse mixing estimators, *IEEE Trans. Image Process.*, vol. 19, no. 11, pp. 2889–2900, Nov. 2010.
16. J. Yang, J. Wright, T. Huang, and Y. Ma, Image super-resolution via sparse representation, *IEEE Trans. Image Process.*, vol. 19, no. 11, pp. 2861–2873, Nov. 2010.
17. K. Kim and Y. Kwon, Single-image super-resolution using sparse regression and natural image prior, *IEEE Trans. Pattern Anal. Mach. Intel.*, vol. 32, no. 6, pp. 1127–1133, Jun. 2010.

18. R. Zeyde, M. Elad, and M. Protter, On single image scale-up using sparse-representations, in *Curves and Surfaces*, pp. 711–730, 2012, Springer Berlin Heidelberg.
19. R. Timofte, V. De, and L. V. Gool, Anchored neighborhood regression for fast example-based super-resolution, *Proceedings of the IEEE International Conference on Computer Vision (ICCV 2013)*, pp. 1920–1927, 2013.
20. C. Dong, C. C. Loy, K. He, and Xiaoou Tang, Image super-resolution using deep convolutional networks, *IEEE Trans. Pattern Anal. Mach. Intel.*, vol. 38, no. 2, February 2016.
21. Y.-L. Chan and W.-C. Siu, New Adaptive pixel decimation for block motion vector estimation, *IEEE Trans. Circuits Syst. Video Technol.*, pp. 113–118, vol. 6, no. 1, February 1996, U.S.A.
22. K.-C. Hui, W.-C. Siu, and Y.-L. Chan, New adaptive partial distortion search using clustered pixel matching error characteristic, *IEEE Trans. Image Process.*, vol. 14, no. 5, pp. 597–607, May 2005.
23. W.-C. Siu, Y.-L. Chan, and K.-T. Fung, On transcoding a B-frame to a P-frame in the compressed domain, *IEEE Trans. Multimedia*, vol. 9, issue 6, pp. 1093–1102, October 2007.
24. K.-W. Hung and W.-C. Siu, Single-image super-resolution using iterative Wiener filter based on nonlocal means, Signal Process. Image Commun., vol. 39, pp. 26–45, Elsevier Science, The Netherlands.
25. K.-W. Hung and W.-C. Siu, A computationally scalable adaptive image interpolation algorithm using maximum-likelihood de-noising for real-time applications, *J. Electron. Imaging*, vol. 22, no. 4, pp. 043006-1:043006_15, October–December 2013.
26. K.-W. Hung and W.-C. Siu, Fast image interpolation using the bilateral filter, *IET Image Process.*, vol. 6, no. 7, pp. 877–90, October 2012, UK.
27. W.-C. Siu and K.-W. Hung, Review of image interpolation and super-resolution, *Proceedings, Paper 337, Invited Paper, Special Session on Sparse and Feature Representation for Image/Video Restoration, 2012 APSIPA Annual Summit and Conference (APSIPA-ASC'2012)*, December 3–6, 2012, Hollywood, California USA.
28. K.-W. Hung and W.-C. Siu, Single image super-resolution using iterative Wiener filter, *Proceedings of the IEEE International Conference on Acoustics, Speech and Signal Processing (ICASSP'2012)*, 25–30 March, 2012, Kyoto Japan, pp. 1269–1272.
29. H. He and W.-C. Siu, Single image super-resolution using Gaussian process regression, *Proceedings of the IEEE Computer Vision and Pattern*

Recognition Conference (CVPR 2011), June 20–24, 2011, Crowne Plaza, Colorado, USA, pp. 449–456.

30. K.-W. Hung and W.-C. Siu, New motion compensation model via frequency classification for fast video super-resolution, *Proceedings of the International Conference on Image Processing (ICIP'2009)*, 7–11 November, 2009, Cairo Egypt, pp. 1193–1196.
31. K.-W. Hung and W.-C. Siu, Novel DCT-based image up-sampling using learning-based adaptive k-NN MMSE estimation, *IEEE Trans. Circuits Syst. Video Technol.*, vol. 24, no. 12, pp. 2018–2033, December 2014.
32. K.-W. Hung and W.-C. Siu, Hybrid DCT-Wiener-based interpolation via learnt Wiener filter, *Proceedings of the IEEE International Conference on Acoustics, Speech and Signal Processing (ICASSP'2013)*, 26–31, May, 2013, Vancouver, Canada, pp. 1419–1423.
33. K.-W. Hung and W.-C. Siu, Learning-based image interpolation via robust k-NN searching for coherent AR parameters, *J. Visual Comm. Image Rep.*, vol. 31, pp. 305–311, Aug. 2015.
34. J.-J. Huang and W.-C. Siu, Practical application of random forests for super-resolution imaging, *Proceedings of the IEEE International Symposium on Circuits and Systems (ISCAS'2015)*, 24–27 May 2015, Lisbon, Portugal, pp. 2161–2164.
35. J.-J. Huang and W.-C. Siu, Fast image interpolation with decision tree, *Proceedings of the 2015 IEEE International Conference on Acoustics, Speech and Signal Processing (ICASSP'2015)*, 19–24 April, 2015, Brisbane, Australia, pp. 1221–1225.
36. J.-J. Huang, K.-W. Hung and W.-C. Siu, Hybrid DCT-Wiener-based interpolation using dual MMSE estimator scheme, *Proceedings of the 19th International Conference on Digital Signal Processing, (19th DSP2014)*, 20–23 August 2014, Hong Kong, pp. 748–753.
37. J.-J. Huang and W.-C. Siu, Learning hierarchical decision trees for single image super-resolution, *IEEE Trans. Circuits Syst. Video Technol.*, pp. 937–950, vol. 27, no. 5, May 2017.
38. Z. S. Liu, W. C. Siu and J. J. Huang, Image super-resolution via weighted random forest, *Proceedings of IEEE International Conference on Industrial Technology (ICIT'2017)*, Toronto, ON, 2017, pp. 1019–1023.
39. Z. S. Liu, W. C. Siu and Y. L. Chan, Fast image super-resolution via randomized multi-split forests, *Proceedings of the IEEE International Symposium on Circuits and Systems (ISCAS'2017)*, Baltimore, MD, 2017, pp. 1–4.

40. Zhi-Song Liu and W.-C. Siu, Cascaded random forests for fast image super-resolution, *Proceedings of the IEEE International Conference on Image Processing (ICIP'18)*, Greece, 2018.
41. J. Kim, J. KwonLee, and K. MuLee, Accurate image super-resolution using very deep convolutional networks, *Proceedings of the IEEE International Conference on Computer Vision and Pattern Recognition (CVPR'2016)*, 2016, Las Vegas, Nevada, pp. 1646–1654.
42. O. Russakovsky, J. Deng, H. Su, J. Krause, S. Satheesh, S. Ma, Z. Huang, A. Karpathy, A. Khosla, M. Bernstein, A. C. Berg, and L. Fei-Fei, ImageNet large scale visual recognition challenge. *Int. J. Comput. Vision*, 2015.
43. C. Ledig, L. Theis, F. Huszar, J. Caballero, A. Cunningham, A. Acosta, A. Aitken, A. Tejani, J. Totz, Z. Wang, et al., Photo-realistic single image super-resolution using a generative adversarial network, *Proceedings of the IEEE International Conference on Computer Vision and Pattern Recognition (CVPR'2017)*, 2017, Honolulu, Hawaii.
44. Z. S. Liu, W. C. Siu and Y. L. Chan, Joint back projection and residual networks for efficient image super-resolution, submitted to *Proceedings of the IEEE International Asia-Pacific Signal and Information Processing Association Annual Summit and Conference (APSIPA'18)*, Honolulu, Hawaii, 2018.
45. R. Timofte et al., NTIRE 2017 challenge on single image super-resolution: Methods and results, *2017 IEEE Conference on Computer Vision and Pattern Recognition Workshops (CVPRW'2017)*, Honolulu, HI, 2017, pp. 1110–1121.
46. Glorot, X., Bordes, A., and Bengio, Y. (2011). Deep sparse rectifier neural networks, *Proceedings of the International Conference on Artificial Intelligence and Statics (AISTATS'2011)*, Ft. Lauderdale, FL, USA, 2011.
47. K. He, X. Zhang, S. Ren, and J. Sun, Deep residual learning for image recognition, *Proceedings of the IEEE International Conference on Computer Vision and Pattern Recognition (CVPR'2016)*, 2016, Las Vegas, Nevada, pp. 770–778.
48. K. Simonyan and A. Zisserman. Very deep convolutional networks for large-scale image recognition. *Proceedings of the International Conference on Learning Representations (ICLR'2015)*, 2015, San Diego, CA.
49. C. Szegedy, W. Liu, Y. Jia, P. Sermanet, S. Reed, D. Anguelov, D. Erhan, V. Vanhoucke, and A. Rabinovich. Going deeper with convolutions. Proceedings, *IEEE International Conference on Computer Vision and Pattern Recognition (CVPR'2015)*, 2015, Boston, Massachusetts.

50. R. M. Mersereau and A. V. Oppenheim, Digital reconstruction of multidimensional signals from their projections, *Proc. IEEE*, vol. 62, no. 10, pp. 1319–1338, Oct. 1974. doi: 10.1109/PROC.1974.9625.
51. K. He, X. Zhang, S. Ren, and J. Sun, Delving deep into rectifiers: Surpassing human-level performance on Imagenet classification, *Proceedings of the IEEE International Conference on Computer Vision (ICCV'2015)*, 2015, Santiago, Chile, pp. 1026–1034.

Chapter 9

Non-Contact Three-Dimensional Measurement Using the Learning Approach

Daniel P. K. Lun and B. Budianto
Department of Electronic and Information Engineering, The Hong Kong Polytechnic University, Hong Kong, China
enpklun@polyu.edu.hk, budianto@ieee.org

Nowadays, non-contact three-dimensional (3D) scanning systems are popularly used in daily applications. Among the various 3D scanning methods, the structured light projection methods such as fringe projection profilometry (FPP) are commonly adopted in the 3D measurement applications that require high resolution and high accuracy. FPP allows full-field measurement of an object's 3D profile based on the triangulation between a projector and a camera located at different viewing angles of the object. For a typical FPP process, the projector first projects a set of fringe patterns onto the target object. Due to the object's height profile, the fringe patterns are deformed as shown on the object surface and are then captured by

Learning Approaches in Signal Processing
Edited by Wan-Chi Siu, Lap-Pui Chau, Liang Wang, and Tieniu Tan

ISBN 978-981-4800-50-1 (Hardcover), 978-0-429-06114-1 (eBook)
www.panstanford.com

a camera. We call the resulting images as fringe images. In principle, the height profile of an object can be estimated by measuring the amount of deformation of the fringe pattern in the fringe images. Since the fringes in a fringe image often appear as a set of sinusoidal signals, the measurement can be achieved by evaluating the phase shift of the sinusoidal fringe signals as compared to a reference obtained during the initial calibration. Similar to many phase-based optical imaging systems, FPP also requires performing a phase unwrapping process in order to obtain the correct phase data for 3D measurement. However, due to the imperfection of the working environment, the captured fringe images often contain different artifacts, which introduce great difficulty to phase unwrapping. To solve the problem, current approaches try to embed additional information into the projected fringe patterns to aid the phase unwrapping process. While the approach is sound in theory, it introduces another problem if the embedded information can be effectively extracted and decoded from the captured fringe image. In this chapter, two learning-based classification methods, namely discriminative dictionary learning (DL) and convolutional neural network (CNN), are introduced for the abovementioned task. While both approaches give superior performance over the traditional methods, the CNN-based approach is more robust and efficient particularly when implementing on a graphics processing unit. These learning methods open up a new research agenda for non-contact 3D measurement and will further improve the performance of existing 3D scanning systems.

9.1 Introduction

Fringe projection profilometry (FPP) is a full-field, high resolution and high accuracy non-contact three-dimensional (3D) measurement method. It has been generally used in industrial inspection [1, 2], 3D face model reconstruction [3], and other 3D scene reconstruction problems [4, 5]. Traditional FPP approaches can be divided into two groups: multiple-frame and single-shot. In general, multiple-frame approaches like phase shifting profilometry (PSP) [6–11] have a better robustness than the single-shot approaches. However, they require the projection of multiple frames of fringe patterns in sequent to the target objects which introduces great difficulty in dynamic applications where the objects are not necessarily static. Single-shot approaches [12–16] like Fourier transform profilometry (FTP) [12] do not have such problem since they only require the projection of a single fringe pattern. However, the robustness of these single-shot approaches is always doubtful. Significant errors will be generated when the measurement is conducted in an adverse condition, such as the object has vivid texture on its surface, or the material of the object has high reflectivity such that some of the fringes are overwhelmed by the reflection of the strong light around.

Both the multiple-frame and single-shot FPP methods also suffer from the ambiguity problem in phase unwrapping. As mentioned above, FPP evaluates the phase shift of the sinusoidal fringe signals to measure the deformation of the fringe pattern. For most phase imaging systems, it is common that only incomplete and noisy phase estimates can be obtained so that the true phase is difficult to be inferred. As in the case of FPP, only a wrapped phase, i.e., a modulo-2π estimation of the phase, can be obtained. Hence, to recover the true phase, additional phase unwrapping procedure needs to be performed. Traditional phase unwrapping algorithm assumes that the Itoh condition is fulfilled. By Itoh condition, it means that the absolute phase difference between two neighboring points is less than π. So, the true phase can be obtained by integrating the phase differences. For FPP, this condition can become invalid when some parts of the fringe are missing due to, for instance, occlusion or sudden jumps in the object's height. In this case, one or more cycles of the fringe will be missing and thus a sudden jump of more than 2π in phase angle will result. In fact, the problem can be solved if the cycle numbers of the fringe signal (we also call it the period order information) are known such that the exact number of 2π jumps can be found and added back to the integration process. Based on how the period order information is obtained, there are three major classes of approach used in FPP. The simplest and intuitive solution is to project additional coded patterns that carry the period order information of the fringe. The gray-code plus phase shifting method [17], the phase code plus phase shifting method [18], the temporal phase unwrapping algorithm [19], and the gray code plus strip-edge shifting method [1] are the methods that belong to this category. Since this approach requires additional fringe projections, the acquisition time will also increase, and thus additional processing time is inevitable. The second approach is by employing another camera located close to the projector [20]. By doing this, all fringes can be found from the captured image of the second camera and the phase unwrapping algorithm can be performed based on this image.

To avoid using additional camera or projecting additional fringe patterns, recent FPP methods embed the period order into the fringe pattern. In these approaches, the period order information is first encoded in various forms: multiple frequencies/wavelengths [21–23], multiple colors [24], random patterns [11, 25], and markers [10, 26–29], etc. However, in practice, due to various distortions, these approaches often end up with requiring additional fringe projections to ensure the correct decoding of the period order information. Besides, the approaches in [10, 21–24, 27–29] can only reconstruct the 3D profile of a simple scene, i.e., only have a single simple object. Otherwise, the accuracy of the period order information will be low and additional refinement procedure such as the voting strategy [11] is required in order to get the correct result.

In recent years, one of the biggest advancements in non-contact 3D measurement is the integration of the learning approaches into the measurement system [30–33]. As to the phase ambiguity problem as mentioned above, a solution using the learning method was proposed and achieved satisfactory results [33]. In this approach, the problem of true phase estimation is tackled by exploiting the sparsity of the fringe pattern and the embedded coded patterns, which encode the period order information. Since the code patterns have different morphological structures from the fringe patterns, they can be separated by performing a modified morphological component analysis (MCA) tailored for FPP. It differs from the traditional MCA [34–36] by considering also the blurring effect and noise in the fringe images. Then, a dictionary is trained to decode the period order information from the extracted code patterns. The dictionary learning and sparse coding techniques have been used successfully for classification tasks such as object classification [37], face recognition [38–40], texture segmentation [40–42], etc. When using in FPP, a dictionary is designed to keep the sparse representation of the embedded code patterns through a procedure known as the sparse coding. When decoding the period order information, the dictionary will be consulted for the classification of the code patterns in the fringe images. Assisted by the estimated period order information, an effective phase unwrapping algorithm is developed for reconstructing the 3D profile of objects.

Different learning algorithms can be used to decode the period order information. Besides the above dictionary learning (DL) method, it can also be achieved by using the convolutional neural network (CNN). Recently, CNN has been popularly used in many classification tasks such as object classification and segmentation [43–45] with great success. Compared to the DL approach, CNN can further improve the estimation of the period order information by massive training process. Furthermore, the regular structure of CNN allows it to be implemented using the GPU and achieves significant improvement in computation speed. In this chapter, the details of these two learning methods as applied to FPP are introduced. Experiments are conducted to evaluate their performance when applying to a real FPP system. They are also compared with other traditional methods. As shown in the experimental results, both the DL and CNN-based FPP methods outperform the traditional approaches in terms of accuracy and robustness. The CNN-based method has a further advantage that it is extremely fast when implementing with graphics processing unit (GPU). Furthermore, it is less sensitive than the DL-based approach to the quality of the code patterns extracted from the fringe images.

This chapter is organized as following. First, the fundamental of FPP is presented in Section 9.2. Then two learning-based methods for FPP are introduced in Section 9.3. Experimental results are reported in Section 9.4 and finally, the conclusion is drawn in Section 9.5.

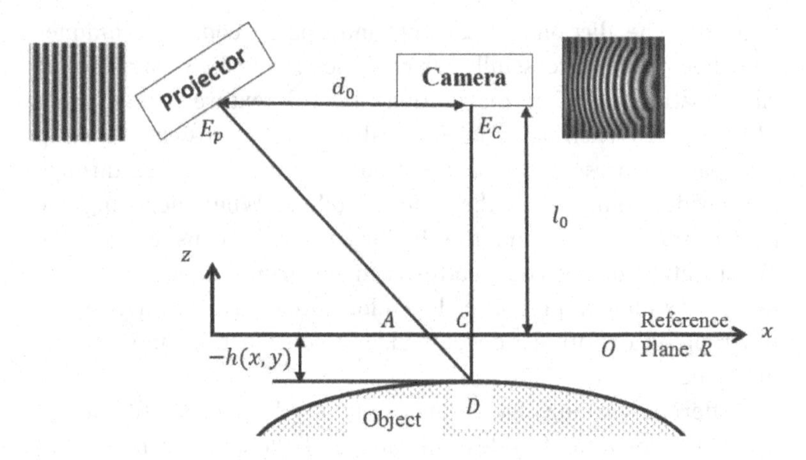

Figure 9.1 Phase-based fringe projection profilometry arrangement in the crossed-optical-axes geometry.

9.2 Fundamental of Fringe Projection Profilometry

For a typical FPP setup, regular fringe patterns are projected onto the target object. Due to the object's height, the fringe patterns as shown on the object surface deform as compared with the original patterns. So, by evaluating the amount of deformation, the height profile of the object can be measured. Figure 9.1 shows a typical FPP setup in the crossed-optical-axis triangulation geometry. For simplicity, it shows only the cross section of the object surface for a given y coordinate. Hence, the object's height can be expressed as a function of a single variable x. As shown in the figure, a projector at E_p projects a fringe pattern onto an object and a camera at E_c captures the deformed fringe pattern as shown on the surface of the object. The distance between the projector and camera is fixed to be d_0 and they are placed at a distance l_0 from a flat reference plane R.

First, let us assume the reference plane is a solid object such that all the light projected onto it will be reflected. In this case, the fringe image captured by the camera can be expressed as

$$G_0(x, y) = a_0(x, y) + b(x, y)\cos[2\pi f_0 x + \phi_0(x, y)] \tag{9.1}$$

where a_0 is the bias of the fringe image due to the texture of the reference plane; b is the amplitude of the fringe pattern; f_0 is carrier frequency of the fringe; and ϕ_0 is the phase shift. Here we assume the intensity of the fringe pattern will change as a cosine function. It is because both the lens of the projector and camera will serve as a low pass filter such that only the fundamental frequency component of the fringe is recorded in the fringe image. Now assume that the reference plane is taken away. The fringe pattern is projected onto the object. The captured fringe image becomes

$$G(x, y) = a(x, y) + b(x, y)\cos[2\pi f_0 x + \phi(x, y)] \quad (9.2)$$

where a is the bias of the fringe image due to the texture of the object; ϕ is the phase shift, which will vary according to the object's height. From Fig. 9.1, it can be seen that the two triangles ACD and $E_p DE_c$ are similar. Then, the object height h is related to the parameters AC, l_0 and d_0 as

$$\frac{AC}{-h(x, y)} = \frac{d_0}{l_0 - h(x, y)} \quad (9.3)$$

The object height can thus be evaluated by

$$h(x, y) = \frac{l_o AC}{AC - d_0} \quad (9.4)$$

Since l_0 and d_0 are known in the calibration process, the key in finding the object height h at point D is to find the distance AC, which is the displacement of the fringe from A to C. Such parameter can be measured as follows: Suppose a light ray from the projector located at E_p is projected onto point A of the reference plane. When the reference plane is removed, the same light ray from E_p will fall onto the object at point D, which is the same point in x-axis as point C. So, AC can be easily measured by comparing the position of the light ray captured by the camera with and without the reference plane. However, the FPP system does not project a single light ray but a fringe pattern. Therefore, to estimate AC, we need to measure the phase difference between the cosine fringe signals in (9.1) and (9.2) at point D captured by the camera. More specifically, the difference of $\phi(xy)$ and $\phi_0(xy)$ should be measured as follows:

$$\Delta\phi(x, y) = \phi(x, y) - \phi_0(x, y) = 2\pi f_0 AC \quad (9.5)$$

Substitute (9.5) to (9.4), we have

$$h(x, y) = \frac{l_o \Delta\phi(x, y)}{\Delta\phi(x, y) - 2\pi f_0 d_0} \tag{9.6}$$

In (9.5), ϕ_0 is the phase shift when the fringe pattern is projected onto the reference plane. It can be obtained during the initial calibration. Hence, if ϕ is known, $\Delta\phi$ and the height h can be determined. However, ϕ cannot be directly measured from the captured fringe image G in (9.2) due to many reasons. First, due to the texture on the object surface, the fringe image G will have a bias term a, which is unknown and object dependent. Particularly for objects with vivid texture, such bias term cannot be removed by a simple low pass filter. Many researches have been conducted for solving this problem. One of them is by using the so-called Phase Shifting Profilometry (PSP) technique. In such technique, multiple fringe patterns with fixed phase shifts are projected onto the object. The resulting fringe images are combined to cancel out the bias term. The technique will be further discussed in Section 9.2.1. The second problem in obtaining ϕ from G is due to the ambiguity of the trigonometric functions. As shown in (9.2) the same G will be obtained for all $\phi + k2\pi$, where k can be any integer (it is also called the period order number). It means that given G, only the wrapped phase $\hat{\phi}$ of the true phase ϕ can be obtained, where $\phi = \hat{\phi} + k2\pi$. Although in theory the problem can be solved by using the phase unwrapping methods, it will be shown in Section 9.2.2 that these methods will fail in many practical situations.

9.2.1 Phase Shifting Profilometry (PSP)

Recall (9.2), the main objective of the fringe analysis procedure is to estimate the phase shift ϕ from the captured fringe image G. In practice, it is a difficult problem because the capture fringe image contains both higher order harmonics and the varied dc bias due to the object texture and noise. They introduce aliasing to the fundamental frequency spectrum. When there is no aliasing, the Fourier transform can be applied to extract the interested fundamental frequency components directly in the Fourier domain. Such approach was first proposed by Takeda and known as the Fourier transform profilometry (FTP) [12]. However, FTP often fails when the object's texture is complex, for instance, having vivid texture or color patterns. They introduce the aliasing problem that makes the FTP method not as popular as other fringe analysis methods.

To avoid the aliasing problem due to the bias, the most common method is the phase shifting profilometry (PSP). In PSP, multiple fringe patterns with fixed phase differences between each other are projected on the object. The bias term can be cancelled by combining the captured fringe images together. More specifically, the captured fringe images in a PSP process can be mathematically described as follows:

$$G^n(x,y) = a(x,y) + b(x,y)\cos\left[2\pi f_0 x + \phi(x,y) + \frac{2\pi n}{N}\right] \quad (9.7)$$

where $n = 1, \ldots, N$ and N is the total number of phase shifted fringe patterns; G^n is the nth fringe image. Since there are three unknown variables in (9.7), the minimum number of projections required is three. Therefore, three-step PSP is the most popular PSP-based fringe analysis method [6, 8, 11, 46–48]. The three fringe images can be expressed as

$$G^1(x,y) = a(x,y) + b(x,y)\cos\left[2\pi f_0 x + \phi(x,y) - \frac{2\pi}{3}\right];$$
$$G^2(x,y) = a(x,y) + b(x,y)\cos\left[2\pi f_0 x + \phi(x,y)\right];$$
$$G^3(x,y) = a(x,y) + b(x,y)\cos\left[2\pi f_0 x + \phi(x,y) + \frac{2\pi}{3}\right] \quad (9.8)$$

The dc bias can be obtained easily by averaging the three fringe patterns

$$a(x,y) = \frac{G^1(x,y) + G^2(x,y) + G^3(x,y)}{3} \quad (9.9)$$

and the resulting phase is

$$\hat{\phi} = \tan^{-1}\left(\sqrt{3}\frac{G^1(x,y) + G^3(x,y)}{2G^2(x,y) - G^1(x,y) - G^3(x,y)}\right) \quad (9.10)$$

As shown in (9.8)–(9.10), the three-step PSP algorithm requires only to project the minimum number of fringe patterns and is simple to apply. It does not suffer from the aliasing problem as in the FTP method. However, similar to the FTP method, this approach can only give modulo-2π wrapped phase $\hat{\phi}$ rather than the true phase ϕ. Hence, additional phase unwrapping procedure is necessary.

9.2.2 Phase Unwrapping in FPP

Any PSP method that attempts to directly retrieve the true phase ϕ from the fringe image G can only have the wrapped version of ϕ since any arc tangent operator in (9.10) will return only the wrapped version of ϕ, i.e., $\hat{\phi} = \Pi(\phi) = \mathrm{mod}(\phi + \pi, 2\pi) - \pi$, where $\mathrm{mod}(xy)$ is the modulus of x with the divisor y. Additional phase unwrapping procedure is necessary to retrieve the true phase ϕ from the wrapped phase $\hat{\phi}$.

$$|\phi(x) - \phi(x-1)| = \Pi\left(\hat{\phi}(x) - \hat{\phi}(x-1)\right) \qquad (9.11)$$

If the condition is fulfilled, the true phase at position x' can be obtained by,

$$\phi(x') = \sum_{x=1}^{x=x'} \Pi\left(\hat{\phi}(x) - \hat{\phi}(x-1)\right) + \phi(0), \qquad (9.12)$$

where $\phi(0)$ is the initial phase value which is assumed to be known. After the true phase for all pixels is known, the height profile of the object can be reconstructed.

Many phase unwrapping algorithms have been developed in last two decades. Most of them rely on the Itoh analysis assumption. Examples of these algorithms includes Goldstein's branch cut algorithm [50], quality-guided phase unwrapping algorithm [51–56], and Flynn phase unwrapping algorithm [57]. All these algorithms are robust and can give accurate results when the Itoh condition is fulfilled. However, as a matter of fact, there are many practical working environments in FPP where the captured fringe images violate this condition.

Figure 9.2 shows two situations. In the upper figure, two objects are positioned at different distances from the camera. As shown in the figure, the distance of the two fringes becomes closer in the fringe image (right) although the period order (the number in the figure) is significantly different. So, the fringes are discontinuous and violate the Itoh condition. Another situation is shown in Fig. 9.2 (lower). As shown in the figure, the middle part of the object is blocked due to the position of the camera. The continuity of the fringes seems preserved in the fringe image (right), but in fact the fringes in region B (period number 2–4) are missing.

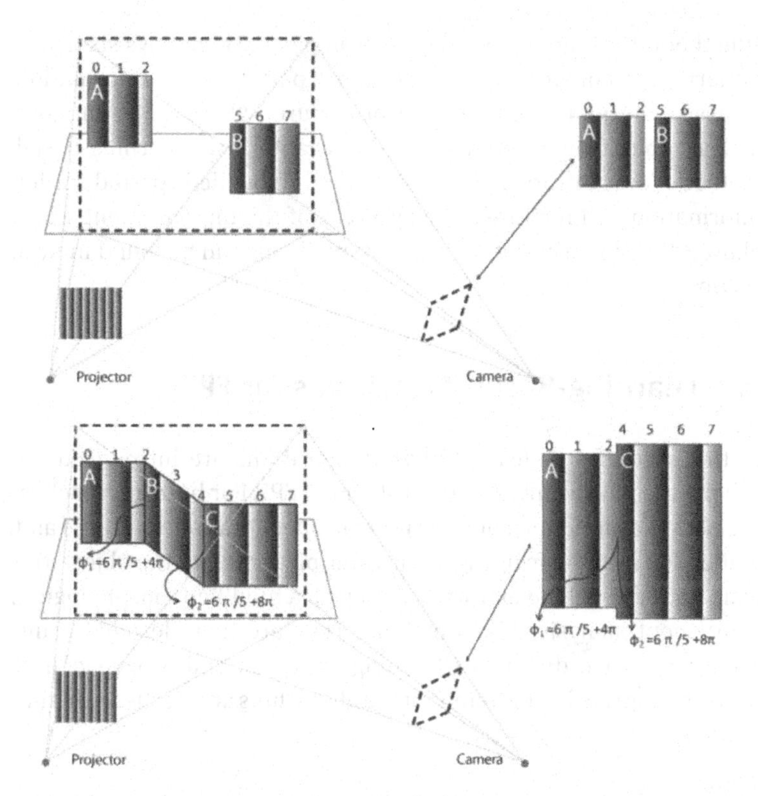

Figure 9.2 Illustration of the ambiguity problem due to isolated objects (upper) and occlusion (lower) [8].

Any conventional phase unwrapping algorithms based on the Itoh condition will fail to recover the true phase in this situation.

To solve the problem, one of the typical methods is to embed the period order numbers into the projected fringe pattern. They may be encoded into the form of markers or texture patterns and added to the fringe pattern. When the embedded fringe pattern is captured, the markers or texture patterns are extracted and then decoded to get back the period order numbers. Different period order encoding strategies have been proposed [8, 11, 23, 24]. However, all these approaches reduce the accuracy of the final 3D measurement since the additional period order information introduces noise to the fringe images and affects the estimation process. Moreover, the approaches in [8, 23] require more than three fringe projections (the theoretical lower bound). For example, [23] employs six fringe patterns for embedding the additional period order information to assist the phase unwrapping procedure, whereas [8] employs at least four fringe patterns. In this chapter, two learning-based methods are introduced so that the embedded period order information will not affect the accuracy of the measurement while allows them to be easily decoded. More details can be found in next section.

9.3 Learning-Based Algorithms for FPP

In this section, two learning-based algorithms are introduced for solving the phase unwrapping problem of FPP. For both approaches, the period order information is first encoded into code patterns and embedded into the fringe pattern. After projecting to the object, the code patterns will be extracted out using a modified morphological component analysis (MCA) method. They are then decoded using a discriminative dictionary learning approach and a deep neural network approach. The details of the algorithms are discussed in this section.

9.3.1 Encoding the Period Order Numbers

As mentioned in the previous section, the height profile of an object is directly related to the true phase shift ϕ of the fringe pattern. To obtain ϕ from the fringe image, the inverse trigonometry method can only provide the wrapped phase $\hat{\phi}$. The relationship between ϕ and $\hat{\phi}$ can be expressed as

$$\phi = \hat{\phi} + k2\pi \tag{9.13}$$

where k is the so-called k-value. It keeps the period order information of $\hat{\phi}$. A map that records all k-value of a fringe image is called K-Map. If this map is available when unwrapping $\hat{\phi}$, we do not need to worry about the ambiguity problem as mentioned in Section 9.2.2. We can make use of the k-value to detect if there is any discontinuity in the phase value as shown in Fig. 9.2. Mathematically, we can express a fringe image embedded with code patterns as a superimposition of two images as follows:

$$G = G_1 + G_2 \tag{9.14}$$

where G_1 is the original sinusoidal fringe pattern, i.e., G in (9.7); and G_2 is the code pattern that encodes the K-Map defined by

$$\begin{aligned} G_2 &= M\left(K(\phi)\right), \\ K &: R \to Z^+ \\ \phi &\to \left\lfloor \frac{\phi + \pi}{N2\pi} - \pi \right\rfloor \end{aligned} \tag{9.15}$$

In (9.15), M is the encoding function for each k-value, $\lfloor \cdot \rfloor$ is the floor function that gives the closest smaller integer number, and N is number of fringes encoded into a single code pattern. The encoding function M should be carefully designed to ensure that (1) M can generate some code patterns G_2 that have different morphological structures from G_1, e.g., different shape, frequencies, colors, etc., so that they can be easily identified; and (2) G_2 should have a unique feature for every N k-values. Figure 9.3 shows an example of five unique textons and their corresponding code pattern. As shown in the figure, each texton has a unique shape. They are concatenated to form a code pattern. The last column of Fig. 9.3 shows a coded fringe pattern with $N = 1$, i.e., each fringe is encoded with a unique code pattern. Such coded fringe pattern will be projected to the object. In practical situation, there is a trade-off between the frequency of the fringe pattern and the number of textons. To achieve an

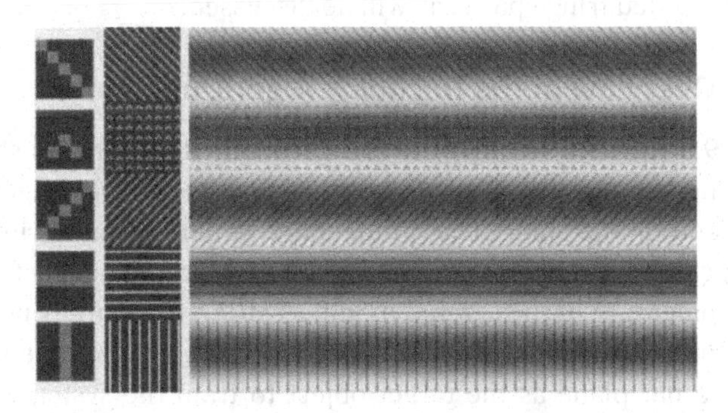

Figure 9.3 5×5 pixels binary textons (most left), the code pattern generated by textons (second column), and the coded fringe pattern with one unique texture assigned for each k-value, i.e., $N = 1$ (column three).

Figure 9.4 The fringe images (with code pattern embedded) captured from a shiny flat board (left) and a jar (right).

accurate 3D measurement, high frequency fringes are preferable (thus more periods), but it requires more textons which will increase the difficulty in designing unique textons, and in turn affect the decoding result. As a compromise, a code pattern can be embedded to more than 1 fringe, i.e., $N > 1$. When the code patterns are decoded, the exact period number of each fringe is not known, but just a rough approximation. However, it is often sufficient for solving the fringe discontinuity problem in fringe unwrapping. Figure 9.4 shows two fringe images in which the coded fringe pattern can be seen. In this figure, each texton is used for three consecutive fringes ($N = 3$) and in total five textons are used. This setting is used for all the experiments mentioned in this chapter. In next subsection, the extraction and decoding techniques for generating the k-values from the coded fringe patterns will be discussed.

9.3.2 Overview of the Decoding Algorithm

Figure 9.5 shows a summary of the decoding algorithm. As shown in the figure, the decoding algorithm consists of two stages: an offline and online stage. The offline stage is performed during the system calibration and the online stage is performed during the 3D measurement. In both stages, a modified MCA procedure is used to separate the fringe pattern and codes pattern. At the offline stage, we use a flat plane as the target object to train the system. Since a flat plane is employed, there is no fringe discontinuity problem in the fringe image. Thus, the period order information (k-values) of the fringes are known. It will be used as the ground truth

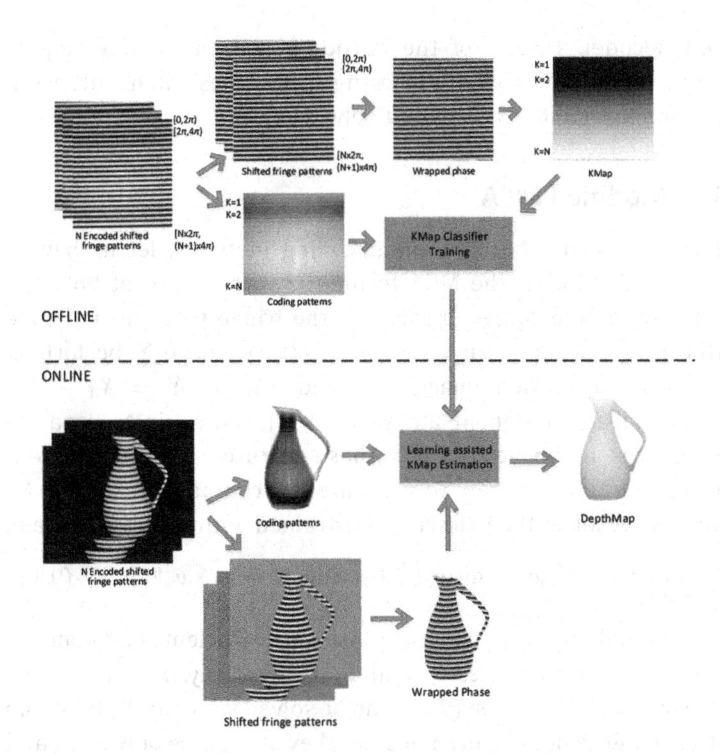

Figure 9.5 A summary of the learning based FPP method for 3D measurement.

to compare with that decoded from the extracted code patterns. Thus, by sampling many training patches of different code patterns and their associated k-values, a supervised training process can be carried out for learning a K-Map Classifier.

At the online stage, the captured encoded fringe images of the testing object are also separated using a modified MCA procedure to get the fringe patterns and coded patterns. The wrapped phase $\hat{\phi}$ is then calculated from the fringe pattern using (9.10) whereas the k-values of $\hat{\phi}$ are decoded from the coded patterns using the trained K-Map Classifier obtained during the offline stage. These k-values are used in the phase unwrapping procedure to evaluate the true phase ϕ. Finally, the 3D profile of the object can be measured. As mentioned above, two types of decoder are introduced in this section: (1) discriminative dictionary-based decoder and (2) CNN-based decoder. Details of the decoding and phase unwrapping processes of using these two learning approaches will be discussed in Section 9.3.4 and 9.3.5, respectively.

9.3.3 Modified MCA

Before discussing the different decoding methods, let us have a brief introduction of the MCA technique that is used at both the online and offline stages to separate the fringe patterns and code patterns in a fringe image. Let an arbitrary image Y be formed by two superimposed images, X_1 and X_2, i.e., $Y = X_1 + X_2$. Morphological component analysis (MCA) allows us to separate the features in image Y when these features present different morphological structures. To estimate the component X_1 and X_2 from Y, MCA solves the following constrained optimization problem,

$$\min_{w_1, w_2} \|Y - \Phi_1 w_1 - \Phi_2 w_2\|_2^2 + \lambda_1 \|w_1\|_1 + \lambda_2 \|w_2\|_1, \qquad (9.16)$$

where w_1 and w_2 are the sparse transform coefficients of X_1 and X_2 under the transform bases Φ_1 and Φ_2, respectively, i.e., $\Phi_1 w_1 = X_1$ and $\Phi_2 w_2 = X_2$. Equation (9.16) can be solved if X_1 and X_2 fulfill the mutual incoherence requirement, i.e., they are sparse only in Φ_1 and Φ_2, respectively, but not vice versa. To implement MCA, [58] employs the rational-dilation wavelet transform (RADWT) [59] and the split augmented Lagrangian shrinkage algorithm (SALSA) developed in [60–62]. The optimization problem in (9.16) can be formulated in the form of SALSA as follows:

$$\arg\min_{W} \|Y - \Phi v\|_2^2 + \lambda_1 \|w_1\|_1 + \lambda_2 \|w_2\|_1 \, s.t. v = w, \qquad (9.17)$$

where

$$\Phi = \begin{bmatrix} \Phi_1 & \Phi_2 \end{bmatrix}, \; v = \begin{bmatrix} v_1 \\ v_2 \end{bmatrix}, \quad w = \begin{bmatrix} w_1 \\ w_2 \end{bmatrix}, \quad \lambda = \begin{bmatrix} \lambda_1 \\ \lambda_2 \end{bmatrix} \qquad (9.18)$$

The problem in (9.17) is then solved by performing the following iterative procedure:

1. $z^{k+1} = S_\lambda \left(v^k + \delta^k\right) - \delta^k$;
2. $\delta^k = \frac{1}{(\mu+2)} \Phi^T \left(Y - \Phi z^{k+1}\right)$;
3. $v^{k+1} = \delta^{k+1} + z^{k+1}$

In the equations above, $z = w - \delta$ and z^{k+1} stands for the z at $k+1$ iterations. $S_\lambda(a)$ is the standard soft thresholding function, i.e., $S_\lambda(a) = \text{sgn}(a)\max(|a| - \lambda, 0)$. The initial v^0 and δ^0 are set to zero. The above procedure is performed repeatedly until it is converged. Then $X_i = \Phi_i \mathrm{w}_i^{\text{converge}}$ where $\mathrm{w}_i^{\text{converge}}$ is the w_i when converged with $i = 1, 2$. It is shown in [58] that the above algorithm can converge to the global minimum.

When applying MCA to FPP, certain modification is needed since the captured fringe image very often is not just the addition of fringe and code patterns. It should be noted that, similar to many optical imaging systems, FPP suffers from the optical aberration problem that the captured fringe images are often noisy and blurred. To model the captured fringe image G accurately, (9.14) should be reformulated as follows:

$$G = \mathcal{K}(G_1 + G_2) + \varepsilon, \tag{9.19}$$

where $\mathcal{K}$ is a matrix that represents the convolution with a blurring kernel, and ε is the Gaussian white noise (usually the variance is assumed to be known). In (9.19), G can be interpreted as the convolution of the captured fringe image with the blurring kernel plus the additive Gaussian white noise. An example of the captured fringe image affected by defocusing is shown in Fig. 9.6. Given $\mathcal{K}$ (can be determined during calibration), the problem in (9.16) needs to be slightly modified as follows:

$$\underset{\alpha_1,\alpha_2}{\arg\min} \|G - \mathcal{K}(\Phi_1\alpha_1 + \Phi_2\alpha_2)\|_2^2 + \lambda_1\|\alpha_1\|_1 + \lambda_2\|\alpha_2\|_1, \tag{9.20}$$

where α_1 and α_2 are the sparse coefficients of G_1 and G_2 in Φ_1 and Φ_2, i.e., $G_1 = \Phi_1\alpha_1$ and $G_2 = \Phi_2\alpha_2$; and λ_1 and λ_2 are the regularization parameters. For the implementation of Φ_1 and Φ_2, the tuned-Q wavelet transform (TQWT) as in [58, 70] can be used. They fulfill the mutual incoherence required by MCA; hence, they can efficiently capture the sparse features of interest. Equation (9.20) can be solved by utilizing the splitting variable approach as follows:

$$f_1(v) = \|G - \mathcal{K}\Phi v\|_2^2, \quad f_2(\alpha) = \lambda_1\|\alpha_1\|_1 + \lambda_2\|\alpha_2\|_1 \tag{9.21}$$

Figure 9.6 An example of the captured fringe image affected by severe defocusing.

with

$$\Phi = \begin{bmatrix} \Phi_1 \ \Phi_2 \end{bmatrix}, \quad v = \begin{bmatrix} v_1 \\ v_2 \end{bmatrix}, \quad \alpha = \begin{bmatrix} \alpha_1 \\ \alpha_2 \end{bmatrix}. \tag{9.22}$$

Finally, the implementation of (9.21) can be carried out using an iterative optimization procedure as follows:

Given the initial v^0 and δ^0,

$$\begin{aligned} \alpha^{k+1} &= \arg\min_{v} \lambda_1 \|\alpha_1\|_1 + \lambda_2 \|\alpha_2\|_1 + \frac{\mu}{2} \left\|\alpha - v^k - \delta^k\right\|_2^2; \\ v^{k+1} &= \arg\min_{v} \|G - \mathcal{K}\Phi v\|_2^2, + \frac{\mu}{2} \left\|\alpha^{k+1} - v - \delta^k\right\|_2^2; \\ \delta^{k+1} &= \delta^k - \left(\alpha^{k+1} - v^{k+1}\right) \end{aligned} \tag{9.23}$$

If Φ_1 and Φ_2 are implemented with a tight-frame TQWT and the matrix $\mathcal{K}$ can be assumed to be a circular convolution operation with a blurring kernel, it is shown in then $\mathcal{K}$, Φ_1 and Φ_2 can be factorized as follows:

$$\mathcal{K} = U^T H U, \quad \Phi_i = U^T C_i U \quad \forall i \in 1, 2, \tag{9.24}$$

where U is a discrete Fourier transform (DFT), $U^T = U^{-1}$ is its inverse (U is unitary, i.e., $U^T U = U^T U = I$), and H and C_i are some diagonal matrices. The Appendix at the end of this chapter shows that (9.23) can be further simplified to **Algorithm I**. The fringe pattern G_1 and the embedded code pattern G_2 can then be obtained from Y

Algorithm I: Modified MCA

Given the initial v^0 and δ^0

1. $\boldsymbol{z^{(k+1)} = S_\lambda \left(v^{(k)} + \delta^{(k)}\right) - \delta^{(k)}}$

2. $\boldsymbol{v^{(k)} = \frac{1}{\mu}\left(I - U^T F U\right)\left(\Phi^T \mathcal{K} G + \mu z^{k+1}\right)}$

3. $\boldsymbol{\delta^{(k+1)} = \delta^{(k+1)} - z^{(k+1)}}$

In **Algorithm I**, a new parameter F is introduced. It is defined as follows:

$$F = C^T H^* \left(\mu + 2\,|H|^2\right)^{-1} H C \qquad (9.25)$$

where H^* is the complex conjugate of H; $|H|^2$ is the squared absolute value of H; $C = [C_1\ C_2]$; and the expression $H^* \left(\mu + 2\,|H|^2\right)^{-1} H$ is in fact a Wiener filter in the frequency domain. It is used as a deblurring operation and enhances the noisy fringe image to facilitate the decomposition. Since the term $\Phi^T \mathcal{K} G$ can be pre-computed prior to the iteration process, the computation speed of **Algorithm I** can be very fast. In addition to that, the computation of the term F is also efficient because it is performed in the Fourier domain. The complexity of the algorithm is only $O(rM \log_2 M)$, where M is the total number of pixels of the fringe image and r is the redundancy factor which is set to 2. The decomposition results of a coded fringe image under different regularization parameter λ_1 are shown in Fig. 9.7. As shown in the figure, the best result can be achieved when λ_1 is set to 0.3 (i.e., $\lambda_2 = 0.7$ since $\lambda_1 + \lambda_2 = 1$). The parameter λ_1 is set at the offline stage and is kept fixed during the online stage.

9.3.4 Dictionary Learning–Based K-Map Classifier

In this section, the design of the dictionary learning-based K-Map Classifier is introduced. First, the training procedure of the dictionary is explained. Then the way how a linear classifier is used in the k-value estimation is discussed.

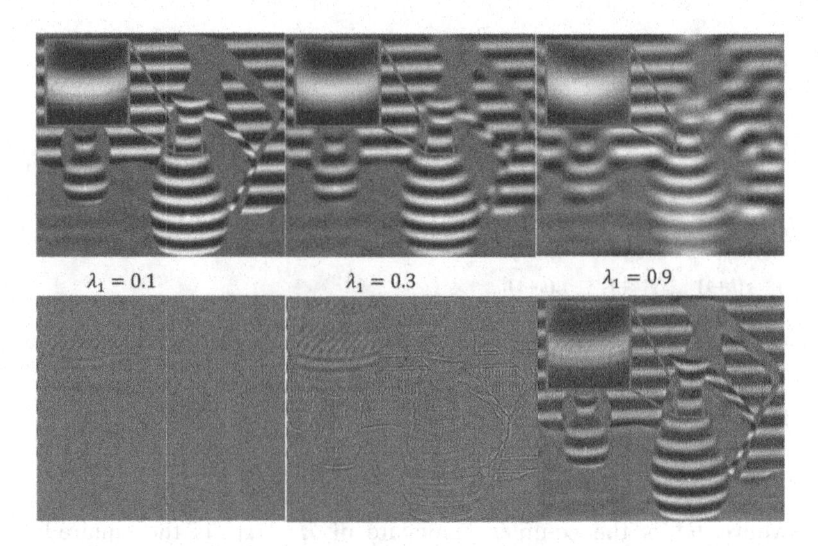

Figure 9.7 The coded fringe pattern decomposition using TQWT. The decomposition results: the fringe pattern G_1 (top row) and the code pattern G_2 (bottom row), under different regularization parameter λ_1.

9.3.4.1 Training dictionary

To train a reliable and robust dictionary for estimating the K-Map, many image patches and their corresponding class labels (k-values) are needed. They are selected randomly during the offline stage by first projecting the coded fringe patterns on a flat board and capturing the reflected fringe image following a standard FPP process. Since the flat board does not introduce sudden jumps or occlusion in the fringe image, the k-values of all fringes can be determined directly from the fringe pattern using traditional phase unwrapping algorithms. Let us denote a region of unique code pattern k as R^k. A set of code pattern patches and their Gabor features can be obtained by

$$P^k = \left\{p_i^k\right\}_{i=1,\ldots,N^k} \quad \text{and} \quad \zeta^k = \left\{\zeta_i | \zeta = \mathcal{G}\left(p_i^k\right)\right\}_{i=1,\ldots,N^k} \tag{9.26}$$

where P^k and ζ^k are a patch set and a patch feature set, respectively. N^k is the number of training patches for each label in set P^k. To allow effective training, the number of patches selected for training cannot be too small. In practice, we may select 256 patches ($N^k = 256$) for each unique k-value. Suppose we have K k-values, i.e., $k = 1, \ldots, K$, the total number of training patches is N^kK. In (9.26), $\mathcal{G}(\cdot)$ is the function to extract the Gabor features of a given patch. Specifically, this feature is obtained after the patch is convolved with the Gabor kernel of $j = 1 \ldots J$; and θ orientations, i.e. $\theta_o = 1, \ldots, \theta$. It gives a complex vector, $\rho^k_{j,\theta} = p^k * G_{j,\theta}$ where '*' denotes the convolution. Finally, a set of features ζ^k which is formed by taking the mean of $\rho^k_{j,\theta}$ of different scales and orientations can be expressed as

$$\zeta^k = \left[\overline{|\rho^k_{1,1}|}, \ldots, \overline{|\rho^k_{1,\theta}|}, \ldots, \overline{|\rho^k_{J,1}|}, \ldots, \overline{|\rho^k_{J,\theta}|}\right] \tag{9.27}$$

where $\overline{|\rho|}$ is the mean of the magnitude of ρ. Note that good results can be obtained by setting $J = 3$ and $\theta = 6$. Given a set of features ζ^k and their corresponding k-values, a discriminative dictionary D can be trained by using a label consistent K-SVD (version 1) algorithm (LC-KSVD) [38]. It can be achieved by minimizing the following optimization problem:

$$\underset{D,A,\gamma}{\arg\min} \|Z - D\gamma\|_2^2 + \|B - A\gamma\|_2^2 \, s.t. \forall i, \|\gamma_i\|_0 \leq T, \tag{9.28}$$

where $Z = \left[\zeta^1, \ldots, \zeta^K\right] \in \mathbb{R}^{\theta J \times N^k K}$ is the training feature set and ζ^k is a patch feature $\zeta^k_i = \mathcal{G}\left(p^k_i\right)$ with $p^k_i \in G_2$. The first term of (9.28) will learn a dictionary D which gives a sparse code $\gamma \in \mathbb{R}^{L \times N^k K}$ of Z, whereas the second term will enhance the discriminability of γ by minimizing the difference between the linear transformation of γ and a discriminative block diagonal binary matrix $B \in \mathbb{Z}^{N^k K \times N^k K}$ defined as follows:

$$B = \begin{bmatrix} 1_{N^k \times N^k} & 0 & 0 & 0 \\ 0 & 1_{N^k \times N^k} & 0 & 0 \\ 0 & 0 & \ddots & 0 \\ 0 & 0 & 0 & 1_{N^k \times N^k} \end{bmatrix} \tag{9.29}$$

where $1_{N^k \times N^k}$ refers to an all-ones matrix of size $N^k \times N^k$. The binary matrix B is used to force the code pattern patches of the same period order k to have the similar sparse representation. This strategy will ensure that the dictionary will have a good discriminative power for classification purpose. When training (9.28), the matrix D and A are initialized using the discrete cosine transform (DCT) and updated using the approach in [38].

At the online stage, given a discriminative dictionary D, a discriminative sparse code of a coded patch can be obtained. It is then used to determine the k-value of the code using a simple classifier. Following the Label Consistent K-SVD method [38–39], a linear classifier (W) can be obtained by solving the following minimization problem:

$$\arg\min_{W} \|H - W\gamma\|_2^2 + \lambda_w \|W\|^2, \tag{9.30}$$

where γ is obtained from (9.28), λ_w is a constant to control the contribution of the corresponding term, and

$$H = \begin{bmatrix} 1_{1\times N^k} & 0 & 0 & 0 \\ 0 & 1_{1\times N^k} & 0 & 0 \\ 0 & 0 & \ddots & 0 \\ 0 & 0 & 0 & 1_{1\times N^k} \end{bmatrix} \tag{9.31}$$

Equation (9.30) has a close form solution as follows:

$$W = K\gamma^T \left(\gamma\gamma^T + \lambda_w I\right)^{-1} \tag{9.32}$$

9.3.4.2 Multi-level scanline phase unwrapping

Using the dictionary D and the linear classifier W obtained during the offline stage, the k-values of the code pattern can be easily determined. To determine the k-value of a code pattern patch, first, their Gabor features are calculated using (9.26) to get its patch feature $\tilde{\xi}$. Then the sparse code $\tilde{\gamma}$ of features vector $\tilde{\xi}$ can be determined by the sparse coding method as

$$\arg\min_{\tilde{\gamma}} \left\|\tilde{\xi} - \mathrm{D}\tilde{\gamma}\right\|_2^2 \; st. \; \forall i, \; \|\tilde{\gamma}\|_0 \leq T_0. \tag{9.33}$$

Equation (9.33) can be solved using the Orthogonal Matching Pursuit (OMP) algorithm [67]. Finally, the k-value of the patch can be determined by the linear classifier W as follows:

$$\hat{k} = \max_i (W\tilde{\gamma}), \tag{9.34}$$

where $\max_i (W\tilde{\gamma})$ returns an index i of the coefficient in the vector $W\tilde{\gamma}$ of which the value is maximum.

Although every k-value of a K-Map can be determined using the above procedure, the process can be rather time consuming if it has to be done for all pixels of a fringe image. To reduce the complexity, the above method can be integrated into the traditional multi-level quality guided phase unwrapping algorithm [53]. To do so, a "good" point based on the quality of the wrapped phase map $\hat{\phi}$ is first chosen. Such "good" point can be determined by making use of the approach in [51, 53] as follows:

$$Q_{\text{map}}(i, j) = \max\left\{\max\left\{\left|\Delta^x_{i,j}\right|, \left|\Delta^x_{i-1,j}\right|\right\}, \max\left\{\left|\Delta^y_{i,j}\right|, \left|\Delta^y_{i-1,j}\right|\right\}\right\} \quad (9.35)$$

where $\Delta^x_{i,j} = \mathcal{W}\left(\hat{\phi}(i+1, j) - \hat{\phi}(i, j)\right)$ and $\Delta^y_{i,j} = \mathcal{W}(\hat{\phi}(i, j+1) - \hat{\phi}(i, j))$ are the wrapped phase differences in the horizontal and vertical directions and $\mathcal{W}(\cdot)$ is the phase wrapping operator such that $Q_{\text{map}}(i, j) \in [0, 2\pi)$. Given the quality map Q_{map}, a "good" point is where its Q_{map} value is lower than a threshold.

The information in Q_{map} is very useful in guiding the traditional phase unwrapping [54, 56]. However, we found that for a particular position where Q_{map} indicates the quality is low (e.g., the Q value is high), the result of k-value classification at that position also will not be good. An example is shown in Fig. 9.8, which depicts the code patterns and Q_{map} obtained in an FPP experiment of an object. Three code pattern patches and Q_{map} patches are extracted. The sparse k-value classification algorithm as mentioned above is applied to each code pattern patch and the resulting K-maps are shown. For the Q_{map} patch centered at position (x, y), the mean $\mu = \overline{Q(x, y)}$ of the patches is also evaluated. In the second row, the code pattern patch and Q_{map} patch are near the boundary of the object. Hence, some of the Q_{map} values (fourth column) are rather high which means that they are not suitable for phase unwrapping. It can be seen that the corresponding Q_{map} patch has a high mean value. One can also see that the K-Map estimated in the corresponding location (third column) is quite far away from the ground truth (second column). The same can be seen in the results in the final row. The code pattern patch and Q_{map} patch are at the location where distortion is found due to the specular reflection of the object. It can be seen that the corresponding Q_{map} patch has a relatively higher mean value. And the K-Map estimated in the corresponding location (third column) is also somewhat different from the ground truth (second column). On the contrary, the code pattern patch and Q_{map} patch in the third

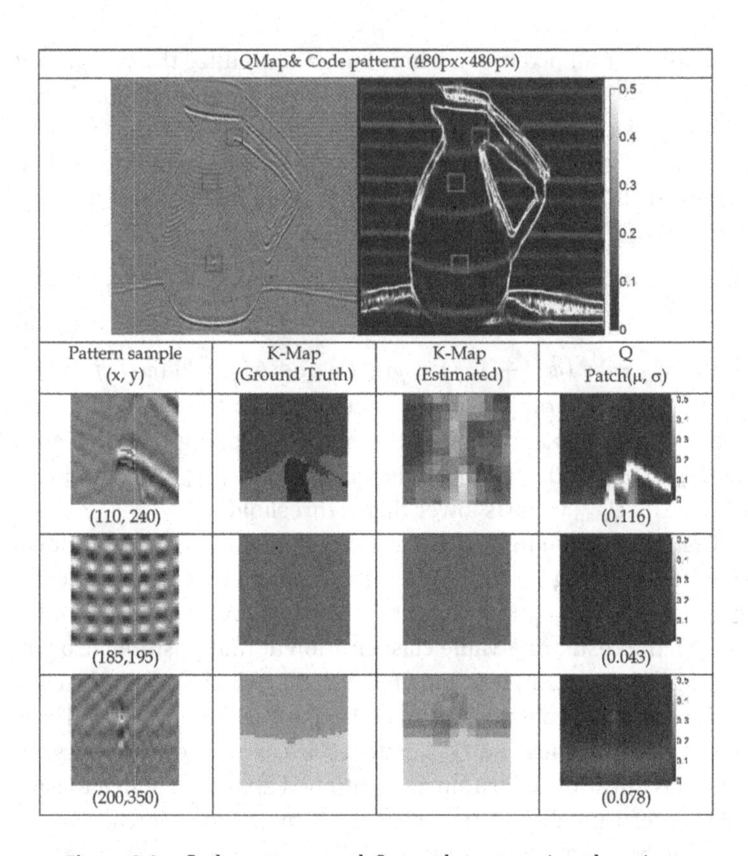

Figure 9.8 Code patterns and Q_{map} values at various locations.

row are at the smooth region of the object. The mean of the Q_{map} patch is low and the K-map estimated is very close to the ground truth. The above observation is expected since for the code pattern patches located at the positions where there are abrupt changes in the fringe pattern, both the MCA and the classification algorithm which heavily rely on the second order statistics in the optimization process will have difficulty to obtain statistically stationary data. Hence, the estimation is prone to error.

The above observation shows that we can make use of Q_{map} to identify the starting points for phase unwrapping. Once a starting point is identified, its k-value can be found using the sparse k-value classification algorithm defined in (9.33) and (9.34). Then the traditional multilevel scanline phase unwrapping algorithm can start based on the k-value until it finds the discontinuity such that the unwrapping cannot continue. Then another starting point can be found based on the same approach. The process repeats until all pixels in the fringe image is unwrapped. **Algorithm II** shows the detailed phase unwrapping procedure.

Algorithm II: Dictionary learning-based phase unwrapping
Inputs: The wrapped phase map $\hat{\phi}$, the quality map Q
Outputs: The unwrapped phase map ϕ

1. **Initialize** *minQThreshold* with a small value
2. **Repeat**
   (a) Randomly select a point (x, y) where $\overline{Q(x, y)} <$ *minQThreshold*
   (b) If the true phase and *k-value* of a neighbor of (x, y) has been obtained, directly compute $\phi(x, y)$ and its *k-value* based on those of its neighbor.
   (c) Otherwise, estimate the *k-value* using the proposed sparse *k-value* classification algorithm (i.e., (9.33) and (9.34)). Then based on the estimated *k-value*, compute $\phi(x, y)$.
   (d) Starting from (x, y), scan all other pixels of which the Q value also smaller than *minQThreshold*. Repeat step (a), (b) and (c) for all these pixels.
   (e) If no more pixel whose Q value has a mean smaller than *minQThreshold*, increase the value of *minQThreshold* by a fixed amount.
3. **Until** no more pixel to be unwrapped

Similar to the traditional multi-level scanline phase unwrapping algorithm, **Algorithm II** is also simple and fast. With the presence of the period order information, it allows true phase estimation even when there are multiple disconnected regions in the wrapped phase, which is common when the scene is complex. In this situation, the traditional algorithm such as [53] will fail. In fact, many traditional algorithms do not have the remedy like Step 2c of **Algorithm II** and hence, errors cannot be avoided. In such complex scene,

Algorithm II provides the *k*-value of a pixel for these regions such that the phase unwrapping algorithm can use them as the reference to unwrap all the pixels in these regions. Some examples will be given in Section 9.4 to illustrate the performance of this algorithm.

9.3.5 Deep Learning–Based K-Map Classifier

For the previous approach, it is assumed that a "good" point can be found and its *k*-value can be determined by the sparse *k*-value classification algorithm. Then the unwrapping procedure can be performed within a particular region. As indicated in **Algorithm II**, these steps are repeated until no more pixels to be unwrapped. However, in the practical environment, finding a good point through random selection (see step 2a of **Algorithm II**) can require many trials and thus delay the whole process. In this section, we introduce another way of decoding the code pattern. It is based on the recent convolutional neural network (CNN), which is more efficient and can provide more accurate pixel-wise prediction due to the massive training process. In this section, the network architecture of the CNN adopted is first discussed. Then the training of the CNN is described in detail. Finally, the K-Map estimation is explained. Unlike the previous method, the CNN-based K-Map classifier does not find the "good" points by trail-and-error method. Furthermore, the K-Map of the whole region can be found directly by CNN; thus the algorithm will be more efficient particularly when implementing using GPU. Finally, it will be shown in Section 9.4 that the CNN approach is sensitive to the quality of the extracted code pattern; hence, the method is more robust.

9.3.5.1 Network architecture

The architecture of the CNN used in this algorithm is designed based on U-Net [45], which can be considered as the concatenation of two smaller networks, namely, encoder network (the contracting path) and decoder network (expansion path). Since K-Map estimation is basically a texture segmentation problem, the network aims to learn from the highly repetitive patterns to segment a particular texture from a code pattern. It is similar to the segmentation problem

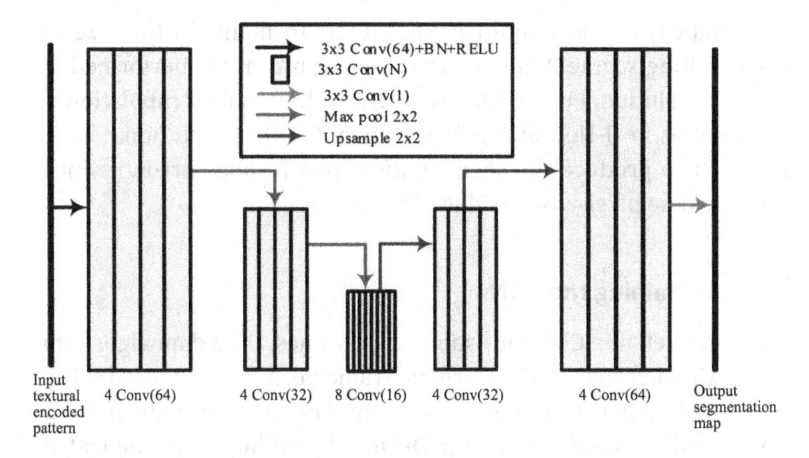

Figure 9.9 The architecture of the CNN adopted in the CNN-based K-Map Classifier.

in [43]. Recall that the embedded code patterns are specifically designed to have highly repetitive structure as shown in Fig. 9.3. They are periodic and its local 'order' is repeated over a small region. By taking account of these characteristics, the architecture of the CNN is designed as in Fig. 9.9. As shown in the figure, the number of contracting paths (red arrows) and expansive paths (green arrows) are reduced into two because of the localized property of the code pattern. Furthermore, the skip connection is also removed from the network since the contracting feature map and the up-sampling map are not directly correlated as in the segmentation problem in [45]. In the figure, each blue rectangle represents a basic building block of the network. It consists of a 3×3 convolution (Conv) followed by batch normalization (BN) and a ReLU. All blocks have the same number of inputs and outputs except the first layer whose input is a single code pattern image.

For the contracting path (encoder) and expansive path (decoder), the 2-dimensional (2D) max pool with size 2×2 for dyadic scale decomposition and the up-sampling operation are employed, respectively. Hence, similar to a multiresolution wavelet decomposition, the encoder is to extract the main features of the code pattern and the decoder is to reconstruct the main features to the k-values. In the figure, a set of four-consecutive basic blocks (blue block) act as a multichannel filter. To maintain the size of the resulting segmented map, the padding process is performed at each convolutional procedure. Hence, no additional extrapolation is required as in U-Net. In the final step, a 3×3 convolutional layer is added to produce the segmentation map (orange arrow) which indicates the pixel-wise k-value of the scene.

9.3.5.2 Training the CNN

To train a reliable CNN for estimating k-values, the Adam algorithm using the Pytorch deep learning framework can be used [71]. The input and the output of the network are the code pattern image and the resulting K-Map. During the offline stage, the coded fringe patterns are projected onto a flat plane as in the dictionary learning mentioned above. The reflected fringe image is captured following a standard FPP procedure. The coded fringe image also goes through the modified MCA to obtain the fringe patterns and coded patterns. One of the difficulties of training CNNs is to collect sufficient number of training samples. As it is time consuming to set up an FPP system to obtain the fringe image and the ground truth of the code patterns, it will take a long time to collect sufficient training samples. To generate more training data, fringe images are first divided in many small patches. They are further slightly modified to let them resemble the real image patches obtained in typical FPP systems. The three types of modification are (1) performing affine transformation to accommodate deformation of fringe due to object's shape; (2) introducing two additional artifacts, namely additive Gaussian noise and Gaussian blur to accommodate artifacts due to lens' distortion or the medium of transmission; and (3) adding and multiplying augmentation to accommodate various changes in intensity due to object's texture.

For training the CNN, the use of smaller patch size makes the training simple and requires less computation time. However, for estimating the K-Map, the effective patch size is highly correlated to the receptive field size of CNN that allows the network to capture accurate context information of the co-occurrence statistics of the texture. To determine the effective patch size at the training stage, we compare the loss when training the CNN with image patches

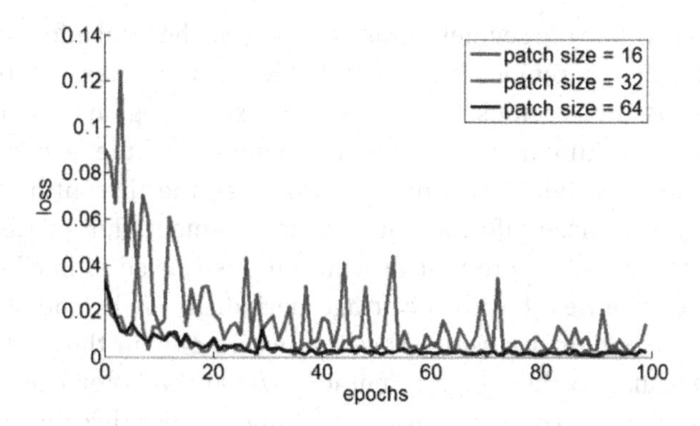

Figure 9.10 Convergence rate of the CNN trained by image patches of different patch sizes.

of the following three patch sizes: 16, 32, and 64, while fixing the learning rate. As shown in the Fig. 9.10, it can be observed that larger patch size can give faster and stable convergence (black line). It can be verified by comparing with the figures in Table 9.1, which shows the percentage of segmentation correctness of the CNN trained with different patch sizes. As shown in the table, using a patch size of 16 × 16 pixels can only give 24.51% correct segmentation while using larger patch sizes such as 32 × 32 and 64 × 64 ixels can give a more than 91% correct segmentation. Thus in all experiments described in this section, the CNN is trained with image patches of size 64 × 64 pixels. 30,000 patches are obtained by sampling the coded images randomly. The learning rate and batch size are set to be 0.001 and 128 respectively.

9.3.5.3 K-Map refinement

Although the segmentation map given by the CNN is a good estimation of the regions of different code patterns, it still has not reached the pixel-wise accuracy as required to generate the K-Map.

Table 9.1 % of segmentation correctness of the CNN trained with various patch sizes (16, 32, and 64)

Patch size	16 × 16	32 × 32	64 × 64
% Correctness	24.51	91.49	93.28

To further refine the segmentation map, we use the local information of the segmented regions to estimate the K-Map as well as to improve the misclassified regions. It consists of two steps: region refinement and k-value estimation. For FPP, the boundary of the k-values in the K-Map can be determined by analyzing the discontinuity of the wrapped phase information as in the segmentation procedure performed in [11]. Note that discontinuity is caused either by the 2π discontinuities due to the fringe periodicity or by the abrupt surface change or occlusion. The first occurs only in the direction perpendicular to the fringe orientation while the latter can occur in any direction. The boundary due to any of these discontinuities can be easily detected. After refining the region boundaries, we can estimate the k-values of each region. Suppose there are N_C regions, $\{R_m\}_{m=1,\ldots,N_C}$ in the extracted code pattern G_2, the k-values for each region R_m can be determined by,

$$k(R_m) = \Theta^{-1}\{G_2(R_m)\} \tag{9.36}$$

where $k(R_m)$ refers to the k-value of region R_m; Θ^{-1} is the decoding function for determining the k-values of a region defined as follows:

$$\Theta^{-1}(X) = V(CNN_{\text{learn}}(X)) \tag{9.37}$$

where $V(\cdot)$ is the voting (or mode in statistics) of all code values in region R and $CNN_{\text{learn}}(X)$ is obtained from the output of the CNN for a given code pattern. The voting is necessary due to the imperfect k-value estimation due to noise, bias, etc. Unlike the previous algorithm, the voting operator is performed locally for each wrapped region, i.e., a region bounded by 2π discontinuity. To further improve the accuracy of the refinement, the sequence of k-values should also be considered. It is noted that the k-value numbers must be monotonically increased. If by accident that the k-value of a region is smaller than that of the region before it, such k-value must have error. It can be replaced by a number just smaller than the next region. It may not be the correct number, but it will minimize the influence to the final 3D measurement.

9.4 Experiments

This section shows the performance of the two learning-based algorithms discussed in this chapter when working in real FPP systems. The system consists of a digital light processing (DLP) projector and a digital single lens reflex (DSLR) camera. The projector has a 2000:1 contrast ratio with light output of 3300 ANSI lumens and the camera has a 22.2 mm × 14.8 mm CMOS sensor and a 17–50 mm lens. They were connected to a personal computer with a 3.4 GHz CPU and 16 GB RAM. The distance between the object and FPP system is approximately 70–120 cm. All programs of were written in MATLAB or Python. The CNN is implemented using the Pytorch framework.

9.4.1 Quantitative Evaluation

9.4.1.1 Distortion introduced by the embedded period order information

While the period order information is extremely helpful to the phase unwrapping of FPP, it may also introduce distortion to the 3D measurement since the period order information are embedded into the fringe patterns, which can be seen by the FPP system as noises. To evaluate how such embedded period order information affects the 3D measurement, an experiment is conducted as following. First, a flat board with size 500 mm × 400 mm is used as the target object. Since its ground truth can be easily measured, it allows an objective comparison of the accuracy of different methods. Several methods were compared including the conventional phase shifting profilometry (PSP) with the Goldstein phase unwrapping algorithm (PSP+Goldstein) [10, 47, 50, 72], the PSP method with speckle-embedded fringe patterns (PSP-Speckle) [11], and the dictionary learning (DL) algorithm discussed in Section 9.3.4 with and without the MCA. PSP+Goldstein is the conventional method. PSP-Speckle was proposed recently which also embeds period order information into the fringe patterns. However, the period order is embedded in the form of random speckles and it does not use the MCA method to separate the code patterns and fringe patterns. It simply sums up all frames to remove the fringe pattern and obtain the code pattern. For the DL method, two methods are used to extract the code patterns. First, the MCA method as discussed in Section 9.3.3 is used; second, the method in PSP-Speckle is used. The objective of the comparison is to find out the performance of the modified MCA method in reducing the error in 3D measurement due to the embedded period order information. To evaluate the performance of these approaches, the board was scanned for 50 times for different wavelengths (i.e., 8, 12, 24, 32, and 48 pixels) and the root means square (RMS) of the reconstructed phase error is obtained by averaging the whole surface. Note that the total number of reconstructed points is 678,665 after removing a few pixels at the boundary which contain some artifacts. Figure 9.11 shows the comparison result. As shown in the figure, PSP-Speckle method shows low accuracy even when comparing with the conventional PSP+Goldstein method. It is obvious since the embedded speckle introduces distortion to the fringe patterns. Meanwhile the accuracy of the DL method without the modified MCA only shows a slightly better performance than the PSP-Speckle method but is worse than the conventional PSP. In fact, the embedded code patterns in the DL method also introduce distortion as the PSP-Speckle method. However, after the modified MCA method is applied, the accuracy of the DL method improves to be slightly better than the conventional PSP+Goldstein method. This result verifies that the modified MCA method can effectively separate the fringe patterns and code patterns.

9.4.1.2 Code pattern classification

As mentioned above, the first step of K-Map classification is similar to an image segmentation process that pixels of the same *k*-value will be segmented into a region. To qualitatively evaluate the performance of the two learning-based K-Map classifiers, the same flat board as in the previous experiment can be used. Since the *k*-values of a flat board can be determinedeasily, they can be used as the labels for training the discriminative dictionary and CNN. In the experiment, fringe patterns are first projected onto a flat board with different patterns as shown in Fig. 9.12 (first and third columns). Using the modified MCA, the code pattern can be extracted

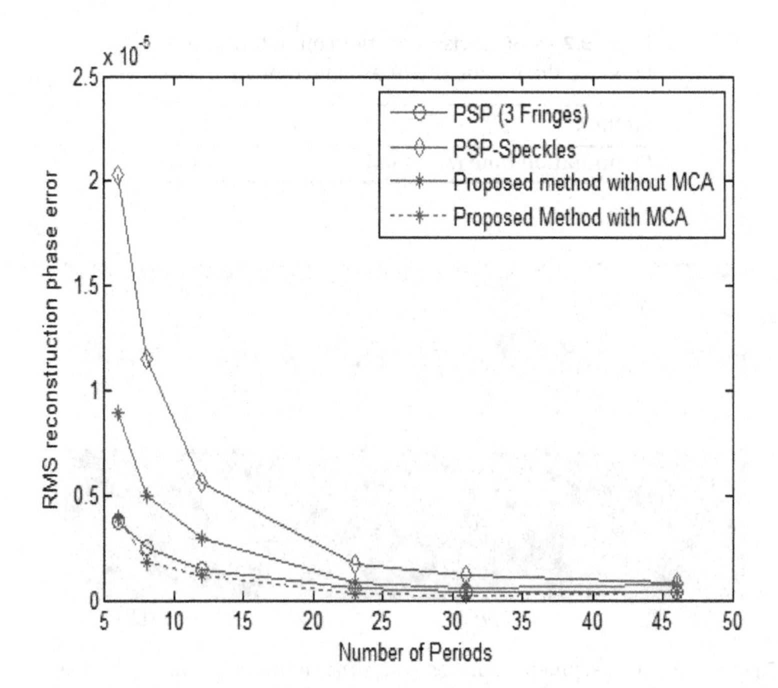

Figure 9.11 3D measurement error in RMS when using different methods with respect to the number of periods in the fringe pattern.

Figure 9.12 Flat boards with vivid texture and checkered texture (first and third columns) used in the experiments and their coded pattern images (second and fourth columns) obtained by the modified MCA method.

from the captured fringe image as depicted in Fig. 9.12 (second and fourth columns). The code patterns are then sent to the DL and CNN K-Map classifiers for code pattern classification. The selection of the parameters of both classifiers have been described in detail in Section 9.3.4.1 and Section 9.3.5.2. The accuracy of both classifiers is measured by the percentage of the correct estimation against

Table 9.2 Comparison on the computation time between the DL and CNN-based methods

Method	DL	CNN
Computation time (seconds)	6.53	0.3283

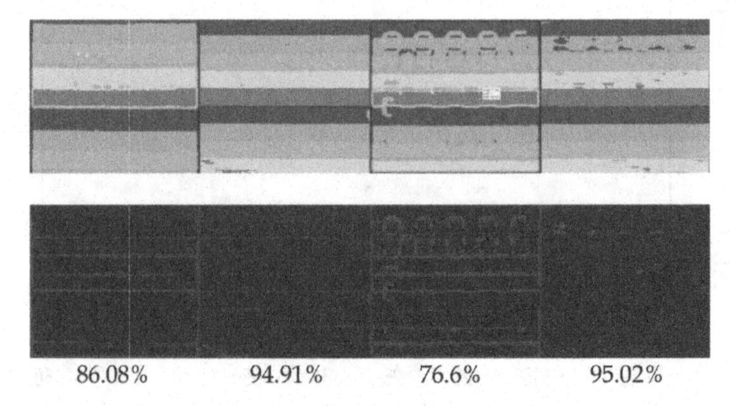

Figure 9.13 The segmentation results. (first and third columns) DL based K-Map Classifier; (second and fourth columns) CNN based K-Map Classifier. The first row is the decoding result and the second row is the error map against the ground truth. The percentages of correct classification are shown at the bottom.

the ground truth. The comparison results are shown in Fig. 9.13, where the segmentation result and error are shown in the first and second row, respectively. As shown in the figure, the CNN approach can accurately segment the code patterns even when the object has vividly changing textures on its surface. For the same texture pattern, the DL approach gives much inferior performance. The run time performance of the DL and CNN approaches are also compared as shown in Table 9.2. It can be seen in the table that the CNN approach is much faster than the DL approach. It is because the CNN method runs on a GPU while the DL approach runs only on a desktop computer. Due to its regular structure, the CNN approach can be easily implemented on a GPU. It is in contrast to the DL approach that much adaptation is needed to port the algorithm to GPU for execution.

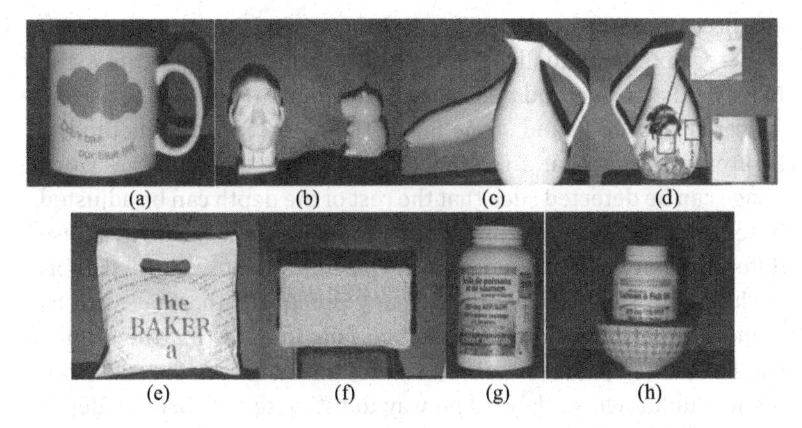

Figure 9.14 Objects that form complex scenes. (a) A cup; (b) A head sculpture and a plastic toy; (c) a plastic banana occluded by a jar; (d) a jar with texture and highlights; (e) a plastic bag with regular textual pattern; (f) a ceramic plate with complex texture covering the whole object; (g) a bottle with complex and strong texture cover almost the whole object; and (h) a bottle and a bowl with complex and strong texture.

9.4.2 Qualitative Evaluation

For the qualitative evaluation, various scenes as shown in Fig. 9.14 are considered. While the first scene is a simple glass with smooth color, the other scenes are more complex. They contain highlight regions, sudden intensity jump, occlusion, or bias due to the object's texture. Specifically, the second scene consists of two isolated objects: a head sculpture and a pink piggy toy. The third scene is a plastic banana that is occluded by a shiny jar. The fourth scene is a jar with vivid texture and highlights. The fifth scene is a plastic bag with regular and smooth texture. The sixth scene is a ceramic plate covered with texture; the seventh scene is a plastic bottle with irregular and strong texture; and the eighth scene is the same bottle inside a bowl with regular texture.

As in Section 9.4.1, the performances of the two learning-based methods are compared with the conventional method (PSP+Goldstein) [10, 47, 50, 72] and PSP-Speckle [11]. Figures 9.15 and 9.16 show the comparison results (together with the signal-to-noise (SNR) compared with the ground truth in each case). In order to generate the ground truths, 30 shifted fringe pattern images are projected onto each scene and the period order information is determined manually. Among the competing algorithms, the PSP+Goldstein method generates erroneous 3D profiles for scene 2, scene 3, scene 7, and scene 8. For the PSP+Goldstein method, it is often assumed that a reference point at the center of the fringe image can be detected such that the rest of the depth can be adjusted based on this reference point. It also requires that the absolute phase difference between two neighboring pixels will not be more than π. Such assumption is not valid in many practical situations such as scene 7, where the Goldstein method fails in detecting the branch cut. For scene 2, since no object locates at the center, the reference point is unknown so there is no way to estimate the absolute depth of the objects. For scene 3 and scene 8, only one of the objects is at the center. Hence, the depth of the object placed at the center can be estimated accurately but not the other which is isolated and has a sudden phase jump at the border. In fact, in practical environment, it is not easy to provide the reference point. The approach such as putting a tiny marker on the object as in [10] will fail if there are other objects located outside the region that is not connected directly to the reference point, such as scenes 2 and 3 in Fig. 9.15 as well as scene 8 in Fig. 9.16.

Unlike the conventional PSP method, the period order information of all pixels is provided in the PSP-Speckle method and the two learning-based algorithms. Hence, true phase can be obtained as shown in Figs. 9.15 and 9.16 (the second and third column). However, the PSP-Speckle method cannot estimate the period order accurately when the scene has complex texture. They are confused with the speckles and lead to the false k-value estimation, and in turn the erroneous 3D profiles as shown in the fourth to eighth scene of Figs. 9.15 and 9.16. To further illustrate this, we show in Fig. 9.17a the speckle map of the fifth scene after the fringe pattern is removed. It can be seen that the texture remains in the speckle map and lets the speckles (the bright dots) be imperceptible. Also, it is noted that the speckles affect the smoothness of the reconstructed 3D surface. Sometimes, they can lead to erroneous reconstruction particularly in the regions where the fringe images have other artifact such as highlights (see the result in the fourth scene in the second column of Figs. 9.15 and 9.16 and its vertical gradient in Fig. 9.18).

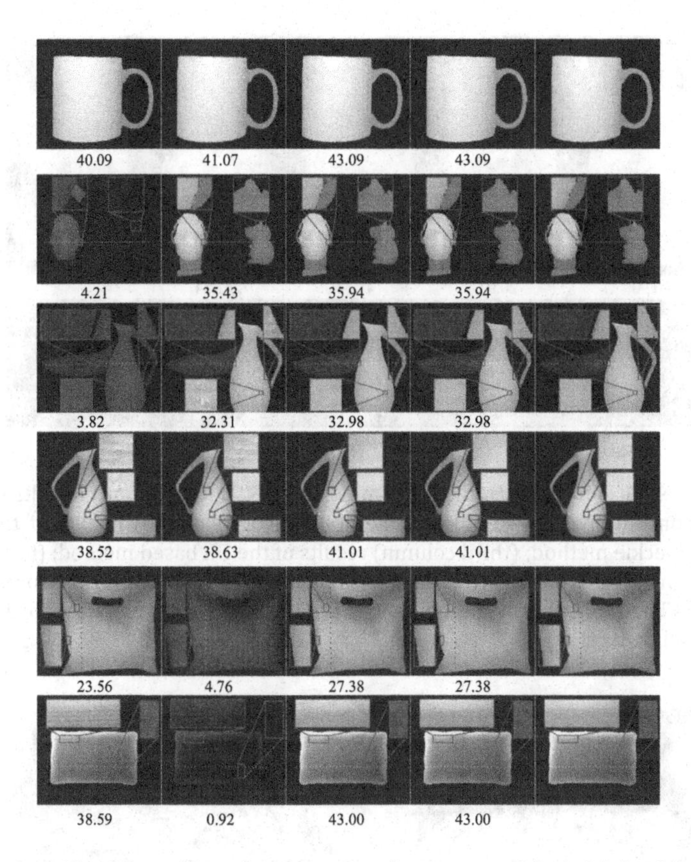

Figure 9.15 3D profiles of various complex scenes with multiple objects, occlusion, highlight regions, and textures. (First column) Results of the conventional PSP+Goldstein method; (second column) results of the PSP-Speckle method; (third column) results of the DL based method; (forth column) results of the CNN based method; and (fifth column) the ground truth. The numbers under the images are the SNR as compared with the ground truth.

The same problem does not exist for the learning-based algorithms because the modified MCA can extract the code pattern from the textural pattern of the object. It is clearly seen in Fig. 9.17b. Together with the period order detection method, the two learning-based algorithms can give the 3D profiles of the objects in different complex scenes. The results are close to the ground truths. However, the DL-based method seems to be more sensitive to the accuracy

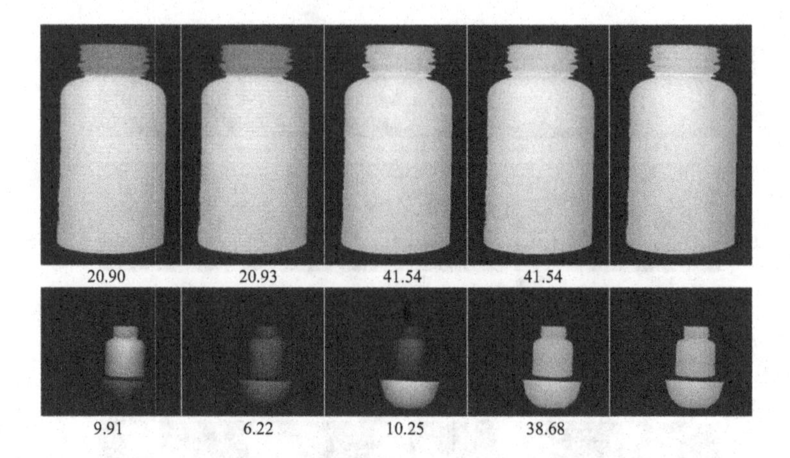

Figure 9.16 More 3D profiles of complex scenes. (First column) Results of the conventional PSP+Goldstein method; (second column) results of the PSP-Speckle method; (third column) results of the DL based method; (forth column) results of the CNN based method; and (fifth column) the ground truth. The numbers under the images are the SNR as compared with the ground truth.

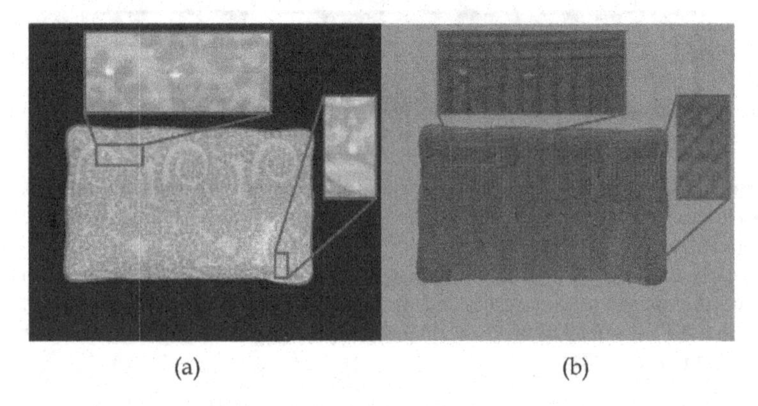

Figure 9.17 (a) The speckle map on the object in scene 6; and (b) the code pattern of the same object resulted from the modified MCA method.

of the code pattern generated by the modified MCA. As shown in Fig. 9.19, the code patterns generated by the MCA is a bit blurry at the top than at the bottom. The K-Map thus obtained by the DL approach contains more incorrect *k*-values. They lead to the

Figure 9.18 The vertical gradient of row 4 in Fig. 9.15; (first column) PSP+Goldstein method; (second column) the PSP-Speckle method; (third column) the DL based method; (fourth column) the ground truth.

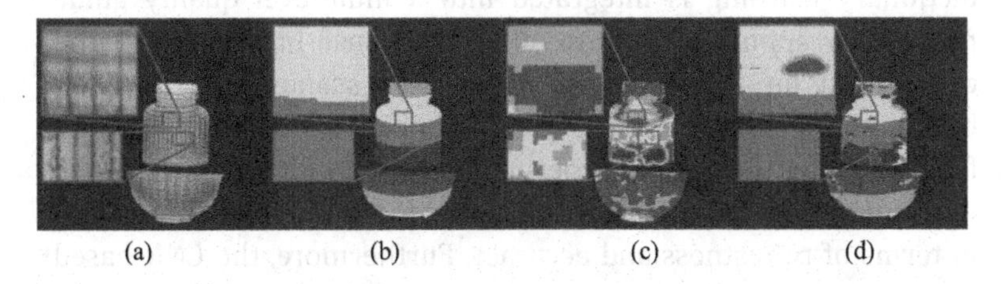

Figure 9.19 (a) The code pattern of the last scene in Fig. 9.16 obtained by the modified MCA (column 1); (b) the K-Map ground truth; (c) the K-Map obtained by the DL method; (d) the K-Map obtained by the CNN method.

erroneous 3D profile in the last scene as shown in Fig. 9.16. It is indicated by the low percentage of correctness, i.e., only 34.53% of the total area. On the other hand, the CNN-based method has a better classification power due to the massive training process. The generated 3D profiles are close to the ground truths. The improvement over the DL-based approach on scene 7 is more than two times, i.e., 76.63%. It is due to the relatively accurate K-Map as shown in Fig. 9.19d.

9.5 Conclusion

In this chapter, robust non-contact three-dimensional measurement algorithms based on the learning method are presented. In particular, the fringe projection profilometry method was introduced and the ambiguity problem due to the discontinuity in the fringe image was explained. It was shown that by embedding codes patterns into the fringe patterns, the ambiguity problem can be solved.

Unlike the conventional methods, the algorithms introduced in this chapter does not require additional hardware setup or additional fringe pattern projections. By adopting the modified morphological component analysis (MCA) method, the code patterns and fringe patterns can be separated. Then a supervised learning is employed to estimate the k-values of the extracted code patterns. Two supervised learning algorithms were introduced in this chapter, namely, the dictionary learning (DL) and convolutional neural network (CNN)-based algorithms. Along with a classification procedure, the dictionary learning is integrated into a multilevel quality guide phase unwrapping process to allow the phase unwrapping to be carried out for fringe images of complex scenes. For the CNN-based approach, it is integrated to a simple refinement procedure for estimating the K-Map. Experimental results have demonstrated the superiority of these algorithms over the traditional approaches in terms of robustness and accuracy. Furthermore, the CNN-based approach can be extremely efficient when implementing on a GPU platform. Comparing between the two learning based approaches, the DL based method is more sensitive to the quality of the code patterns extracted by the modified MCA. Hence, the CNN-based approach will have a higher robustness provided that sufficient number of training samples are available.

Acknowledgments

This research work is supported by the Hong Kong Polytechnic University under research grant number 4-ZZHM.

Appendix: Proof of Algorithm I

Consider the minimization problem as stated in (9.20),

$$\underset{\alpha_1,\alpha_2}{\arg\min} \|G - \mathcal{K}(\Phi_1\alpha_1 + \Phi_2\alpha_2)\|_2^2 + \lambda_1 \|\alpha_1\|_1 + \lambda_2 \|\alpha_2\|_1$$

As shown in (9.23), the above can be solved by the following iterative procedure:

Initialize $k \leftarrow 0, \mu > 0, \delta^0$, and$v^0$

Repeat

$$
\begin{aligned}
& z^{k+1} \leftarrow S_{\lambda/\mu}\left(v^k + \delta^k\right) - \delta^k \\
& v^{k+1} \leftarrow \left(\Phi^T \mathcal{K}^T \mathcal{K}\Phi + \mu I\right)^{-1}\left(\Phi^T \mathcal{K}^T G + \mu z^{k+1}\right) \\
& \delta^{k+1} = v^{k+1} - z^{k+1} \\
& k \leftarrow k+1
\end{aligned}
$$

Until meeting the stopping criteria

Now let us focus on the first term in line 4 of the above procedure. Since we adopt a tunable Q-factor wavelet transform (TQWT), which has a tight frame ($\Phi_i \Phi_i^T = I$) for both Φ_1 and Φ_2, we have

$$
\Phi\Phi^T = \left[\Phi_1\ \Phi_2\right]\begin{bmatrix}\Phi_1 \\ \Phi_2\end{bmatrix} = 2I
$$

Thus, applying the Sherman–Morrison–Woodbury matrix inversion lemma to the first term in line 4, we have

$$
\left(\Phi^T \mathcal{K}^T \mathcal{K}\Phi + \mu I\right)^{-1} = \frac{1}{\mu}\left(I - \Phi^T \mathcal{K}^T\left(\mu I + 2\mathcal{K}\mathcal{K}^T\right)^{-1}\mathcal{K}\Phi\right)
$$

Assume that $\mathcal{K}$ can be approximated as a circular convolution operator. Then $\mathcal{K}$, Φ_1, and Φ_2 can be factorized as

$$
\mathcal{K} = U^T H U,\quad \Phi_i = U^T C_i U,\quad \Phi_i^T = U^T C_i^T U,\quad \forall i \in 1, 2,
$$

where U represents the discrete Fourier transform (DFT), $U^T = U^{-1}$ is its inverse; H and C_i are some diagonal matrices. Therefore, the above terms can be written as,

$$
\begin{aligned}
& \Phi^T \mathcal{K}^T\left(\mu I + 2\mathcal{K}\mathcal{K}^T\right)^{-1}\mathcal{K}\Phi \\
& = \mathrm{U}^{\mathrm{T}} C_i^T H^* U\left(\mu U^T U + 2U^T H H^* U\right)^{-1} U^T H C_i U \\
& = \mathrm{U}^{\mathrm{T}} \underbrace{\left(C_i^T H^* (\mu + 2|H|^2)^{(-1)} H C_i\right)}_{F}
\end{aligned}
$$

where H^* is the complex conjugate of H^*; $|H|^2$ is the squared absolute of the entries of H; and $C = \left[C_1\ C_2\right]$. By substituting the above terms to the algorithm, **Algorithm I** is obtained as follows:

Initialize $k \leftarrow 0$, $\mu > 0$, δ^0, and v^0

Repeat

$$z^{k+1} \leftarrow S_{\lambda/\mu}\left(v^k + \delta^k\right) - \delta^k$$

$$v^{k+1} \leftarrow \frac{1}{\mu}\left(I - U^T F U\right)\left(\Phi^T K^T G + \mu z^{k+1}\right)$$

$$\delta^{k+1} = v^{k+1} - z^{k+1}$$

$$k \leftarrow k+1$$

Until meeting the stopping criteria

(Q.E.D.)

References

1. Z. Song, R. Chung, and X.-T. Zhang (2013). An accurate and robust strip-edge-based structured light means for shiny surface micromeasurement in 3-D, *IEEE Trans. Ind. Electron.*, **60**(3), pp. 1023–1032.
2. T.-W. Hui and G. K.-H. Pang (2008). 3-D Measurement of solder paste using two-step phase shift profilometry, *IEEE Trans. Electron. Packag. Manuf.*, **31**(4), pp. 306–315.
3. S. Zhang (2010). High-resolution, high-speed 3-D dynamically deformable shape measurement using digital fringe projection techniques, in *Advances in Measurement Systems*, eds. M. K. Sharma (InTechOpen), pp. 29–50.
4. R. R. Garcia and A. Zakhor (2012). Consistent stereo-assisted absolute phase unwrapping methods for structured light systems, *IEEE J. Sel. Topics Signal Process.*, **6**(5), pp. 411–424.
5. F. Sadlo, T. Weyrich, R. Peikert, and M. Gross (2005). A practical structured light acquisition system for point-based geometry and texture, *Proceedings Eurographics/IEEE VGTC Symposium on Point-Based Graphics.*
6. P. S. Huang and S. Zhang (2006). Fast three-step phase-shifting algorithm, *Appl. Opt.*, **45**(2), pp. 5086–5091.
7. S. K. Nayar (2012). Micro phase shifting, *Proceedings of the IEEE Conference on Computer Vision Pattern Recognition (CVPR).*

8. Y. Wang, K. Liu, Q. Hao, D. L. Lau, and L. G. Hassebrook (2011). Period coded phase shifting strategy for real-time 3-D structured light illumination, *IEEE Trans. Image Process.*, **20**(11), pp. 3001–3013.
9. F. Yang and X. He (2007). Two-step phase-shifting fringe projection profilometry: Intensity derivative approach, *Appl. Opt.*, **46**(29), pp. 7172–7178.
10. S. Zhang and S.-T. Yau (2006). High-resolution, real-time 3D absolute coordinate measurement based on a phase-shifting method, *Opt. Express*, **14**(7), pp. 2644–2649.
11. Y. Zhang, Z. Xiong, and F. Wu (2013). Unambiguous 3D measurement from speckle-embedded fringe, *Appl. Opt.*, **52**(32), pp. 7797–7805.
12. M. Takeda and K. Mutoh (1983). Fourier transform profilometry for the automatic measurement of 3-D object shapes, *Appl. Opt.*, **22**(24), pp. 3977–3982.
13. T.-C. Hsung and D. P.-K. Lun (2010). On optical phase shift profilometry based on dual tree complex wavelet transform, *Proceedings of the IEEE International Conference on Image Processing (ICIP) 2010*, pp. 337–340.
14. T.-C. Hsung, D. Pak-Kong Lun, and W. W. L. Ng (2011). Efficient fringe image enhancement based on dual-tree complex wavelet transform, *Appl. Opt.*, **50**(21), pp. 3973–3986.
15. Q. Kemao (2007). Two-dimensional windowed Fourier transform for fringe pattern analysis: Principles, applications and implementations, *Opt. Lasers Eng.*, **45**(2), pp. 304–317.
16. Q. Kemao, H. Wang, and W. Gao (2008). Windowed Fourier transform for fringe pattern analysis: theoretical analyses, *Appl. Opt.*, **47**(29), pp. 5408–5419.
17. J. Salvi, J. Pagès, and J. Batlle (2004). Pattern codification strategies in structured light systems, *Pattern Recognition*, **37**(4), pp. 827–849.
18. Y. Wang and S. Zhang (2012). Novel phase-coding method for absolute phase retrieval, *Opt. Lett.*, **37**(11), pp. 2067–2069.
19. J. M. Huntley and H. Saldner (1993). Temporal phase-unwrapping algorithm for automated interferogram analysis, *Appl. Opt.*, **32**(17), pp. 3047–3052.
20. Y. Wang, K. Liu, Q. Hao, X. Wang, D. L. Lau, and L. G. Hassebrook (2012). Robust active stereo vision using Kullback–Leibler divergence, *IEEE Trans. Pattern Anal. Mach. Intell.*, **34**(3), pp. 548–563.
21. J. Gass, A. Dakoff, and M. K. Kim (2003). Phase imaging without 2π ambiguity by multiwavelength digital holography, *Opt. Lett.*, **28**(13), pp. 1141–1143.

22. W.-H. Su and H. Liu (2006). Calibration-based two-frequency projected fringe profilometry: A robust, accurate, and single-shot measurement for objects with large depth discontinuities, *Opt. Express*, **14**(20), pp. 9178–9187.
23. K. Liu, Y. Wang, D. L. Lau, Q. Hao, and L. G. Hassebrook (2010). Dual-frequency pattern scheme for high-speed 3-D shape measurement, *Opt. Express*, **18**(5), pp. 5229–5244.
24. Y. Wang, S. Yang, and X. Gou (2010). Modified Fourier transform method for 3D profile measurement without phase unwrapping, *Opt. Lett.*, **35**(5), pp. 790–792.
25. Y. Zhang, Z. Xiong, Z. Yang, and F. Wu (2014). Real-time scalable depth sensing with hybrid structured light illumination, *IEEE Trans. Image Process.*, **23**(1), pp. 97–109.
26. P. Cong, Z. Xiong, Y. Zhang, S. Zhao, and F. Wu (2015). Accurate dynamic 3D sensing with Fourier-assisted phase shifting, *IEEE J. Sel. Topics Signal Process.*, **9**(3), pp. 396–408.
27. H. Cui, W. Liao, N. Dai, and X. Cheng (2012). A flexible phase-shifting method with absolute phase marker retrieval, *Measurement*, **45**(1), pp. 101–108.
28. S. Gai and F. Da (2010). A novel phase-shifting method based on strip marker, *Opt. Lasers Eng.*, **48**(2), 205–211.
29. B. Budianto, D. P. K. Lun, and T.-C. Hsung (2014). Marker encoded fringe projection profilometry for efficient 3D model acquisition, *Appl. Opt.*, **53**(31), pp. 7442–7453.
30. W. Luo, A. G. Schwing, and R. Urtasun (2016). Efficient deep learning for stereo matching, *Proceedings of the 2016 IEEE Conference on Computer Vision and Pattern Recognition (CVPR)*, pp. 5695–5703.
31. Z. Chen, X. Sun, L. Wang, Y. Yu, and C. Huang (2015). A deep visual correspondence embedding model for stereo matching costs, *Proceedings of the 2015 IEEE International Conference on Computer Vision (ICCV)*.
32. S. R. Fanello et al. (2016). HyperDepth: Learning depth from structured light without matching, *Proceedings of the 2016 IEEE Conference on Computer Vision and Pattern Recognition (CVPR)*, pp. 5441–5450.
33. Budianto and D. P. K. Lun (2016). Robust fringe projection profilometry via sparse representation, *IEEE Trans. Image Process.*, **25**(4), pp. 1726–1739.

34. J. Bobin, J. L. Starck, J. M. Fadili, Y. Moudden, and D. L. Donoho (2007). Morphological component analysis: An adaptive thresholding strategy, *IEEE Trans. Image Process.*, **16**(11), pp. 2675–2681.
35. J. Bobin, J. L. Starck, J. Fadili, and Y. Moudden (2007). Sparsity and morphological diversity in blind source separation, *IEEE Trans. Image Process.*, **16**(11), pp. 2662–2674.
36. L.-W. Kang, C.-W. Lin, and F. Yu-Hsiang (2012). Automatic single-image-based rain streaks removal via image decomposition, *IEEE Trans. Image Process.*, **21**(4), pp. 1742–1755.
37. J. Mairal, F. Bach, and J. Ponce (2012). Task-driven dictionary learning, *IEEE Trans. Pattern Anal. Mach. Intell.*, **34**(4), pp. 791–804.
38. Z. Jiang, Z. Lin, and L. S. Davis (2013). Label consistent K-SVD: Learning a discriminative dictionary for recognition, *IEEE Trans. Pattern Anal. Mach. Intell.*, **35**(11), pp. 2651–2664.
39. Q. Zhang and B. Li (2010). Discriminative K-SVD for dictionary learning in face recognition, *Proceedings of the IEEE Conference on Computer Vision and Pattern Recognition (CVPR)*, 2010.
40. S. Yubao, L. Qingshan, T. Jinhui, and T. Dacheng (2014). Learning discriminative dictionary for group sparse representation, *IEEE Trans. Image Process.*, **23**(9), pp. 3816–3828.
41. J. Mairal, F. Bach, J. Ponce, G. Sapiro, and A. Zisserman (2008). Discriminative learned dictionaries for local image analysis, *Proceedings of the IEEE Conference on Computer Vision and Pattern Recognition (CVPR)*.
42. I. Ramirez, P. Sprechmann, and G. Sapiro (2010). Classification and clustering via dictionary learning with structured incoherence and shared features, *Proceedings of the IEEE Conference on Computer Vision and Pattern Recognition (CVPR)*.
43. V. Andrearczyk and P. F. Whelan (2017). Texture segmentation with Fully Convolutional Networks, *arXiv preprint arXiv:1703.05230*.
44. E. Shelhamer, J. Long, and T. Darrell (2017). Fully convolutional networks for semantic segmentation, *IEEE Trans. Pattern Anal. Mach. Intel.*, **39**(4), pp. 640–651.
45. O. Ronneberger, P. Fischer, and T. Brox (2015). U-Net: Convolutional networks for biomedical image segmentation, *Proceedings of Medical Image Computing and Computer-Assisted Intervention (MICCAI 2015)*, pp. 234–241.
46. T. Chen, H. P. A. Lensch, C. Fuchs, H.-p. Seidel, and M. Informatik (2007). Polarization and phase shifting for 3D scanning of translucent objects,

Proceedings of the IEEE Conference on Computer Vision and Pattern Recognition (CVPR), 2007.

47. D. Malacara (2007). *Optical Shop Testing* (Wiley Series in Pure and Applied Optics).
48. T. Chen (2008). Modulated phase-shifting for 3D scanning, *Proceedings of the IEEE Conference on Computer Vision and Pattern Recognition (CVPR)*.
49. K. Itoh (1982). Analysis of the phase unwrapping algorithm, *Appl. Opt.*, **21**(14), p. 2470.
50. R. M. Goldstein, H. A. Zebker, and C. L. Werner (1998). Satellite radar interferometry: Two-dimensional phase unwrapping, *Radio Sci.*, **23**(4), pp. 713–720.
51. D. C. Ghiglia and M. D. Pritt (1998). *Two-Dimensional Phase Unwrapping: Theory, Algorithms, and Software* (John Wiley & Sons).
52. L. Meng, S. Fang, P. Yang, L. Wang, M. Komori, and A. Kubo (2012). Image-inpainting and quality-guided phase unwrapping algorithm, *Appl. Opt.*, **51**(13), pp. 2457–2462.
53. S. Zhang, X. Li, and S.-T. Yau (2007). Multilevel quality-guided phase unwrapping algorithm for real-time three-dimensional shape reconstruction, *Appl. Opt.*, **46**(1), pp. 50–57.
54. K. Chen, J. Xi, and Y. Yu (2013). Quality-guided spatial phase unwrapping algorithm for fast three-dimensional measurement, *Opt. Commun.*, **294**, pp. 139–147.
55. M. Zhao, L. Huang, Q. Zhang, X. Su, A. Asundi, and Q. Kemao (2011). Quality-guided phase unwrapping technique: comparison of quality maps and guiding strategies, *Appl. Opt.*, **50**(33), pp. 6214–6224.
56. T. Flynn (1996). Consistent 2-D phase unwrapping guided by a quality map, *Proceedings of the International Symposium on Geoscience and Remote Sensing (IGARSS)*.
57. T. J. Flynn (1997). Two-dimensional phase unwrapping with minimum weighted discontinuity, *J. Opt. Soc. Am. A*, **14**(10), p. 2692.
58. I. W. Selesnick (2011). Resonance-based signal decomposition: A new sparsity-enabled signal analysis method, *Signal Process.*, **91**(12), pp. 2793–2809.
59. İ. Bayram and I. W. Selesnick (2011), A dual-tree rational-dilation complex wavelet transform, *IEEE Trans. Signal Process.*, **59**(12), pp. 6251–6256.

60. M. V. Afonso, J. M. Bioucas-Dias, and M. A. T. Figueiredo (2010). Fast image recovery using variable splitting and constrained optimization, *IEEE Trans. Image Process.*, **19**(9), pp. 2345–2356.
61. M. V. Afonso, J. M. Bioucas-Dias, and M. A. T. Figueiredo (2011). An augmented Lagrangian approach to the constrained optimization formulation of imaging inverse problems, *IEEE Trans. Image Process.*, **20**(3), pp. 681–695.
62. M. A. T. Figueiredo, J. M. Bioucas-Dias, and M. V. Afonso (2009). Fast frame-based image deconvolution using variable splitting and constrained optimization, *Proceedings of the IEEE/SP 15th Workshop on Statistical Signal Processing*.
63. M. Aharon, M. Elad, and A. Bruckstein (2006). K-SVD: An algorithm for designing overcomplete dictionaries for sparse representation, *IEEE Trans. Signal Process.*, **54**(11), pp. 4311–4322.
64. J. Mairal, F. Bach, J. Ponce, and G. Sapiro (2009). online dictionary learning for sparse coding, *Proceedings of the International Conference on Machine Learning (ICML)*.
65. S. Chen, D. Donoho, and M. Saunders (1998). Atomic decomposition by basis pursuit, *SIAM J. Sci. Comput.*, **20**(1), pp. 33–61.
66. R. Tibshirani (2011). Regression shrinkage and selection via the lasso: a retrospective, *Journal of the Royal Statistical Society: Series B (Statistical Methodology)*, **73**(3), pp. 273–282.
67. S. G. Mallat and Z. Zhang (1993). Matching pursuits with time-frequency dictionaries, *IEEE Trans. Signal Process.*, **41**(12), pp. 3397–3415.
68. D.-S. Pham and S. Venkatesh (2008). Joint learning and dictionary construction for pattern recognition, *Proceedings of the IEEE Conference on Computer Vision and Pattern Recognition (CVPR)*.
69. M. Yang, L. Zhang, X. Feng, and D. Zhang (2014). Sparse representation based fisher discrimination dictionary learning for image classification, *Int. J. Comput. Vision*, **109**(3), pp. 209–232.
70. I. W. Selesnick (2011). Wavelet transform with tunable Q-factor, *IEEE Trans. Signal Process.*, **59**(8), pp. 3560–3575.
71. A. Paszke, G. Sam, S. Chintala, G. Chanan, E. Yang, Z. DeVito, Z. Lin, A. Desmaison, L. Antiga, and A. Lerer (2017). Automatic differentiation in PyTorch, *Proceedings Thirsty-First Conference on Neural Information Processing Systems (NIPS 2017)*.
72. P. Huang (2007). A fast three-step phase shifting algorithm for real-time three-dimensional shape measurement, *Proceedings of Frontiers in Optics 2007*.

Part IV

Biometrics and Health Applications

Chapter 10

Computational and Learning Aspects of DNA Sequences

Ngai-Fong Law
Department of Electronic and Information Engineering,
The Hong Kong Polytechnic University, Hung Hom, Hong Kong
ennflaw@polyu.edu.hk

DNA is an extremely long chain of molecules that contains all the information necessary for the life functions of a cell. Through representing DNA as a character string, computational methods can be used to study DNA sequences. In this chapter, we have discussed the problem of identifying protein-coding regions called exons in a DNA sequence. Various computational measures have been used to study the bias distribution of base compositions in exons, such as position asymmetry and three-periodicity characterization through discrete Fourier transform. These statistical features can then be fed into machine learning methods such as *k*-nearest neighbor and neural networks to obtain reliable identification.

10.1 Introduction

DNA contains the genetic composition of every living organism in this planet. Its study is vital to facilitate our understanding

Learning Approaches in Signal Processing
Edited by Wan-Chi Siu, Lap-Pui Chau, Liang Wang, and Tieniu Tan

ISBN 978-981-4800-50-1 (Hardcover), 978-0-429-06114-1 (eBook)
www.panstanford.com

of the biological and genetic information encoded inside DNA. Recent advancement in technologies enables individual genome to be sequenced in a fast, accurate and affordable manner. Through sequencing, DNA can be represented as a long sequence composing of four nucleotide bases: namely A (Adenine), C (Cytosine), G (Guanine) and T (Thymine). Therefore, in addition to carrying out experiments, computational methods can be developed based on this representation for extracting knowledge within the DNA sequences.

One important problem in DNA sequence analysis is to identify the function in different parts of a DNA sequence [1]. In particular, genes are specific parts that are of interest because genes can act as instructions to make proteins and to determine our physical traits. Along the DNA sequence, genes appear at different regions in a distributed manner. For example, human is estimated to have between 20,000 and 25,000 genes. These genes may not be located next to each other. In fact, most genes are separated with a few hundred or thousand bases from each other. Genes in eukaryotes further contains two regions called exons and introns. Exons correspond to the protein coding region while introns are often called non-coding region. It is important to know their exact locations within a gene so that we can understand how different proteins are formed from a gene.

In the past decades, many computational methods have been proposed to distinguish exons from introns in a DNA sequence. All these computational methods are based on finding features that show different statistics in the two regions. Example features include the different bases compositions [2], the different profiles of the occurrence of bases [3, 4], the different distribution of co-occurrence frequencies [5, 6], and the periodicity information of the two regions. Through using these features, exons can be identified. To further improve the identification accuracy, learning approach can be performed. In other words, these features are used to train the learning method so that the learned model can identify potential exon regions in a new DNA sequence. Typical learning methods include clustering-based approaches [8], support vector machine [9–11], neural network [12–14] and hidden Markov model [15, 16].

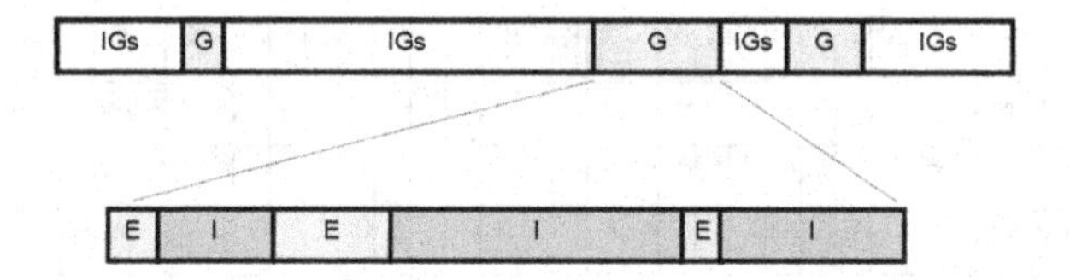

Figure 10.1 Structure of a DNA sequence of eukaryotes. A DNA sequence is made up from genes (G) and inter-genic spaces (IGs). Genes are further divided into two regions: exons (E) and introns (I).

Recently, deep models such as convolutional neural network [17] and recurrent neural networks [19] have been proposed for exon identification.

In this chapter, computational methods for learning-based approaches for exon and intron identification are summarized. Our objective is to explain how these learning-based methods can be applied in the biological area. We will first give a background on DNA sequence. Features about exons and introns are discussed. Details about inclusion of biological knowledge into signal processing methods will be described. Then machine learning approach for exon identification using these features is summarized. Finally, the use of deep model is discussed.

10.2 Background on DNA Sequences

DNA sequence can be considered to be a character string consisting of four bases: A, C, G and T. Despite that there are only 4 distinct bases, DNA is a very long sequence. For example, human genome contains around 3 billion bases and the yeast contains about 12 million bases. The length of a DNA sequence, however, may not reflect the complexity of the living organism. The single cell organism Amoeba genome is 100 times larger than the human genome [20]. Along a DNA sequence of eukaryotes, genes are specific parts of the long string that encode instructions for proteins synthesis. The segment separating two genes is called inter-genic spaces. Figure 10.1 shows a schematic diagram of a DNA sequence.

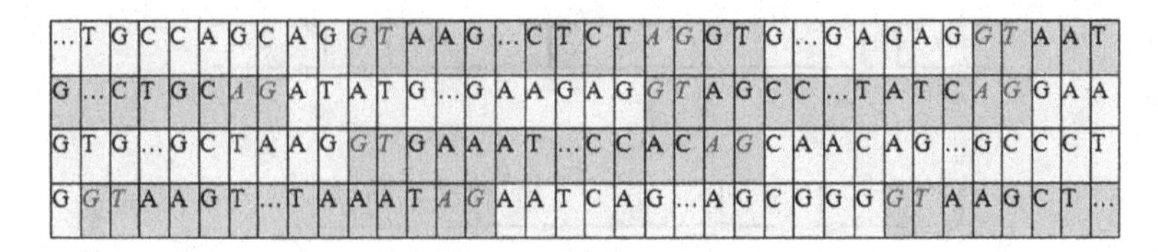

Figure 10.2 Genes are composed of two regions called exon and intron which are interleaved with each other. Exon is highlighted with blue color while intron is with orange color.

Figure 10.1 illustrates that genes can appear in multiple random parts of a DNA sequence with variable size of inter-genic spaces in between genes. These inter-genic spaces may be involved in regulations or providing structural integrity [21]. Genes can have diverse lengths but all of them are further sub-divided into two regions called exon and intron. Exon corresponds to protein coding regions while introns are non-coding regions. Generally, the length of the exons is shorter than introns. An example showing the base composition in a gene is given in Fig. 10.2 where exon and intron are highlighted with blue color and orange color, respectively. We can see that exons and introns are interleaved with each other and their lengths are not fixed.

Splice junction refers to the boundary between exon and intron. In a process called splicing, introns are removed from the gene sequence [22]. Various combinations of alterative exons would remain to construct different types of proteins from a single gene. Therefore, an important problem in DNA sequence analysis is to identify the location of exons and introns so that an understanding of how the different proteins are produced can be achieved. It is, however, a very difficult problem because genes are relatively few in number in many genomes. For example, around 3% of human genome corresponds to genes only [1]. Thus, many techniques have high number of false splice junction identifications. Besides, the length of exons can be very short. This makes their identification challenging.

To identify splice junctions, an experimental technique called RNA-seq [23] can be performed. This is usually combined with alignment-based method for identifying the locations of exons and introns. Majority of splice sites contain special canonical splicing patterns [24, 25]. The most frequent patterns are "GT" (called donor) and "AG" (called acceptor) at intron and exon boundaries. For example in Fig. 10.2, we can see that most introns start with "AG" and end with "GT." Hence, splice junctions can be identified by finding "AG" and "GT" patterns. However, using only these canonical splicing patterns for splice junction identification is not reliable. While some junctions may not have these patterns, these patterns can also occur elsewhere in the sequence as there are only four distinct bases constituting the DNA. Thus, a large number of false positives would result. Other features in the exons and introns should be explored to achieve a reliable splice junction identification.

10.3 Features about Exons (Coding Regions)

A way to identify exon and intron regions is to find out features that show different behavior in these two regions. For example, exon and intron regions often have different base compositions and arrangement. This base compositional difference can then be used to infer gene structure and distinguish exon region from intron region. Let S denote a DNA sequence containing bases S_i at the i-th position, i.e.,

$$S = S_1 S_2 S_3 S_4 S_5 S_6 \cdots \tag{10.1}$$

Codon, which is three consecutive bases in a DNA sequence, forms an amino acid. Different combinations of amino acids then form proteins. Because of this structure, the sequence S sometimes is sub-divided into three sub-sequences as follows,

$$\begin{aligned} SubS_1 &= S_1 S_4 S_7 \cdots \\ SubS_2 &= S_2 S_5 S_8 \cdots \\ SubS_3 &= S_3 S_6 S_9 \cdots \end{aligned} \tag{10.2}$$

The following summarizes different features that have been proposed to characterize exon (coding region) and intron (non-coding region).

- **Compositional bias:** compositional bias refers to the uneven distribution of bases in the three subsequences [3]. For

Table 10.1 The normalized occurrence frequency of different bases $F(b, SubS_j)$ in coding and non-coding regions

	Coding region			Non-coding region		
	$SubS_1$	$SubS_2$	$SubS_3$	$SubS_1$	$SubS_2$	$SubS_3$
A	0.2728	0.3211	0.3397	0.2994	0.2958	0.2955
C	0.2044	0.3091	0.1612	0.1760	0.1948	0.1908
G	0.3946	0.1516	0.1021	0.1941	0.1768	0.2045
T	0.1283	0.2181	0.3970	0.3304	0.3326	0.3092

example, in subsequence $SubS_3$ (i.e., the third position of a codon), it is more common to find bases G/C than A/T. This forms the compositional bias towards G and C content. Mathematically, these compositional bias is obtained through finding the occurrence frequencies of different bases at $SubS_1$, $SubS_2$ and $SubS_3$. Let $F(b, SubS_j)$ be the normalized occurrence frequency of the base b in $SubS_j$ where $b \in \{A, C, G, T\}$ and $j = 1, 2, 3$. In non-coding region, the base is considered to appear randomly. Thus, the occurrence frequencies $F(b, SubS_j)$ would be uniformly distributed and close to 0.25 for all bases b in $SubS_j$. However, exon contains particular structure of genetic coding. Hence, the coding region has uneven occurrence frequencies among the four bases. Table 10.1 shows an example of the normalized occurrence frequencies in a coding and a non-coding regions.

To quantify the difference in the occurrence frequency of bases in the coding and non-coding region, the square sum difference from the uniform distribution can be calculated. For the example in Table 10.1, the values for the coding and non-coding regions are 0.1170 and 0.0460, respectively. This shows that the bias inside coding region is larger than that inside non-coding region. Thus, the base compositional bias can be used to distinguish coding and non-coding regions.

- **Position Asymmetry:** the position asymmetry can be measured by considering the variance of $F(b, SubS_j)$ from its average values [5]. The average frequencies of the base b over the three sub-sequences can be obtained as

$$AveF(b) = \frac{\sum_{j=1}^{3} F(b, SubS_j)}{3}. \qquad (10.3)$$

Then the position asymmetry in the distribution of the bases can be obtained as

$$PA = \sum_{b \in \{A,C,G,T\}} Var(b), \qquad (10.4)$$

where

$$Var(b) = \sum_{j=1}^{3} \{F(b, SubS_j) - AveF(b)\}^2. \qquad (10.5)$$

If there is no bias in the distribution of the occurrence frequency (i.e., in the non-coding region), PA will be close to zero. In contrast in coding region, the bias will make the value of PA deviate from zero. As an example, the position asymmetries of the coding and the non-coding regions for the set of data considered in Table 10.1 are, respectively, 0.1004 and 0.00093. We can see the PA value in coding region deviates from 0 but the value in non-coding region is close to 0.

- **Periodic correlation between base positions:** a way to quantify the correlation information between base positions is to compute the co-occurrence frequency of a particular base b and another base d when b and d are separated by k bases [4, 6]. Consider an example of a simple DNA sequence $S = ACTTAGGTACT$. The co-occurrence of "A" and "A" separated by 3 bases is two while the occurrence of "A" and "G" separated by 0 base is 1.

  Studies found that the base "A" is more likely to be found at $k = 2, 5, 8, \ldots$ than at other k values. This forms a periodic pattern reflecting the strong correlation at certain base positions. Figure 10.3 shows an example in which the periodic pattern can be observed for coding region while the pattern looks random for non-coding region.

  To quantify the periodic pattern in Fig. 10.3 using a single measure, probabilities ratio can be calculated. Let P_1, P_2

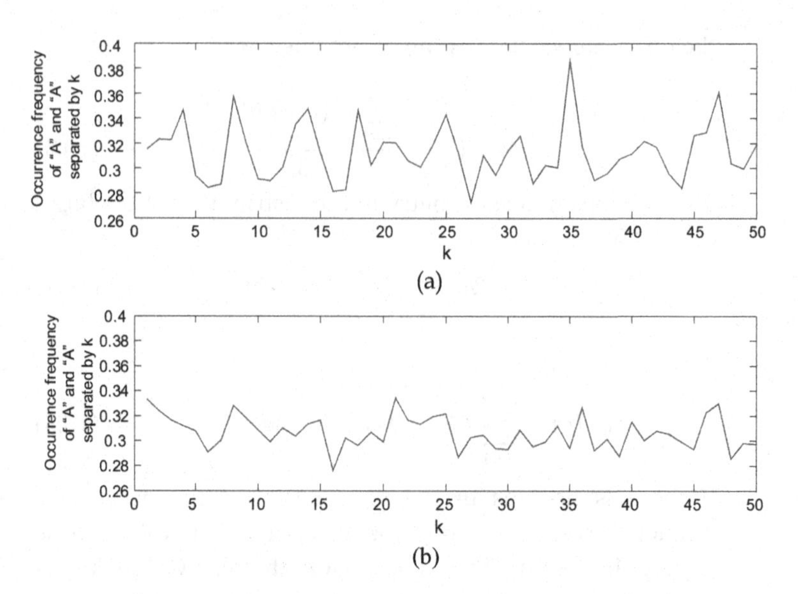

Figure 10.3 The normalized frequency of the pair "AA" separated by k number of bases in (a) a coding and (b) a non-coding region.

and P_3 be the probabilities of finding pairs of the same base at distance $k = 0, 3, 6, \ldots$, $k = 1, 4, 7, \ldots$ and $k = 2, 5, 8, \ldots$, respectively. Because of the periodic pattern, P_3 should be much larger than P_1 and P_2. The ratio between $\max(P_1, P_2, P_3)$ and $\min(P_1, P_2, P_3)$ can be used to distinguish coding and non-coding regions. In particular, the ratio in coding region would be large while it is close to 1 for non-coding region.

- **Mutual information:** It is a way to quantify the dependency of two bases along a DNA sequence [26]. Let $P_{b,d}(k)$ be the probability of finding bases b and d that are separated by k bases. Let P_b be the probability of base b in the DNA sequence. Mutual information is defined as

$$I(k) = \sum_{b,d \in \{A,C,G,T\}} P_{b,d}(k) \log \left(\frac{P_{b,d}(k)}{P_b P_d} \right). \qquad (10.6)$$

The mutual information shows the amount of information that can be observed from one base about another base

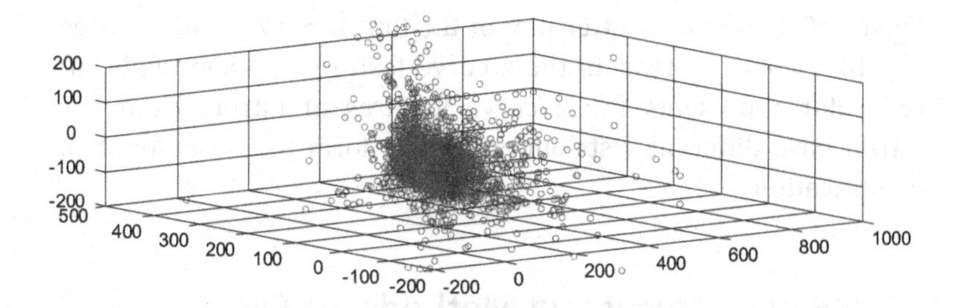

Figure 10.4 The scatter plot of the first three z curve features of exon (in blue color) and intron (in red color).

at a distance of k bases. $I(k)$ does not vary with k in non-coding region but varies periodically with k in coding regions. Essentially, mutual information measures the periodic correlation pattern between base positions.

- **Z curve:** the Z curve approach attempts to find statistical information about the cumulative frequencies of the occurrence of individual bases [27, 28]. There are nine features in Z curve which are defined as follows,

$$\begin{aligned}
zc_{3(i-1)} &= F\left(A, SubS_i\right) + F\left(G, SubS_i\right) - F\left(C, SubS_i\right) \\
&\quad - F\left(T, SubS_i\right) \\
zc_{3(i-1)+1} &= F\left(A, SubS_i\right) + F\left(C, SubS_i\right) - F\left(G, SubS_i\right) \\
&\quad - F\left(T, SubS_i\right) \qquad i = 1, 2, 3 \\
zc_{3(i-1)+2} &= F\left(A, SubS_i\right) + F\left(T, SubS_i\right) - F\left(C, SubS_i\right) \\
&\quad - F\left(G, SubS_i\right) \qquad (10.7)
\end{aligned}$$

The z curve features measure the uneven distribution of certain biological properties. In particular, zc_0, zc_3 and zc_6 measure the relative distribution of purine (A or G) and pyrimidine (C or T) types for bases at positions $\{0, 3, 6, 9, \cdots\}$, $\{1, 4, 7, 10, \cdots\}$ and $\{2, 5, 8, 11, \cdots\}$, respectively. The features zc_1, zc_4 and zc_7 measure the relative distribution of amino (A or C) and keto (G or T) types for bases at position $\{0, 3, 6, 9, \cdots\}$, $\{1, 4, 7, 10, \cdots\}$ and $\{2, 5, 8, 11, \cdots\}$, respectively. Lastly, the features zc_2, zc_5 and zc_8 measure the relative distribution of the weak H-bond group (A or T) and the strong H-bond (G or C) types for bases at position $\{0, 3, 6, 9, \cdots\}$, $\{1, 4, 7, 10, \cdots\}$ and $\{2, 5, 8, 11, \cdots\}$, respectively.

Figure 10.4 shows a scatter plot of the first three z curve features. The blue color represents the z curve features of exon while the red color represents the z curve features of intron. Exon and Intron have different distributions which form the basis for their identification.

10.4 Signal Processing Methods for Exon Identification

In Section 10.3, features that show difference in coding and non-coding regions are discussed. All these features attempt to characterize the unbalanced base composition in coding regions in which periodicity is observed in exon while absent in intron. In fact, a natural way to obtain periodicity information in a sequence is to use the discrete Fourier transform (DFT). If a signal has a period of 3, the power spectrum is expected to show a peak at the normalized frequency $f = 2\pi/3$. Thus, a peak at $f = 2\pi/3$ should be present in coding region and absent in non-coding region [7, 29, 30].

Before DFT can be used to check the periodicity in a DNA sequence, the DNA sequence must be converted into a numerical sequence. A simple way to obtain the numerical sequence is to use the one-hot encoding scheme [31–35]. A binary indicator sequence $u_b[n]$ is defined for base b at the base position n. The value of $u_b[n]$ is 1 if the base b is present at the position n and 0 otherwise. Then the DNA sequence can be written as

$$S = w_A u_A[n] + w_C u_C[n] + w_G u_G[n] + w_T u_T[n] \qquad (10.8)$$

where w_b is the weighting associated with base b. In this way, DFT can be applied to the binary indicator sequences $u_b[n]$ and the power spectrum can be obtained as

$$PS[k] = |U_A[k]|^2 + |U_C[k]|^2 + |U_G[k]|^2 + |U_T[k]|^2, \qquad (10.9)$$

where

$$U_b[k] = DFT\{u_b[n]\}. \qquad (10.10)$$

Figure 10.5 shows the power spectra of different sequences. Figure 10.5a shows the spectrum of a long exon consisting of 39168 bases. Peak at 13056 ($f = 2\pi/3$) can clearly be seen. When the exon sequence is shorter as in Fig. 10.5b, the peak at $f = 2\pi/3$ can still be seen, despite that the power spectrum is noiser than that in Fig. 10.5a. Figure 10.5c shows the spectrum of an intron with 3429 bases. Peak at $f = 2\pi/3$ is missing. Thus, Fig. 10.5 shows that value at $f = 2\pi/3$ of the power spectrum can be used to identify an exon.

The use of DFT in periodicity characterization of long sequences is good. However, for short sequences, the periodicity characterization may not be good. Unfortunately, length of exon sequence can be rather short. Approximately 80% of exons have fewer than 200 bases [36]. Figure 10.6 shows the power spectra for short sequences with 300 bases. In Fig. 10.6a, the peak at $f = 2\pi/3$ (i.e., $k = 100$) is not obvious for an exon sequence. Comparing the spectrum of an intron in Fig. 10.6b, it is difficult to distinguish the two sequences using DFT. Figure 10.6c shows the power spectrum of a random sequence. In fact, all the three sequences in Fig. 10.6 show similar characteristics. This example shows the limitation of using DFT on short exons.

To improve the robustness of DFT-based periodicity characterization, biological properties should be incorporated into the DFT framework. Our numerical sequence representation should be modified to incorporate biological knowledge. Z curve considers three kinds of biological knowledge: the purine (A/G) and pyrimidine (C/T) periodicity, the amino (A/C) and keto (G/T) periodicity and the weak H-bond (A/T) and strong H-bond (G/C) periodicity. Thus, three modified sequences can be obtained as

$$\begin{aligned} zs_1[n] &= u_a[n] + u_g[n] - u_c[n] - u_t[n] \\ zs_2[n] &= u_a[n] + u_c[n] - u_g[n] - u_t[n]. \\ zs_3[n] &= u_a[n] + u_t[n] - u_g[n] - u_c[n] \end{aligned} \tag{10.11}$$

DFT can then be applied to these modified sequences [37, 38]. For the same short exon sequence considered in Fig. 10.6a, the spectrum of $zs_1[n]$ shown in Fig. 10.7 clearly shows a peak at $f = 2\pi/3$. Thus, the modified sequence should be able to characterize exon better than the original sequence.

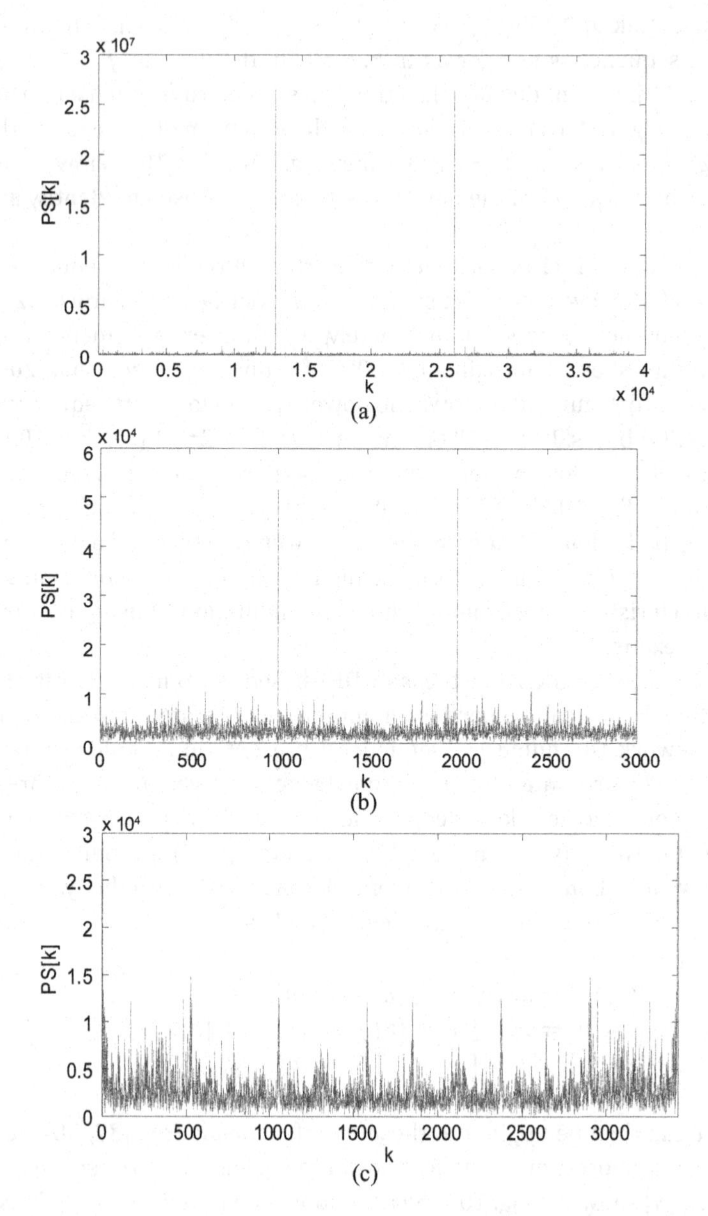

Figure 10.5 Power spectra for three different sequences, (a) a long exon with 39168 bases, (b) an exon with 3000 bases and (c) an intron with 3429 bases.

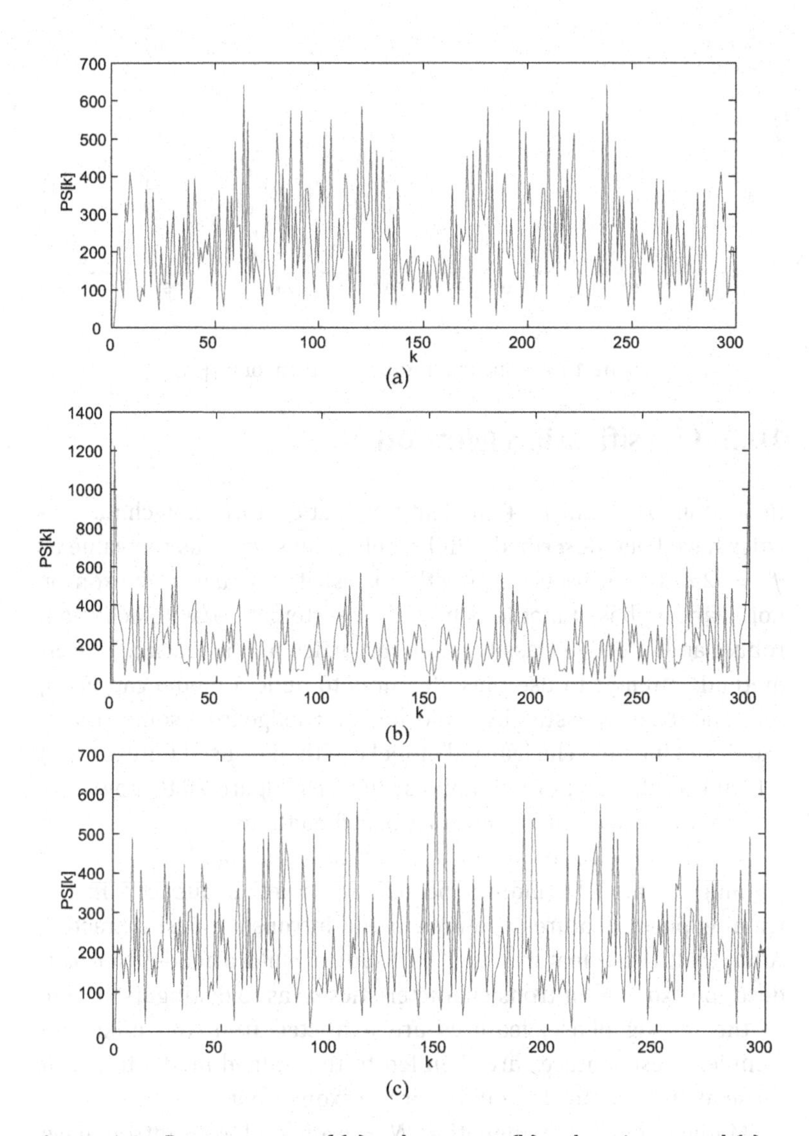

Figure 10.6 Power spectra of (a) a short exon, (b) a short intron and (c) a random sequence. All of them have 300 bases.

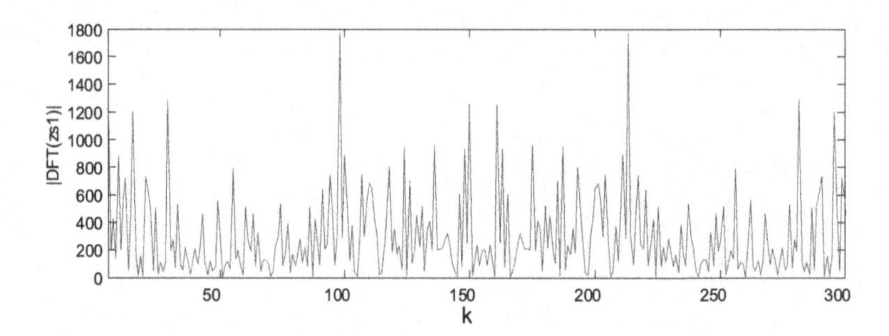

Figure 10.7 The magnitude spectrum of $zs_1[n]$.

10.5 Classification Methods

In Sections 10.3 and 10.4, the features that capture exon characteristics have been described. While their values (such as the value at $f = 2\pi/3$) can be used directly to test if a sequence is exon or not, learning-based methods provide a systematic way to achieve a robust and effective classification. In particular, the learning-based methods attempt to use these features to build a model capturing sequence characteristics by exploring knowledge from some known exons and introns. The learned model can then be used for exon and intron classification of unknown sequences. Figure 10.8 shows the general framework of the learning-based methods.

There are two phrases in learning-based methods, namely training for model building and testing of new sequences. In the training phase, features about exon and introns are first extracted. After that, these features will be used to build a model that can describe exon and introns characteristics so as to distinguish them. In the testing phase, features are extracted from the unknown sample. These features are then fed to the trained model to test if those features match the model for the exon or not.

Mathematically, assume that *N/2* exon and *N/2* intron have been collected. These sequences will be used to build a model for describing their characteristics. The first step is to extract features from these sequences. Let the features for the i-th sequence be denoted as

$$FeatV_i = \left\{ f_{i,1}, f_{i,2}, \cdots, f_{i,m} \right\}, \tag{10.12}$$

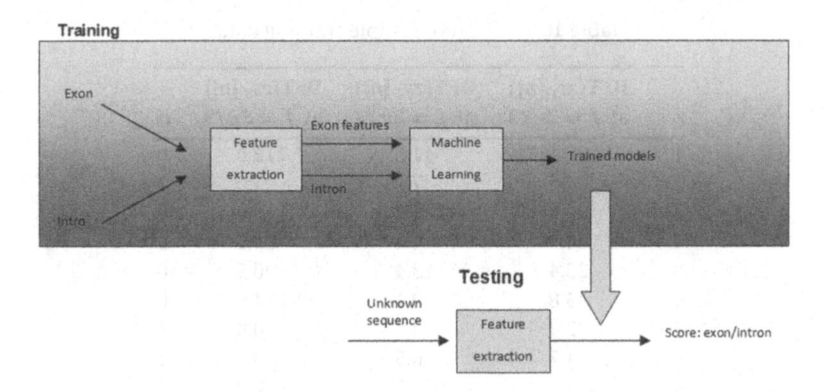

Figure 10.8 General machine learning framework.

where $i = 1, \cdots, N$ and m is the total number of features values. These features are labeled according to the prior knowledge of whether $FeatV_i$ is from an exon or not, i.e.,

$$Label\{FeatV_i\} = d_i \tag{10.13}$$

where $d_i = 0$ if it is an exon and $d_i = 1$ if it is an intron. Consider that DFT is applied to the three modified sequences $zs_1[n]$, $zs_2[n]$ and $zs_3[n]$. Three features, namely the DFT values at $f = 2\pi/3$ of these three sequences are extracted. Table 10.2 shows an example of these three feature values of exons and introns. In this case, we have $FeatV_1 = \{82.1, 41.5, 116.0\}$ and $Label\{FeatV_1\} = d_1 = 0$ for the first exon sequence. The second sequence is an intron with $FeatV_2 = \{11.7, 20.2, 1.3\}$ and $Label\{FeatV_2\} = d_2 = 1$. Figure 10.9 shows a scatter plot of the feature vectors $FeatV_i$ for 100 sequences of *C. Elegans* where blue color denotes the exon and red color denoted the intron. Machine learning methods are then applied to these features so that a model is built for distinguishing exons and introns, i.e.,

$$\text{Model} = \text{Machine Learning} (FeatV_1, FeatV_2, \cdots, FeatV_N; d_1, d_2, \cdots d_N) \tag{10.14}$$

Over the year, various machine learning methods have been developed, such as k-nearest neighbor (kNN), neural network (NN) and support vector machine. These methods are based on different ideas to build the learned model. Brief discussions on kNN and NN are given in the following two sub-sections.

Table 10.2 A set of sample training data

i	DFT(zs_1 [n]) at $f = 2\pi/3$	DFT(zs_2 [n]) at $f = 2\pi/3$	DFT(zs_3 [n]) at $f = 2\pi/3$	d_i
1	82.1	41.5	116.0	0
2	4.5	7.0	1.0	1
3	94.8	3.3	146.9	0
4	84.0	4.4	93.1	0
5	23.4	13.4	90.5	0
6	45.8	8.1	114.9	0
7	2.9	0.4	0.0	1
8	4.7	6.5	0.1	1
9	6.0	2.5	0.0	1
10	11.7	20.2	1.3	1

10.5.1 *k*-Nearest Neighbor (kNN) Method

kNN is one of the simplest methods although it is also one of the popular methods. kNN does not make any assumption on the underlying data distribution. It simply performs classification by finding the most similar data points in the training data and making a prediction using those similar points only. "k" in kNN means the number of similar data points that will be considered in prediction. Hence, $k = 1$ means only the most similar data point is considered while $k = 3$ means three most similar data points.

Consider the example given in Table 10.2 and its scatter plot in Fig. 10.9a. The two unknown sequences with features U1 = $\{30.1, 28.3, 87.9\}$ and U2 = $\{3.1, 10.5, 9.6\}$ are located in two different parts in the scatter plot. kNN is based on the similarity checking. One popular similarity measure is the Euclidean distance. Consider $k =$ 1 in kNN for simplicity. By obtaining the Euclidean distance between U1 and all other 10 sequences, the smallest distance is obtain from $I = 6$. Thus, U1 will have the same label as the 6th sequence, i.e., U1 is predicted to be an exon. Similarity, U2 is found to have the smallest Euclidean distance with $I = 2$. Hence, U2 is predicted to be an intron.

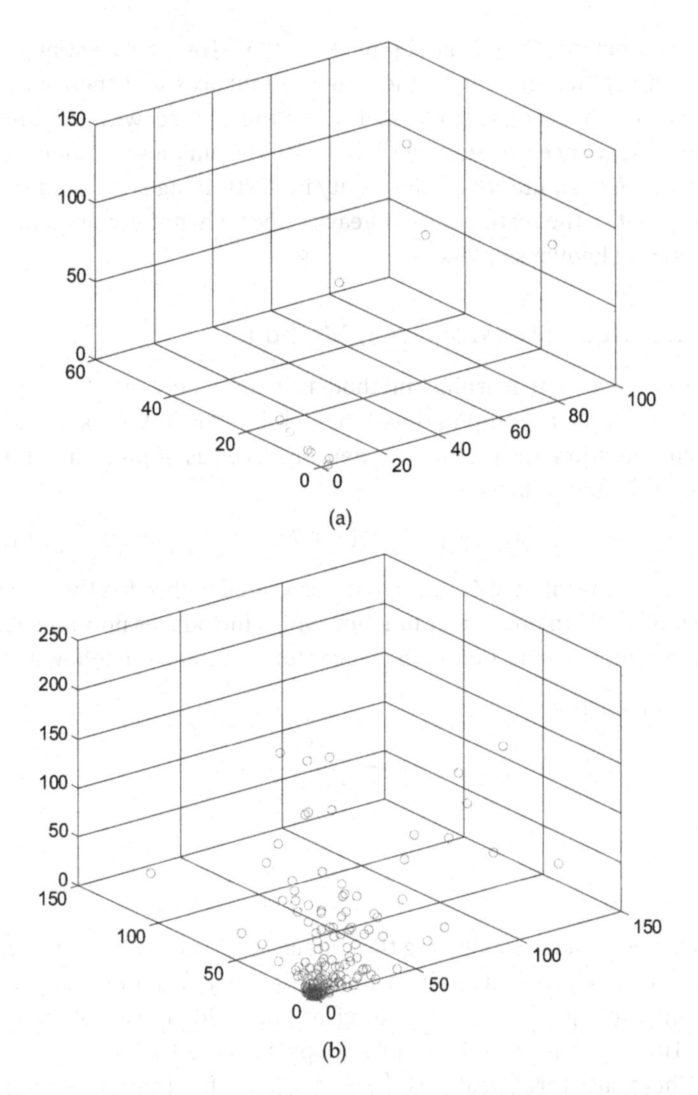

Figure 10.9 The scatter plot of the features of (a) the 10 sequences in Table 10.2 and (b) 100 sequences of *C. Elegans*. The blue color represents features from exons while the red color represents features from introns. The green color denotes features from two unknown features with features $\{30.1, 28.3, 87.9\}$ and $\{3.1, 10.5, 9.6\}$.

In summary, the training part in the kNN is essentially to "remember" features of all the known sequences and their labels. Similarity checking is done in the testing phrase with all these known sequences to find the labels for the unknown sequences. Hence, the computational complexity in the training part is minimal while that in the testing part is heavy, especially if there are a large number of known sequences.

10.5.2 Neural Network (NN) Method

Another machine learning method is neural network (NN) [13, 14, 39]. It is a computational model in which the basic unit of computation is neurons. A neuron receives inputs and then computes an output as

$$y = F\left(wx + b\right), \tag{10.15}$$

where x is the input, w is the weight associated with x, b is the bias in the model, F is a non-linear function which introduces non-linearity to the computations. Common choices for F include the following:

sigmoid function:

$$F\left(z\right) = \frac{1}{1 + e^{-z}}$$

ReLU:

$$F\left(z\right) = \begin{cases} 0 & z < 0 \\ z & z \geq 0 \end{cases}$$

A NN was built by connecting these neurons together. Usually there is an input layer, a number of hidden layers and an output layer. To illustrate, a network consisting of one hidden layer shown in Fig. 10.10 is constructed for our example in Table 10.2.

There are three features $\left\{f_{i,1}, f_{i,2}, f_{i,3}\right\}$ for the i-th training sequence. For simplicity, let's denote x_j as $f_{i,j}$. There is a weight $w^1_{j,p}$ connecting the input node x_j and the hidden node h_p. As there are three inputs and 2 hidden nodes, altogether we have six weights. Hence, the output for the two hidden nodes can be expressed as

$$\begin{aligned} &\text{Output of h1} : h_1 = F\left(w^1_{11}x_1 + w^1_{12}x_2 + w^1_{13}x_3 + b^1_1\right) \\ &\text{Output of h2} : h_2 = F\left(w^1_{21}x_1 + w^1_{22}x_2 + w^1_{23}x_3 + b^1_2\right) \end{aligned} . \tag{10.16}$$

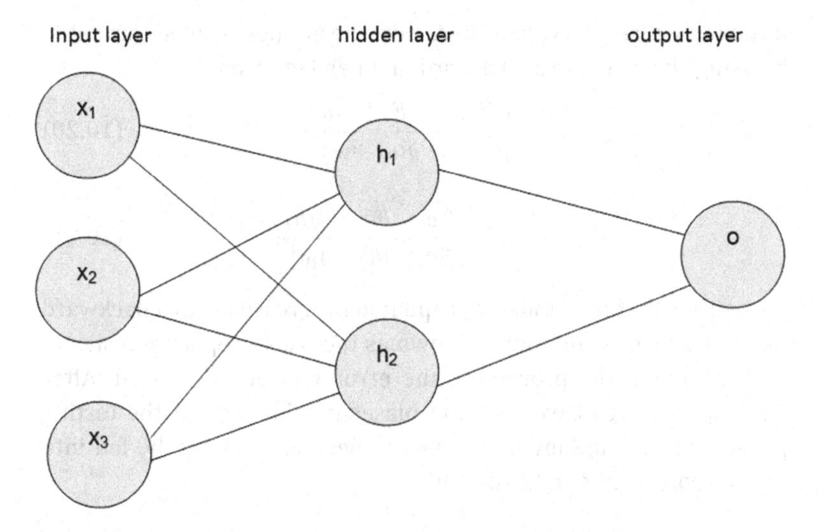

Figure 10.10 A simple neural network consisting of three inputs in the input layer, two nodes in the hidden layer and one node in the output layer.

The output layer contains one node only because our problem is a binary classification problem. The output label is either an exon or an intron. Using similar notation, the output of the NN can be written as

$$o = F\left(w_{11}^{2} h_1 + w_{12}^{2} x_2 + b^2\right). \tag{10.17}$$

Our goal is to find the best weights and biases that fit the training data. This involves the definition of an error function. We may consider the sum of the square error to be the error one wants to minimize, i.e.,

$$E = \frac{1}{2} \sum_{N} (d - o)^2 \tag{10.18}$$

where d is the desired output (as defined in Eq. 10.13) and the sum is over all the training sequences N. A way to find the weights and the biases is through the calculation of the gradient and update the weights iteratively as

$$w_{j,p}^{q,L+1} = w_{j,p}^{q,L} - \alpha \frac{\partial E}{\partial w_{j,p}^{q,L}}, \quad \text{where } q = 1, 2. \tag{10.19}$$

This means that the weight at the $(L + 1)$ iteration is updated by modifying the weight at the L-th iteration along the gradient direction with α as a weighting factor. The gradient can be computed by using chain rule in a backward manner. For example,

$$\frac{\partial E}{\partial w_{j,p}^{2,L}} = \frac{\partial E}{\partial o} \cdot \frac{\partial o}{\partial w_{j,p}^{2,L}} \tag{10.20}$$

$$\frac{\partial E}{\partial w_{j,p}^{1,L}} = \frac{\partial E}{\partial o} \cdot \frac{\partial o}{\partial h_j} \cdot \frac{\partial h_j}{\partial w_{j,p}^{1,L}} \tag{10.21}$$

We can see that the gradient is updated progressively in a backward manner which is commonly known as the back-propagation strategies. Through this procedure, the error will be minimized. After training, the set of weights and biases can be used in the testing phase. For any unknown sequences, their features can be fed into the network to obtain the output.

10.6 Classification Experiments

Three datasets were considered, namely Yeast, *C. Elegans* and Human. The Yeast, *C. Elegans* and Human datasets have 12000, 3000, and 6000 sequences, respectively. Half of them are exon and half are intron. We considered four experimental settings:

- Setting 1: Only one feature is used. It is the value of $PS[k]$ (Eq. 10.9) at normalized frequency of $2\Pi/3$
- Setting 2: Three features are used. They are DFT($zs_1[n]$), DFT($zs_2[n]$) and DFT($zs_3[n]$) at normalized frequency of $2\Pi/3$.
- Setting 3: Six features are used. The first three features are same as those in Setting 2. The remaining three features are their DC values.
- Setting 4: all features in setting 3 together with features such as compositional bias, position asymmetry, periodic correlation between base positions, mutual information and the z-curve features as described in Section 10.3.

In kNN, K is set to be 10, i.e., 10 most similar data points are used in making prediction. In performing experiments, data in each dataset is randomly partitioned into two parts. The first part is

Table 10.3 The classification performance of using different features for three datasets

	Yeast		*C. Elegans*		Human	
	Exon	**Intron**	**Exon**	**Intron**	**Exon**	**Intron**
Numbers:	6000	6000	3000	3000	1500	1500
Setting 1						
Predicted to be exon	5148	647	2690	373	1294	1069
Predicted to be intron	852	5353	310	2627	206	431
Accuracy	87.5%		93.8%		64.4%	
False positive	10.8%		12.4%		71.3%	
False negative	14.2%		10.3%		13.7%	
Setting 2						
Predicted to be exon	5293	284	2609	311	1056	301
Predicted to be intron	707	5716	391	2689	444	1199
Accuracy	91.7%		88.3%		75.2%	
False positive	4.7%		10.4%		20.1%	
False negative	11.8%		13.0%		29.6%	
Setting 3						
Predicted to be exon	5164	265	2511	205	1141	238
Predicted to be intron	836	5735	489	2795	359	1262
Accuracy	90.8%		88.4%		80.1%	
False positive	4.4%		6.8%		15.7%	
False negative	13.93%		16.3%		23.9%	
Setting 4						
Predicted to be exon	5602	230	2860	228	1369	157
Predicted to be intron	398	5770	140	2772	131	1343
Accuracy	94.8%		93.9%		90.4%	
False positive	3.8%		7.6%		10.5%	
False negative	6.6%		4.7%		8.7%	

the training data while the second part is the testing data. The procedure is repeated six times and an average performance over these six trials was obtained. Table 10.3 shows the results for the four experimental settings.

In setting 1, the value of $PS[k]$ (Eq. 10.9) at normalized frequency of $2\Pi/3$ is used as the feature. For the first two datasets: Yeast and *C. Elegans*. They are less challenging for the DFT-based method as the sequence length is long. In both cases, the accuracy is high with false positive and false negatives are around 10%. However, human exon sequence is rather challenging because of its short length. The accuracy drops to 64%. The main problem is, however, the true positive where 71.3% of intron was classified as exons.

In settings 2 and 3, DFT is applied to the three modified sequences. Setting 2 considers the DFT values at normalized frequency of $2\Pi/3$ while Setting 3 considers 3 more DC values. The overall performance for the long sequence was more or less similar to setting 1. However, for short sequences, the performance was significantly improved. The accuracies for settings 1, 2 and 3 were 64.4%, 75.2% and 80.1%, respectively. The false positives dropped from 71.3% to 15.7% from setting 1 to setting 3.

If the features from the three modified sequences were combined with statistical features such as position asymmetry as in setting 4, the accuracies for all sequences improved to be over 90%. The false positive dropped to be less than 11%. Hence, the results showed that the use of appropriate features to build the model has significant effect on the overall performance in machine learning-based methods.

10.7 Deep Learning Methods

Performance of the traditional machine learning method relied on the use of appropriate features. If the chosen features could not capture important input characteristics, the trained model would have limited capability. In recent years, deep learning method has been applied to many different classification problems. In contrast to the general machine learning framework, features are not extracted for training. Rather, the raw data is usually fed into the deep learning model. With a number of non-linear transforming layers, important features from the raw data are learned progressively. One recent attempt for exon and intron classification is to use Convolutional Neural Network (CNN) [17]. Figure 10.11 shows the schematic diagram of the CNN-based exon intron classification.

In the pre-processing layer, the input is the character-based DNA sequence while the output is a numerical sequence which can be processed by the CNN. It is possible to use the one-hot encoding scheme to convert the base in each position into a 4D binary vector [31–34]. Suppose that the sequence is "AAACAT," then the numerical

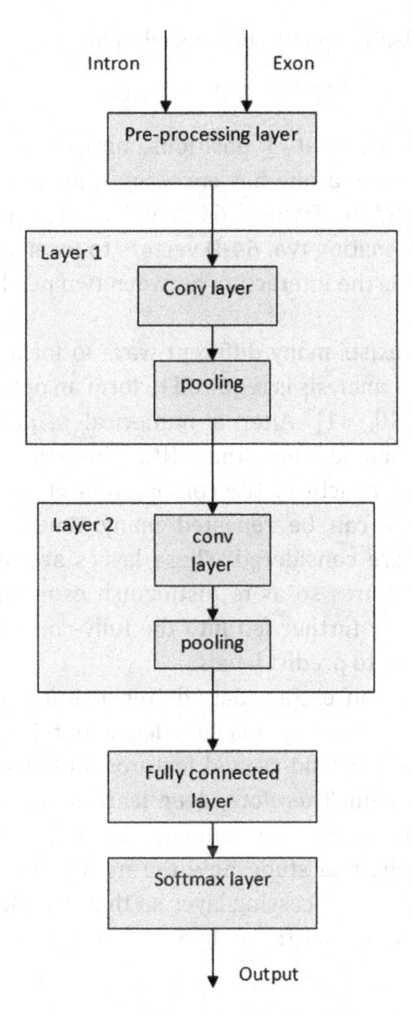

Figure 10.11 Schematic diagram of the CNN-based exon intron classification.

sequence becomes

$$1000 \quad 1000 \quad 1000 \quad 0100 \quad 1000 \quad 0001.$$

Note that the space introduced above is for clarity only. Another way to produce the numerical sequence is to form a 2D matrix which would have spatial relationship between different bases. For example, an amino acid is formed from three consecutive bases, thus, the sequence "AAACA" can be considered to be

"AAA," "AAC," "ACA."

As there are 4 bases at 3 positions, altogether there are 64 combinations. Hence, a one-hot encoding scheme can be applied to the amino acid to form a 64-D vector. The output can be formed by concatenating two 64-D vectors to form a 64×2 matrix which can describe the interaction between two neighboring amino acids [17].

In fact, there exists many different ways to form the numerical sequence. Further analysis is required to form an optimal numerical representation [40, 41]. After a numerical sequence/matrix is obtained, it can be fed into the CNN. There are a number of layers in the CNN structure. The combination of convolution layer and pooling layer can be repeated many times. In Fig. 10.11, only two layers are considered. These layers are used to extract characteristics features so as to distinguish exon and intron. The features are usually further fed into the fully-connected layer and a softmax classifier to predict labels.

Deep learning can extract data-driven features and deal with high-dimensional data, while machine learning relies on the expert knowledge to extract hand-crafted features and deals mostly with low-dimensional data. Therefore, deep learning is becoming more and more popular in genomic sequence analysis. Further analysis is, however, required to study how the numerical sequences can be formed in the pre-processing layer so that deep learning can be used effectively to extract those data-driven features for sequence classification.

10.8 Conclusions

Due to the recent progress in human genome sequencing, there is a huge surge of demand and interest in analyzing DNA sequence. Its analysis plays a vital role in the interpretation of biological results. In this chapter, an important problem of identifying protein coding region called exons has been discussed. We have described various approaches that were used to characterize the bias base distribution in exons. We have also described how biological property of exon can be integrated with the discrete Fourier transform to give an effective exon feature characterization. These features can be fed into learning-based methods such as k-nearest neighbor and neural network to achieve a robust and effective sequence identification. In the future, deep learning methods can be explored. Instead of using the hand-crafted features as in the traditional learning-based methods, DNA sequences can be fed directly to the deep learning framework to extract data-driven features for sequence classification.

References

1. Cooper, G. M. (2000), *The Cell: A Molecular Approach,* 2nd ed. (Sunderland MA: Sinauer Associates).
2. Gautier, C. (2000), Compositional Bias in DNA, *Current Opinion in Genetics and Development,* 10(6), pp. 656–661.
3. Fickett, J. W. (1982), Recognition of Protein Coding Regions in DNA Sequences, *Nucleic Acids Research,* 10, pp. 5303–5318.
4. Fickett, J. W., and Tung, C. S. (1992), Assessment of Protein Coding Measures, *Nucleic Acids Research,* 20, pp. 6441–6450.
5. Guigo, R. (1999), DNA Composition, Codon Usage and Exon Prediction, *Genetic Databases,* pp. 53–80.
6. Konopka, A. K. (1994), *Structure and Methods: VI, Human Genome Initiative and DNA Recombination, Chapter Towards Mapping Functional Domains in Indiscriminantly Sequenced Nucleic Acids: A Computational Approach* (Adenine Press, Guilderland).
7. Tiwari, S., Ramachandran, S., Bhattacharya, A., Bhattacharya, S., and Ramaswamy, R. (1997), Prediction of Probable Genes by Fourier Analysis of Genomic Sequences, *Computer Applications in Biosciences,* 13, pp. 263–270.
8. Liew, A. W. C., Wu, Y., Yan, H., and Yang, M. (2005), Effective Statistical Features for Coding and Non-coding DNA Sequence Classification for Yeast, *C. Elegans* and Human, *International Journal on Bioinformatics Research and Applications,* 1(2), pp. 181–201.
9. Huang, J., Li, T., Chen, K., and Wu, J. (2006), An Approach of Encoding for Prediction of Splice Sites using SVM, *Biochimie,* 88, pp. 923–929.

10. Sonnenburg, S., Schweikert, G., Philips, P., Behr, J., and Ratsch, G. (2007), Accurate Splice Site Prediction using Support Vector Machines, *BMC Bioinformatics*, 8 (Suppl 10):S7.
11. Goel, N., Singh, S., and Aseri, T. C. (2015), An Improved Method for Splice Site Prediction in DNA Sequences Using Support Vector Machines, *Procedia Computer Science*, 67, pp. 358–367.
12. Meher, P. K., Sahu, T. K., and Rao, A. R. (2016), Performance Evaluation of Neural Network, Support Vector Machine and Random Forest for Prediction of Donor Splice Sites in Rice, *Indian J. Genet.*, 76(2), pp. 173–180.
13. Noordewier, M., Towell, G., and Shavlik, J. (1991), Training Knowledge-Based Neural Networks to Recognize Genes in DNA Sequences, *Advances in Neural Information Processing Systems*, pp. 530–536.
14. Brunak, S., Engelbrecht, J., and Knudsen, S. (1991), Prediction of Human mRNA Donor and Acceptor Sites from the DNA Sequence, *Journal Molecular Biology*, 220(1), pp. 49–65.
15. Borodovsky, M., and McIninch, J. (1993), Genmark: Parallel Gene Recognition for both DNA Strands, *Comput. Chem*, 17, pp. 123–133.
16. Salzberg, S. L., Pertea, M., Delcher, A. L., Gardner, M. J., and Tettelin, H. (1999), Interpolated Markov Models for Eukaryotic Gene Finding, *Genomics*, 59, pp. 24–31.
17. Nguyen, N. G., Tran, V. A., Ngo, D. L., Phan, D., Lumbanraja, F. R., Faisal, M. R., Abapihi, B., Kubo, M., and Satou, K. (2016), DNA Sequence Classification by Convolutional Neural Network, *J. Biomedical Science and Engineering*, 9, pp. 280–286.
18. Hill, S. T., Kunitzle, R., Teegarden, A., Merrill, E., Danaee, P., and Hendrix, D. A. (2018), A Deep Recurrent Neural Network Discovers Complex Biological Decipher RNA Protein-coding Potential, *Nucleic Acids Research*, gky567, https://doi.org/10.1093/nar/gky567.
19. Lee, B., Lee, T., Na, B., and Yoon, S. (2015), DNA-Level Splice Junction Prediction Using Deep Recurrent Neural Networks. (https://arxiv.org/abs/1512.05135)
20. http://www.genomenewsnetwork.org/articles/02_01/Sizing_genomes.shtml.
21. Nelson, C. E., Hersh, B. M., and Carroll, S. B. (2004), The Regulatory Content of Intergenic DNA Shapes Genome Architecture, *Genome Biology*, 5(4): R25.
22. Clancy, S. (2008), RNA Splicing: Introns, Exons and Spliceosome, *Nature Education*, 1(1): 31.

23. Wang, Z., Gerstein, M., and Snyder, M. (2009), RNA-Seq: a Revolutionary Tool for Transcriptomics, *Nature Reviews Genetics*, 10, pp. 57–63.
24. Xiong, F., Gao, J., Li, J., Liu, Y., Feng, G., Fang, W., Chang, H., Xie, J., Zheng, H., Li, T., and He, L. (2009), Noncanonical and Canonical Splice Sites: a Novel Mutation at the Rare Noncanonical Splice-Donor Cut Site (IVS4+1A>G) of SEDL Causes Variable Splicing Isoforms in X-Linked Spondyloepiphyseal Dysplasia Tarda, *European Journal of Human Genetics*, 17.4, pp. 5190–516.
25. Burset, M., Seledtsov, I. A., and Solovyev, V. V. (2000), Analysis of Canonical and Non-Canonical Splice Sites in Mammalian Genomes, *Nucleic Acids Research*, 28.21, pp. 4364–4375.
26. Grosse, I., Buldyrev, S. V., Stanley, H. E., Holste, D., and Herzel, H. (2000), Average Mutual Information of Coding and Noncoding DNA, *Pacific Symposium on Biocomputing*, 5, pp. 611–620.
27. Zhang, R., and Zhang, C. T. (1994), Z Curves, an Intuitive Tool for Visualizing and Analyzing DNA Sequences, *Journal Biomolecular Structure Dynamics*, 11, pp. 767–782.
28. Zhang, C. T., and Wang, J. (2000), Recognition of Protein Coding Genes in the Yeast Genome at Better than 95% Accuracy Based on the Z Curve, *Nucleic Acids Research*, 28, pp. 2904–2814.
29. Berger, J. A., Mitra, S. K., and Astola, J. (2003), Power Spectrum Analysis for DNA Sequences, *Seventh International Symposium on Signal Processing and Its Applications*, 2, pp. 29–32.
30. Tsonis, A. A., Kumar, P., Elsner, J. B., and Tsonis, P. A. (1996), Wavelet Analysis of DNA Sequences, *Physical Review E*, 53(2), pp. 1828–1834.
31. Vaidyanathan, P. P., and Yoon, B. J. (2004), The Role of Signal-Processing Concepts in Genomics and Proteomics, *Journal of the Franklin Institute* (Invited paper), Special Issue on Genomics.
32. Wang, W., and Johnson, D. H. (2002), Computing Linear Transforms of Symbolic Signals, *IEEE Transactions on Signal Processing*, 50(3), pp. 628–634.
33. Anastassiou, D. (2000), Frequency-domain analysis of biomolecular sequences, *Bioinformatics*, 16 (12), pp. 1073–1081.
34. Vaidyanathan, P. P., and Yoon, B. J. (2002), Digital Filters for Gene Prediction Applications, *Thirty-Sixth Asilomar Conference on Signals, Systems and Computers*, 1, pp. 306–310.
35. Zhao, J., Yang, X. W., Li, J. P., and Tang, Y. Y. (2001), DNA Sequences Classification Based on Wavelet Packet Analysis, *WAA 2001, LNCS 2251*, pp. 424–429 (Springer-Verlag Berlin Heidelberg).

36. Sakharkar, M. K., Chow, V. T., and Kangueane, P. (2004), Distributions of Exons and Introns in the Human Genome, *In Silico Biology*, 4(4), pp. 387–393.
37. Law, N. F., Cheng, K. O., and Siu, W. C. (2006), On Relationship of Z-curve and Fourier Approaches for DNA Coding Sequence Classification, *Bioinformation*, 1(7), pp. 242–246.
38. Law, N. F., Cheng, K. O., and Siu, W. C. (2007), Spectral Approaches for DNA Sequence Classification, *International Conference in IT and Application*, 2, pp. 541–544.
39. Arniker, S. B., Kwan, H. K., Law, N. F., and Lun, D. P. (2011), Promoter Prediction Using DNA Numerical Representation and Neural Network: Case study with three organisms, *2011 Annual IEEE India Conference.*
40. Berger, J. A., Mitra, S. K., and Astola, J. (2003), Power Spectrum Analysis for DNA Sequences, *Seventh International Symposium on Signal Processing and Its Applications*, 2, pp. 29–32.
41. Cristea, P. (2002), Real and Complex Genomic Signals, *DSP*, pp. 543–546.

Chapter 11

Visual Food Recognition for Dietary Logging and Health Monitoring

Sharmili Roy,[a] Zhao Heng,[a] Kim-Hui Yap,[a] Alex Kot,[a] and Lingyu Duan[b]

[a] *Rapid-Rich Object Search Lab, Nanyang Technological University, Singapore*
[b] *School of Electronics Engineering and Computer Science, Peking University, China*
ekhyap@ntu.edu.sg

Visual food recognition has gained popularity in recent years due to its key roles in monitoring an individual's diet and public health management. The aim of mobile visual food recognition is to recognize food dishes from pictures taken using portable devices such as mobile phones/tablets, and then analyze and monitor the dietary intakes of the users. Existing solutions for visual food recognition employ Deep Neural Networks (DNNs) that are trained as dish classification engines. These DNNs are typically deployed in back-end servers. When an individual snaps the photo of a food dish, the image is sent to this DNN-based classifier for recognition and analysis. The results from this analysis are sent back to the mobile end-user and logged in the end-user's food journal for monitoring and health management. This chapter gives an overview of recent advances in visual food recognition systems and the state-of-the-art solutions for dietary logging and management. We

Learning Approaches in Signal Processing
Edited by Wan-Chi Siu, Lap-Pui Chau, Liang Wang, and Tieniu Tan

ISBN 978-981-4800-50-1 (Hardcover), 978-0-429-06114-1 (eBook)
www.panstanford.com

explore various challenges in designing such systems and highlight some solutions that address these challenges. In addition, we introduce a Personalized Compact Network that incorporates user-specific dietary habits/preferences into the DNNs and achieves high recognition accuracy with low memory and energy foot prints.

11.1 Introduction

Unhealthy diet is a primary cause of various serious illnesses such as cardiovascular diseases, diabetes and obesity. About 30% to 40% of human cancers are related to improper diets. A healthy dietary structure is beneficial to improving people's health and physical fitness and is of great significance to the country and individuals.

Large-scale, cumulative, and traceable dietary data can be used to establish national level continuous monitoring of people's health conditions, guide people's dietary behaviors, and deliver early warning of diet-related diseases. However, current dietary data are collected mainly via online or offline questionnaire surveys and interviews. These methods are inefficient, and make it difficult to trace and monitor people's dietary behaviors. As a result, the data collected cannot comprehensively and timely reflect people's dietary habits.

With recent development in e-health, significant progress has been made in the fields of health monitoring, disease prevention, medical diagnosis, and medical care. Due to their portability and efficiency, smart terminals, in particular smart mobile phones, have become powerful tools for health information collection, transmission, processing, and analysis. There are nowadays more than 120,000 health-related applications (apps). Most of the health-related apps, however, target recording and analysis of human physiological data. Some food logging apps such as MyFitnessPal (MyFitnessPal, 2018) and LoseIt (LoseIt, 2018) allow users to manually log their diets. Manual data entry is tedious and time consuming and research shows that such applications are unable to retain their users in the long run (Cordeiro et al., 2015b). With the proliferation of digital cameras in smart phones, increasingly more people like to record and share their food via images in the Internet. Some solutions have been proposed to leverage on the mobile camera to either just log the meal without any diet analysis (Cordeiro et al., 2015a) or rely on expert nutritionists (Martin et al., 2008) or crowd sourcing (Noronha et al., 2011) to analyze the images offline. One fundamental problem with these approaches is that they do not provide immediate feedback to the users about their diet. Since the analysis is done offline, the feedback is delayed and often loses its effectiveness.

11.2 Visual Food Recognition and Analysis Solutions

For immediate analysis of an individual's diet, different aspects of automated food/diet recognition and analysis have been investigated in the published literature. Some works have focused on food recognition strategies and evaluated various features and classification methods. Handcrafted global and local features, SIFT, local binary patterns (LBP) (Nguyen et al., 2014) and some contextual information such as where the image was taken have been evaluated for dish recognition (Bettadapura et al., 2015). k-nearest neighbor (k-NN) (He et al., 2014), support vector machines (SVMs) (Nguyen et al., 2014), artificial neural networks, and random forest classification methods (Anthimopoulos et al., 2014) are amongst the widely used classifiers in the context of food classification. Recently, convolutional neural networks (CNNs) based food recognition engines have achieved high classification accuracies (Hassannejad et al., 2016; Pouladzadeh and Shirmohammadi, 2017; Zhang et al., 2016).

Design of complete systems for dietary monitoring for real end-users is another focus of the current literature. These systems leverage mobile computing devices such as mobile phones and tablets for food logging, recognition, and analysis. Some examples of such systems are FoodLog (Kitamura et al., 2009), DietCam (Kong and Tan, 2012), Menu-Match (Beijbom et al., 2015), and Food-Cam (Kawano and Yanai, 2015).

Calorie estimation is another research focus in the context of diet analysis and assessment. Calorie estimation is a very hard problem because calorie not only depends on the volume of the food but also on the food density. Assuming that the density information for standard food items is available, some works have attempted to solve the portion size estimation problem. Portion size estimation attempts to determine how much food is present in an image in terms of cm^3 or grams. Aizawa et al. (2013) proposed a solution to estimate food portion using pre-determined serving size classification. A collection of portion estimation solutions require reference information in the form of markers, tokens, camera calibration, size of reference objects such as thumb or eating tools or geographical locations of the dishes (Akpro Hippocrate et al., 2016; Mariappan et al., 2009; Pouladzadeh et al., 2016, 2014, 2012; Puri et al., 2009; Sun et al., 2008; Villalobos et al., 2012). Volume estimation from a single image/view is an ill-posed problem. Using 3D information from multi-view images, some template matching and shape reconstruction based volume estimation has also been proposed in the literature (Chae et al., 2011; He et al., 2013; Puri et al., 2009). With the spirit of estimating how much food a person has consumed, some works have also explored left over estimation in a curated task-specific dataset (Ciocca et al., 2015).

More recently, food attributes have attracted a lot of research attention. Attributes such as ingredients, cooking methods, chopping methods, cuisine and course are being classified from food images using multi-task CNN training. Some of the most notable works in this field are Chen and Ngo (2016); Chen et al. (2017); Min et al. (2018, 2017); Zhang et al. (2016). The goal of such solutions is to identify various attributes of a dish from its image and recommend/retrieve recipes of the corresponding dish.

For the various tasks of food recognition and analysis discussed above, food datasets on which recognition and analytics engines can be trained and evaluated form a crucial part of the solution. The following section discusses and compares various food datasets available in the literature.

11.3 Food Datasets

The first food dataset, named Food50, was introduced in Joutou and Yanai (2009). This dataset contains 50 dishes from Japanese cuisine. The images are crawled from the Internet where each image shows a single dish. This dataset was further expanded to 85 dishes in a subsequent work (Hoashi et al., 2010). These two datasets, however, are proprietary. Another proprietary dataset, TADA, was introduced by Bosch et al. (2011); Mariappan et al. (2009), where images were acquired in a lab and markers were placed alongside the dishes to facilitate color calibration and dish recognition. TADA dataset contains images of real food dishes (256 images) and images of food replica (50 images). Each image may have multiple food dishes making the dataset more realistic and more challenging.

One of the first large-scale public food database, UECFOOD-100, was published by Kawano and Yanai (2013). This dataset contains 9,000 images of 100 food dishes. Ground truth bounding boxes surrounding the food are provided for each image. This dataset was further expanded to 256 dish categories in Kawano and Yanai (2014). Another dataset of 50 Chinese dishes with a total of 5,000 images was published by Chen et al. (2012) for the purpose of food identification and quantity estimation.

Currently the largest publicly available food dataset is Food-101 (Bossard et al., 2014). It contains 101,000 images of 101 food categories. Food-101 primarily focuses on western cuisine. A twin dataset of Food-101 called UPMC Food-101 is published in Wang et al. (2015). UPMC Food-101 has the same dish categories as Food-101. For most images, they also provide the HTML webpage from where the image was downloaded. This makes the dataset multi-modal, offering both dish images and dish description in the form of the text available in the webpage. Another large database is UNICT-FD889 which contains 889 dish categories with a total of 3,583 images (Farinella et al., 2014). Although a large variety of dishes are represented in UNICT-FD889, but the number of examples for each category is comparatively low. This dataset is designed for the purpose of near duplicate food retrieval as opposed to the standard task of dish classification. Addressing another interesting

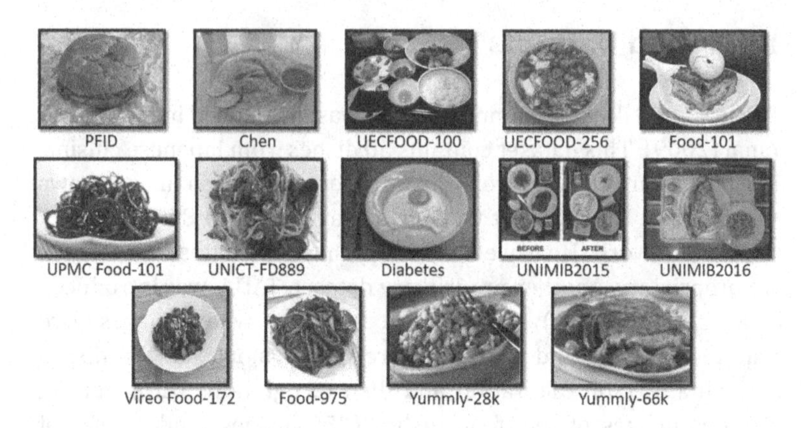

Figure 11.1 Sample food images from the non-proprietary datasets published in the literature.

task of leftover estimation, two canteen datasets UNIMIB2015 and UNIMIB2016 have been published in Ciocca et al. (2015, 2017) that contain images of canteen food trays where images are taken in University canteens and the dish items are arranged in an unconstrained manner on the food tray. Each tray contains multiple instances of a dish. A collection of 1,027 canteen trays, 3,616 dish instances belonging to 73 dish categories are published.

Dish attributes such as ingredients, cooking and chopping methods have gained increasing interest in the past few years. New datasets have been published for recognition of such attributes. Vireo Food-172 is one of the first such datasets (Chen and Ngo, 2016). Focusing on Chinese cuisine, Vireo Food-172 provides dish ingredients for each food image. The dataset comprises of 172 Chinese dishes with a total of 110,241 images. A few other large-scale datasets that also provide ingredient and other attribute information such as cuisine, course, restaurant, etc., are published in Min et al. (2018, 2017); Zhang et al. (2016). Table 11.1 gives a comprehensive overview of the food datasets available in the literature. For each dataset, we compare its size, number of dish categories, whether or not a single image depicts multiple dishes, the available annotation and the availability of the dataset (i.e., public, proprietary, or unknown). Figure 11.1 shows some examples of food images taken from the non-proprietary datasets.

Table 11.1 Overview of the food datasets published in the literature

Name	Year	#Images	#Classes	Type	Annotation	Availability	Reference
Food50	2009	5,000	50	Single	Dish label	Proprietary	Joutou and Yanai (2009)
PFID	2009	1,098	61	Single	Dish label	Public	Chen et al. (2009)
TADA	2009	50/256	—	Multi	—	Proprietary	Bosch et al. (2011) Mariappan et al. (2009)
Food85	2010	8,500	85	Single	Dish label	Proprietary	Hoashi et al. (2010)
Chen	2012	5,000	50	Single	Dish label	Public	Chen et al. (2012)
UEC-Food100	2012	9,060	100	Multi	Dish label, bounding box	Public	Kawano and Yanai (2013)
UEC-Food256	2014	31,397	256	Multi	Dish label, bounding box	Public	Kawano and Yanai (2014)
Food-101	2014	101,000	101	Single	Dish label	Public	Bossard et al. (2014)
UPMC Food-101	2015	90,840	101	Single	Dish label, html text	Public	Wang et al. (2015)
UNICT-FD889	2014	3,583	889	Single	Dish label	Public	Farinella et al. (2014)
Diabetes	2014	4,868	11	Single	Dish label	Public	Anthimopoulos et al. (2014)
UNIMIB2015	2015	1,000 × 2	15	Multi	Polygonal segmentation	Public	Ciocca et al. (2015)
UNIMIB2016	2016	1,027	73	Multi	Polygonal segmentation	Public	Ciocca et al. (2017)
Vireo Food-172	2016	110,241	172	Single	Dish label, ingredient list	Public	Chen and Ngo (2016)
FoodCat	2016	46,000	115	Single	Dish label, hierarchical category	Unknown	Herruzo et al. (2016)
Food-975	2017	37,885	975	Single	Dish label, restaurant, ingredients	Unknown	Zhou and Lin (2016)
Zhang et al.	2017	250,000	360	Single	Dish label, ingredients, cooking method	Unknown	Zhang et al. (2016)
Yummly-28k	2017	27,638	—	Single	Dish label, ingredients, cuisine, course	Public	Min et al. (2017)
Yummly-66k	2017	66,615	—	Single	Dish label, ingredients, cuisine, course	Public	Min et al. (2018)

11.4 Vision-Based Food Recognition

Early solutions to food recognition problem employed handcrafted features and vision-based methods. On the very first dataset, Food50, multiple kernel learning based feature fusion was used to obtain a recognition accuracy of 61.34% on Japanese food. On the larger Japanese food dataset of UECFOOD-100, SVM classifiers with color histogram and SURF features were used to achieve a classification accuracy of 81.55% for the top-5 classes when the ground-truth bounding boxes were given. On the larger version of this dataset, UECFOOD-256, such a system obtained a classification accuracy of only 74.4% on top-5 dish classes. On the dataset by Chen, a multi-label SVM trained on SIFT, LBP, color, and Gabor features achieved an overall recognition accuracy of 68.3%.

On the large Food-101 dataset, random forest was used to mine discriminant superpixel-grouped parts in the food images. SVM was then used to classify these parts with an average accuracy of 50.76% (Bossard et al., 2014). On a diabetic dataset meant to help diabetic patients control their daily carbohydrate consumption, Anthimopoulos et al. (2014) designed a food recognition system based on the bag-of-features model. Various visual features and classifiers were evaluated and a best classification accuracy of close to 78% was reached using a dictionary of 10,000 words.

Vision-based solutions work best in laboratory conditions where the picture backgrounds are clean and the food components are well demarcated. In real-life situations, however, the food components may be occluded or often mixed and placed at different locations owing to the varying cooking, chopping and mixing methods involved. This is particularly true for Asian food such as Chinese and Indian cuisine. This decreases the reliability of local descriptive features thereby making traditional vision-based solutions unsuitable in such settings.

11.5 CNN-Based Food Recognition

Deep convolutional neural network (DCNN) is currently the state-of-the-art technique for image recognition problems. DCNNs can

Table 11.2 Comparison of classification accuracy of various CNN architectures on the Food-101 dataset

CNN architecture	Top-1 (%)	Top-5 (%)	Reference
AlexNet	56.40	—	Martinel et al. (2016)
Modified AlexNet	70.41	—	Wang et al. (2015)
Modified GoogLeNet	77.40	93.70	Liu et al. (2016)
VGG16	79.17	—	—
Inception V3	88.28	96.88	Hassannejad et al. (2016)
ResNet-200	88.38	97.85	Martinel et al. (2016)
Wide Residual Network	88.72	97.92	Martinel et al. (2016)
Wide-slice Residual Network	90.27	98.71	Martinel et al. (2016)

estimate optimal feature representations from the data adaptively as against the traditional handcrafted features. Recent research on DCNN has shown that deep architectures can outperform traditional vision-based methods in food recognition problems (Hassannejad et al., 2016; Pouladzadeh and Shirmohammadi, 2017; Tanno et al., 2016).

Most existing methods use standard convolutional neural network (CNN) architectures and employ deep features extracted directly from the neural networks for image-level food classification (Bossard et al., 2014; Liu et al., 2016; Yanai and Kawano, 2015). Various studies (Kawano and Yanai, 2015; Liu et al., 2016) show that deeper networks such as VGGNet (Simonyan and Zisserman, 2014), GoogleNet (Szegedy et al., 2015) and ResNet (He et al., 2016) generate better food features than AlexNet (Ioffe and Szegedy, 2015). Table 11.2 compares various CNN architectures in terms of their classification accuracy on the benchmark Food-101 dataset.

Food-101 dataset is primarily based on western cuisine. Although high classification accuracy has been achieved on the Food-101 dataset, it is still challenging to accurately classify Asian food, especially Chinese and Indian food because of the way ingredients are chopped and mixed together. Contextual information such as geographical location (Ioffe and Szegedy, 2015), dish ingredients (Chen and Ngo, 2016), textual information (Wang et al., 2015), etc., can be used to improve the classification accuracy in

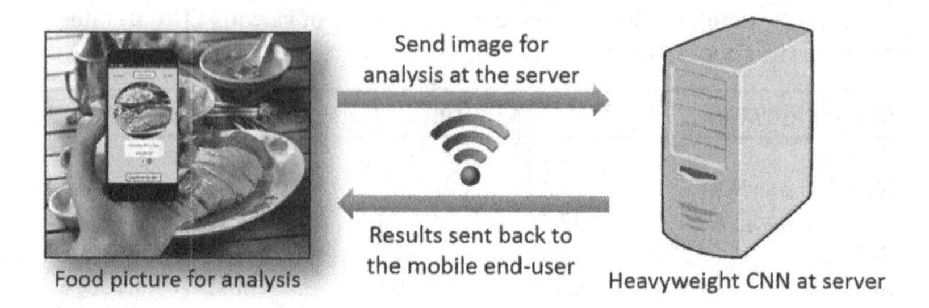

Figure 11.2 An overview of traditional food recognition engines based on deep learning methods.

such cases. Though these attributes are not always available, they are generally effective in boosting the recognition performance.

A typical deep learning based food recognition solution would take the following approach. The user would snap a picture of the food and the picture would be sent to a central server possibly with some contextual information such as geo-location. The classification would happen at the central server based on a food classification engine and the results would be sent back to the local mobile device. The food classifiers could be based on VGGNet (Chen and Ngo, 2016), ResNet (Aguilar et al., 2017) or InceptionNet (Hassannejad et al., 2016) frameworks which usually have high memory, computation, and energy footprints. Figure 11.2 gives an overview of such a solution.

There are two key limitations of such solutions. First, the classification engine at the server does not take into consideration the mobile user's personal food preferences, dietary habits, restrictions and dietary history. Second, since the classifier is not compact enough to fit on a mobile device, the users are required to send every picture to the server for analysis. This may not be practically feasible since the user may not have access to the Internet at all times or may not be willing to transfer pictures over the network due to privacy concerns.

In the next section, we propose a personalized and lightweight solution to food recognition where we integrate an end-user's dietary preferences in the learning process and train a compact network to achieve high classification accuracy. This compact network is more amenable for deployment on mobile devices as

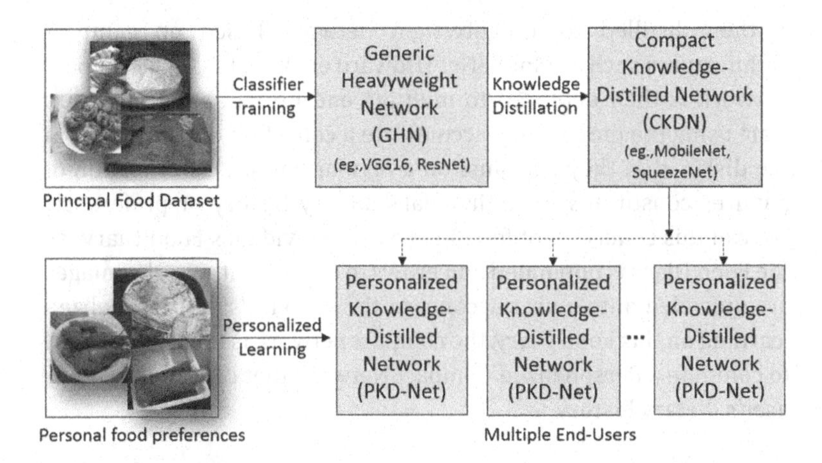

Figure 11.3 Overview of the proposed Personalized Compact Network. A generic heavyweight network is trained on the principal food database. This network then distills its knowledge to a compact network. The compact network is then personalized based on the end-user's dietary preferences.

opposed to heavyweight architectures such as VGGNet, ResNet and Inception module.

11.6 Personalized Food Recognition Engine

Typically, most people have preferences on what they like to eat. Often they eat their favorite dishes more frequently and visit their favorite restaurants time and again. This, we believe, presents key information that can be leveraged to greatly improve the accuracy of the food classification engine. Leveraging on (i) personal dietary history and (ii) knowledge distillation from a big trainer network, we train a compact neural network architecture that achieves high classification accuracy. Figure 11.3 provides an overview of the proposed framework.

We train a generic heavyweight network based on VGG16 architecture for the classification task on the principal food dataset. The principal food dataset is typically a wide collection of food images crawled from the Internet image search engines such as Google and Yahoo. The knowledge of this heavyweight network is then distilled to a lightweight network based on compact architectures such as MobileNet (Howard et al., 2017). This compact network is then deployed to multiple end users. As mobile users start using the model, they accumulate a collection of pictures from the dishes that they consume on a regular basis. This collection of pictures constitutes the individual's dietary history or preference. We call this collection of food images an individual's Food Diary. As the Food Diary is populated and grows to a sufficient size, the images are uploaded onto a user account in the server. Using personalized learning on this Food Diary, the compact network is then customized to compute a Personalized Compact Network that captures the end-user's dietary history.

11.6.1 Knowledge Distillation

To design a compact recognition engine with low computational and memory requirements, we distill the knowledge of the heavyweight VGG16 network, which has a large number of parameters per layer to a compact network based on MobileNet-224 (Howard et al., 2017) architecture. MobileNets need only about 10 to 12 MB of storage as opposed to 510 MB required by a full weight VGG16. In addition, MobileNets use depth-wise separable convolutions to achieve 8 to 9 times less computation than standard convolutions with a small drop in accuracy (Howard et al., 2017). To boost the accuracy of the MobileNet-224 model, we use knowledge distillation from the VGG16 classifier.

The goal of knowledge distillation is to transfer the knowledge of a large trainer network to a smaller trainee network such that the trainee network approximates the trainer network accurately. The trainer network, in our case, is the heavyweight VGG16 network and the trainee network is the compact MobileNet model (Figure 11.3). If we represent the trainer with $\phi_{tr}(x)$ and the trainee with $\phi_{te}(x)$, then the goal of knowledge distillation is to achieve $\forall_{(x_i \in X)} \phi_{tr}(x_i) = \phi_{te}(x_i)$ where X is the set of training images from the principal food dataset and $|X| = N$.

In order to achieve this goal, the mapping from input vectors to output vectors learned by $\phi_{tr}(x)$ should be distilled into $\phi_{te}(x)$. Trainer networks such as $\phi_{tr}(x)$ learn to discriminate between large number of classes using the usual objective of maximizing the average log probability of the correct outcome. As a result of the learning, the trained model also assigns probabilities to all the incorrect answers/classes even though these probabilities are very small. This relative probability of incorrect answers signifies how the trainer model tends to generalize (Hinton et al., 2015). When distilling knowledge to a trainee network, it is important to train the trainee to generalize in the same way as the trainer model.

A simple way to transfer the generalization ability of the trainer to the trainee is to use the class probabilities produced by the trainer model as "soft targets" during the training process of the trainee. Hinton et al. proposed to use a parameter called temperature in the softmax function to generate these soft targets in the trainer model (Hinton et al., 2015). The following paragraphs explain this process in detail with respect to the trainer model $\phi_{tr}(x)$ and the trainee model $\phi_{te}(x)$.

For each input image, the output of the trainer $\phi_{tr}(x)$ is a probability distribution over all the dish categories. This probability is generated by the softmax layer from logits. Logits represent the output of the last fully connected layer. The dimension of the logits vector for both the trainer and the trainee network are equal to the number of classes/dish categories. For an input food image x_i, the logits vector generated by $\phi_{tr}(x)$, can be denoted by $\boldsymbol{v_i}$, where the dimension of vector $\boldsymbol{v_i} = (v_i^1, v_i^2, \ldots, v_i^C)$ is the number of dish categories C. A generalized softmax layer converts the logits vector $\boldsymbol{v_i}$ to a probability distribution $\boldsymbol{q_i}$ as follows:

$$M_T(\boldsymbol{v_i}) = \boldsymbol{q_i}, \text{ where } q_i^j = \frac{exp(\frac{v_i^j}{T})}{\sum_k exp(\frac{v_i^k}{T})}, \tag{11.1}$$

where T is the temperature parameter which is normally set to 1. Using a higher value of T produces the softer probability distribution over the C classes.

To distill the knowledge, the trainee model $\phi_{te}(x)$ is trained on a set of training images, also called the transfer set, using soft target distribution. For each image in the transfer set, the soft target distribution generated by $\phi_{tr}(x)$ is used as the target for training $\phi_{te}(x)$. The temperature of the trainee and the trainer models are kept equal during this distillation process. The trainee model $\phi_{te}(x)$ generates a student logits vector $\boldsymbol{w}_i$ and the corresponding probability distribution $M_T(\boldsymbol{w}_i)$.

For knowledge distillation, we minimize the Kullback–Leibler divergence between the trainer and the trainee probability distributions as defined below (Hinton et al., 2015):

$$L_{KD}(\phi_{tr}(x), \phi_{te}(x)) = \frac{1}{N}\sum_{i=1}^{N} KL(M_T(\boldsymbol{v}_i)||M_T(\boldsymbol{w}_i)), \qquad (11.2)$$

where $KL(\boldsymbol{x}||\boldsymbol{y})$ represents the Kullback–Leibler divergence between vectors $\boldsymbol{x}$ and $\boldsymbol{y}$.

When a set of hard image-label training pairs $\{(x_i, \boldsymbol{l}_i)\}$ are provided, the trainee $\phi_{te}(x)$ learns a standalone pure supervised classification task using the traditional cross-entropy loss which is defined as

$$L_S(\phi_{te}) = \frac{1}{N}\sum_{i=1}^{N} \mathcal{H}(\boldsymbol{l}_i, M_{T=1}(\boldsymbol{w}_i)), \qquad (11.3)$$

where $\mathcal{H}$ is the entropy function. We use a weighted combination of Kullback–Leibler loss (Eq. 11.2) and the standalone cross-entropy loss (Eq. 11.3) to train $\phi_{te}(x)$. The combined loss is defined as follows:

$$L(\phi_{tr}, \phi_{te}) = \alpha L_S(\phi_{te}) + (1-\alpha)T^2 L_{KD}(\phi_{tr}, \phi_{te}), \qquad (11.4)$$

where α is the weighting factor. To make the distillation process as effective as possible, more emphasis should be given to $L_{KD}(\phi_{tr}(x), \phi_{te}(x))$ while keeping the emphasis on $L_S(\phi_{te}(x))$ small. Hence, we set the weight α to 0.1.

It can be shown that when the temperature T is high, the distillation process is equivalent to minimizing $\frac{1}{2}(\boldsymbol{w}_i - \boldsymbol{v}_i)^2$. At low temperature, however, most of the logits that are more negative than the average are neglected even though they may convey some useful information acquired by the trainer model (Hinton et al., 2015). As a result, we choose an intermediate value for the temperature, $T = 4$, in order to let the compact network capture most of the knowledge from the heavyweight network while ignoring the most negative logits. The proposed system is implemented in the Caffe deep learning framework (Jia et al., 2014) on NVIDIA Titan Xp

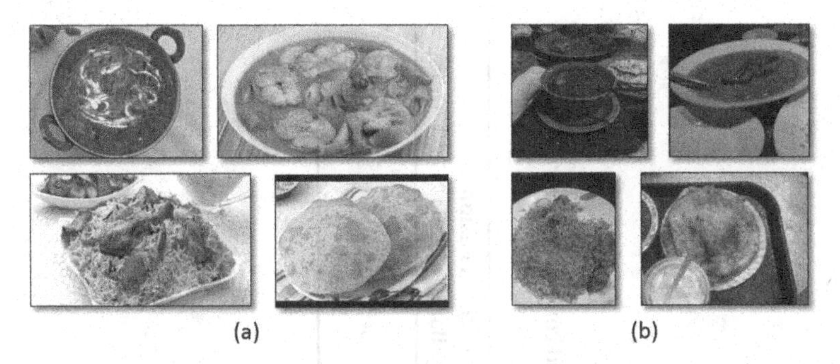

Figure 11.4 Sample images from (a) NTUIndianFood50 dataset and (b) the Food Diary. Four dishes are shown in clockwise direction: "Dal Makhani, " "Fish Curry," "Bhatura," and "Chicken Biryani."

graphics processing unit (GPU). To realize the distillation process, we implemented a new Caffe layer which takes logits from the trainer (v_i) and the trainee (w_i) as input and computes the loss function as formulated in Eq. 11.4.

11.6.2 Personalized Learning

For personalized learning of the compact network on an end-user's Food Diary, we created a new dataset called NTUIndianFood50. NTUIndianFood50 comprises of 50 Indian dishes and a total of 30,378 images. These images are crawled from the popular image search engines Google and Yahoo. In addition, for 24 dishes, images are crawled from a restaurant review website called www.zomato.com (Zomato, 2018) to represent the Food Diary. Since the Food Diary images are user-posted pictures from real restaurant settings, they are much poorer in illumination and quality and have much more clutter in the background when compared to the NTUIndianFood50 database images (Figure 11.4).

Classification on the Food Diary images is challenging but more realistic. When the knowledge distilled compact network is tested for classification accuracy on images from the Food Diary, the accuracy was found to be only 74.3% as opposed to 86.4% when test images were selected from the NTUIndianFood50 dataset crawled from Google and Yahoo.

Table 11.3 Comparison of classification accuracy, architecture and model size of the proposed compact network with other methods on Food-101 dataset

	DCNN-Food (Yanai and Kawano, 2015)	**DeepFood (Liu et al., 2016)**	**Classifiers Fusion (Aguilar et al., 2017)**	**Classifiers Fusion (Aguilar et al., 2017)**	**Ours**
Top-1(%)	70.4	77.4	82.3	83.8	84.0
Architecture	Modified AlexNet	Modified GoogLeNet	ResNet-50	Inception-V3	MobileNet
Model Size	425 MB	51 MB	98 MB	91 MB	**12 MB**

Table 11.4 Comparison of classification accuracy between compact network and the Personalized Compact Network

Model	Test Case	Top-1 Accuracy (%)
Compact Network	Baseline test	74.3
Compact Network	Personalized test	82.6
Personalized Compact Network	Personalized test	**95.6**

In order to boost the classification accuracy of the compact network and integrate the end-user's dietary preferences, we perform classifier personalization on the Food Diary images by optimizing the cross entropy loss function described in Eq. 11.3 where now the image-label pairs $\{(x_i, \boldsymbol{l}_i)\}$ come from the training images of the end-user's Food Diary and N is the size of this training set. Since each end user has a different set of images in the Food Diary, $\{(x_i, \boldsymbol{l}_i)\}$, the resulting network parameters would be different and personal to each user. The resulting optimized network is the Personalized Compact Network.

11.6.3 Performance Evaluation

Table 11.3 summarizes the top-1% classification performance of the compact network when compared with other networks on a popular Food-101 benchmark dataset. For fair comparison, all experiments used the same partition of data. Our compact network only used knowledge distillation from VGG16 network and did not leverage on personalized learning. The architecture information and the model size are also included for comparison.

Our proposed compact network clearly outperforms other deep learning based methods such as DCNN-Food (Yanai and Kawano, 2015) and DeepFood (Liu et al., 2016). In addition, slightly higher classification accuracy is achieved when compared to recent network architectures in Aguilar et al. (2017), but with a much smaller memory storage requirement, which makes it more suitable for mobile deployment.

Table 11.4 shows the classification performance of the compact network and the Personalized Compact Network on two test cases, (i) baseline test, where the network was trained and knowledge

Table 11.5 Comparison of heavyweight VGG16 model and the Personalized Compact Network in terms of model size and number of parameters to optimize

Model	Model size	Number of parameters
VGG16	510 MB	138.0 Million
Personalized Compact Network	12 MB	4.2 Million

distilled using NTUIndianFood50 dataset only and all test images came from the Food Diary; (ii) personalized test, where the network was trained and knowledge distilled as same as the baseline test, but test images came from both NTUIndianFood50 dataset and the Food Diary. This represents the scenario where the user regularly visits his favorite restaurants and eats his favorite dishes but from time to time the user may also visit new restaurants and try new dishes. To simulate this scenario, we selected 360 test images from the Food Diary and 1,000 test images from the NTUIndianFood50 dataset. These test images were not used for training or personalization of the networks.

The Personalized Compact Network which takes advantage of the Food Diary outperforms the non-personalized compact network by a significant margin, where the network has no knowledge about the Food Diary. The improvement in classification accuracy achieved by classifier personalization is 13%.

Table 11.5 provides the model size and total number of network parameters for the Personalized Compact Network and compares these parameters with those of the teacher network VGG16 on the server. The proposed Personalized Compact Network provides more than 40 fold reductions in model size and the number of network parameters. Inference time of the Personalized Compact Network was found to be 23 milliseconds per image.

11.7 Conclusion

In this chapter, we introduced the concept of visual diet recognition and analysis. We reviewed a number of works that attempt to solve various aspects of diet analysis such as dish recognition, end-to-end dietary logging systems, quantity estimation, leftover estimation and dish attribute recognition. Current approach towards diet logging and analytics focus on deep neural networks where an analytics engine is trained on a server and the end-user sends images over the network for interpretation at the server. Such systems do not focus on reducing the computation and memory requirement of the recognition engines. In addition, they do not leverage on end-user's dietary preferences. We propose a Personalized Compact Network that integrates end-user's dietary preferences into the classification process and trains from a heavyweight trainer network to achieve high classification accuracy on a compact neural network architecture. Such compact recognition engines are more amenable for deployment on mobile devices.

Acknowledgments

This research was carried out at the Rapid-Rich Object Search (ROSE) Lab at the Nanyang Technological University, Singapore. The ROSE Lab is supported by the Infocomm Media Development Authority, Singapore. This research is supported by the National Research Foundation, Prime Minister's Office, Singapore, under the NRF-NSFC grant NRF2016NRF-NSFC001-098. We gratefully acknowledge the support of NVIDIA Corporation for the donation of the Titan X Pascal GPU used for this research.

References

Aguilar, E., Bolaños, M., and Radeva, P. (2017). Food recognition using fusion of classifiers based on CNNs, in *International Conference on Image Analysis and Processing* (Springer), pp. 213–224.

Aizawa, K., Maruyama, Y., Li, H., and Morikawa, C. (2013). Food balance estimation by using personal dietary tendencies in a multimedia food log. *IEEE Trans. Multimedia* **15**, 8, pp. 2176–2185.

Akpro Hippocrate, E. A., Suwa, H., Arakawa, Y., and Yasumoto, K. (2016). Food weight estimation using smartphone and cutlery, in *Proceedings of the First Workshop on IoT-enabled Healthcare and Wellness Technologies and Systems* (ACM), pp. 9–14.

Anthimopoulos, M., Gianola, L., Scarnato, L., Diem, P., and Mougiakakou, S. G. (2014). A food recognition system for diabetic patients based on an optimized bag-of-features model. *IEEE J. Biomedical and Health Informatics* **18**, 4, pp. 1261–1271.

Beijbom, O., Joshi, N., Morris, D., Saponas, S., and Khullar, S. (2015). Menu-match: Restaurant-specific food logging from images, in *Applications of Computer Vision (WACV), 2015 IEEE Winter Conference on* (IEEE), pp. 844–851.

Bettadapura, V., Thomaz, E., Parnami, A., Abowd, G. D., and Essa, I. (2015). Leveraging context to support automated food recognition in restaurants, in *Applications of Computer Vision (WACV), 2015 IEEE Winter Conference on* (IEEE), pp. 580–587.

Bosch, M., Zhu, F., Khanna, N., Boushey, C. J., and Delp, E. J. (2011). Combining global and local features for food identification in dietary assessment, in *Image Processing (ICIP), 2011 18th IEEE International Conference on* (IEEE), pp. 1789–1792.

Bossard, L., Guillaumin, M., and Van Gool, L. (2014). Food-101–mining discriminative components with random forests, in *European Conference on Computer Vision* (Springer), pp. 446–461.

Chae, J., Woo, I., Kim, S., Maciejewski, R., Zhu, F., Delp, E. J., Boushey, C. J., and Ebert, D. S. (2011). Volume estimation using food specific shape templates in mobile image-based dietary assessment, in *Computational Imaging IX*, Vol. 7873 (International Society for Optics and Photonics), p. 78730K.

Chen, J., and Ngo, C.-W. (2016). Deep-based ingredient recognition for cooking recipe retrieval, in *Proceedings of the 2016 ACM on Multimedia Conference* (ACM), pp. 32–41.

Chen, J.-j., Ngo, C.-W., and Chua, T.-S. (2017). Cross-modal recipe retrieval with rich food attributes, in *Proceedings of the 2017 ACM on Multimedia Conference* (ACM), pp. 1771–1779.

Chen, M., Dhingra, K., Wu, W., Yang, L., Sukthankar, R., and Yang, J. (2009). Pfid: Pittsburgh fast-food image dataset, in *2009 16th IEEE International Conference on Image Processing (ICIP)*, pp. 289–292, doi: 10.1109/ICIP.2009.5413511.

Chen, M.-Y., Yang, Y.-H., Ho, C.-J., Wang, S.-H., Liu, S.-M., Chang, E., Yeh, C.-H., and Ouhyoung, M. (2012). Automatic chinese food identification and quantity estimation, in *SIGGRAPH Asia 2012 Technical Briefs*, SA '12 (ACM, New York, NY, USA), ISBN 978-1-4503-1915-7, pp. 29:1–29:4, doi:10.1145/2407746.2407775, http://doi.acm.org/10.1145/2407746.2407775.

Ciocca, G., Napoletano, P., and Schettini, R. (2015). Food recognition and leftover estimation for daily diet monitoring, in *International Conference on Image Analysis and Processing* (Springer), pp. 334–341.

Ciocca, G., Napoletano, P., and Schettini, R. (2017). Food recognition: A new dataset, experiments, and results, *IEEE Journal of Biomedical and Health Informatics* **21**, 3, pp. 588–598, doi:10.1109/JBHI.2016.2636441.

Cordeiro, F., Bales, E., Cherry, E., and Fogarty, J. (2015a). Rethinking the mobile food journal: Exploring opportunities for lightweight photo-based capture, in *Proceedings of the 33rd Annual ACM Conference on Human Factors in Computing Systems* (ACM), pp. 3207–3216.

Cordeiro, F., Epstein, D. A., Thomaz, E., Bales, E., Jagannathan, A. K., Abowd, G. D., and Fogarty, J. (2015b). Barriers and negative nudges: Exploring challenges in food journaling, in *Proceedings of the 33rd Annual ACM Conference on Human Factors in Computing Systems* (ACM), pp. 1159–1162.

Farinella, G. M., Moltisanti, M., and Battiato, S. (2014). Classifying food images represented as bag of textons, in *2014 IEEE International Conference on Image Processing (ICIP)*, pp. 5212–5216, doi:10.1109/ICIP.2014.7026055.

Hassannejad, H., Matrella, G., Ciampolini, P., De Munari, I., Mordonini, M., and Cagnoni, S. (2016). Food image recognition using very deep convolutional networks, in *Proceedings of the 2nd International Workshop on Multimedia Assisted Dietary Management* (ACM), pp. 41–49.

He, K., Zhang, X., Ren, S., and Sun, J. (2016). Deep residual learning for image recognition, in *2016 IEEE Conference on Computer Vision and Pattern Recognition (CVPR)*, pp. 770–778, doi:10.1109/CVPR.2016.90.

He, Y., Xu, C., Khanna, N., Boushey, C. J., and Delp, E. J. (2013). Food image analysis: Segmentation, identification and weight estimation, in *Multimedia and Expo (ICME), 2013 IEEE International Conference on* (IEEE), pp. 1–6.

He, Y., Xu, C., Khanna, N., Boushey, C. J., and Delp, E. J. (2014). Analysis of food images: Features and classification, in *Image Processing (ICIP), 2014 IEEE International Conference on* (IEEE), pp. 2744–2748.

Herruzo, P., Bolaños, M., and Radeva, P. (2016). Can a CNN recognize catalan diet? *CoRR* **abs/1607.08811**, http://arxiv.org/abs/1607.08811.

Hinton, G., Vinyals, O., and Dean, J. (2015). Distilling the knowledge in a neural network, *arXiv preprint arXiv:1503.02531* .

Hoashi, H., Joutou, T., and Yanai, K. (2010). Image recognition of 85 food categories by feature fusion, in *Multimedia (ISM), 2010 IEEE International Symposium on* (IEEE), pp. 296–301.

Howard, A. G., Zhu, M., Chen, B., Kalenichenko, D., Wang, W., Weyand, T., Andreetto, M., and Adam, H. (2017). Mobilenets: Efficient convolutional neural networks for mobile vision applications, *arXiv preprint arXiv:1704.04861* .

Ioffe, S., and Szegedy, C. (2015). Batch normalization: Accelerating deep network training by reducing internal covariate shift, *CoRR* **abs/1502.03167**, http://arxiv.org/abs/1502.03167.

Jia, Y., Shelhamer, E., Donahue, J., Karayev, S., Long, J., Girshick, R., Guadarrama, S., and Darrell, T. (2014). Caffe: Convolutional architecture for fast feature embedding, *arXiv preprint arXiv:1408.5093* .

Joutou, T., and Yanai, K. (2009). A food image recognition system with multiple kernel learning, in *Image Processing (ICIP), 2009 16th IEEE International Conference on* (IEEE), pp. 285–288.

Kawano, Y., and Yanai, K. (2013). Real-time mobile food recognition system, in *2013 IEEE Conference on Computer Vision and Pattern Recognition Workshops*, pp. 1–7, doi:10.1109/CVPRW.2013.5.

Kawano, Y., and Yanai, K. (2014). Automatic expansion of a food image dataset leveraging existing categories with domain adaptation, in *Proc. of ECCV Workshop on Transferring and Adapting Source Knowledge in Computer Vision (TASK-CV)*.

Kawano, Y., and Yanai, K. (2015). Foodcam: A real-time food recognition system on a smartphone, *Multimedia Tools and Applications* **74**, 14, pp. 5263–5287.

Kitamura, K., Yamasaki, T., and Aizawa, K. (2009). Foodlog: Capture, analysis and retrieval of personal food images via web, in *Proceedings of the ACM Multimedia 2009 Workshop on Multimedia for Cooking and Eating Activities* (ACM), pp. 23–30.

Kong, F., and Tan, J. (2012). Dietcam: Automatic dietary assessment with mobile camera phones, *Pervasive and Mobile Computing* **8**, 1, pp. 147–163.

Liu, C., Cao, Y., Luo, Y., Chen, G., Vokkarane, V., and Ma, Y. (2016). Deepfood: Deep learning-based food image recognition for computer-aided dietary assessment, in *International Conference on Smart Homes and Health Telematics* (Springer), pp. 37–48.

LoseIt! (2018). Weight loss that fits, Www.loseit.com.

Mariappan, A., Bosch, M., Zhu, F., Boushey, C. J., Kerr, D. A., Ebert, D. S. and Delp, E. J. (2009). Personal dietary assessment using mobile devices, in *Computational Imaging VII*, Vol. 7246 (International Society for Optics and Photonics), p. 72460Z.

Martin, C. K., Han, H., Coulon, S. M., Allen, H. R., Champagne, C. M., and Anton, S. D. (2008). A novel method to remotely measure food intake of free-living individuals in real time: The remote food photography method, *British Journal of Nutrition* **101**, 3, pp. 446–456.

Martinel, N., Foresti, G. L., and Micheloni, C. (2016). Wide-slice residual networks for food recognition, *CoRR* **abs/1612.06543**, http://arxiv.org/abs/1612.06543.

Min, W., Bao, B.-K., Mei, S., Zhu, Y., Rui, Y., and Jiang, S. (2018). You are what you eat: Exploring rich recipe information for cross-region food analysis, *IEEE Transactions on Multimedia* **20**, 4, pp. 950–964.

Min, W., Jiang, S., Sang, J., Wang, H., Liu, X., and Herranz, L. (2017). Being a supercook: Joint food attributes and multimodal content modeling for recipe retrieval and exploration, *IEEE Transactions on Multimedia* **19**, 5, pp. 1100–1113.

MyFitnessPal (2018). Free calorie counter, diet and exercise journal, www.myfitnesspal.com.

Nguyen, D. T., Zong, Z., Ogunbona, P. O., Probst, Y., and Li, W. (2014). Food image classification using local appearance and global structural information, *Neurocomputing* **140**, pp. 242–251.

Noronha, J., Hysen, E., Zhang, H., and Gajos, K. Z. (2011). Platemate: crowd-sourcing nutritional analysis from food photographs, in *Proceedings of the 24th Annual ACM Symposium on User Interface Software and Technology* (ACM), pp. 1–12.

Pouladzadeh, P., Kuhad, P., Peddi, S. V. B., Yassine, A., and Shirmohammadi, S. (2016). Food calorie measurement using deep learning neural network, in *IEEE International Instrumentation and Measurement Technology Conference Proceedings (I2MTC)*, pp. 1–6.

Pouladzadeh, P., and Shirmohammadi, S. (2017). Mobile multi-food recognition using deep learning, *ACM Transactions on Multimedia Computing, Communications, and Applications (TOMM)* **13**, 3s, p. 36.

Pouladzadeh, P., Shirmohammadi, S., and Al-Maghrabi, R. (2014). Measuring calorie and nutrition from food image, *IEEE Transactions on Instrumentation and Measurement* **63**, 8, pp. 1947–1956.

Pouladzadeh, P., Villalobos, G., Almaghrabi, R., and Shirmohammadi, S. (2012). A novel svm based food recognition method for calorie measurement applications, in *Multimedia and Expo Workshops (ICMEW), 2012 IEEE International Conference on* (IEEE), pp. 495–498.

Puri, M., Zhu, Z., Yu, Q., Divakaran, A., and Sawhney, H. (2009). Recognition and volume estimation of food intake using a mobile device, in *Applications of Computer Vision (WACV), 2009 Workshop on* (IEEE), pp. 1–8.

Simonyan, K., and Zisserman, A. (2014). Very deep convolutional networks for large-scale image recognition, *CoRR* **abs/1409.1556**, http://arxiv.org/abs/1409.1556.

Sun, M., Liu, Q., Schmidt, K., Yang, J., Yao, N., Fernstrom, J. D., Fernstrom, M. H., DeLany, J. P., and Sclabassi, R. J. (2008). Determination of food portion size by image processing, in *Engineering in Medicine and Biology Society, 2008. EMBS 2008. 30th Annual International Conference of the IEEE* (IEEE), pp. 871–874.

Szegedy, C., Liu, W., Jia, Y., Sermanet, P., Reed, S., Anguelov, D., Erhan, D., Vanhoucke, V., and Rabinovich, A. (2015). Going deeper with convolutions, in *2015 IEEE Conference on Computer Vision and Pattern Recognition (CVPR)*, pp. 1–9, doi:10.1109/CVPR.2015.7298594.

Tanno, R., Okamoto, K., and Yanai, K. (2016). Deepfoodcam: A dcnn-based real-time mobile food recognition system, in *Proceedings of the 2nd International Workshop on Multimedia Assisted Dietary Management* (ACM), pp. 89–89.

Villalobos, G., Almaghrabi, R., Pouladzadeh, P., and Shirmohammadi, S. (2012). An image procesing approach for calorie intake measurement, in *Medical Measurements and Applications Proceedings (MeMeA), 2012 IEEE International Symposium on* (IEEE), pp. 1–5.

Wang, X., Kumar, D., Thome, N., Cord, M., and Precioso, F. (2015). Recipe recognition with large multimodal food dataset, in *Multimedia & Expo Workshops (ICMEW), 2015 IEEE International Conference on* (IEEE), pp. 1–6.

Yanai, K., and Kawano, Y. (2015). Food image recognition using deep convolutional network with pre-training and fine-tuning, in *Multimedia & Expo Workshops (ICMEW), 2015 IEEE International Conference on* (IEEE), pp. 1–6.

Zhang, X.-J., Lu, Y.-F., and Zhang, S.-H. (2016). Multi-task learning for food identification and analysis with deep convolutional neural networks, *Journal of Computer Science and Technology* **31**, 3, pp. 489–500.

Zhou, F., and Lin, Y. (2016). Fine-grained image classification by exploring bipartite-graph labels, in *2016 IEEE Conference on Computer Vision and Pattern Recognition (CVPR)*, pp. 1124–1133, doi:10.1109/CVPR.2016.127.

Zomato (2018). Indian restaurant search and discovery service, www.zomato.com.

Part V

Motions in Videos, Pose Recognition, and Human Activity Analysis

Chapter 12

Learning Randomized Decision Trees for Human Behavior Capture

Zhen-Peng Bian,[a] Cheen-Hau Tan,[b] Junhui Hou,[c] and Lap-Pui Chau[b]

[a]*Singapore Telecommunications Limited, Singapore*
[b]*Nanyang Technological University, Singapore*
[c]*City University of Hong Kong, Kowloon Tong, Hong Kong*

This chapter focuses on recent research in Randomized Decision Tree (RDT) algorithm; in particular, how to classify human face and body using pixelwise classification of depth image. Similar to popular machine learning algorithms, training of RDT is also computation intensive. This chapter shows a more efficient technique to reduce the training time of the RDT algorithm. Hence, it is suitable for power-constrained devices. Besides, two applications are presented in this chapter to show the efficiency of the technique: (i) a fall detection system that monitor human fall down; (ii) a human–computer interface system that enable human to use nose and mouth to control computer mouse. The applications end with experimental results and performance evaluations.

Learning Approaches in Signal Processing
Edited by Wan-Chi Siu, Lap-Pui Chau, Liang Wang, and Tieniu Tan

ISBN 978-981-4800-50-1 (Hardcover), 978-0-429-06114-1 (eBook)
www.panstanford.com

12.1 Introduction

Motion capture is the process of capturing and representing human motion in a form that is suitable for further processing. Typically, human motion capture data is represented as a sequence of poses, where each pose is a data sample at a point in time, and each pose is represented as a skeletal structure that is composed of a set of connected joints. Capturing and representing human motion in such a simplified form provides a powerful tool for analyzing human behavior. Hence, motion capture has been applied in many fields, including surveillance, human–computer interfacing, movies, games, robotics, military and so on.

Traditional motion capture systems are marker-based systems, where a number of markers are attached to the subject's body (Tanie et al., 2005). The positions of these markers are captured and then used to deduce the locations of the subject's joints. The main drawbacks of these systems are the requirement of specialized and expensive equipment, inconvenience to the subject, and difficultly in operation.

More recently, the availability of inexpensive depth cameras (Fanello et al., 2014, Microsoft, 2018, SoftKinetic, 2018) allow a promising new approach for motion capture. Unlike traditional RGB cameras that capture light and color information, depth cameras capture depth information from a scene. Hence, a single depth camera can be used to capture the three dimensional structure of a surface of an object or a person.

Recently, a motion capture algorithm (Shotton et al., 2011) that utilizes cameras has been proposed. The algorithm uses the Randomized Decision Tree (RDT) algorithm, to rapidly classify depth pixels into segments of the human body, before using the segmentation information to derive the positions of body joints. As with other machine learning methods, the RDT algorithm has to be trained before use. However, the training of this algorithm is computationally complex. In Shotton et al. (2011), the training process needs a computer system with a cluster with 1000 cores for one day. Hence, to be able to use the algorithm in practical applications, it is very important to reduce the complexity of the training phase.

In Section 12.2, we present both the RDT algorithm and an improved version of the RDT algorithm, which has significantly reduced computational complexity for the training phase, for motion tracking. Although the RDT based motion tracking method is fast enough to run on standard hardware, it is beneficial to reduce the computation complexity of the testing phase for use on low power or mobile devices. Hence, we also propose a method to meet this requirement.

We also present two motion capture applications that utilize the proposed improved RDT (Bian, 2015). In Section 12.3, we describe a fall detection system. This system uses a depth camera to monitor human subjects for falls; if the monitored subject falls down and fails to get up, an alert will be triggered. For the second application, we describe a human–computer interface system in Section 12.4. This system allows a disabled user to use his head and mouth motions to control the mouse pointer.

12.2 Randomized Decision Trees in Motion Capture

A number of motion capture algorithms based on the depth camera have been proposed (Baak et al., 2011, Buys et al., 2014, Ganapathi et al., 2010, Grest et al., 2007, Knoop et al., 2006, Lehment et al., 2010, Plagemann et al., 2010, Shotton et al., 2011, Suau et al., 2012, Wei et al., 2012, Zhu et al., 2008), including an influential motion tracking method by Shotten (Shotton et al., 2011). In this method, depth images, with pixels containing depth information, of a human subject are captured. The RDT algorithm is then used to classify foreground pixels to a body part label. After that, a mean-shift algorithm extracts the 3D joint position from each group of pixels. Thus, a set of joints that indicates the location of each body part is obtained. This motion tracking system runs at 200 fps on the Xbox GPU and extract highly accurate positions of human body joints from a single depth image.

The RDT algorithm has to be trained beforehand using depth images with labeled pixels. In order to address the variation of poses, a large and highly varied training dataset of depth images is required. Since depth images do not contain color and illumination information, it is straightforward to synthesize a large depth image dataset. However, the training process of the RDT is very computational intensive; in Shotton et al. (2011), the training time is cited as one day for training three trees with twenty levels from one million images on a cluster with 1000 cores. The training time in Buys et al. (2014) is seven days for 80,000 images on a sixteen nodes cluster in one tree with sixteen levels, where each node is a Xeon X5570 CPU with 48 GB RAM and two GTX550ti GPUs.

This exceptionally high training complexity limits the practical applicability of the RDT for motion capture. Hence, in the following subsections, after describing the RDT algorithm, we will present a RDT-based algorithm that has a much lower computation requirement during the training process.

12.2.1 The Randomized Decision Tree Algorithm

12.2.1.1 Overview of the RDT

Figure 12.1 shows a RDT, a tree structure that contains leaf nodes and split (intermediate) nodes. Each split node extracts a feature that coarsely describes the neighborhood of a tested pixel; this feature is expressed as a number. The feature is parameterized, so that features extracted at different split nodes describe different aspects of the neighborhood of the test pixel. Each leaf node, on the other hand, contains a label probability distribution.

To classify a pixel, we first evaluate its feature at the root node of the RDT and compare it to a threshold. If the feature value f is larger or equals to a threshold τ, the left child node is evaluated next; if not, the right child node is evaluated. This process is repeated with the child node, and we progress down the RDT until a leaf node is reached. A leaf node contains label probability distributions that are assigned to the pixels that reach it; in the simplest case, a leaf node contains a single label to be assigned to the pixel.

In the paper by Shotten (Shotton et al., 2011), the RDT feature describes the depth difference between two neighboring pixels of

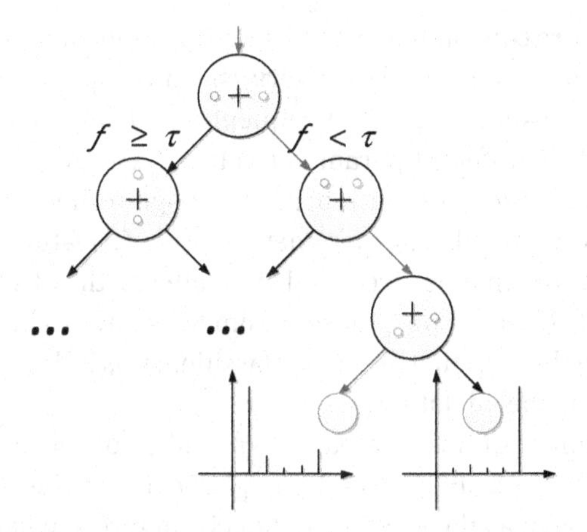

Figure 12.1 Illustration of a RDT. Each node contains an evaluation function f. In Shotten's work (Shotton et al., 2011), f compares the depth of two neighbors (circles) of the tested pixel (cross). The blue arrow shows an example of the sequence of evaluations a tested pixel might undergo. The leaf nodes show the label probability distributions for pixels that reach them.

the tested pixel,

$$f((x_0, y_0)|(\Delta x_1, \Delta y_1), (\Delta x_2, \Delta y_2)) = z\left((x_0, y_0) + \frac{(\Delta x_1, \Delta y_1)}{z(x_0, y_0)}\right) - z\left((x_0, y_0) + \frac{(\Delta x_2, \Delta y_2)}{z(x_0, y_0)}\right), \quad (12.1)$$

where (x_0, y_0) is the location of the test pixel, $(\Delta x_1, \Delta y_1)$ and $(\Delta x_2, \Delta y_2)$ are the location offsets of the two neighboring pixels relative to the test pixel, and function $z(x, y)$ denotes the depth of pixel (x, y). The location offset is normalized with respect to $z(x_0, y_0)$ to ensure that the feature is invariant to the depth of the tracked surface; a tracked surface should produce the same feature values regardless of its depth. We can express the feature comparison function at each node as

$$E((x_0, y_0); p_n) = B(f((x_0, y_0)|(\Delta x_{1,n}, \Delta y_{1,n}), (\Delta x_{2,n}, \Delta y_{2,n})) - \tau_n), \quad (12.2)$$

where parameter $p_n = ((\Delta x_{1,n}, \Delta y_{1,n}), (\Delta x_{2,n}, \Delta y_{2,n}), \tau_n)$. When the value of f is greater than or equals to τ, $B(f) = 1$, and the left child node is evaluated next; otherwise $B(f) = 0$, and the right child node is evaluated next.

Each feature contributes a small amount of information about the patch around the test pixel at a very cheap computational cost. In a RDT, a test pixel is evaluated sequentially using a large number of features with different parameter values $(\Delta x_1, \Delta y_1)$, $(\Delta x_2, \Delta y_2)$, and τ, so accurate information of the neighborhood around the test pixel is obtained. The parameters of each feature determine the information that is extracted and affects the discrimination ability of the feature. Thus, these parameters have to be tailored to discriminate the data that are to be classified; this is done by training the RDT with labeled data.

If the number of possible pairs of neighbor offsets is very large, building a tree with all these nodes might be intractable. In this case, a method known as the randomized decision forest, where multiple trees containing different sets of evaluation function parameters can be used. The results from each tree are then aggregated to obtain the final result.

12.2.1.2 Training the RDT

The purpose of the training step is to build the RDT and to obtain feature parameters for the split nodes and the labels for the leaf nodes. A set of labeled pixels from example depth images, the training set, is used for training. If we evaluate the training set at a node with candidate parameter $p = ((\Delta x_1, \Delta y_1), (\Delta x_2, \Delta y_2), \tau)$ (Equation 12.2), we can partition the training set Q into two subsets Q_L and Q_R, where

$$Q_L((\Delta x_1, \Delta y_1), (\Delta x_2, \Delta y_2), \tau) = \\ \{E((x_0, y_0); p((\Delta x_1, \Delta y_1), (\Delta x_2, \Delta y_2), \tau)) = 1\}, \tag{12.3}$$

$$Q_R((\Delta x_1, \Delta y_1), (\Delta x_2, \Delta y_2), \tau) = \\ \{E((x_0, y_0); p((\Delta x_1, \Delta y_1), (\Delta x_2, \Delta y_2), \tau)) = 0\}. \tag{12.4}$$

with E as defined in Equation 12.2. The aim is to find the parameter $p = ((\Delta x_1, \Delta y_1), (\Delta x_2, \Delta y_2), \tau)$ for each node that generates a partitioning with the lowest Shannon entropy.

Starting from the root node, for each node, the set (or subset) of all candidate neighbor pixel offsets $((\Delta x_1, \Delta y_1), (\Delta x_2, \Delta y_2))$ is evaluated. For each offset $(\Delta x_1, \Delta y_1)$, $(\Delta x_2, \Delta y_2)$, we find the τ that produces the partitioning with the smallest Shannon entropy:

$$\tau^* = \underset{\tau \in T}{\arg\min} S(Q((\Delta x_1, \Delta y_1), (\Delta x_2, \Delta y_2), \tau)), \tag{12.5}$$

$$S(Q((\Delta x_1, \Delta y_1), (\Delta x_2, \Delta y_2), \tau)) = \sum_{sub \in L, R} \frac{|Q_{sub}((\Delta x_1, \Delta y_1), (\Delta x_2, \Delta y_2), \tau)|}{Q} H(Q_{sub}((\Delta x_1, \Delta y_1), (\Delta x_2, \Delta y_2), \tau)), \tag{12.6}$$

where $H(Q)$ is the Shannon entropy of set Q, and $S(Q((\Delta x_1, \Delta y_1), (\Delta x_2, \Delta y_2), \tau))$ is the Shannon entropy of the partitioning generated by $((\Delta x_1, \Delta y_1), (\Delta x_2, \Delta y_2), \tau)$. Equation 12.5 is solved by brute force search on a set of candidate values for τ. The Shannon entropy for each candidate $((\Delta x_1, \Delta y_1), (\Delta x_2, \Delta y_2))$ with its optimal τ is compared; the chosen parameter is the one with the smallest Shannon entropy, i.e.,

$$\begin{aligned} & p((\Delta x_1^*, \Delta y_2^*), (\Delta x_2^*, \Delta y_2^*), \tau^*) \\ &= \underset{(\Delta x, \Delta y) \in (\Delta X, \Delta Y), \tau \in T}{\arg\min} S(Q((\Delta x_1, \Delta y_1), (\Delta x_2, \Delta y_2), \tau)) \\ &= \underset{(\Delta x_1, \Delta y_1), (\Delta x_2, \Delta y_2) \in (\Delta X, \Delta Y)}{\arg\min} \\ & \{\min_{\tau \in T} S(Q((\Delta x_1, \Delta y_1), (\Delta x_2, \Delta y_2), \tau))\}. \end{aligned} \tag{12.7}$$

With the chosen parameter $p((\Delta x_1^*, \Delta y_2^*), (\Delta x_2^*, \Delta y_2^*), \tau^*)$, the training set is split into subsets Q_L and Q_R (Eqs. 12.3, 12.4); two child nodes are created for the processed node, and the training algorithm repeats for two child nodes with subsets Q_L and Q_R. Training stops when the desired tree depth is reached, and label probabilities are assigned to the leaf nodes. The probability of label c for a leaf is the ratio of the number of pixels with label c and the total number of pixels that reach the leaf.

The number of possible pairs of candidate offsets are $(2M + 1)^4$ pairs, where the range of M is Δx, Δy, i.e., $\Delta x, \Delta y \in [-M, M]$. For example, the number of possible pairs of candidate offsets is around 1.6×10^9 for $M = 100$. As a result, training could take a very long time.

12.2.2 The Compact Feature RDT

In this subsection, we propose the Compact Feature RDT (CF-RDT) algorithm, which can be trained at a much faster rate than the standard RDT algorithm.

12.2.2.1 The compact feature

The feature of a RDT algorithm is designed to extract information about the geometric surface around a test pixel. As shown in Equation 12.1, this is done by comparing the depth values between two neighbor pixels of the test pixel. Here, we propose that a similar amount of information can be obtained by comparing the depth values between the test pixel and a neighbor pixel:

$$f((x_0, y_0)|(\Delta x, \Delta y)) = z(x_0, y_0) - z\left((x_0, y_0) + \frac{(\Delta x, \Delta y)}{z(x_0, y_0)}\right) \tag{12.8}$$

We call this new feature the compact feature, and its corresponding algorithm the Compact Feature RDT(CF-RDT) algorithm. Compared with RDT's feature, the compact feature has higher computational efficiency: there is only one addition, one subtraction, and one division, and also indexes two depth pixels. The number of candidate offsets is $(2M+1)^2$ for $\Delta x, \Delta y \in [-M, M]$ for CF-RDT which is much fewer than RDT. For example, for $M = 100$, the number of candidate offsets for CF-RDT is around 4×10^4, compared with 1.6×10^9 for RDT. Using this new feature, the evaluation function of a node is

$$E((x_0, y_0); p_n) = B(f((x_0, y_0)|(\Delta x_n, \Delta y_n)) - \tau_n), \tag{12.9}$$

where parameter $p_n = ((\Delta x_n, \Delta y_n), \tau_n)$

12.2.2.2 Training of the compact feature RDT

By using the compact feature, we only need to evaluate a single neighbor pixel, thus reducing the number of candidate offsets to be evaluated. For each candidate parameter $p((\Delta x, \Delta y), \tau)$, by evaluating the training samples using Equation 12.9, we can partition the training set Q into two sets Q_L and Q_R, where

$$Q_L((\Delta x, \Delta y), \tau) = \{E((x_0, y_0); p((\Delta x, \Delta y), \tau)) = 1\}, \tag{12.10}$$

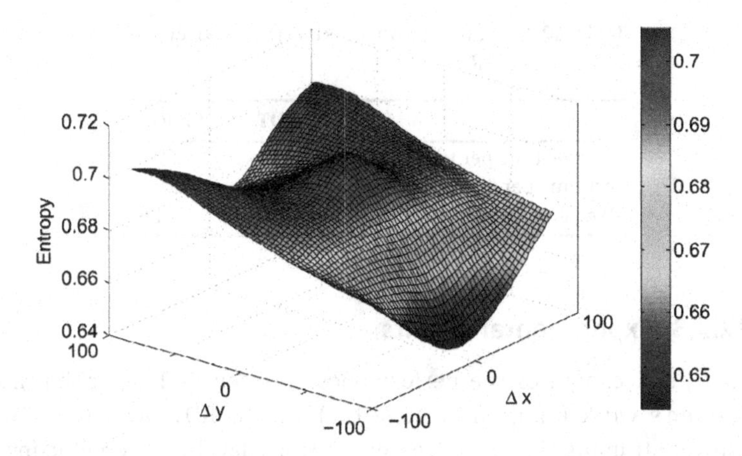

Figure 12.2 An entropy map corresponding to different offsets $(\Delta x, \Delta y))$ with their optimal τ^*.

$$Q_R((\Delta x, \Delta y), \tau) = \{E((x_0, y_0); p((\Delta x, \Delta y), \tau)) = 0\}. \qquad (12.11)$$

As before, we compute the parameter p that produces the partition with the lowest entropy

$$\begin{aligned} p((\Delta x^*, \Delta y^*), \tau^*) &= \underset{(\Delta x, \Delta y) \in (\Delta X, \Delta Y), \tau \in T}{\arg\min} S(Q((\Delta x, \Delta y), \tau)) \\ &= \underset{(\Delta x, \Delta y) \in (\Delta X, \Delta Y)}{\arg\min} \{\min_{\tau \in T} S(Q((\Delta x, \Delta y), \tau))\} \end{aligned} \qquad (12.12)$$

where

$$S(Q((\Delta x, \Delta y), \tau)) = \sum_{sub \in L, R} \frac{|Q_{sub}((\Delta x, \Delta y), \tau)|}{Q} H(Q_{sub}((\Delta x, \Delta y), \tau)), \qquad (12.13)$$

and $H(Q)$ is the Shannon entropy of set Q, and $S(Q((\Delta x, \Delta y), \tau))$ is the Shannon entropy of the partitioning generated by $((\Delta x, \Delta y), \tau)$.

To further improve training speed, we propose that it is not necessary to evaluate all the possible values of $(\Delta x, \Delta y)$. Figure 12.2 shows an example of the entropy corresponding to different offsets $(\Delta x, \Delta y))$ with their optimal τ^*. The entropy changes smoothly with varying $(\Delta x, \Delta y)$; this is true for all datasets. Thus, instead of the prior brute force approach, we use a suitable search algorithm to find the $p((\Delta x, \Delta y), \tau)$ with the lowest entropy. This dramatically reduces the number of evaluated offsets and results in faster training.

Table 12.1 Comparison of RDT (Shotton et al., 2011) and CF-RDT

	RDT	CF-RDT
Training time per tree (hour)	310	3.9
Test time per frame (ms)	5.0	2.8
Mean error (cm)	3.2	3.1

12.2.3 Experimental Results

Table 12.1 compares the performance between RDT algorithm in Shotten's work (Shotton et al., 2011) and CF-RDT algorithm. To evaluate training time, one tree with twenty levels is trained using 10,000 images (with 1000 pixels each). The Matlab program is run on a single core standard PC. Using the compact feature and a search algorithm, the CF-RDT takes less than 1 h to train, while the RDT requires 310 h.

To compare the performance of the RDT based and CF-RDT based motion tracking algorithms, we evaluate the speed and accuracy of head joint extraction using 2000 depth images with ground truths. The depth images are first classified with the RDT or CF-RDT algorithms, before the mean shift algorithm is applied for joint extraction. In this experiment, the algorithms are implemented in C++. The tracking time per frame for the CF-RDT based method is 2.8 ms, while the time for the RDT based method is 5.0 ms, which is 79% higher. The RDT based method and the CF-RDT based method have errors of 3.2 cm and 3.1 cm, respectively; both errors are not significantly different.

12.2.4 Efficient Feature Evaluation

There are situations where the speed of CF-RDT classification can be improved. In the feature evaluation function for CF-RDT (Equation 12.8), the offset $(\Delta x_n, \Delta y_n)$ is normalized with respect to $z(x_0, y_0)$. In certain applications, the differences in depth between all pixels of a tracked surface is very small relative to their depth. In this case, it is reasonable to normalize the offsets of all the pixels on the surface with the same depth value. We can exploit this by shifting the normalization operation to the training phase. During training, we normalize all pixels using a set of depth values, and store the set of corresponding RDTs. During the test phase, we check the depth of the tracked surface and retrieve the corresponding RDT. Thus, by eliminating the division operation, we can evaluate the feature more efficiently.

The efficiently evaluated feature is formulated as

$$f((x_0, y_0)|(\Delta x, \Delta y)) = z(x_0, y_0) - z\left((x_0, y_0) + \frac{(\Delta x, \Delta y)}{\bar{z}/R}\right), \tag{12.14}$$

where, $\bar{z}/R$ replaces $z(x_0, y_0)$ in Equation 12.8. $\bar{z}$ is a normalized variable related to depth, and R is a constant.

In face tracking (see Section 12.4), $\bar{z}$ is set to the depth value of nose during training phase. The offset parameter $(\Delta x, \Delta y)$ is normalized with respect to $\bar{z}/R$ after training, which kept as $\text{RDT}_{\bar{z}}$. Thus, for a given $\text{RDT}_{\bar{z}}$, the feature becomes

$$f((x_0, y_0)|(\Delta x_{\bar{z}}, \Delta y_{\bar{z}})) = z(x_0, y_0) - z((x_0, y_0) + (\Delta x_{\bar{z}}, \Delta y_{\bar{z}})), \tag{12.15}$$

where $(\Delta x_{\bar{z}}, \Delta y_{\bar{z}})$ is normalized in advance. $\bar{z}$ is the index to the corresponding $\text{RDT}_{\bar{z}}$ in test phase.

12.3 Fall Detection Using Motion Capture

Fall detection is primarily used as a monitoring system for elderly people who are prone to falls; if the monitored person accidentally falls down and fails to get up, an alert will be triggered to request for help. Fall detection systems can be classified into non-vision-based and vision-based. Non-vision-based methods typically require the monitored subject to wear inertial sensors, such as accelerometers; this creates an inconvenience for the subject.

Vision-based methods monitor the subject using cameras, and are non-intrusive for the subject. They not only capture the subject's movement, but also identify the background and analyze the interaction between the subject and the background, such as recognizing that the subject lying on the floor. Vision-based methods monitor specific features of the subject, such as posture (Brulin et al., 2012, Thome et al., 2008, Yu et al., 2012, 2013) changes in shape (Foroughi et al., 2008a,b, Rougier et al., 2011b), 3D head position (Rougier and Meunie, 2006, Rougier et al., 2013), and 3-D silhouette (Auvinet et al., 2011). Reviews of fall detection methods can be found in Mubashir et al. (2013), Noury et al. (2007), Yu (2008).

For vision-based methods, systems utilizing depth cameras are more practical than systems utilizing RGB cameras. This is because RGB cameras do not work well in poorly illuminated environments, while depth cameras are not affected by environmental illumination. RGB camera-based systems require multiple cameras to extract 3D information of a scene; only areas that are visible to every camera can be monitored. Thus, the monitored subject must not be occluded from any of the cameras; this limitation poses challenges in proper positioning of the cameras, especially if there are furniture in the room. This issue affects systems using a depth camera to a much lesser extent since the monitored subject only needs to be visible to one camera.

12.3.1 Overview of Proposed Approach

In this section, we develop a fall detection system (Bian et al., 2015) that utilizes the Microsoft Kinect (Microsoft, 2018) depth camera and the Microsoft Kinect Software Development Kit (SDK). The SDK provides an implementation of the motion capture framework (Shotton et al., 2011) described in the previous section; it retrieves depth images from the Kinect sensor and produces the captured subject's pose in the form of a skeleton, which consists of a set of joints and their connectivity information. Instead of using this implementation in its entirety, we utilize the Compact Feature RDT instead of the standard RDT. This allows for higher computational efficiency. We also added a pose correction step to improve tracking robustness.

Figure 12.3 shows the block diagram of the fall detection system. Using the SDK, we first extract a depth image from the sensor.

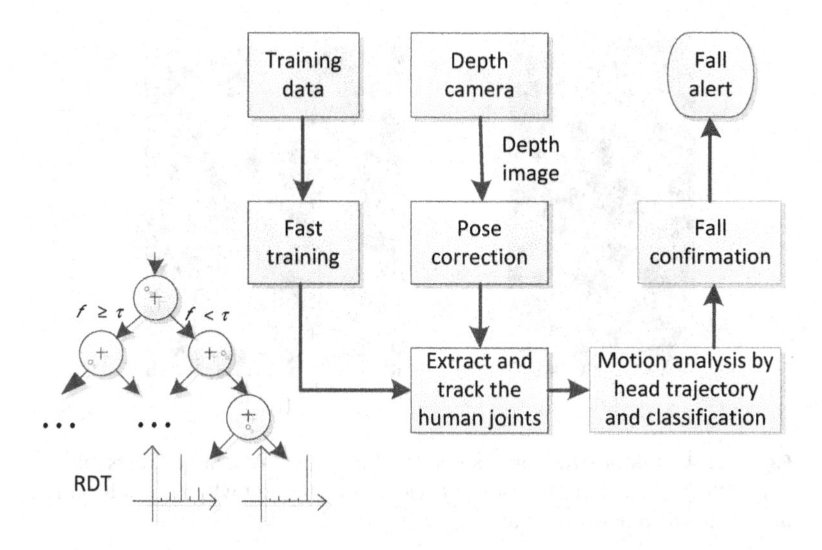

Figure 12.3 Block diagram of the fall detection framework.

From the depth image, the SDK can be used to extract the skeleton; however, this operation is performed reliably when the subject is in the vertical orientation (Fig. 12.4a), but not when the subject is in the horizontal orientation (Fig. 12.4b). Hence, we first apply pose correction—we rotate the depth image so that the subject is always in the vertical orientation. After that, we extract the skeleton from the depth image using the SDK with the Compact Feature RDT algorithm. Our system detects falls by analyzing the trajectory of the head joint. Thus, we track this trajectory, and apply Support Vector Machine (SVM) classifier on the trajectory to decide whether a fall has occurred.

This motion capture method is highly suitable for fall detection because: (1) It is a markerless method; hence, monitored subjects would not be encumbered by any equipment, (2) It uses a depth camera to monitor subjects; hence, it can work well in poorly illuminated rooms, (3) It has low computational complexity during operation; hence, it can operate in real-time. The results of our experiments for proposed fall detection framework is more robust and accurate when compared with existing approaches.

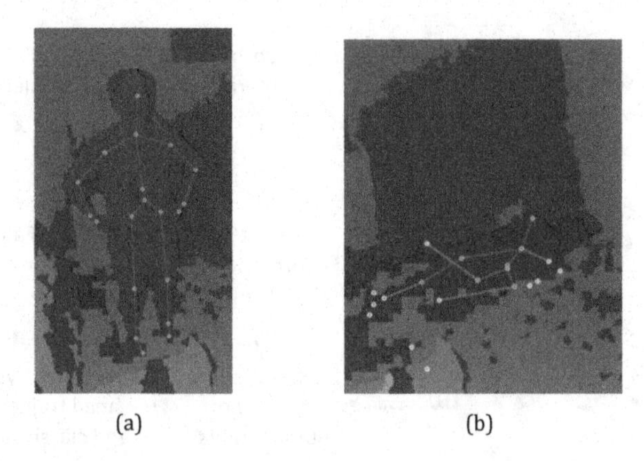

Figure 12.4 Joint extraction results. The Kinect SDK reliably extracts joints when the subject is in a (a) vertical orientation, but not when the subject is in a (b) horizontal orientation.

12.3.2 Pose Correction

As mentioned earlier, we rotate the depth image so that the subject's pose is in the vertical orientation. First, we need to determine the subject's pose. To do this, we require the skeleton information. Since we do not have the skeleton for the current frame yet, and since the change in pose between successive frames is small, we use the skeleton from the previous frame as an estimate. We define the orientation of the subject's pose as the orientation of the straight line from the head joint to the hip center. To correct the current pose, we rotate the current depth image by the angle that vertically aligns the orientation of the previous pose. Figure 12.5 demonstrates the images before and after pose correction. Figure 12.5a demonstrates the images of a simulated person fall. Figure 12.5b displays the pose-corrected images. Pose correction ensures that the orientation of the subject's torso in the depth image is vertical.

There are times when we need to do pose correction without relying on the previous frame. If tracking of the subject is lost, we will not have pose information from the previous frame. To apply pose correction in this case, we first rotate the depth image by two angles, 120° and 240°, as shown in Fig. 12.6. We assume that skeleton extraction for each image is reliable. We then rotate the

Figure 12.5 A simulated fall poses. (a) The sequence of original poses. (b) Poses after correction.

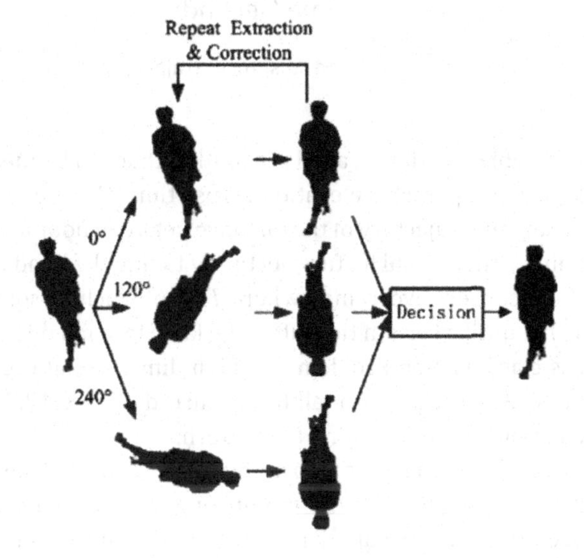

Figure 12.6 Pose correction without prior frames.

image so that the pose is in the vertical orientation. We apply this operation to each image individually for several iterations. The final corrected pose is then selected from one of the three rotated depth images.

12.3.3 Fall Detection Using SVM

The World Health Organization defined (WHO, 2007) a fall as "Inadvertently coming to rest on the ground, floor or other lower level, excluding intentional change in position to rest in furniture,

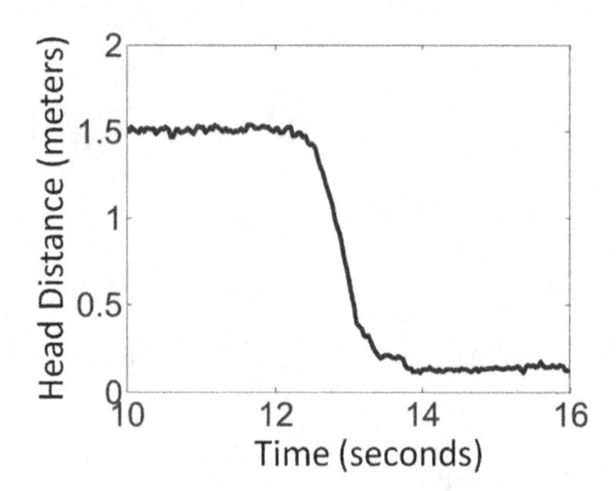

Figure 12.7 Fall pattern of head distance trajectory from an image sequence.

wall or other objects." Hence, a fall is a motion that can be measured by the change of a person's elevation across time. We thus detect a fall by tracking the trajectory of the distance between floor and head.

We form a dimensional feature vector by taking the head to floor distance for d consecutive frames, where d should include some time before the fall until a certain time after the fall. Since fall detection is a two-class classification problem of a high dimensional vector, we use SVM to classify whether a fall has occurred. Figure 12.7 shows the head distance trajectory for a fall pattern.

After a fall, a person might recover by himself, so a fall alert does not need to be triggered by the monitoring system. Thus, after a fall has been detected, we apply recovery detection to confirm this event. There are two measures for recovery detection: (a) head and hip are higher than a recovery threshold T_{recover1}; (b) head is higher than another threshold T_{recover2}. If one of these two measures is satisfied, it signifies that the person has recovered. In our work, we set $T_{\text{recover1}} = 0.5$ m and $T_{\text{recover2}} = 0.8$ m according to references from anthropometry studies (Motmans and Ceriez, 2005). After a fall is detected, if the subject cannot stand up within a specific time, a fall alert will be triggered.

In order to train the SVM classifier, we need a set of fall and non-fall examples. Instead of obtaining samples from actual human fall motions, which is impractical for large data sets, we decide to synthesize the training samples. To do so, we model the head motion during a fall using a free fall body model. According to the laws of physics, if a body starts to fall freely from a standstill, its height is

$$h(t) = h_0 + \frac{1}{2}g(t - t_0)^2, \tag{12.16}$$

where $h(t)$ is the height at time t, g is the acceleration due to gravity, which is 9.81 ms^{-2}, h_0 is the height at time t_0, which is the time at start of the fall.

12.3.4 Experimental Results

Before describing the experiments, we first explain the performance evaluation metrics used. The metrics, as suggested by (Noury et al., 2007), are:

(1) True positives (TP): falls detected correctly.
(2) True negatives (TN): non-falls detected correctly.
(3) False positives (FP): non-falls detected as falls.
(4) False negatives (FN): falls detected as non-falls.
(5) Sensitivity (SE): ability to detect falls $SE = \frac{TP}{TP+FN}$.
(6) Specificity (SP): ability to detect non-falls $SP = \frac{TN}{TN+FP}$.
(7) Accuracy (AC): correct classification rate $AC = \frac{TP+TN}{TP+TN+FP+FN}$.
(8) Error rate (ER): incorrect classification rate $ER = \frac{FP+FN}{TP+TN+FP+FN}$.

A high specificity implies that majority of non-falls are correctly classified. A high sensitivity signifies that majority of falls are detected. A reliable fall detection system requires high specificity and high sensitivity. Furthermore, for an ideal system, we expect high accuracy and low error rate.

To evaluate the performance of proposed method, we have a few test subjects perform several motions that might happen in everyday life, such as walking, sitting down, and falling. We apply motions and their ground truth classification outcomes that are suggested in Noury et al. (2007). Half of the motions are classified as "Positive," which means that they are classified as a fall event, while the other motions are classified as "Negative," which corresponds to a non-fall event. Each type of motion is performed several times, so we get 380 samples in total. There are four subjects, three male and one female, and their heights and ages are: 159–182 cm and 24–31 years

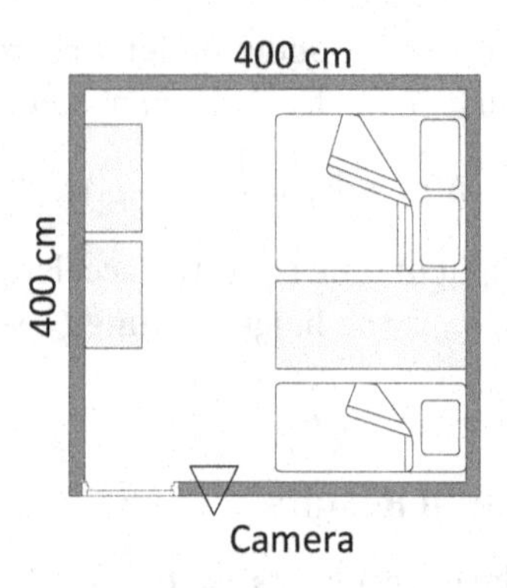

Figure 12.8 Room plan for the evaluation of fall detection.

old. The experiments are conducted in a bedroom, the camera is mounted on a wall 2.3 m above the floor. The layout is shown in Fig. 12.8.

For performance validation, we first train the SVM classifier using a set of 100,000 synthesized non-fall and fall patterns. After training, the classifier is used to determine the motions performed by human test subjects as fall or non-fall events. The experimental results are compared with two latest approaches that are based on a monocular depth camera, as shown in Table 12.2.

For the proposed method, there are only nine incorrect classification results, which are all false negatives, i.e., the system omits nine fall alerts, but there is no false alert. Eight incorrect results are due to incorrect classification of falls, and one incorrect result is due to the loss of tracking the subject during fall confirmation stage. In

Table 12.2 The results of different approaches for fall detection

	(Rougier et al., 2011a)	(Planinc and Kampel, 2012)	Proposed
TP	**188**	175	181
TN	104	151	**190**
FP	86	39	**0**
FN	**2**	15	9
SE(%)	**98.9**	92.1	95.3
SP(%)	54.7	79.5	**100**
AC(%)	76.8	85.8	**97.6**
ER(%)	23.2	14.2	**2.4**

this case, the fall motion is detected by SVM classifier, but the system failed to confirm it in the fall confirmation.

The method (Rougier et al., 2011a) detects the most fall events, but with many false positives (FP). In this method, a fall is detected when the height of the center point of a human silhouette is lower than a threshold. Thus, the method cannot distinguish a fall event from intentional activities, such as lying down or sitting on the floor. Furthermore, the silhouette center is derived using foreground pixels, and the foreground segmentation algorithm used in this method is not very robust. The approach of (Planinc and Kampel, 2012) detects falls by thresholding the orientation of body and the height information of spine using predefined empirical values. The orientation and height information are computed from the joint positions extracted from the Kinect SDK. However, as mentioned earlier, the Kinect SDK joint extraction is inaccurate when the body is horizontally oriented. Overall, the proposed method performs the best compared to existing methods, as can be seen from the superior values in terms of specificity, accuracy and error rate.

12.4 Human–Computer Interface Using Motion Capture

Assistive Technologies (AT) have been developed to help people who have lost the use of their hands interface with computers, by allowing them to use other voluntary signals and motions. ATs can be classified into four categories, based on different types of interfaces (Music et al., 2009, Yousefi et al., 2011), i.e.,

(1) Physiological Signal-Based ATs (Mak and Wolpaw, 2009, McCool et al., 2014, Mihajlovic et al., 2014, Nam et al., 2014, Williams and Kirsch, 2008).
(2) Voice Command-Based ATs (Huo et al., 2013, Park et al., 2013).
(3) Mechanical Motion-Based ATs (Jose and Lopes, 2014, Lau and O'Leary, 1993, Mazo, 2001).
(4) Motion Tracking-Based ATs (track motions of human body), such as eye (MacKenzie, 2012, Zhang and MacKenzie, 2007), tongue (Huo et al., 2013, Yousefi et al., 2011), head (Betke et al.,

2002, Guness et al., 2012), and face (Epstein et al., 2014, Morris and Chauhan, 2006, Tu et al., 2007).

In this section, we propose a human–computer interface that utilizes motion tracking based on a depth camera. This system allows a user to use his facial position and expression to replace the use of a mouse. The user moves the on-screen mouse pointer by turning his head, and activates mouse clicks by opening and closing his mouth. We call this proposed system Facial position and expression Mouse (FM) system. The FM system addresses several challenging issues faced by other ATs, i.e.,

(1) FM is not affected by illumination changes in the environment, such as the presence or absence of lighting, that affects RGB camera based methods. This is due to the use of a depth camera.
(2) The use of RDT for position and expression detection from single image avoids the "feature drift" problem that appears in many of the existing tracking algorithms.
(3) The small range of head motion means that the system has to either sacrifice motion accuracy or motion range of the on-screen mouse pointer. We solve the problem by using a non-linear function to map the nose motion to the motion of the pointer. We also utilize mouth status commands to allow the user to return his head to the natural position without affecting the pointer.

When compared with other ATs, the advantage of FM system is handiness (no guardian, no accessory on body, no calibration for the individual user, and no initialization required) and robustness (not affected by illumination or color). It demonstrates the FM system is superior to other ATs.

12.4.1 Overview of FM System

The human–computer interface (Bian et al., 2016) employs a depth camera, such as SoftKinetic (SoftKinetic, 2018) and Kinect (Microsoft, 2018) to track the user's face. Specifically, we track the position of nose, and the status of mouth (open or close). We choose the position of nose as tracked point for controlling the on-screen

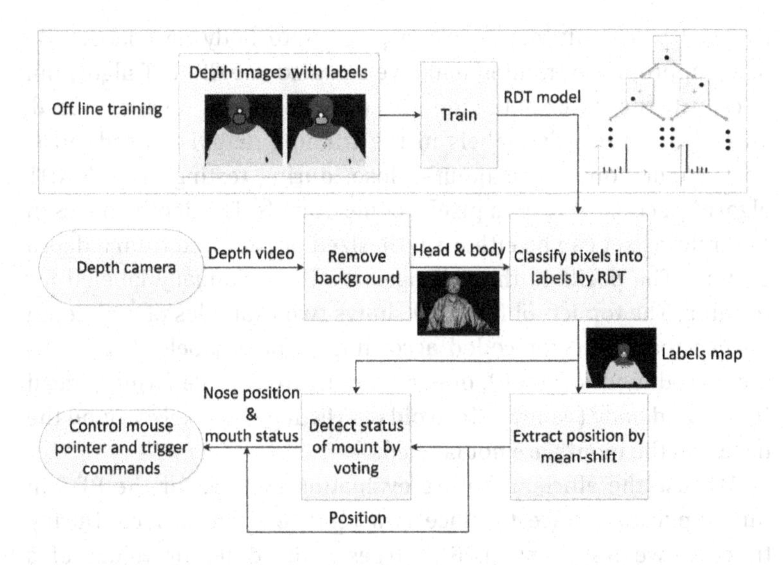

Figure 12.9 Block diagram of the human–computer interface.

pointer. The reason for using the nose is that it, unlike the eyes for example, is a rigid part of the face; hence, it is stable and will not move unintentionally. It is also not occluded by glasses.

The CF-RDT algorithm is employed to label the nose and mouth pixels on the depth image, after which the mean-shift algorithm is applied on the nose pixels to compute the position of the nose. With appropriate training, the CF-RDT algorithm is not only able to identify mouth pixels, but also determine whether those pixels correspond to an opened mouth or a closed mouth. Figure 12.9 shows the flow chart of the proposed approach.

12.4.2 Detection of Facial Position and Expression

As shown in Fig. 12.9, the first step of the interface is to separate the background from the human user. If the depth value of pixel $z(x, y)$ is larger than $T_{\max}$ or smaller than $T_{\min}$, this pixel is considered as the background and is excluded from further processing. We empirically set $T_{\min} = 50$ and $T_{\max} = 150$.

The next step is to classify pixels into labels using the CF-RDT algorithm. We have explained previously that CF-RDT algorithm can be used to classify depth image pixels into body part labels. By using appropriate training data, we can also use CF-RDT algorithm to classify the status of the mouth, i.e., whether it is open or closed. To do so, we use five labels in the training data, i.e., head, nose, body, mouth open, and mouth close; during testing, the CF-RDT algorithm classifies depth pixels to these labels. The depth images in the training set can be either synthesized or captured from a depth camera. The pixels in these images are then manually labeled for training. The top left of Fig. 12.9 shows two examples of real depth images that are color coded according to pixel labels. The labels are closed mouth (green), opened mouth (red), nose (white), head (blue), and body (yellow). To avoid overfitting, the pixels around the nose and the mouth are not labeled.

We use the efficient feature evaluation method for the RDT in this application since the face is a relatively flat surface. During training, we train several RDT trees using different values of $\overline{z}$ (Equation 12.14, 12.15), where $\overline{z}$ is the depth of nose in training images. In the test phase, the depth of user's nose $\overline{z}$ is used to index the corresponding $\mathrm{RDT}_{\overline{z}}$.

12.4.3 Decision Approach of Nose Position and Mouth Status

The RDT classifier output is a distribution D_{il} over the labels l for each test pixel (x_i, y_i). After the pixels classification, we use a mean-shift clustering approach (Shotton et al., 2011) to extract the position of nose. To extract the status of mouth $S_m \in$ {mouth-close, mouth-open}, we use a voting method. Each pixel contributes a vote with a magnitude that is dependent on its label probability, its distance from the mouth position, and the world surface area of the pixel. The voting metric is

$$S_m^* = \arg\max_{S_m}(w_{S_m} \sum_{i=1}^{M} w_{iS_m} k(P_m - L_i) z(L_i)^2), \qquad (12.17)$$

where w_{S_m} denotes the probability of label S_m, which is learned from training data. w_{iS_m} denotes the probability of label S_m for test pixel (x_i, y_i); it is the result of RDT classification, D_{il}. To assign higher importance to pixels nearer to the mouth position, we apply a kernel function $k(\bullet)$, e.g., a Gaussian kernel, on the distance between pixel location L_i and mouth position P_m. The surface area of pixel (x_i, y_i) is accounted for by its depth $z(L_i)$. The mouth status S_m^* is decided by voting value.

12.4.4 Determining User Input Based on Nose Position and Mouth Status

In this section, we show how the nose position and mouth status are utilized together to control the computer mouse pointer position and trigger mouse button.

12.4.4.1 Mapping function of motion

In Tu et al. (2007), the authors showed that among joystick mode, direct mode and location mode, the location mode offers the best performance. Thus, we use the location mode, where the operation is very similar to the mouse operation. However, we are faced with a problematic trade-off due to the low amplitude of head motion and the comparatively low resolution of the depth image. Since a small mapping gain leads to a small range of pointer motion, if we linearly map nose motion to pointer motion, pointer cannot cover the whole screen. On the other hand, a large mapping gain will reduce pointer precision and is susceptible to noise. Therefore, we need to trade-off the precision of pointer motion with the range of pointer motion.

To address this problem, we use a non-linear mapping function, i.e.,

$$D = Gd^{p_1}\left(1 - \exp\left(-\left(\frac{d}{n}\right)^{p_2}\right)\right), \tag{12.18}$$

where d represents the distance between current and last sampled nose position, G is mapping gain, D is distance between current and last sampled pointer position, and p_1 and p_2 and n are coefficients. The term $(1 - \exp(-(\frac{d}{n})^{p_2}))$ forces $\lfloor D \rfloor = 0$ when d is a small value; this eliminates mouse pointer jitter, making the interface more accurate. The non-linear mapping function increases mapping gain with increased nose movement speed; this behavior exploits the user's tendency to move his head faster when pointer is far away from its target. When pointer is close to its target location, the user slows down his head movement; thus, mapping gain is reduced, and pointer precision is increased.

12.4.4.2 Commands triggered by mouth status

A user can simulate mouse clicks or trigger other commands by closing and opening his mouth in a sequence. For example, a mouse click can be triggered by an open-close-open sequence.

We also utilize the mouth status to improve the ease of use the system. The open and close of mouth are used to disable and enable the movement of pointer. This enables the user to expediently adjust the head pose without moving the pointer. After moving pointer for a large distance, the user can disable pointer movement by opening his mouth, return his head back to a comfortable position, and then continue operating the computer. Besides improving the comfort for operating the interface, utilizing the mouth status also increases the range of pointer motion.

12.4.5 Experiments

In this section, we validate the performance of the detection method together with the computer interface operation.

12.4.5.1 Validation of detection algorithm

To verify the performance of our algorithm, we conduct experiments using the Bosphorus 3D face database (Savran et al., 2012). Twenty challenging pose types, i.e., Lower-Face-Action-Units (LFAU), are used as training and testing samples. In total, there are 105 users, with 1876 negative samples and 105 positive samples.

The 10-fold cross validation method is used to estimate the performance. The dataset is randomly split into ten approximately same size subsets, where each user only includes in one subset. The training and testing are conducted ten times. In each experiment, nine subsets are used for training and the remaining subset is used for testing. The results from ten rounds are averaged to estimate classification accuracy.

Table 12.3 Mouth status detection and nose tracking results

	RDT	CF-RDT
False Negative of mouth	2	2
False Positive of mouth	6	5
Total errors of mouth	8	7
Error rate of mouth (%)	0.40	0.35
Mean error of nose (mm)	3.2	1.9
Time (ms)	5.4	1.5
Frame rate (fps)	185	667

In Table 12.3, we show the mouth status detection and nose tracking results for methods based on RDT and CF-RDT with efficient feature evaluation. In this table, False Negative (FN) refers to positive events appeared as negative. False Positive (FP) refers to negative events appeared as positive. The mean error is the error of the extracted position of the nose. The algorithms are written in C++ and tested on an Intel Core i7 with 8 GB RAM.

The results show that, in this application, CF-RDT with efficient feature evaluation outperforms RDT in all metrics, except for FN, where both algorithms are even. The mouth status error rate is improved by 12.5% while the nose error is improved by 40% for the CF-RDT. Using the CF-RDT instead of RDT also allows the application to run 3.6× faster.

12.4.5.2 Evaluation of computer interface operation

We first explain the procedure for evaluating the interface operation. We evaluate based on the ISO/TS 9241-411 standard (ISO/TS, 2012), in that the multi-directional tapping task is used. In this task, a circular target appears on the screen one at a time; users have to move the pointer to the center of a target as soon as possible. A new target, chosen from the set of targets shown in Fig. 12.10, appears after the current one has been selected, and the user continues the test by selecting the new target. After this task, each user is given a questionnaire to evaluate the comfort of using the interfaces based on ISO/TS 9241-411 standard.

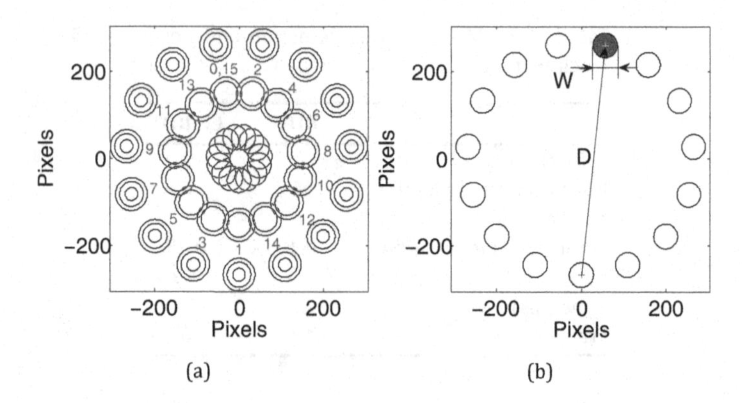

Figure 12.10 The on-screen tapping task on Graphical User Interface (GUI). (a) 90 possible targets and their orders. (b) Targets appear one at a time. Arrow shows path of pointer from location of previous target to newly appeared target.

According to the ISO/TS 9241-411 standard, throughput (*TP*) is the metric to show the user performance in terms of speed and accuracy. *TP* shows data rate that users pass to computer via Graphical User Interface (GUI). Higher *TP* indicates better performance for the interface. *TP* can be formulated as

$$TP = \frac{ID}{MT}, \tag{12.19}$$

where MT is movement time, time taken for the pointer to a target, averaged over all trials for the same target conditions, and ID is the index of difficulty of the target. ID is defined using Shannon's formula, i.e.,

$$ID = \log_2\left(\frac{D}{W} + 1\right), \tag{12.20}$$

where D is the distance from the previous target to the current target, and W is the width of the current target, as shown in Fig. 12.10b. In our experiments, we generate and use six target conditions based on their D and W values. These target conditions, along with their corresponding D, W and ID values, are shown in Table 12.4. The mean *TP* is calculated as

$$TP = \frac{1}{S}\sum_{i=1}^{S}\left(\frac{1}{C}\sum_{j=1}^{C}\frac{ID_{ij}}{MT_{ij}}\right), \tag{12.21}$$

Table 12.4 Design parameters of tapping task

D(pixels)	W(pixels)	ID(bits/s)
100	57	1.46
305	76	2.33
305	57	2.67
534	76	3.00
534	57	3.37
534	30	4.23

where S is the number of users, and C is the number of target conditions.

Additionally, we also evaluate Task Completion Time (TCT), as used in Huo et al. (2013), Jose and Lopes (2014), Yousefi et al. (2011). TCT is the total time used to complete each experiment, starting from display of first target, and ending at selection of last target. Clearly, lower TCT indicates better performance.

Next, we describe the experimental setup. Three interfaces are experimentally compared, i.e., FM, Camera Mouse and a standard optical mouse. For FM, the Kinect v2 depth sensor with resolution of 512×424 pixels is used. Camera Mouse is a popular interface (Camera Mouse, 2018). The input RGB camera for Camera Mouse is a 1280×720 resolution HD webcam. The standard optical mouse is used as a baseline to evaluate our methods.

For each user, the test order of the three interfaces is randomized. The user has to finish all tests of one interface before starting another test. Before their respective tests, the users get familiar with using the Camera Mouse and FM for longer than 1 h, and using the standard optical mouse for longer than 10 min. For each interface, we conducted 90 trials each round, and twenty rounds for each user, for 16 users.

12.4.5.3 Results of interface operation

The interface evaluations are shown in Table 12.5. The throughput of standard optical mouse is 4.39 bps, which is similar to other experiments (Huo et al., 2013, Jose and Lopes, 2014, Soukoreff and MacKenzie, 2004, Yousefi et al., 2011). *TP* and TCT of FM achieve

Table 12.5 Experimental results of operation evaluation

	TP(bps)	TP_n(%)	TCT(minutes)	TCT_n(%)
Camera Mouse	1.52±0.12	100	4.68±0.35	100
FM	3.26±0.26	215	2.88±0.25	61
Mouse	4.39±0.36	289	1.53±0.12	33

115% and 39% improvement when compare with Camera Mouse. In short, Table 12.5 demonstrates that FM outperform Camera Mouse. There are four main reasons for the improved performance:

(1) The head position relative to the pointer position is fixed for the Camera Mouse at first. Our approach allows users to comfortably adjust their head poses at any time.
(2) The mapping function for Camera Mouse is linear. The FM mapping function accelerates pointer movement when target position is far from pointer position, and improves precision when target is nearby.
(3) The Camera Mouse is susceptible to "feature drift," which makes the users feel inconvenient. Our detection method can detect position and expression from an image while resisting "feature drift" that is a common issue for many tracking algorithms.
(4) The Camera Mouse takes a longer time to stop the pointer movement when pointer moves quickly. The FM mouth status enables the users to stop the pointer movement rapidly.

To compare the performance of FM with other types of ATs, we normalize with respect to mouse *TP*, and show them in Table 12.6. FM provides the best performance among all the ATs.

Figure 12.11 shows the results of the questionnaire for Camera Mouse and FM. The responses of questions are rated from 1 (the least favorable response) to 7 (the most favorable response). One-way analysis of variance with pairwise comparison is used to analyze the outcomes of Camera Mouse and FM. The comparisons that have significant difference with $p < 0.5$ are denoted with "*" on the vertical axis.

Except the "Mouth fatigue," FM receives better scores in all aspects, as mouth motion are not required for Camera Mouse. In

Table 12.6 Benchmarking ATs: normalized *TP*

AT	Normalized *TP*
Mouse	100
FM (face)	74
EyeTracker (eye) (MacKenzie, 2012)	65
SmartNav (head) (Guness et al., 2012)	56
Camera Mouse (face) (Camera Mouse, 2018)	35
dTDS (tongue+speech) (Huo et al., 2013)	28
TDS (tongue) (Yousefi et al., 2011)	27
HeadTracker (head) (Guness et al., 2012)	22
LCS (lip) (Jose and Lopes, 2014)	21
Head Orientation (head) (Williams and Kirsch, 2008)	20
EMG (neck muscle) (Williams and Kirsch, 2008)	16

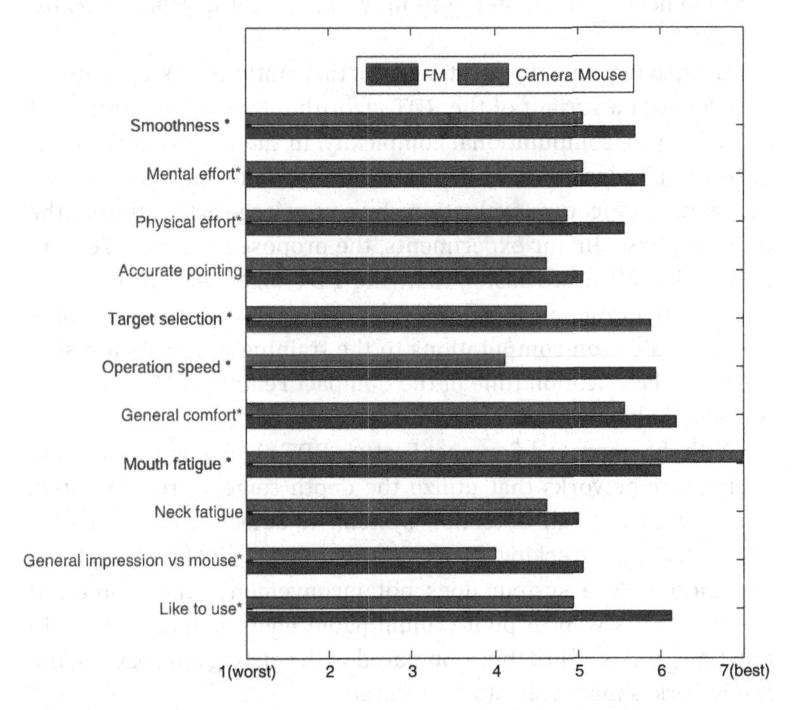

Figure 12.11 Assessment of the questionnaire for Camera Mouse and FM. A higher score indicates a favorable response. The results that are significantly different, with $p < 0.5$, are denoted with "*" on the vertical axis.

terms of "Operation speed" and "Like to use", the users prefer FM over Camera Mouse significantly. In addition, the average score for FM and Camera Mouse are 5.7 and 5.0, respectively. FM outperforms Camera Mouse by 14%.

12.5 Conclusion

Single depth camera based motion capture systems provide a compact, convenient, and affordable motion capture solution for various applications in modern daily life. One such system, the RDT based motion tracking method, achieves high accuracy and robustness at a low computational complexity. However, for motion tracking, the RDT algorithm is very computational intensive. Training needs a computer system with a cluster of 1000 cores for one day.

Thus, to improve the practicality of this motion tracking method, we proposed a variant of the RDT algorithm that can be trained at a much lower computational complexity. In the proposed Compact Feature RDT algorithm, by using an alternative feature, we reduced the number of feature evaluations by a very large extent during the training phase. In our experiments, the proposed Compact Feature RDT trains 79 times faster than the RDT algorithm. In order to improve operational speed, we also showed how to shift some feature evaluation computations to the training phase. As a result, the pixel classification time of the Compact Feature RDT is reduced to nearly half of the time for RDT.

With the improved Compact Feature RDT algorithm, two motion analysis frameworks that utilize the depth camera are presented. The first one is fall detection system. It detects the event of a person falling by tracking the motion of his head joint and classifying this motion. The system does not inconvenience the monitored person, works well in poorly illuminated environments, works in real time, and is affordable. Compared with existing approaches, the framework is more robust and accurate.

The second framework is a human–computer interface for persons who are unable to use their hands to operate the computer. In proposed Facial position and expression Mouse (FM) system, the user controls mouse pointer by rotating his head and triggers commands by opening and closing his mouth. The system tracks nose of user and the status of user's mouth from a depth image. A nonlinear nose-pointer mapping function is used to improve precision and speed of pointer navigation, and mouth status commands are utilized to improve ease of use for mouse pointer navigation. The proposed human–computer interface system is robust, convenient to use, affordable, and works in real time. The proposed framework achieves the best user operation performance among alternative human–computer interface systems, ranking just behind the mouse input.

References

Auvinet, E., Multon, F., Saint-Arnaud, A., Rousseau, J. and Meunier, J. (2011). Fall detection with multiple cameras: An occlusion-resistant method based on 3-D silhouette vertical distribution, *IEEE Transactions on Information Technology in Biomedicine* **15**, 2, pp. 290–300.

Baak, A., Muller, M., Bharaj, G., Seidel, H.-P. and Theobalt, C. (2011). A data-driven approach for real-time full body pose reconstruction from a depth camera, in *Computer Vision (ICCV), 2011 IEEE International Conference on*, pp. 1092–1099, doi:10.1109/ICCV.2011.6126356.

Betke, M., Gips, J. and Fleming, P. (2002). The Camera Mouse: visual tracking of body features to provide computer access for people with severe disabilities, *IEEE Transactions on Neural Systems and Rehabilitation Engineering* **10**, 1, pp. 1–10.

Bian, Z. (2015). *Markerless Motion Capture and Analysis Based on Depth Images*, Ph.D. thesis, Nanyang Technological University.

Bian, Z., Hou, J., Chau, L. and Magnenat-Thalmann, N. (2015). Fall detection based on body part tracking using a depth camera, *IEEE Journal of Biomedical and Health Informatics* **19**, 2, pp. 430–439, doi:10.1109/JBHI.2014.2319372.

Bian, Z., Hou, J., Chau, L. and Magnenat-Thalmann, N. (2016). Facial position and expression-based human–computer interface for persons with tetraplegia, *IEEE Journal of Biomedical and Health Informatics* **20**, 3, pp. 915–924, doi:10.1109/JBHI.2015.2412125.

Brulin, D., Benezeth, Y. and Courtial, E. (2012). Posture recognition based on fuzzy logic for home monitoring of the elderly, *IEEE*

Transactions on Information Technology in Biomedicine **16**, 5, pp. 974–982.

Buys, K., Cagniart, C., Baksheev, A., Laet, T. D., Schutter, J. D. and Pantofaru, C. (2014). An adaptable system for RGB-D based human body detection and pose estimation, *Journal of Visual Communication and Image Representation* **25**, 1, pp. 39–52.

Epstein, S., Missimer, E. and Betke, M. (2014). Using kernels for a video-based mouse-replacement interface, *Personal and Ubiquitous Computing* **18**, 1, pp. 47–60.

Fanello, S. R., Keskin, C., Izadi, S., Kohli, P., Kim, D., Sweeney, D., Criminisi, A., Shotton, J., Kang, S. B. and Paek, T. (2014). Learning to be a depth camera for close-range human capture and interaction, *ACM Trans. Graph.* **33**, 4, pp. 86:1–86:11.

Foroughi, H., Aski, B. S. and Pourreza, H. (2008a). Intelligent video surveillance for monitoring fall detection of elderly in home environments, in *11th IEEE International Conference on Computer and Information Technology (ICCIT)*, pp. 219–224.

Foroughi, H., Naseri, A., Saberi, A. and Yazdi, H. S. (2008b). An eigenspace-based approach for human fall detection using integrated time motion image and neural network, in *9th International Conference on Signal Processing, 2008. ICSP 2008* (IEEE), pp. 1499–1503.

Ganapathi, V., Plagemann, C., Koller, D. and Thrun, S. (2010). Real time motion capture using a single time-of-flight camera, in *2010 IEEE Conference on Computer Vision and Pattern Recognition (CVPR)*, pp. 755–762.

Grest, D., Krüger, V. and Koch, R. (2007). Single view motion tracking by depth and silhouette information, in *Image Analysis* (Springer), pp. 719–729.

Guness, S., Deravi, F., Sirlantzis, K., Pepper, M. and Sakel, M. (2012). Evaluation of vision-based head-trackers for assistive devices, in *2012 Annual International Conference of the IEEE Engineering in Medicine and Biology Society (EMBC)*, pp. 4804–4807.

Huo, X., Park, H., Kim, J. and Ghovanloo, M. (2013). A dual-mode human computer interface combining speech and tongue motion for people with severe disabilities, *IEEE Transactions on Neural Systems and Rehabilitation Engineering* **21**, 6, pp. 979–991.

ISO/TS (2012). Ergonomics of human-system interaction - part 411 – evaluation methods for the design of physical input devices, ISO/TS 9241-411:2012, *ISO/TS*.

Jose, M. and Lopes, R. (2014). Human-computer interface controlled by the lip, *IEEE Journal of Biomedical and Health Informatics* **PP**, 99, pp. 1–1.

Knoop, S., Vacek, S. and Dillmann, R. (2006). Sensor fusion for 3d human body tracking with an articulated 3d body model, in *Robotics and Automation, 2006. ICRA 2006. Proceedings 2006 IEEE International Conference on* (IEEE), pp. 1686–1691.

Lau, C. and O'Leary, S. (1993). Comparison of computer interface devices for persons with severe physical disabilities, *Amer. J. Occupat. Therapy* **47**, 11, pp. 1022–1030.

Lehment, N. H., Kaiser, M., Arsić, D. and Rigoll, G. (2010). Cue-independent extending inverse kinematics for robust pose estimation in 3d point clouds, in *Image Processing (ICIP), 2010 17th IEEE International Conference on* (IEEE), pp. 2465–2468.

MacKenzie, I. S. (2012). Evaluating eye tracking systems for computer input, in *Gaze Interaction and Applications of Eye Tracking: Advances in Assistive Technologies* (IGI Global), pp. 205–225.

Mak, J. and Wolpaw, J. (2009). Clinical applications of brain-computer interfaces: Current state and future prospects, *IEEE Reviews in Biomedical Engineering* **2**, pp. 187–199.

Mazo, M. (2001). An integral system for assisted mobility automated wheelchair, *IEEE Robotics Automation Magazine* **8**, 1, pp. 46–56.

McCool, P., Fraser, G., Chan, A., Petropoulakis, L. and Soraghan, J. (2014). Identification of contaminant type in surface electromyography (EMG) signals, *IEEE Transactions on Neural Systems and Rehabilitation Engineering* **22**, 4, pp. 774–783.

Microsoft (2018). http://www.microsoft.com/.

Mihajlovic, V., Grundlehner, B., Vullers, R. and Penders, J. (2014). Wearable, wireless EEG solutions in daily life applications: What are we missing? *IEEE Journal of Biomedical and Health Informatics* **PP**, 99, pp. 1–1.

Morris, T. and Chauhan, V. (2006). Facial feature tracking for cursor control, *Journal of Network and Computer Applications* **29**, 1, pp. 62–80.

Motmans, R. and Ceriez, E. (2005). Anthropometry table, in *Ergonomie RC, Leuven, Belgium.*

Mubashir, M., Shao, L. and Seed, L. (2013). A survey on fall detection: Principles and approaches, *Neurocomputing* **100**, pp. 144–152.

Music, J., Cecic, M. and M., B. (2009). Testing inertial sensor performance as hands-free human–computer interface, *WSEAS Trans. Comput.* **8**, pp. 715–724.

Nam, Y., Koo, B., Cichocki, A. and Choi, S. (2014). GOM-Face: GKP, EOG, and EMG-based multimodal interface with application to humanoid robot control, *IEEE Transactions on Biomedical Engineering* **61**, 2, pp. 453–462.

Noury, N., Fleury, A., Rumeau, P., Bourke, A., Laighin, G., Rialle, V. and Lundy, J. (2007). Fall detection-principles and methods, in *Engineering in Medicine and Biology Society, 2007. EMBS 2007. 29th Annual International Conference of the IEEE* (IEEE), pp. 1663–1666.

WHO (2007). WHO global report on falls prevention in older age, *World Health Organization (WHO) Library Cataloguing-in-Publication Data.*

Park, J.-S., Jang, G.-J., Kim, J.-H. and Kim, S.-H. (2013). Acoustic interference cancellation for a voice-driven interface in smart TVs, *IEEE Transactions on Consumer Electronics* **59**, 1, pp. 244–249.

Plagemann, C., Ganapathi, V., Koller, D. and Thrun, S. (2010). Real-time identification and localization of body parts from depth images, in *Robotics and Automation (ICRA), 2010 IEEE International Conference on* (IEEE), pp. 3108–3113.

Planinc, R. and Kampel, M. (2012). Introducing the use of depth data for fall detection, *Personal Ubiquitous Computing.*

Rougier, C., Auvinet, E., Rousseau, J., Mignotte, M. and Meunier, J. (2011a). Fall detection from depth map video sequences, in *Proceedings of the 9th international conference on Toward useful services for elderly and people with disabilities: smart homes and health telematics*, ICOST'11 (Springer-Verlag, Berlin, Heidelberg), ISBN 978-3-642-21534-6, pp. 121–128, URL http://dl.acm.org/citation.cfm?id=2026187.2026206.

Rougier, C. and Meunie, J. (2006). Fall detection using 3D head trajectory extracted from a single camera video sequence, in *First International Work-shop on Video Processing for Security(VP4S).*

Rougier, C., Meunier, J., St-Arnaud, A. and Rousseau, J. (2011b). Robust video surveillance for fall detection based on human shape deformation, *IEEE Transactions on Circuits and Systems for Video Technology* **21**, 5, pp. 611–622.

Rougier, C., Meunier, J., St-Arnaud, A. and Rousseau, J. (2013). 3D head tracking for fall detection using a single calibrated camera, *Image Vision Comput.* **31**, 3, pp. 246–254.

Savran, A., Sankur, B. and Taha Bilge, M. (2012). Comparative evaluation of 3D vs. 2D modality for automatic detection of facial action units, *Pattern Recogn.* **45**, 2, pp. 767–782.

Shotton, J., Fitzgibbon, A., Cook, M., Sharp, T., Finocchio, M., Moore, R., Kipman, A. and Blake, A. (2011). Real-time human pose recognition in parts from single depth images, in *2011 IEEE Conference on Computer Vision and Pattern Recognition (CVPR)*, pp. 1297–1304.

SoftKinetic (2018). http://www.softkinetic.com/.

Soukoreff, R. W. and MacKenzie, I. S. (2004). Towards a standard for pointing device evaluation, perspectives on 27 years of Fitts' Law research in HCI, *Int. J. Hum.-Comput. Stud.* **61**, 6, pp. 751–789.

Suau, X., Ruiz-Hidalgo, J. and Casas, J. R. (2012). Real-time head and hand tracking based on 2.5 d data, *Multimedia, IEEE Transactions on* **14**, 3, pp. 575–585.

Tanie, H., Yamane, K. and Nakamura, Y. (2005). High marker density motion capture by retroreflective mesh suit, in *Proceedings of the 2005 IEEE International Conference on Robotics and Automation, 2005. ICRA 2005*, pp. 2884–2889.

Camera Mouse (2018). http://www.cameramouse.org/.

Thome, N., Miguet, S. and Ambellouis, S. (2008). A real-time, multiview fall detection system: A LHMM-based approach, *IEEE Trans. Circuits Syst. Video Technol.(TCSVT)* **18**, 11, pp. 1522–1532.

Tu, J., Tao, H. and Huang, T. (2007). Face as mouse through visual face tracking, *Computer Vision and Image Understanding* **108**, 1–2, pp. 35–40, special Issue on Vision for Human-Computer Interaction.

Wei, X., Zhang, P. and Chai, J. (2012). Accurate realtime full-body motion capture using a single depth camera, *ACM Trans. Graph.* **31**, 6, pp. 188:1–188:12.

Williams, M. and Kirsch, R. (2008). Evaluation of head orientation and neck muscle EMG signals as command inputs to a human-computer interface for individuals with high tetraplegia, *IEEE Transactions on Neural Systems and Rehabilitation Engineering* **16**, 5, pp. 485–496.

Yousefi, B., Huo, X., Veledar, E. and Ghovanloo, M. (2011). Quantitative and comparative assessment of learning in a tongue-operated computer input device, *IEEE Transactions on Information Technology in Biomedicine* **15**, 5, pp. 747–757.

Yu, M., Rhuma, A., Naqvi, S., Wang, L. and Chambers, J. (2012). A posture recognition-based fall detection system for monitoring an elderly person in a smart home environment, *IEEE Transactions on Information Technology in Biomedicine* **16**, 6, pp. 1274–1286.

Yu, M., Yu, Y., Rhuma, A., Naqvi, S., Wang, L. and Chambers, J. (2013). An online one class support vector machine based person-specific fall detection system for monitoring an elderly individual in a room environment, *IEEE Journal of Biomedical and Health Informatics*.

Yu, X. (2008). Approaches and principles of fall detection for elderly and patient, in *10th International Conference on e-health Networking, Applications and Services, 2008. HealthCom 2008* (IEEE), pp. 42–47.

Zhang, X. and MacKenzie, I. S. (2007). Evaluating eye tracking with ISO 9241 – part 9, in *Proceedings of HCI International* (Springer), pp. 779–788.

Zhu, Y., Dariush, B. and Fujimura, K. (2008). Controlled human pose estimation from depth image streams, in *Computer Vision and Pattern Recognition Workshops, 2008. CVPRW'08. IEEE Computer Society Conference on* (IEEE), pp. 1–8.

Chapter 13

Deep Learning in Gesture Recognition Based on sEMG Signals

Panagiotis Tsinganos,[a] Athanassios Skodras,[a] Bruno Cornelis,[b] and Bart Jansen[b]

[a]*Department of Electrical and Computer Engineering, University of Patras, Greece*
[b]*Department of Electronics and Informatics, Vrije Universiteit Brussel, Belgium*
{panagiotis.tsinganos, skodras}@ece.upatras.gr, {bcorneli, bjansen}@etrovub.be

Over the past years, Deep Learning methods have shown promising results to a wide range of research fields including image classification and natural language processing. Their increased success rates have drawn the attention of many researchers from various domains. This chapter investigates the application of Deep Learning methods to the problem of electromyography-based gesture recognition. A signal processing pipeline based on Deep Learning is presented through examples taken from the literature, whereas the details of state-of-the-art neural network architectures are discussed. In addition, this chapter illustrates a few ways adopted from image classification tasks that visualize what the neural network learns. Finally, new approaches are proposed and evaluated with publicly available datasets.

Learning Approaches in Signal Processing
Edited by Wan-Chi Siu, Lap-Pui Chau, Liang Wang, and Tieniu Tan

ISBN 978-981-4800-50-1 (Hardcover), 978-0-429-06114-1 (eBook)
www.panstanford.com

13.1 Introduction

Over the past decades, there has been particular interest in gesture recognition for human-computer interaction (HCI). This particular combination finds many applications, including sign language recognition, robotic equipment control, virtual reality gaming, and prosthetics control [1]. Among the various sensor modalities that have been used to capture hand gesture information, electromyography (EMG) is considered more appropriate since it captures the muscle's electrical activity; the physical phenomenon that results in hand gestures. EMG data can be recorded either with invasive or non-invasive methods. Surface electromyography (sEMG) is a technique that measures muscle's action potential from the surface of the skin, contrary to invasive methods that penetrate the skin to reach the muscle.

A popular approach to sEMG-based gesture recognition consists of using pattern recognition methods derived from Machine Learning (ML) [2]. Conventional ML pipelines include data acquisition, feature extraction, model definition, and inference. Acquisition of sEMG signals involves attaching one or more electrodes around the target muscle group. The features used for classification are usually hand-crafted by human experts and capture the temporal and frequency characteristics of the data. They serve as the input to ML classifiers, such as k-Nearest Neighbors (kNN), Support Vector Machines (SVM), Multi-Layered Perceptron (MLP), Linear Discriminant Analysis (LDA), and Random Forests (RF), where the classifier's parameters are adjusted towards accurate classification.

Deep Learning (DL) is a class of ML algorithms that has revolutionized many fields of data analysis [3]. For example, Convolutional Neural Networks (CNNs) and Recurrent Neural Networks (RNNs) were successfully deployed for image classification and speech recognition tasks respectively. DL methods differ from conventional ML approaches in that feature extraction is part of the model definition, therefore obviating the need for hand-crafted features. Although these methods are not new [3], they recently gained more attention due to the increased availability of abundant data

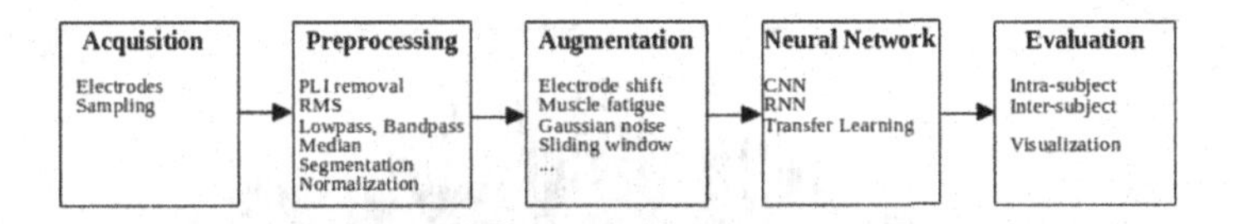

Figure 13.1 Flowgraph of an EMG-based gesture recognition system.

and vast improvements in computing hardware allowing these computationally demanding methods to be executed in less time.

In view of these advancements, the problem of sEMG-based gesture recognition has been formulated as a DL classification task. The main steps in a DL pipeline are shown in Fig. 13.1. The acquisition and preprocessing methods applied in this methodology are quite similar to classical EMG signal processing techniques. Compared to other methods, DL requires many data, which in the case of sEMG can be difficult to acquire. Therefore, an augmentation step is usually employed to create synthetic data from a small dataset. The most significant part in a DL pipeline is the model definition. Two types of neural networks are mostly used, CNNs and RNNs, from which different variants can be created or improved with Transfer Learning approaches. Finally, once the model is specified, the last step is the analysis of the network performance using evaluation metrics, whereas there exist methods that try to explain how neural networks make their decision.

This chapter discusses techniques of approaching the task of gesture recognition based on sEMG signals using DL. Firstly, data acquisition methods and parameters that affect the quality of the recorded signal are briefly presented. In Section 13.3 common practices for data preparation, including signal processing and data augmentation are discussed. Then, state-of-the-art neural network architectures are introduced, followed by the presentation of evaluation and network visualization techniques. Finally, future directions are discussed.

13.2 Data Acquisition

The detection of bioelectric signals, including EMG, is mostly based on bipolar electrodes placed over the region of interest.

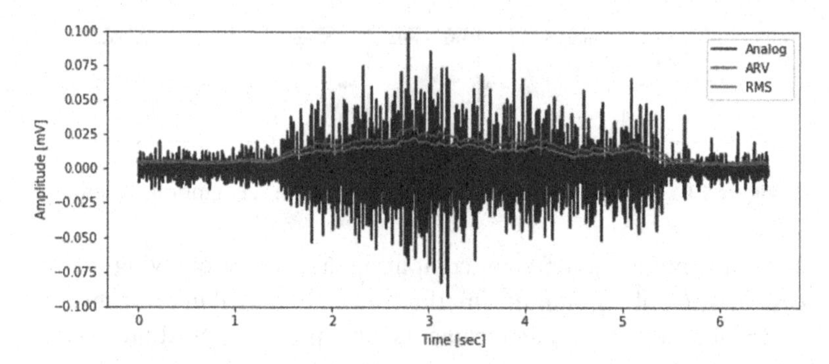

Figure 13.2 Example of a raw sEMG signal and its estimated amplitude envelope with ARV and RMS.

In the case of myoelectric signals, the electrodes are attached above the target muscle group and measure the generated signals while a movement is performed. Under this setup, the recorded signal provides information about the electrical activity of the muscle over time. However, when performing a gesture more than one muscles participate in different time instants, therefore multichannel detection is used, based on 1D or 2D electrode arrays that cover the forearm muscles. The spatially sampled signal can either be the analog surface potential or EMG features such as, Average Rectified Value (ARV), and Root Mean Square (RMS), estimated over a time interval [4]. An example of a single channel EMG surface potential and extracted features are shown in Fig. 13.2. Since the EMG frequency content lies in the range of 5–500 Hz, a sampling frequency of at least 1 kHz is required – yet, most frequency power is contained between 20 and 150 Hz

In the literature of EMG gesture recognition with Deep Learning, the analog surface potential or the RMS envelope are mostly preferred. The authors of [5] record data from a 1D electrode array that reports the RMS signal, whereas in [6, 7] a high-density electrode grid measures the instantaneous image of the analog myoelectric activity. Details on the acquisition practices found in literature are shown in Table 13.1.

From the generation in the muscle fibers to the detection by the electrodes, the EMG signal can be influenced by many factors. These are either related to the detection system or the

Table 13.1 Acquisition practices in literature

Ref.	Dataset	Sampling (Hz)	Electrodes	Feature
[5]	Ninapro-DB1	100	$1 \times 8 + 2$	RMS
[5]	Ninapro-DB2	2000	$1 \times 8 + 4$	Analog
[6]	CapgMyo	1000	8×16	Analog
[7, 8]	CSL-HDEMG	2000	7×24	Analog
[9]	Myo	200	1×8	Analog

physiological characteristics of the muscles. In the first category, the environmental noise (e.g., power line interference) and the noise induced by the electronics are the most important parameters that affect the quality of the recorded signal. To lessen their effect, proper skin preparation (e.g., rubbing with abrasive conductive paste) and careful instrumentation amplifier design are usually proposed. In addition, capacitive electrodes have shown to reduce the recorded noise in long-term monitoring applications, compared to bipolar electrodes that require a galvanic contact with the body [4]. The most important factors from the second category are the thickness of the tissue separating the muscle from the electrode, the cross-talk from other muscles and biomedical signals (e.g., ECG), as well as changes to muscle geometry due to movement [10]. Some of the induced artifacts by these inherent factors can be removed either with careful electrode arrangement (e.g., taking into consideration the possible muscle translation due to dynamic joint movement) or signal processing methods (e.g., ECG cross-talk reduction) [10].

13.3 Data Preparation

13.3.1 EMG Preprocessing

Basic filtering methods applied in amplitude and spectral characteristics of muscle activity are part of the data preparation step of a DL pipeline as well. These steps are modified depending on which EMG signal characteristic is used for the classification task. Common preprocessing steps for sEMG signals are shown in Table 13.2.

When the EMG amplitude is used for the gesture classification, the main steps are noise reduction, rectification, and smoothing.

Table 13.2 Data preparation practices in literature

Ref.	Preprocessing	EMG characteristic
[5]	PLI[1] removal, RMS, low-pass	RMS
[6, 7]	PLI removal, median, normalization	raw
[9]	(not mentioned)	spectrogram
[12]	PLI removal	spectrogram

[1] PLI: Power-Line Interference.

The aim of noise reduction is to eliminate additive noise, artifacts, and power-line interference that contaminate the EMG signal [4]. The next step computes the absolute value of each EMG sample, which makes the computation of amplitude parameters, like the mean, feasible (raw EMG signals have an average value of zero) [10]. Finally, smoothing is applied to extract the amplitude estimate. There are mainly two methods encountered in the literature. The first algorithm simply computes the moving average of the signal (i.e., the ARV), whereas the second is the RMS. Both methods require the selection of an appropriate window length over which the signal is considered stationary. Typically, windows of 100 ms to 200 ms are used [4, 10]. Smoothing can also be performed by low-pass filtering, where second-order or higher Butterworth filters with cutoff frequencies between 1 Hz and 6 Hz are common [10]. In [5], the RMS value is calculated from 200 ms windows and an additional low-pass filtering is performed with a second-order Butterworth filter with 1 Hz cut-off, whereas the authors of [7] employ a non-linear low-pass filter (median filtering) to the instantaneous EMG maps.

Apart from EMG amplitude estimates, spectral representations based on the Short Time Fourier Transform (STFT) and Wavelet transforms have also been used. In the first case, the spectrogram of EMG segments is computed using the STFT, so that from a single EMG channel a 2D signal is generated with the two axes representing the time and the frequency. The window length and overlap used for the calculation of the STFT are important parameters that affect the time and frequency resolution. On the other hand, utilizing the Wavelet Transform allows for a multi-resolution analysis which is well suited for non-stationary signals. In that case, the Mexican hat wavelet is

Table 13.3 Data augmentation practices in literature

Ref.	Augmentation	Dataset[1]
[13]	Time warping	ECG
[5]	Additive noise	EMG
[6, 7]	Electrode shift	EMG
[14]	Permutation, Scaling, Magnitude warping, Sliding window	IMU
[9]	Muscle Fatigue, Electrode Displacement	EMG

[1] ECG: Electrocardiogram, EMG: Electromyogram, IMU: Inertial Measurement Unit.

preferred since it best approximates the action potential generated by the muscles [11]. The authors of [12] calculate the spectrogram of each 200 ms (400 samples) segment with a 256 samples window, whereas in [9] it is calculated with a 28 samples window for EMG segments of 260 ms (52 samples) duration.

13.3.2 Data Augmentation

One of the prerequisites for training a Deep Neural Network with good generalization properties is the availability of a huge dataset. Although there are a few publicly available sEMG datasets, direct use of these data is not always feasible since the acquisition system and the preprocessing steps vary. Therefore, data augmentation techniques have been applied to increase the size of the training set. In the domain of image classification, additive Gaussian noise and affine transformations, like rotation and translation, have resulted in better generalization [3]. However, these methods are not directly applicable to sEMG signals.

In order to augment time-series signals, a few data augmentation approaches have been proposed (Table 13.3). These include additive noise, sliding window, permutation, shift, warping, and scaling. Additive noise and sliding window have successfully been applied to sEMG data [5, 9], whereas the applicability of permutations and warping to EMG signals has not been verified. These methods are presented next and their relation to EMG data is briefly discussed. Examples of data augmentation applied to sEMG signals can be seen in Fig. 13.3.

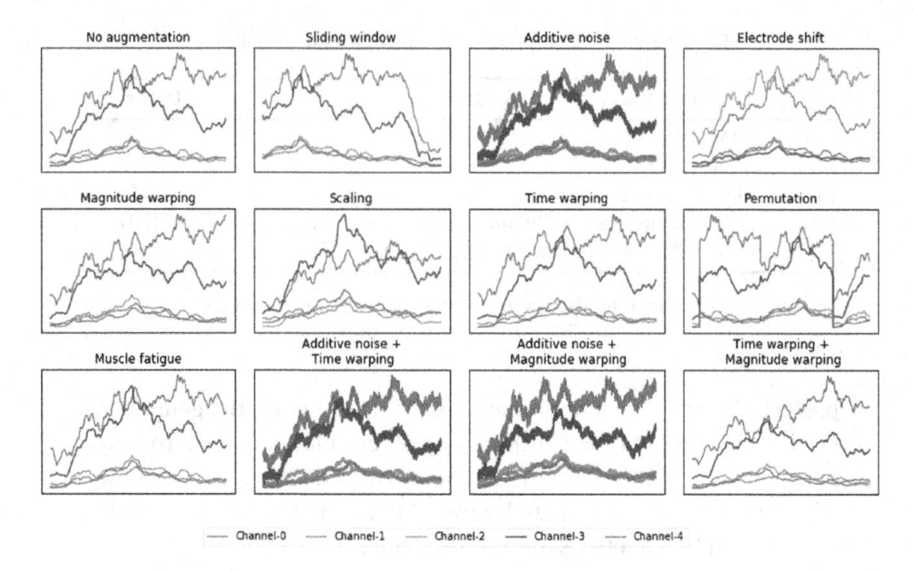

Figure 13.3 Examples of data augmentation techniques applied on sEMG signal.

Considering that training examples need to have a set duration, a *sliding window* can be applied to the EMG signal to generate these examples. Using overlapping windows (with constant or random overlap) is similar to the translation operation applied to images [9, 14].

Similarly to noise augmentation found in image classification, adding Gaussian noise with a defined signal to noise ratio has been used. This *additive noise* can simulate the noise found in EMG recordings without altering the class of the performed gesture.

Augmenting the training data by randomly shifting the channels of the EMG signal can simulate *electrode shift* that may happen between recording sessions [6]. Additionally, electrode displacement due to muscle movement can be emulated by shifting part of the power spectrum from one channel to an adjacent one [9].

Accounting for the fact that during movement repetitions the EMG amplitude and duration cannot be replicated (due to the stochastic generation of action potentials) [10], perturbing the location of sEMG samples (*time-warping*) or their value (*magnitude-warping*) is a way to augment sEMG data. To this effect, a smooth random curve can be used to either distort the time intervals between the samples or change their value through multiplication [14]. In addition, the magnitude of the sEMG samples can be altered by multiplication with a random scalar (*scaling*).

Another augmentation technique is to simulate *muscle fatigue*. The main effect of muscular fatigue on the EMG signal is the compression of the power spectrum. As a result, the amplitude of lower frequencies is increased, whereas higher frequencies are attenuated. Therefore, this method can be easily implemented in the frequency domain, where part of the power of a frequency bin is redistributed to adjacent lower frequency bins [9].

Finally, perturbations to the location of the samples can be performed by dividing the signal into smaller slices and then permuting them [14]. However, the *permutation* method cannot be considered appropriate for sEMG data, since the generated signal does not resemble a "true" sEMG signal.

13.4 Network Architectures

sEMG-based hand gesture recognition can be formulated either as an image classification problem using Convolutional Neural Networks (CNNs), where the input sEMG image has a size of $H \times W \times 1$ (height×width×depth) or as a sequence classification task using Recurrent Neural Networks (RNNs). In the first case, various approaches have been employed to construct an sEMG image, among which (i) the instantaneous sEMG signals from a 2D electrode array (where the width and the height of the array match the dimensions of the image) [6, 7, 15], (ii) segments of sEMG signals using time-windows (where the width matches the number of electrodes and the height is equal to the window length) [5], and (iii) spectrograms or Wavelet transformations of sEMG segments (where for each channel of the EMG a transformation is applied resulting in a time-frequency-space representation) [9, 12]. On the other hand, there are no works utilizing RNNs, probably due to the extensive use of CNNs, which makes the application of recurrent networks a possible research subject.

13.4.1 CNN Architectures

CNN models have been widely used for image classification problems, since the neurons (filters) are arranged in three dimensions (height, width, depth) similar to images. A typical CNN architecture consists of convolutional layers followed by a non-linear activation function that are stacked many times. The operation performed by a convolutional layer (assuming zero bias) can be described by the equation:

$$z[i, j, n] = \sum_{m=0}^{N_{in}-1} \sum_{l=\lfloor -f_h/2 \rfloor}^{\lfloor f_h/2 \rfloor} \sum_{k=\lfloor -f_w/2 \rfloor}^{\lfloor f_w/2 \rfloor} x[i-k, j-l, m] \times h_n[k, l, m], \tag{13.1}$$

where $n \in [0, N_{out} - 1]$, z is the output feature map, x the input, h_n the nth filter of the layer, N_{in} the depth of the input, N_{out} the number of filters, and f_h / f_w the height/width dimension of the filter. Many activation functions have been proposed in the literature, yet the rectified linear unit (ReLU) and its variants are preferred. The equation for this function is given by

$$ReLU(x) = \max\{0, x\}. \tag{13.2}$$

In addition, pooling layers are employed to reduce the spatial dimensions and therefore decrease the number of learned parameters. Pooling operates independently on every depth slice of the input and resizes it spatially, using either the *max* or *average* operation. Finally, a few fully connected layers compute the class scores, while a softmax activation function, $\sigma(x)$, normalizes these values into class probabilities as follows:

$$\sigma(\mathbf{z})_n = \frac{e^{z_n}}{\sum_{k=0}^{N-1} e^{z_k}}, \quad n \in [0, N-1], \tag{13.3}$$

where $\mathbf{z} = [z_0, \ldots, z_{N-1}]$ is the output of the last fully connected layer, and N the number of the output units. CNNs are trained similar to regular neural networks. A loss function is defined and the neuron parameters are optimized using the back-propagation algorithm such that the value of the loss function is minimized. Stochastic gradient descent (SGD) is the basic algorithm used for the optimization, while many improved versions of it have been proposed. The choice of the loss function to minimize depends on

Table 13.4 CNN architectures used in EMG-based gesture recognition

Ref.	Input dimensions (height×width×depth)[1]	Layers	Layer types[2]	Parameters
[5]	15×10 $(t \times c)$	5	CONV, POOL	84,853[3]
[6, 7]	1×10 $(t \times c)$	8	CONV, LC, FC	555,897[3]
[15]	20×10 $(t \times c)$	44	CONV, LC, FC	2,717,877[3]
[9]	$8 \times 14 \times 4$ $(c \times f \times t)$	8	CONV, FC	71,825[4]

[1] t: time, c: electrode channels, f: frequency.
[2] CONV: Convolutional, POOL: Average Pooling, FC: Fully connected, LC: Locally connected.
[3] Calculated for the Ninapro DB1 with $\#G = 53$ gestures.
[4] Calculated for the Ninapro DB5 with $\#G = 53$ gestures.

the problem: For the classification between two (many) classes, the binary (categorical) cross-entropy is used, whereas for regression problems the mean squared error is preferred.

In the literature of EMG-based gesture recognition, the above principles are followed. What affects the number of the layers and the size of the filters the most is the dimensions of the input image. Usually, the first two dimensions correspond to the size of the 2D electrode grid, whereas in the case of 1D electrode array, one of the width and height corresponds to time and the other to the size of the array. Regarding the depth dimension, it is typically set to 1. Table 13.4 summarizes the main characteristics of these architectures. The filter sizes and parameter values refer to the publicly available Ninapro dataset [16], which contains sEMG recordings of subjects performing more than 50 gestures (Table 13.6). In the subsequent analysis, in case the depth dimension is omitted it is considered to be equal to 1 and the symbol $\#G$ denotes the number of class labels.

The network architecture used in [5] comprises five convolutional blocks, where each one contains a convolutional layer and an activation function. The activation function for the last block is a softmax, whereas for the rest a ReLU is used. In addition, an average pooling layer is used after the second and third blocks. The filter sizes of the convolutions are 1×10, 3×3, 5×5, 5×1 and 1×1, while the pooling is performed with 3×3 filters. Regarding the number of filters in each block, the model computes 32, 32, 64, 64 and $\#G$ filters respectively. In this architecture, the convolution at the output layer is equivalent to the dot product performed by a fully connected layer. The output of the fourth convolution has a size of (1, 1, 64). Thus, substituting $N_{in} = 64$, $N_{out} = \#G$, $f_h = f_w = 1$ into Equation (13.1) results in

$$y[0, 0, n] = \sum_{m=0}^{d-1} x[0, 0, m] \times h_n[0, 0, m] = \boldsymbol{x} \cdot \boldsymbol{h_n}, \quad n \in [0, N_{out} - 1],$$

which is equal to the dot product performed by a fully connected layer.

The convolutional network of [7] and [6] consists of eight blocks. The first two are convolutional layers, each of which consists of 64 3×3 filters. The next two blocks use locally connected layers with 64 filters of size 1×1. The next three are fully connected layers consisting of 512, 512, and 128 units, respectively. The network ends with a $\#G$-way fully connected layer and a softmax function. After each layer a batch normalization [17] and a ReLU non-linearity are used. In addition, dropout [18] with a probability of 0.5 is inserted after the fourth, fifth and sixth blocks to reduce overfitting. An improvement on this architecture is presented in [15], where the input image is divided into M equally sized patches and processed separately. Each one of these segments is the input to one of M subnetworks consisting of the first four blocks of [7]. The output feature maps of these networks are concatenated into a single feature map and the last fully connected layers follow. Among the various configurations used to divide the initial image, the strategy that creates patches equal to the number of electrode channels (i.e., an $L \times 10$ $(t \times c)$ image divided into 10 patches of size $L \times 1$) performs best.

Finally, the model proposed in [9] is similar to the improvement proposed in [15] since the input is divided into smaller patches. However, the input is a time-frequency representation of sEMG segments arranged into an image of size $(c \times f \times t)$. Another difference is that a parametric ReLU (PReLU) activation function is applied to the initial input before the separation. Each patch is processed by two blocks consisting of a convolutional layer, a batch normalization [18] and a parametric exponential linear unit (PELU) activation. The convolutional layers use 12 filters with a size of 4×3 and 24 filters with a size of 3×3, respectively. The output feature maps are concatenated in an element-wise summation fashion and a convolutional block of 24 filters of size 3 × 3, follows. Then, three fully connected layers of 100, 100 and $\#G$ units come before the final softmax activation. This model is also augmented by Transfer Learning techniques since the trained network exploits information learned during a pre-training stage.

13.4.2 RNN Architectures

RNNs have recently become popular due to successful applications in problems related to sequential data [19]. Unlike feedforward networks, RNNs exhibit memory properties, which makes them suitable for processing sequence of values. Among the different variants of RNNs, Long Short Term Memory (LSTM) [20] networks are widely used as a result of their capability to learn long-term dependencies without suffering from vanishing/exploding gradient problems.

LSTMs belong to the category of gated RNNs which use gates (a sigmoid activation function followed by pointwise multiplication) to create paths through time that have derivatives that neither vanish nor explode [3]. The basic model of an LSTM cell is shown in Figure 13.4. Compared to an RNN, it contains three gating units ($\boldsymbol{f_t}$, $\boldsymbol{i_t}$, and $\boldsymbol{o_t}$) that control the flow of information as follows [21]: (i) the forget gate $\boldsymbol{f_t}$ decides how much of the previous cell state ($\boldsymbol{C_{t-1}}$) information to retain, (ii) the input gate $\boldsymbol{i_t}$ determines the amount of new information ($\boldsymbol{\tilde{C}_t}$) to store into the cell state ($\boldsymbol{C_t}$), and (iii) the output gate $\boldsymbol{o_t}$ decides what part of the cell state will go through the output ($\boldsymbol{h_t}$).

Although the temporal nature of sEMG signals suggests that an LSTM network should be able to map a sequence of sEMG values to a gesture label, no approach in the literature exploits these properties. A possible explanation is that recurrent networks require a higher memory bandwidth compared to CNN models. In addition, it has been shown recently [22] that convolutional networks that take into account temporal information can surpass LSTMs in a wide range of tasks. Nevertheless, in this section we try to approach the problem of sEMG-based gesture recognition using a simple LSTM architecture. This consists of one LSTM layer with 128 cells followed

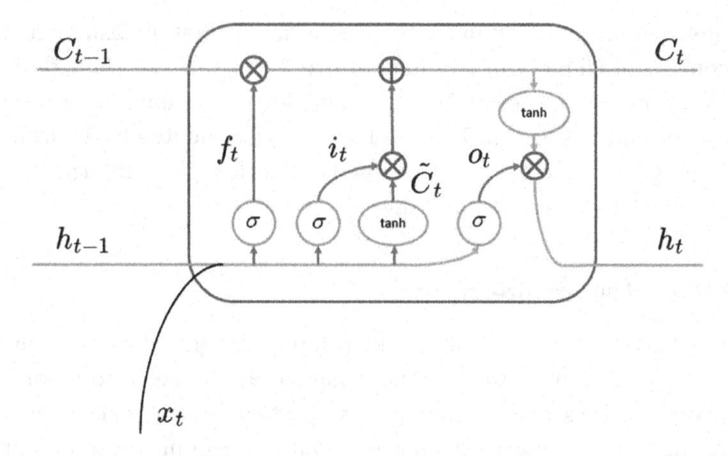

Figure 13.4 Basic model of an LSTM cell (image taken from https://chunml.github.io/ChunML.github.io/project/Creating-Text-Generator-Using-Recurrent-Neural-Network/).

Table 13.5 Simple LSTM architecture for EMG-based gesture recognition

Input dimensions (time × channels)	LSTM cells	Layer types[1]	Parameters
200 × 10	128	LSTM, FC	78,005[2]
200 × 10	128, 128	LSTM, FC	209,589[2]

[1]FC: Fully connected.
[2]Calculated for the Ninapro DB1 with $\#G = 53$ gestures.

by a fully connected layer with $\#G$ units. To investigate whether the addition of more LSTM layers improves performance, another variant of this architecture with two stacked LSTM layers, having 128 cells each, is evaluated. In addition, considering that all the training examples within a batch should have the same dimensions, the sEMG sequences are zero-padded to the longest gesture duration of the dataset. Table 13.5 summarizes the main characteristics of these architectures.

Similar to CNNs, after model definition, a loss function is defined and the RNN parameters are optimized using back-propagation algorithm.

13.5 Network Evaluation

13.5.1 Evaluation Schemes

Once the neural network is defined, the capacity of the model to learn from the data needs to be evaluated. At this step, the training dataset is partitioned into train and validation sets according to an evaluation scheme. In the literature of sEMG-based gestures recognition, two approaches are followed [6]: (i) intra-subject, and (ii) inter-subject evaluation.

In the first case, data recorded from one subject are partitioned into train and validation sets. The network is optimized on the train data and the accuracy on the validation set is measured. This procedure is repeated for all the subjects available in the dataset and an average performance is reported. Usually, during a recording session, each gesture is repeated more than once, therefore the data can be partitioned based on the gesture repetitions. For example, in [5] repetitions 2, 5, and 7 comprise the validation set, while the rest (repetitions 1, 3, 4, 6, 8, 9, and 10) are used for training. Additionally, a cross-validation scheme can be applied for each subject, where one repetition is used for validation and the rest for training [6]. On the other hand, the inter-subject evaluation follows a leave-one-subject-out cross-validation, where the validation set consists of all the repetitions of one subject, while for training the repetitions of all the other subjects are used [6]. This is repeated until the recordings of all the subjects have been in the validation set.

These evaluation approaches are not substitute of one another, rather they should both be applied since they provide different insights about the network. With an intra-subject evaluation, one obtains a metric that quantifies the ability of a network to generalize to unseen data from the same subject it was trained on. On the other hand, an inter-subject evaluation is useful for the assessment of methods that tackle subject variability problems (e.g., electrode shift, fatigue), as well as measure the ability of the network to generalize when the source and target distributions differ (Domain Adaptation, Transfer Learning).

In Section 13.5.3 an application of these schemes is presented with respect to the Ninapro-DB1 dataset.

13.5.2 Learning Visualization

Although deep neural networks exhibit state-of-the-art performance in many tasks, it is difficult to understand why some models perform better than others or how one architecture can be improved. Recently, many researchers have made progress in opening the black box and various neural network interpretation methods have been proposed [23–26]. For example, the authors of [23] create representations that show what input yields a given activation, while by occluding parts of the input they reveal which regions are important for the classification. While these methods concentrate on explaining what a CNN has learned, similar techniques can be applied to RNNs. In [19], it is demonstrated that LSTM cells can learn to count the number of characters in a sentence, whereas cells that retain their state for long time periods can be located by examining the activation of the forget gates.

Considering how valuable these insights are for image classification and language processing tasks, it is expected that similar outcomes can be achieved for the classification of sEMG recordings of hand movements. Therefore, in this section an attempt is made to explain the results of some of the aforementioned methods with regard to sEMG data and gesture recognition.

The simplest way to interpret what a neural network has learned is to visualize the network weights and particularly the filters of a CNN (Figure 13.5). This helps to identify interesting patterns that the network tries to match in the input. Another benefit of this visualization is that it shows whether the network has converged (there is some structure in the filters) or not (the filters look random). Considering sEMG signals, the interpretation depends on what each input dimension corresponds to. For example, if the width and height dimensions are matched to a 2D electrode grid, then a

Figure 13.5 The 32 learned 1×10 filters of the first convolutional layer of AtzoriNet [5]. Lighter color shades correspond to higher values. From bottom to top, the values correspond to channels 1 to 10.

3×3 filter may indicate the direction of muscle action potentials. In case of $t \times c$ input, a filter can detect the electrodes' transition from negative to positive values.

Another visualization is occlusion sensitivity [23], which consists of observing classification scores while occluding (i.e., setting the values to a constant) parts of the input. As a result, important regions for the classification of a specific image, yield a lower accuracy when occluded. In the case of sEMG inputs, an occlusion map can be generated if each electrode channel ($t \times c$ input) or rows/columns of electrodes (2D grid input) are occluded one at a time. This can reveal whether for a given gesture the network assigns more weight to a specific electrode. In addition, for gestures based on the activation of a few muscles, this method can verify whether the network has learned this correlation or not. An example of occlusion map can be seen in Figure 13.6.

Sensitivity analysis based on saliency maps [27] has been widely used for interpreting neural networks. In this method, partial derivatives measure how much small local changes in the pixel value affect the network output. Therefore, it can be used to highlight input regions that cause the most change in the output. It is expected that these pixels correspond to the object location in the image.

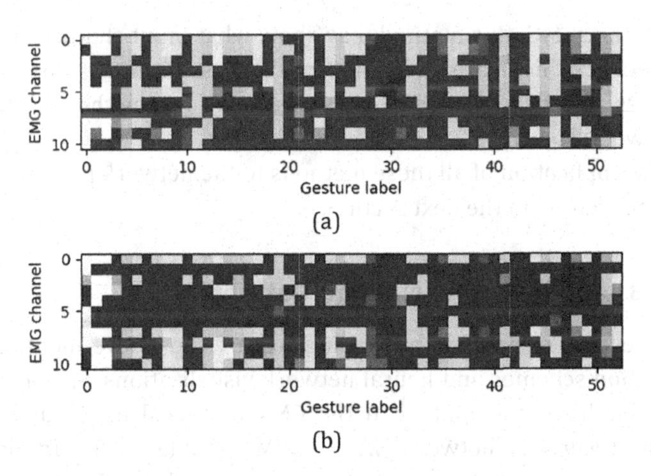

Figure 13.6 Occlusion sensitivity for the AtzoriNet [5]. The first row is the accuracy per gesture without occlusion. Subsequent rows correspond to the occlusion of each channel. Lighter color shades denote higher accuracy.

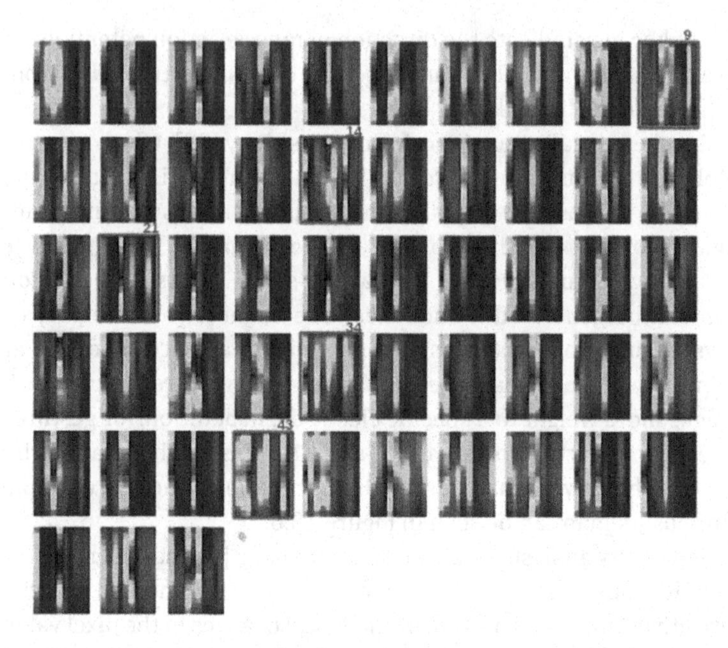

Figure 13.7 Saliency maps for the middle segment of each gesture. From top to bottom and left to right, the images correspond to gesture labels 0 ("rest") to 52.

Regarding sEMG, this method may show what input characteristics the network thinks are important for the classification of the input sEMG as a specific gesture. The application of this method to sEMG is shown in Figure 13.7.

The application of all these methods to the network proposed in [5] is presented in the next section.

13.5.3 Application to Ninapro Dataset

This section comprises an application of the above-mentioned evaluation schemes and neural network visualizations. Specifically, these methods are applied to the CNN proposed in [5] and the simple recurrent network with LSTMs (Table 13.5). In addition, the augmentation techniques presented in Section 13.3 are compared using a Wilcoxon signed-rank test. The neural networks are implemented using the Keras [28] API for

Table 13.6 The Ninapro dataset

Dataset	Subjects	Movements (#G)	Electrodes	Sampling (Hz)
DB1	27	53	$1 \times 8 + 2$	100
DB2	40	50	$1 \times 8 + 4$	2000
DB3	11	50	$1 \times 8 + 4$	2000
DB4	10	53	$1 \times 8 + 4$	2000
DB5	10	53	2×8	200

Python, whereas the CNN visualizations are made with the keras-vis [29] tool. The implementation code is available at https://github.com/DSIP-UPatras/DSP_EMGDL_Chapter

The dataset used for the following experiments is the Ninapro-DB1 (Table 13.6). In this dataset, sEMG signals are captured with an 1×10 array of electrodes, where the first 8 electrodes are equally spaced around the forearm, and the other two are placed on the main activity spots of finger flexor and extensor muscles respectively [30]. For the intra-subject evaluation of the network proposed in [5] (AtzoriNet) and the simple LSTM of Table 13.5 (LstmNet), repetitions 1, 2, 4, 6, 8, and 9 are used for training, while repetitions 3, and 10 for validation. On the other hand, the inter-subject evaluation is performed only for the AtzoriNet using repetitions 4, and 6 for training, while repetitions 1, 2, 3, 4, 6, 8, 9, and 10 for validation. In addition, the following augmentation techniques are compared: additive noise (AN), time-warping (TW), magnitude-warping (MW), and electrode shift (ES).

Table 13.7 shows the intra-subject evaluation for the AtzoriNet. In addition to the classification accuracy of a single sEMG segment (i.e., $15 \times 10 \times 1$ image), the accuracy of classifying a complete gesture, measured as the majority voting of the corresponding segment predictions, is reported in parentheses. As we can see there is a big difference between the two classification metrics. The reason is that during the recording there is a gradual transition between rest, gesture and rest, in contrast to the discrete changes of the gesture labels. Consequently, accuracy is lower during the transition periods where the change in movement is not yet clearly evident from the input EMG signal [16]. In addition, the small temporal length of the input (150 ms) contributes to this discrepancy, since

Table 13.7 Intra-subject evaluation for the network of Atzori [5]

Augmentation	Validation accuracy[1]	Wilcoxon test[2]
none	0.6027 (0.8637)	—
AN	0.6200 (0.8812)	$p < 0.01$
TW	0.6320 (0.8756)	$p < 0.01$
MW	0.6331 (0.8735)	$p < 0.01$
AN-TW-MW	0.6196 (0.8753)	$p < 0.01$
		$(p > 0.01, p < 0.01, p < 0.01)$

[1] The value in parentheses is the average accuracy using voting over the duration of gesture.
[2] p-value of the Wilcoxon test. The values in parentheses correspond to the following comparisons: AN and AN-TW-MW, TW and AN-TW-MW, MW and AN-TW-MW.

during the transitions many of these segments look similar between different gestures.

Regarding the comparison of the augmentation schemes, a simple additive Gaussian noise improves the accuracy by almost 2%, whereas the addition of TW and MW do not provide any statistical improvement. However, when only TW or MW is applied, the accuracy improves by 3% compared to training without augmentation. This indicates that augmentation used in other related tasks such as, motion recognition with IMUs, may be helpful for the classification of sEMG. Therefore, different combinations of these methods without AN should be examined.

More insight about the network is gained with the following visualizations. A filter visualization of the first layer can show how the network looks at the raw data. In AtzoriNet the first layer contains 1×10 filters, i.e., filters that look at the relation between the sEMG channels. From Figure 13.5, we can see that there are filters that favor specific electrode combinations, whereas others attribute almost equal weights to each electrode channel. In addition, the structure of the filters shows that the network has converged.

While the filter weights show what the network searches for in the input, an analysis of the occlusion sensitivity should reveal which portion of the input contains important information for a specific class. Figure 13.6 shows the occlusion sensitivity map generated by occluding (setting to '0' or '1') the sEMG values of each channel one at a time. Occlusion with '0' of channels 9 and 10 has a negative impact to the classification of almost all the gestures, whereas less gestures are affected when the occlusion is performed with '1'. This behavior is expected since the sEMG signal values in these regions indicate a high muscle activity for a wide range of gestures. An interesting observation can be made about the occlusion of electrodes 1-6. An occlusion with '0' of these electrodes has a smaller effect than the occlusion with '1'. This indicates that the values from these electrodes are close to zero.

The last visualization we will investigate is the saliency analysis. This method highlights the input regions that when their values is slightly changed, the class labels is affected the most. Figure 13.7 shows the average saliency map for the middle 1500 ms segment (i.e., average of 10 segments) of each gesture. Only the middle segment of the gestures is used in order to avoid any misinterpretations from the analysis of transient periods. We can observe that almost all the gestures are influenced by the changes of the values of the first electrodes, whereas there are hand movements (e.g., '14') where for adjacent channels the behavior of the network is different. Furthermore, only a few gestures (those with labels '9', '21', '34', '43') are greatly affected by changes in the last two channels. This means that the electrodes at the main activity spots of the finger flexors/extensors capture useful information for the correct classification of these gestures.

The inter-subject evaluation for the AtzoriNet is given in Table 13.8. It is obvious from the results that training a single network with data from different subjects is indeed very difficult. Therefore, techniques based on transfer learning and domain adaptation can improve the performance of the network in this scenario. For example, in [6] the parameters of the Batch Normalization layers are adapted to each specific user during a pre-training phase with unlabeled data, whereas in [9], a source network trained with all available data shares information with user specific networks.

The evaluation of the LstmNet is shown in Table 13.9. The intra-subject validation accuracy for two versions of the LstmNet and with/without augmentation, is given. In every case, the classification accuracy is pretty low compared to what is achieved by a CNN. The main reason why the accuracy of the LstmNet is low, is the fact that the classification is applied only after the last timestep. Therefore, architectures that take into account the output at each

Table 13.8 Inter-subject evaluation for the network of Atzori [5]

Augmentation	Validation accuracy[1]	Wilcoxon test[2]
AN	0.0754 (0.0995)	—
AN-ES	0.0848 (0.1068)	$p < 0.01$

[1] The value in parentheses is the average accuracy using voting over the duration of gesture.
[2] p-value of the Wilcoxon test.

Table 13.9 Intra-subject evaluation for the simple RNN of Table 13.5

Model	Augmentation	Validation accuracy	Wilcoxon test[1]
LSTM - 1 layer	none	0.0855	—
LSTM - 1 layer	AN-TW-MW	0.1879	$p < 0.05$
LSTM - 2 layers	none	0.1109	—
LSTM - 2 layers	AN-TW-MW	0.1028	$p < 0.05$

[1] p-value of the Wilcoxon test.

timestep (e.g., attention model) should perform better. Nevertheless, the augmentation has a positive impact on the single layer LSTM, by almost 10%, which should be expected considering the small size of the training dataset. In addition, there is a small performance improvement when using a deeper LSTM. However, the performance is reduced when the same augmentation is applied.

13.6 Conclusions

In this chapter, we have presented a basic Deep Learning methodology for sEMG-based gesture recognition. We started from the basics of sEMG acquisition and the parameters that influence the quality of the signal, to data preparation techniques, including signal processing and data augmentation. Typical neural network architectures based on convolutional and recurrent layers were presented, while their application to gesture recognition was discussed. Methods to evaluate the performance of a neural network were explained. Based on results from the domain of image classification, network visualization techniques that try to shed light on the inner-workings of neural networks were proposed. Finally, the steps of this methodology were illustrated by means of two examples applied on a publicly available dataset.

The evaluation of the CNN and the RNN models in Section 13.5 provided useful insights into the performance of these two types of networks. In particular, it was experimentally shown that under the same optimization settings, CNNs are more suitable for sEMG-based gesture recognition. In addition, the visualizations for the CNN model have proven to be a useful tool, since they help explain the decisions made by the network during inference. Therefore, we believe they should become an integral part of the DL pipeline for hand movement classification.

References

1. M. J. Cheok, Z. Omar, and M. H. Jaward, A review of hand gesture and sign language recognition techniques, *International Journal of Machine Learning and Cybernetics*, Aug 2017.
2. E. Scheme and K. Englehart, Electromyogram pattern recognition for control of powered upper-limb prostheses: State of the art and challenges for clinical use, *The Journal of Rehabilitation Research and Development*, vol. 48, no. 6, pp. 643–659, 2011.
3. I. Goodfellow, Y. Bengio, and A. Courville, *Deep Learning*, MIT Press, Cambridge, MA, 2016.
4. R. Merletti and D. Farina, *Surface Electromyography: Physiology, Engineering, and Applications*, John Wiley & Sons, Inc., Hoboken, New Jersey, Apr 2016.
5. M. Atzori, M. Cognolato, and H. Müller, Deep learning with convolutional neural networks applied to electromyography data: A resource for the classification of movements for prosthetic hands, *Frontiers in Neurorobotics*, vol. 10, Sep 2016.
6. Y. Du, W. Jin, W. Wei, Y. Hu, and W. Geng, Surface EMG-based inter-session gesture recognition enhanced by deep domain adaptation, *Sensors*, vol. 17, no. 3, Feb 2017.
7. W. Geng, Y. Du, W. Jin, W. Wei, Y. Hu, and J. Li, Gesture recognition by instantaneous surface EMG images, *Scientific Reports*, vol. 6, no. 36571, Nov 2016.

8. C. Amma, T. Krings, J. Böer, and T. Schultz, Advancing muscle-computer interfaces with high-density electromyography, in *Proceedings of the 33rd Annual ACM Conference on Human Factors in Computing Systems*, New York, NY, USA, 2015, CHI '15, pp. 929–938, ACM.
9. U. Côté-Allard, C. Latyr Fall, A. Drouin, A. Campeau-Lecours, C. Gosselin, K. Glette, F. Laviolette, and B. Gosselin, Deep learning for electromyographic hand gesture signal classification by leveraging transfer learning, *ArXiv e-prints*, Jan 2018.
10. P. Konrad, The ABC of EMG, *A practical introduction to kinesiological electromyography*, vol. 1, pp. 30–35, 2005.
11. M. B. I. Reaz, M. S. Hussain, and F. Mohd-Yasin, Techniques of EMG signal analysis: detection, processing, classification and applications, *Biological Procedures Online*, vol. 8, no. 1, pp. 11–35, Dec 2006.
12. X. Zhai, B. Jelfs, R. Chan, and C. Tin, Self-recalibrating surface EMG pattern recognition for neuroprosthesis control based on convolutional neural network, *Frontiers in Neuroscience*, vol. 11, pp. 379–390, Jul 2017.
13. A. Le Guennec, S. Malinowski, and R. Tavenard, Data augmentation for time series classification using convolutional neural networks, in *Proceedings of 2nd ECML/PKDD Workshop on Advanced Analytics and Learning on Temporal Data*, Riva del Garda, Italy, Sep 2016.
14. T. T. Um, F. M. J. Pfister, D. Pichler, S. Endo, M. Lang, S. Hirche, U. Fietzek, and D. Kulić, Data augmentation of wearable sensor data for Parkinson's disease monitoring using convolutional neural networks, in *Proceedings of the 19th ACM International Conference on Multimodal Interaction*, New York, NY, USA, 2017, ICMI 2017, pp. 216–220, ACM.
15. W. Wei, Y. Wong, Y. Du, Y. Hu, M. Kankanhalli, and W. Geng, A multi-stream convolutional neural network for sEMG-based gesture recognition in muscle-computer interface, *Pattern Recognition Letters*, Dec 2017.
16. M. Atzori, A. Gijsberts, I. Kuzborskij, S. Elsig, A. G. M. Hager, O. Deriaz, C. Castellini, H. Müller, and B. Caputo, Characterization of a benchmark database for myoelectric movement classification, *IEEE Transactions on Neural Systems and Rehabilitation Engineering*, vol. 23, no. 1, pp. 73–83, Jan 2015.
17. S. Ioffe and C. Szegedy, Batch Normalization: Accelerating deep network training by reducing internal covariate shift, *ArXiv e-prints*, Feb 2015.
18. N. Srivastava, G. Hinton, A. Krizhevsky, I. Sutskever, and R. Salakhutdinov, Dropout: A simple way to prevent neural networks from overfitting, *Journal of Machine Learning Research*, vol. 15, pp. 1929–1958, 2014.

19. A. Karpathy, J. Johnson, and L. Fei-Fei, Visualizing and understanding recurrent networks, *ArXiv e-prints*, Jun 2015.
20. S. Hochreiter and Schmidhuber U., Long short-term memory, *Neural Computation*, vol. 9, no. 8, pp. 1735–1780, 1997.
21. C. Colah, Understanding LSTM networks, http://colah.github.io/posts/2015-08-Understanding-LSTMs/, 2015, Accessed: 2018-06-30.
22. S. Bai, J. Z. Kolter, and V. Koltun, An empirical evaluation of generic convolutional and recurrent networks for sequence modeling, *ArXiv e-prints*, Mar 2018.
23. R. Zeiler, M.and Fergus, Visualizing and understanding convolutional networks, *ArXiv e-prints*, Nov 2013.
24. G. Montavon, S. Bach, A. Binder, W. Samek, and Müller. K., Explaining nonlinear classification decisions with deep Taylor decomposition, *Pattern Recognition*, vol. 65, pp. 211–222, May 2017.
25. W. Samek, A. Binder, G. Montavon, S. Lapuschkin, and Müller. K., Evaluating the visualization of what a deep neural network has learned, *IEEE Transactions on Neural Networks and Learning Systems*, vol. 28, no. 11, pp. 2660–2673, Nov 2017.
26. Q. Zhang and S. Zhu, Visual interpretability for deep learning: A survey, *ArXiv e-prints*, Feb 2018.
27. K. Simonyan, A. Vedaldi, and A. Zisserman, Deep inside convolutional networks: visualising image classification models and saliency maps, *ArXiv e-prints*, Dec 2013.
28. F. Chollet et al., Keras, https://keras.io, 2015, Accessed: 2018-06-30.
29. R. Kotikalapudi et al., keras-vis, https://github.com/raghakot/keras-vis, 2017, Accessed: 2018-06-30.
30. M. Atzori, A. Gijsberts, C. Castellini, B. Caputo, A. G. M. Hager, S. Elsig, G. Giatsidis, F. Bassetto, and H. Müller, Electromyography data for non-invasive naturally-controlled robotic hand prostheses, *Scientific Data*, vol. 1, no. 140053, 2014.

Chapter 14

Measuring Precise Inter-Person Physiological Synchrony and Its Trends through Adaptive, Data-Driven Algorithms: Combining NA-MEMD and the Synchrosqueezing Transform to Identify Synchronised Respiratory and HRV Frequencies

Apit Hemakom,[a,d] Katarzyna Powezka,[b] Valentin Goverdovsky,[a] Usman Jaffer,[b] Jonathon Chambers,[c] and Danilo P. Mandic[a]

[a] *Department of Electrical and Electronic Engineering, Imperial College London, London SW7 2AZ, UK*
[b] *Department of Vascular Surgery, Imperial College London, London SW7 2AZ, UK*
[c] *Department of Engineering, University of Leicester, Leicester LE1 7RH, UK*
[d] *National Electronics and Computer Technology Center (NECTEC), Pathumthani 12120, Thailand*
apit.hemakom@nectec.or.th, d.mandic@imperial.ac.uk

Learning Approaches in Signal Processing
Edited by Wan-Chi Siu, Lap-Pui Chau, Liang Wang, and Tieniu Tan

ISBN 978-981-4800-50-1 (Hardcover), 978-0-429-06114-1 (eBook)
www.panstanford.com

14.1 Introduction

Cooperative human activities require high degrees of mental and physical synchronisation among multiple participants, to the extent that synchrony underpins performance level in activities such as choir singing, playing music in an ensemble, rowing, flying an airplane with a co-pilot, or performing surgical procedures. When it comes to quantifying the degree of synchronisation among participants, synchrony in physiological responses has been reported in respiration and heart rate variability (HRV) among the choir members (Hemakom et al., 2016; Vickhoff et al., 2013).

Synchrony in respiration among the choral singers is a result of their breathing rhythm being dictated by the tempo and demands of a musical score, that is, they typically perform the short inhalation and long exhalation in unison. In addition to the voluntarily controlled breathing, the respiration is also involuntarily controlled by the autonomic nervous system (ANS), which comprises the sympathetic (SNS) and parasympathetic (PNS) nervous subsystems. The SNS also accelerates other functions, such as the arterial blood pressure and heart rate (Hlastala and Berger, 2001; Jänig, 2006; Silverthorne, 2009), by dilating bronchioles in the lungs, and by regulating neuronal and hormonal responses to stimulate the body. The PNS, on the other hand, slows down physiological functions when the body is at rest.

The interplay between the SNS and PNS, among other factors, manifests itself in variations of the timing of the cardiac cycle—heart rate variability (HRV) —in response to both external and internal factors. Changes in HRV are commonly evaluated in two frequency bands: (i) the low frequency (LF) band, 0.04–0.15 Hz, which is linked to the interaction of the SNS and PNS, and (ii) the high frequency (HF) band, 0.15–0.4 Hz, which primarily reflects the activity of the PNS (Billman, 2013; von Rosenberg et al., 2017). In addition, it is well understood that breathing modulates HRV via a phenomenon referred to as the respiratory sinus arrhythmia (RSA), whereby the heart rate accelerates during inspiration and decelerates during expiration. The RSA is usually attributed to the activity of the PNS, so that the HF component of HRV is dominated by the changes in heart rate induced by breathing.

In an attempt to quantify the degrees of synchronisation in the singers' physiological responses (respiration and HRV), a quantitative measure of the level of cooperation has been recently proposed in Hemakom et al. (2016). This is achieved via the assessment of phase relationship between multiple physiological responses, based on the intrinsic phase synchrony (IPS) and intrinsic coherence (ICoh) measures proposed in our recent work (Hemakom et al., 2016; Looney et al., 2014) under a framework referred to as *intrinsic multi-scale analysis*. The algorithms under this framework are capable of quantifying intra- and inter-component dependence of a complex system, such as multiple synchronies and causalities. The IPS and ICoh are implemented through a combination of the novel data-driven multivariate signal decomposition algorithm called *noise-assisted multivariate empirical mode decomposition* (NA-MEMD) (Mandic et al., 2013; ur Rehman and Mandic, 2011) and two standard data-association measures, phase synchrony and coherence. As desired, the IPS accounts only for phase information between dependent signals, regardless of the differences in the magnitude of intrinsic oscillations between data channels obtained using NA-MEMD. The time and frequency aspects of synchrony, however, are not highly localised using this algorithm. Conversely, using ICoh, the degree of signal dependence can be quantified as a function of frequency, at a loss of the time dimension, because ICoh is computed over the whole data set.

To mitigate the aforementioned problems posed by both the intrinsic data association measures—poor frequency localisation in IPS and loss of time information in ICoh—a highly localised time-frequency data association measure was therefore proposed in Hemakom et al. (2017). It is achieved based on the combination of NA-MEMD, and short-time Fourier transform (STFT)-based synchrosqueezing transform (SST) (Daubechies et al., 2011; Oberlin et al., 2014) and multivariate SST (MSST) (Ahrabian et al., 2015) algorithms. The NA-MEMD is first employed to obtain physically meaningful intrinsic oscillations of a given multivariate signal. The STFT-based SST and MSST (F-M/-SST) algorithms are next employed to generate highly localised time-frequency representations of signal dependence between the intrinsic oscillations produced by the NA-MEMD. This procedure is referred to as the *intrinsic synchrosqueezing coherence* (ISC). The F-M/-SST algorithms are employed, instead of the continuous wavelet transform (CWT)-based SST and MSST (W-M/-SST) algorithms, by the ISC algorithm, since the performance of the CWT-based MSST (W-MSST) degrades with noise power (noticeable spurious harmonics are produced as the noise power increases). This is because: (i) the joint instantaneous frequency estimator is sensitive to noise, and (ii) wavelet functions are time limited—when performing time convolution (i.e. spectral windowing), accurate localisation of both time and frequency by the CWT cannot be achieved at the same time, the CWT thus produces mathematical artefacts (additional noise), that is, wavelet coefficients which correspond to undesired frequency components. The additional noise (artefacts) generated by the W-MSST can be reduced by STFT-based MSST (F-MSST).

However, for certain scenarios, in the presence of

- collaborative tasks which take place over a long period of time,
- physically meaningful and straightforward to interpret trends in the level of cooperation during long and complex events being of interest,
- prior knowledge on periods of trends for physically meaningful and straightforward interpretation being unavailable,

it is imperative to have a data-driven, data-association measure which empirically quantifies *intrinsic trends*, whereby only those components which contain physically meaningful and straightforward interpretation can be combined. To this end, an extension to the standard IPS was proposed in Hemakom et al. (2017) and is referred to as the *nested intrinsic phase synchrony* (N-IPS), which further decomposes time series of synchrony between data channels calculated using sliding windows in the standard IPS into multiple scales of synchrony; then only certain scales which admit straightforward and meaningful physical interpretation are combined, without any prior knowledge on the frequencies of such scales.

The aim of this chapter is to illustrate the enhanced discrimination capability of the intrinsic synchrosqueezing coherence and nested intrinsic phase synchrony data association metrics, to precise identification of physiological synchrony in frequency and time, using the ISC, and to empirical quantification of physically meaningful intrinsic trends in the level of cooperation associated with *events* during the course of long cooperative tasks, using the N-IPS. The intrinsic synchrosqueezing coherence is employed to precisely reveal the synchronised respiratory and HRV frequencies among a subset of bass singers of the Eric Whitacre Choir during 4-minute and 40-second rehearsal and performance of the same musical score. The nested intrinsic phase synchrony is employed to empirically estimate physically meaningful trends in the levels of cooperation of: (i) subsets of soprano and bass singers of the Imperial College Chamber Choir during a 1-hour evensong performance through trends of synchrony in their respiratory and HRV signals, and (ii) three pairs of catheterisation laboratory team members (cardiology consultant, cardiology registrar and physiologist) during a 2-hour invasive coronary procedure through trends of synchrony in their HRV signals.

14.2 Algorithms

All the relevant algorithms (NA-MEMD, SST, MSST, ISC, and N-IPS) are described below.

14.2.1 Noise-Assisted Multivariate Empirical Mode Decomposition (NA-MEMD)

It has been shown in Flandrin et al. (2004) and Wu and Huang (2004) that empirical mode decomposition (EMD) (Huang et al., 1998) acts essentially as a dyadic filter when decomposing white Gaussian noise (WGN), enforcing the dyadic filterbank structure onto intrinsic mode function (IMFs)—Fourier spectra of the IMFs are all identical and cover the same area on a semi-logarithmic period scale. Therefore, the aim of simultaneously decomposing added separate artificial WGN channels and the input signal using noise-assisted MEMD (NA-MEMD) is to enforce the dyadic filterbank structure within the IMFs by using WGN as a decomposition reference and to reduce mode mixing. *While this may hinder the data-driven operation of MEMD, it is essential in multi-channel operations where the requirement is to compare IMFs with similar centre frequencies and bandwidths in order to preserve the physical meaning of the analysis.* It should also be noted that the most important aspect of an MEMD approach, via either standard MEMD or NA-MEMD, is that even if mode mixing occurs it is likely to happen simultaneously in every channel. In this way, multi-component comparisons at the IMF level are matched and physically meaningful. The work in Looney et al. (2014) has shown that using NA-MEMD to decompose WGN using 10 *different* artificial noise channels reduced the degree of overlap between power spectral densities of the resulting IMFs, compared to using NA-MEMD to decompose the same WGN using 10 *identical* artificial noise channels. It was therefore suggested that the number of WGN channels should be as large as possible—computation time permitting—in order to reduce the degree of overlap between IMF spectra, that is, imposes a more pronounced dyadic filterbank structure within the IMFs.

In terms of phase responses, NA-MEMD, which acts as a dyadic filterbank, can be deemed having a linear phase response. This is because of the computational time required by the sifting process to decompose a given input signal into a set of IMFs, thus introducing the same delay to all frequency components of the input signal.

Algorithm 14.1. Noise-assisted multivariate empirical mode decomposition (NA-MEMD)

(1) Given a P-variate signal $\mathbf{x}(t)$, construct an N-variate WGN signal $\mathbf{z}(t)$, $\mathbf{z}(t) = [z_1(t), z_2(t), \ldots, z_N(t)]^T$.
(2) Treat both signals $\mathbf{x}(t)$ and $\mathbf{z}(t)$ as a single (P+N)-variate signal $\mathbf{y}(t)$,
$\mathbf{y}(t) = [x_1(t), x_2(t), \ldots, x_P(t), z_1(t), z_2(t), \ldots, z_N(t)]^T$.
(3) Let multivariate proto-IMF $\tilde{\mathbf{y}}(t) = \mathbf{y}(t)$, and $R = P + N$.
(4) Choose a weighting scheme based on a suitable pointset for sampling on a (R)-hypersphere,
$\mathbf{w}_q = \{w_{\{q,1\}}, w_{\{q,2\}}, \ldots, w_{\{q,R\}}\}_{q=1}^{Q}$.

(5) Calculate a projection, denoted by $\Upsilon\left(\tilde{\mathbf{y}}(t)\right)_{\mathbf{w}_q}$, of the proto-IMF $\tilde{\mathbf{y}}(\mathbf{t})$ along the direction vector $\mathbf{w}_q$, for all q (the whole set of direction vectors), giving $\{\Upsilon\left(\tilde{\mathbf{y}}(t)\right)_{\mathbf{w}_q}\}_{q=1}^{Q}$ as the set of projections.

(6) Find the $j = 1, 2, \ldots, J$ time instants $\{t_{j,\min}^{\mathbf{w}_q}\}$ and $\{t_{j,\max}^{\mathbf{w}_q}\}$ corresponding, respectively, to the minima and maxima of the set of projected signals $\{\Upsilon\left(\tilde{\mathbf{y}}(t)\right)_{\mathbf{w}_q}\}_{q=1}^{Q}$.

(7) Interpolate $\tilde{\mathbf{y}}(t_{j,\min}^{\mathbf{w}_q})$ and $\tilde{\mathbf{y}}(t_{j,\max}^{\mathbf{w}_q})$ to obtain, respectively, multivariate minima and maxima envelope curves $\{\mathbf{e}_{\min}^{\mathbf{w}_q}(t)\}_{q=1}^{Q}$ and $\{\mathbf{e}_{\max}^{\mathbf{w}_q}(t)\}_{q=1}^{Q}$.

(8) For a set of Q direction vectors, the mean $\mathbf{m}(t)$ of the envelope curves is calculated as:

$$\mathbf{m}(t) = \frac{1}{2Q}\sum_{q=1}^{Q} \mathbf{e}_{\min}^{\mathbf{w}_q}(t) + \mathbf{e}_{\max}^{\mathbf{w}_q}(t) \tag{14.1}$$

(9) Extract the detail $\mathbf{d}(t)$ using $\mathbf{d}(t) = \tilde{\mathbf{y}}(t) - \mathbf{m}(t)$.

(10) If the detail $\mathbf{d}(t)$ fulfils the stoppage criterion for a multivariate IMF (the difference between two consecutive sifting results is below a certain threshold), $\mathbf{d}(t)$ becomes a multivariate IMF, $\mathbf{c}_k(t)$, and $\tilde{\mathbf{y}}(t) = \tilde{\mathbf{y}}(t) - \mathbf{c}_k(t)$. Otherwise, $\tilde{\mathbf{y}}(t) = \mathbf{d}(t)$.

(11) Repeat Steps (5)–(10) until $\tilde{\mathbf{y}}(t)$ is a multivariate monotonic residue or trend.

14.2.2 Intrinsic Phase Synchrony (IPS)

The degree of phase synchronisation between data channels can be measured through phase synchrony, which quantifies only the phase relationship between two signals without accounting for amplitude information. It is defined in terms of the deviation from perfect synchrony via the phase synchronisation index (PSI). The PSI values range from 0 to 1, with 1 indicating the perfect phase locking and 0 non-phase-synchronous relationship.

Intrinsic phase synchrony (IPS) was originally proposed in the so-called intrinsic multi-scale analysis framework (Looney et al., 2014), and generalises standard phase synchrony by equipping it with the ability to operate at the intrinsic scale level. It employs NA-MEMD to decompose a given multivariate signal into its narrowband intrinsic oscillations (IMFs), which makes it possible to quantify the temporal locking of the phase information in IMFs using the standard phase synchronisation index (PSI), as outlined in Algorithm 14.2.

Algorithm 14.2. Intrinsic phase synchrony (IPS)

(1) Given a P-variate signal $\mathbf{x}(t)$, obtain multivariate IMFs via NA-MEMD, $\mathbf{c}_k(t)$, where $k = 1, 2, \dots, K$, and K denotes the number of IMFs.
(2) For each channel p, calculate the instantaneous phases $\phi_p(t)$ for the IMFs through the analytic signals generated using the Hilbert transform.
(3) Calculate phase difference between the instantaneous phases for the IMFs of channels i and j, $\Delta\phi_{i,j}(t)$, where $i = 1, 2, \dots, P$, $j = 1, 2, \dots, P$, and $i \neq j$.
(4) Phase synchrony between channels i and j is then defined in terms of the deviation from perfect synchrony via the phase synchronisation index (PSI) (Tass et al., 1998), given by

$$\gamma_{i,j}(t) = \frac{H_{\max} - H}{H_{\max}}, \tag{14.2}$$

where $H = -\sum_{m=1}^{M} pr_m \ln pr_m$ is the Shannon entropy of the distribution of phase differences $\Delta\phi_{i,j}\left(t - \frac{W}{2} : t + \frac{W}{2}\right)$ within a window of length W, M is the number of bins within the distribution of phase differences, and pr_m is the probability of $\Delta\phi_{i,j}\left(t - \frac{W}{2} : t + \frac{W}{2}\right)$ within the mth bin. The maximum entropy $H_{\max}$ is given by

$$H_{\max} = 0.626 + 0.4 \ln(W - 1). \tag{14.3}$$

14.2.3 Synchrosqueezing Transform (SST)

The synchrosqueezing transform (SST or WSST) (Daubechies et al., 2011) is a post-processing technique originally applied to the continuous wavelet transform (CWT) in order to generate highly localised time-frequency representations of non-stationary signals. The CWT is a projection based algorithm that identifies oscillatory components of interest through a series of time-frequency filters known as wavelets. A wavelet $\psi(t)$ is a finite oscillatory function, which when convolved with a signal $x(t)$, in the form

$$W(a,b) = \int x(t)a^{(-1/2)}\psi\left(\frac{t-b}{a}\right)dt \tag{14.4}$$

gives the wavelet coefficients $W(a, b)$, for each scale-time pair (a, b). In this way, the wavelet coefficients can be seen as the outputs of a set of scaled bandpass filters. The scale factor a shifts the bandpass filters in the frequency domain, and also changes the bandwidth of the bandpass filters. Therefore, while the energy of a sinusoid at a frequency ω_s is spread across the scale factor a_s, the energy of the wavelet transform of a sinusoid at a frequency ω_s will spread out around the scale factor $a_x = \omega_\psi/\omega_s$, where ω_ψ is the centre frequency of a wavelet. Thus, the estimated frequency present in those scales is equal to the original frequency ω_s. Consequently, the instantaneous frequency $\Omega(a, b)$ of the wavelet coefficients $W(a, b)$ can be estimated as

$$\Omega(a,b) = -i(W(a,b))^{-1}\frac{\partial}{\partial b}W(a,b). \tag{14.5}$$

The wavelet coefficients that contain the same instantaneous frequencies can then be combined using a procedure referred to as synchrosqueezing (SST). For a set of wavelet coefficients $W(a, b)$ (Daubechies et al., 2011), the CWT-based synchrosqueezing transform coefficient is given by

$$T_{\mathrm{CWT}}(\omega_m, b) = (\Delta\omega)^{(-1)} \sum_{a_l:|\Omega(a_l,b)-\omega_m|\leq\Delta\omega/2} W(a_l,b)a_l^{-3/2}\Delta a_l, \tag{14.6}$$

where ω_m are the frequency bins with a resolution of $\Delta\omega$. A summary of the SST is outlined in Algorithm 2.4.

Algorithm 14.3. Synchrosqueezing transform (SST or WSST)

(1) Given a univariate signal $x(t)$, convolve a wavelet $\psi(t)$ with the signal in order to obtain the wavelet coefficients $W(a, b)$, given by Eq. 14.4.
(2) Estimate the instantaneous frequency $\Omega(a, b)$ of the wavelet coefficients $W(a, b)$, given by Eq. 14.5.
(3) Combine the wavelet coefficients which contain the same instantaneous frequency in order to obtain the CWT-based synchrosqueezing transform coefficients $T_{\mathrm{CWT}}(\omega_m, b)$, given by Eq. 14.6.

The synchrosqueezing transform has been extended to the short-time Fourier transform (STFT) setting in Oberlin et al. (2014), and is referred to as STFT-based SST (FSST). It performs STFT, instead of CWT, on a given univariate signal $x(t)$ to obtain STFT coefficients $U(\eta, t)$, given by

$$U(\eta, t) = \int_{\Re} x(\tau)g(\tau - t)e^{-2i\pi\eta(\tau-t)}d\tau, \tag{14.7}$$

where g is a sliding window, and τ time lag. The STFT coefficients, $U(\eta, t)$, are next relocated to the instantaneous frequency, Ω, according to the map $(\eta, t) \mapsto (\Omega(\eta, t))$, where Ω is defined by

$$\Omega(\eta, t) = \frac{1}{2\pi}\partial_t \arg U(\eta, t) = \mathrm{Re}\left(\frac{1}{2i\pi}\frac{\partial_t U(\eta, t)}{U(\eta, t)}\right). \tag{14.8}$$

The STFT-based synchrosqueezing transform coefficient is then given by

$$T_{\mathrm{STFT}}(f, t) = \frac{1}{g(0)}\int_{\Re} U(\eta, t)\delta(f - \Omega(\eta, t))d\eta. \tag{14.9}$$

The FSST has been shown to exhibit better localisation, that is, sharper representation of instantaneous frequencies in low frequencies of linear, exponential and hyperbolic chirps, compared to the SST (Oberlin et al., 2014). A summary of the FSST is outlined in Algorithm 14.4.

Algorithm 14.4. STFT-based synchrosqueezing transform (FSST)

(1) Given a univariate signal $x(t)$, perform STFT on the signal in order to obtain the STFT coefficients $U(\eta, t)$, given by Eq. 14.7.
(2) Estimate the instantaneous frequency $\Omega(\eta, t)$ of the STFT coefficients $U(\eta, t)$, given by Eq. 14.8.
(3) Combine the STFT coefficients in order to obtain the STFT-based synchrosqueezing transform coefficients $T_{\mathrm{STFT}}(f, t)$ given by Eq. 14.9.

14.2.4 Multivariate Synchrosqueezing Transform (MSST)

The multivariate synchrosqueezing transform (MSST or WMSST) (Ahrabian et al., 2015) is an extension of the CWT-based SST to multivariate cases. It was proposed in order to identify oscillations common to multiple data channels. It first partitions the time-frequency domain into M frequency bands $\{\omega_m\}_{m=1,2,\ldots,M}$. This makes it possible to identify a set of matched monocomponent signals from a given multivariate signal. The instantaneous amplitudes and frequencies present within those frequency bands can then be determined.

For a multivariate signal $\mathbf{x}(t)$ with the corresponding CWT-based SST coefficients of each channel p, $T^p_{\mathrm{CWT}}(\omega, b)$, and a set of frequency bands $\{\omega_m\}_{m=1,2,\ldots,M}$, the instantaneous frequency $\Omega(\omega_m, b)$ for each channel p and frequency band m is given by

$$\Omega(\omega_m, b) = \frac{\sum_{\omega\in\omega_m} |T^p_{\mathrm{CWT}}(\omega, b)|^2 \omega}{\sum_{\omega\in\omega_m} |T^p_{\mathrm{CWT}}(\omega, b)|^2} \tag{14.10}$$

and the instantaneous amplitude $A(\omega_m, b)$ for each channel and frequency band as

$$A(\omega_m, b) = \sqrt{\sum_{\omega\in\omega_m} |T^p_{\mathrm{CWT}}(\omega, b)|^2}. \tag{14.11}$$

In order to estimate the multivariate instantaneous frequency $\Omega(\omega_m, b)$ for a given frequency band m, the instantaneous frequencies across the P channels are then combined using

$$\Omega(\omega_m, b) = \frac{\sum_{p=1}^{P} (A(\omega_m, b))^2 \Omega(\omega_m, b)}{\sum_{p=1}^{P} (A(\omega_m, b))^2}, \tag{14.12}$$

while the multivariate instantaneous amplitude $\mathbf{A}(\omega_m, b)$ for each frequency band is given by

$$\mathbf{A}(\omega_m, b) = \sqrt{\sum_{p=1}^{P} (A(\omega_m, b))^2}. \tag{14.13}$$

The multivariate CWT-based synchrosqueezing transform coefficient $\mathbf{T}_{\mathrm{CWT}}(\omega_m, b)$ for each frequency band m is given by

$$\mathbf{T}_{\mathrm{CWT}}(\omega_m, b) = \mathbf{A}(\omega_m, b)\delta(\omega - \Omega(\omega_m, b)) \tag{14.14}$$

where $\delta(\cdot)$ is the Dirac delta function. The multivariate CWT-based synchrosqueezing transforms for all frequency bands are then given by $\mathbf{T}_{\mathrm{CWT}}(\omega, b) = \mathbf{T}_{\mathrm{CWT}}(\omega_m, b)|_{m=1,2,\ldots,M}$. A summary of the MSST is outlined in Algorithm 14.5.

Algorithm 14.5. Multivariate synchrosqueezing transform (MSST or W-MSST)

(1) Given a P-variate signal $\mathbf{x}(t)$, apply the SST channel-wise in order to obtain the univariate SST coefficients $T^p_{\text{CWT}}(\omega, b)$.
(2) Determine a set of partitions along the frequency axis for the time-frequency domain, and calculate the instantaneous frequency $\Omega(\omega_m, b)$ and amplitude $A(\omega_m, b)$ for each frequency band m and channel p, given by Eqs. 14.10 and 14.11, respectively.
(3) Calculate the multivariate instantaneous frequency $\Omega(\omega_m, b)$ and amplitude $\mathbf{A}(\omega_m, b)$ for each channel m, using the Eqs. 14.12 and 14.13, respectively.
(4) Determine the multivariate CWT-based synchrosqueezing coefficients $\mathbf{T}_{\text{CWT}}(\omega, b)$.

14.2.5 Intrinsic Synchrosqueezing Coherence (ISC)

The intrinsic synchrosqueezing coherence (ISC) (Hemakom et al., 2017) is a data association measure which exhibits precise time and frequency localisation for the analysis of coupled non-linear and non-stationary multivariate signals. This is achieved through the combination of NA-MEMD, FSST and F-MSST algorithms. The NA-MEMD is first employed to decompose a given multivariate signal into a set of physically meaningful narrowband intrinsic oscillations (IMFs). The FSST is next employed channel-wise to generate multiple univariate multicomponent time-frequency (TF) planes with high localisation in both time and frequency. The F-MSST is then used to construct multivariate highly-localised TF representations. These univariate and multivariate TF representations are subsequently employed to generate TF representations of signal dependence (synchrony) in IMFs; see Algorithm 14.6 for detail of the proposed intrinsic synchrosqueezing coherence (ISC). The synchrosqueezing coherence index (SCI) ranges between 0 and 1, with 1 indicating perfect synchrony and 0 a non-synchronous state. For the implementation of F-MSST, FSST is employed instead of SST in Step (1) of Algorithm 14.5.

Algorithm 14.6. Intrinsic synchrosqueezing coherence (ISC)

(1) Given a P-variate signal $\mathbf{x}(t)$, obtain multivariate IMFs via NA-MEMD, $\mathbf{c}_k(t)$, where $k = 1, 2, \ldots, K$, and K is the number of IMFs.
(2) For each channel p, combine the IMFs corresponding to the frequency band of interest, IMF_p.
(3) Apply FSST channel-wise to the combined IMFs, IMF_p, to generate P univariate STFT-based synchrosqueezing transform coefficients, $T^p_{\text{STFT}}(f, t)$, where $p = 1, 2, \ldots, P$.
(4) Apply F-MSST to the combined IMFs between channels i and j, IMF_i and IMF_j, where $i = 1, 2, \ldots, P$, $j = 1, 2, \ldots, P$, and $i \neq j$, to generate multivariate STFT-based synchrosqueezing transform coefficients of the two channels, $T^{i,j}_{\text{STFT}}(f, t)$.
(5) Time-frequency representation of the degree of signal dependence (synchrony) in frequency and time between channels i and j is then determined via the synchrosqueezing coherence index (SCI), given by

$$SCI_{i,j}(f, t) = \frac{\sqrt{\frac{|T^i_{\text{STFT}}(f,t)|\cdot|T^j_{\text{STFT}}(f,t)|}{|T^{i,j}_{\text{STFT}}(f,t)|}}}{\max\limits_{f,t}\left(\sqrt{\frac{|T^i_{\text{STFT}}(f,t)|\cdot|T^j_{\text{STFT}}(f,t)|}{|T^{i,j}_{\text{STFT}}(f,t)|}}\right)}; \forall f, \forall t \quad (14.15)$$

(6) Perform 30 realisations of NA-MEMD,[1] repeat steps (1)–(5) for each realisation, and average the 30 time-frequency representations of signal dependence between channels i and j,

---

[1] In order to reduce the 'leakage' effect in the input channels, which is caused by artificial WGN channels while performing NA-MEMD, additional multiple realisations of WGN are recommended and must be averaged out across the realisations to obtain the 'true' IMFs (ur Rehman and Mandic, 2011). This study has empirically found that performing 30 realisations of NA-MEMD is computationally time permitted and is sufficient to reduce the leakage effect. To exemplify computational time required to perform a single realisation of NA-MEMD, a single realisation of NA-MEMD was performed on a personal computer with an 8-core processor running at the speed of 3.4 GHz, with 32 GB of RAM, to decompose 2-channel 1-minute-long synthetic data with a sampling frequency of 1 kHz, where both the channels consisted of 100 Hz sinusoids, using 5 additional WGN channels. The computational time required for this NA-MEMD operation was 11 minutes. In general, computational time required for each realisation of NA-MEMD depends on the numbers of data channels and data points.

$SCI_{i,j}(f, t)$, in order to obtain a highly localised representation of synchrony.

The rationale behind Eq. 14.15 is as follows: The multiplication of univariate TF representations, that is, instantaneous amplitudes and frequencies, of channels i and j, which are, respectively, represented by $|T^i_{\mathrm{STFT}}(f, t)|$ and $|T^j_{\mathrm{STFT}}(f, t)|$, is performed in order to reveal instantaneous frequencies which exist in both the channels *at the same time instants*. However, if there exists smearing in both the univariate TF representations, the resulting multiplication also exhibits such smearing. This can be removed by dividing the resulting multiplication with the multivariate TF representation of both the channels, $|T^{i,j}_{\mathrm{STFT}}(f, t)|$. If the resulting division is not a number (infinity) due to a value of zero of the denominator, $|T^{i,j}_{\mathrm{STFT}}(f, t)|$, which indicates the non-existence of smearing, it must be replaced by zero. Finally, SCI values in the range 0 to 1 are obtained by dividing the numerator with its maximum value across the whole time and frequency considered.

14.2.6 Nested Intrinsic Phase Synchrony (N-IPS)

The intrinsic phase synchrony (IPS) was originally proposed in the so-called intrinsic multi-scale analysis framework in Looney et al. (2014) and generalises standard phase synchrony by equipping it with the ability to operate at the intrinsic scale level. It employs NA-MEMD to decompose a given multivariate signal into its narrowband intrinsic oscillations (IMFs), which makes it possible to quantify the temporal locking of the phase information in IMFs using the standard phase synchronisation index (PSI).

An extension of the IPS was introduced in Hemakom et al. (2017) and is referred to as nested intrinsic phase synchrony (N-IPS), for the empirical quantification of physically meaningful and straightforward to interpret trends in phase synchrony. The N-IPS first employs the conventional IPS to quantify intrinsic phase relationship in IMFs, and further decomposes the resulting time series of intrinsic phase synchrony into multiple scales, whereby only certain scales which contain physically meaningful and straightforward to interpret information are then combined, as outlined in Algorithm 14.7. Note that trends in synchrony obtained using the N-IPS algorithm can be negative, since IMFs of synchrony which contain a positive offset in the raw PSI values can be neglected. In such circumstances, the baseline for the trends in synchrony obtained using N-IPS is an imperative, since it is used to judge whether the trends in synchrony are significant or not—they are deemed significant if above the baseline.

Algorithm 14.7. Nested intrinsic phase synchrony (N-IPS)

(1) Given a P-variate signal $\mathbf{x}(t)$, obtain multivariate IMFs via NA-MEMD, $\mathbf{c}_{k_x}(t)$, where $k_x = 1, 2, \ldots, K_x$, and K_x is the number of IMFs.
(2) Calculate phase synchrony between channels i and j, $\gamma_{i,j}(t)$, using the IPS algorithm (see Algorithm 14.2), where $i = 1, 2, \ldots, P$, $j = 1, 2, \ldots, P$, and $i \neq j$.
(3) Obtain IMFs of the phase synchrony between channels i and j, $\gamma_{i,j}(t)$, via NA-MEMD, $c_{\gamma_{i,j},k_\gamma}$, $k_\gamma = 1, 2, \ldots, K_\gamma$, and K_γ is the number of IMFs of the phase synchrony.
(4) Combine certain IMFs of the phase synchrony, $c_{\gamma_{i,j},k_\gamma}$, which contain physically meaningful and straightforward interpretation.

14.3 Applications

Applications of the proposed ISC algorithm shall now be demonstrated, which include: (i) estimating degrees of synchrony in a linear bivariate signal; (ii) estimating degrees of synchrony in a non-linear narrowband bivariate signal; and (iii) the precise identification of physiological synchrony among choir members. Applications of the N-IPS algorithm in large scale analyses for the empirical quantification of the levels of cooperation in (i) choir singing and (ii) performing a surgical procedure are also demonstrated.

14.3.1 Estimating Degrees of Synchrony in a Linear Bivariate Signal

The utility of the proposed ISC algorithm, a combination of NA-MEMD and STFT-based M/-SST (F-M/-SST) algorithms, is demonstrated over the task of estimating degrees of synchrony in a linear bivariate signal against the standard intrinsic phase synchrony (IPS) algorithm, which is a combination of NA-MEMD and phase synchrony, and other five combinations of algorithms for performing Steps (1)–(4) in Algorithm 14.6. These combinations are:

- FIR band-pass filter (BPF) & CWT-based M/-SST (W-M/-SST)
- FIR BPF & STFT-based M/-SST (F-M/-SST)
- IIR BPF & CWT-based M/-SST (W-M/-SST)
- IIR BPF & STFT-based M/-SST (F-M/-SST)
- NA-MEMD & CWT-based M/-SST (W-M/-ST)

To resemble real-world physiological signals, which typically contain $1/f$ noise and a low-frequency trend (Radeka, 1969), the linear bivariate signal consisted of sinusoidal oscillations corrupted by additive $1/f$ noise and a low-frequency trend of magnitude A oscillating at 0.01 Hz, $A_{0.01}$, given by

$$
\begin{aligned}
x_1(t) &= A_{0.01} + k\cos(2\pi f(t)t + \phi) + n_1(t),\\
f(t) &= \begin{cases} f_1; t = 0s, \ldots, 1.667s \\ f_2; t = 1.667s, \ldots, 3.334s \\ f_3; t = 3.334s, \ldots, 5s \end{cases}\\
k &= \begin{cases} k_1; t = 0s, \ldots, 1.667s \\ k_2; t = 1.667s, \ldots, 3.334s \\ k_3; t = 3.334s, \ldots, 5s \end{cases}\\
x_2(t) &= A_{0.01} + k\cos(2\pi f(t)t + \phi) + n_2(t),\\
f(t) &= \begin{cases} f_1; t = 0s, \ldots, 1.667s \\ f_3; t = 1.667s, \ldots, 5s \end{cases}\\
k &= \begin{cases} k_1; t = 0s, \ldots, 1.667s \\ k_4; t = 1.667s, \ldots, 5s \end{cases}
\end{aligned}
\tag{14.16}
$$

where $A_{0.01} = 5$, $f_1 = 5$ Hz, $f_2 = 11$ Hz, $f_3 = 18$ Hz, $k_1 = 1$, $k_2 = 0.5$, $k_3 = 2$, $k_4 = 0.1$ and the sampling frequency $f_s = 200$ Hz. The SNR of the first channel corrupted by additive $1/f$ noise n_1 was 10 dB, while the SNR of the second channel governed by $1/f$ noise n_2 was set at 0 dB, 5 dB and 10 dB. For each combination of the algorithms 30 realisations were performed, whereby different values of phase ϕ were set in different realisations, ranging from 0 to 2π, that is, $\phi = \frac{q-1}{30}(2\pi)$, where q is the realisation number. It must be noted that this simulation aimed to demonstrate straightforward comparisons between the combinations of algorithms in: (i) removing $1/f$ noise and trend, which typically govern physiological signals, using the standard filtering techniques and NA-MEMD, essentially acting as a dyadic filterbank; and (ii) most importantly, estimating degrees of synchrony. This simulation, therefore, did not introduce other non-biological or physiological artefacts, such as electrical motor noise, mechanical ventilator noise, respiration, eye blinks, and muscle movement, which often overlap with physiological signals of interest in both spectral and temporal domains such that it becomes difficult to use a simple filtering technique for the removal (Islam et al., 2016). Efficient detection and removal of such artefacts require further sophisticated signal processing schemes, such as independent component analysis (ICA) (Klados et al., 2011; Soomro et al., 2013; Zou et al., 2014), artificial neural network (ANN) (Burger and van den Heever, 2015; Hu et al., 2015; Jafarifarmand and Badamchizadeh, 2013), and support vector machines (SVM) (Bhattacharyya et al., 2013; Lawhern et al., 2012; Shao et al., 2008), which are beyond the scope of this chapter.

Figure 14.1 shows the spectrograms of the clean linear bivariate signal without the low-frequency trend estimated using the standard STFT. In this bivariate signal, there was strong synchrony between the two channels at 5 Hz from $0s$ to $1.667s$, and weak synchrony at 18 Hz from $3.334s$ to $5s$. The bivariate signal was decomposed using NA-MEMD with 10 additional WGN channels. Three IMFs of each of the two channels which governed the frequency band of 4–19 Hz were combined and averaged over 30 realisations of NA-MEMD. Alternatively, the bivariate signal was band-pass filtered using an FIR BPF of order 50 with the passband set to 4–19 Hz, a bandwidth similar to the combined IMFs. A

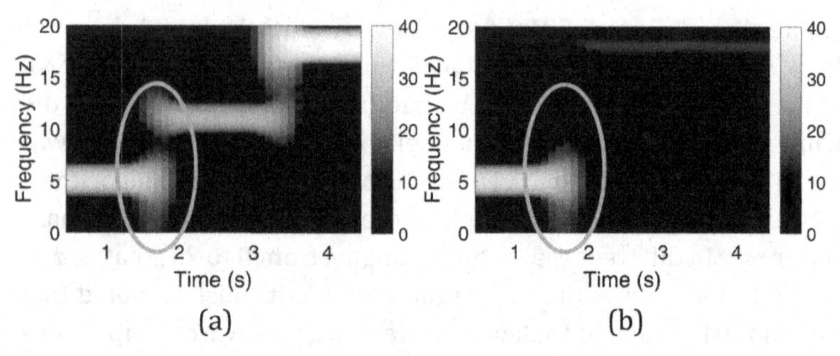

Figure 14.1 Standard TF spectrograms of the clean linear bivariate signal. (a) First channel consisted of 3 sinusoids of frequencies 5 Hz, 11 Hz, and 18 Hz. (b) Second channel consisted of 2 sinusoids of frequencies 5 Hz and 18 Hz. There were therefore both strong and weak signal dependence at 5 Hz and 18 Hz at different time instants. Observe smearing due to frequency switches from 5 Hz to 11 Hz in the first channel, and from 5 Hz to 18 Hz in the second channel (see the ellipses).

Butterworth IIR BPF of order 3 with the same passband was also applied to the bivariate signal. The different filtered bivariate signals obtained using the NA-MEMD, FIR BPF, and IIR BPF were then fed into the W-M/-SST and F-M/-SST algorithms to generate TF representations of signal dependence (synchrony) calculated using Eq. 14.15 in Algorithm 14.6. Phase synchrony estimation was also employed to estimate phase relationship in the filtered bivariate signal obtained using NA-MEMD, that is, intrinsic phase synchrony, and TF representations of phase synchrony were obtained using the Hilbert transform.

Figures 14.2 and 14.3 show TF representations of degrees of synchrony in the linear bivariate signal estimated by averaging over 30 realisations of the different combinations of algorithms considered. The W-M/-SST combined with FIR BPF, IIR BPF, or NA-MEMD, produced noticeable spurious synchrony (see first and third rows in Fig. 14.2, and first row in Fig. 14.3), since CWT introduced mathematical artefacts (noise),[2] and consequently they were

[2]Mathematical artefacts (additional noise) are produced by CWT, because wavelet functions are time limited, thus when performing time convolution (i.e. spectral windowing), accurate localisation of both time and frequency by the CWT cannot be achieved at the same time.

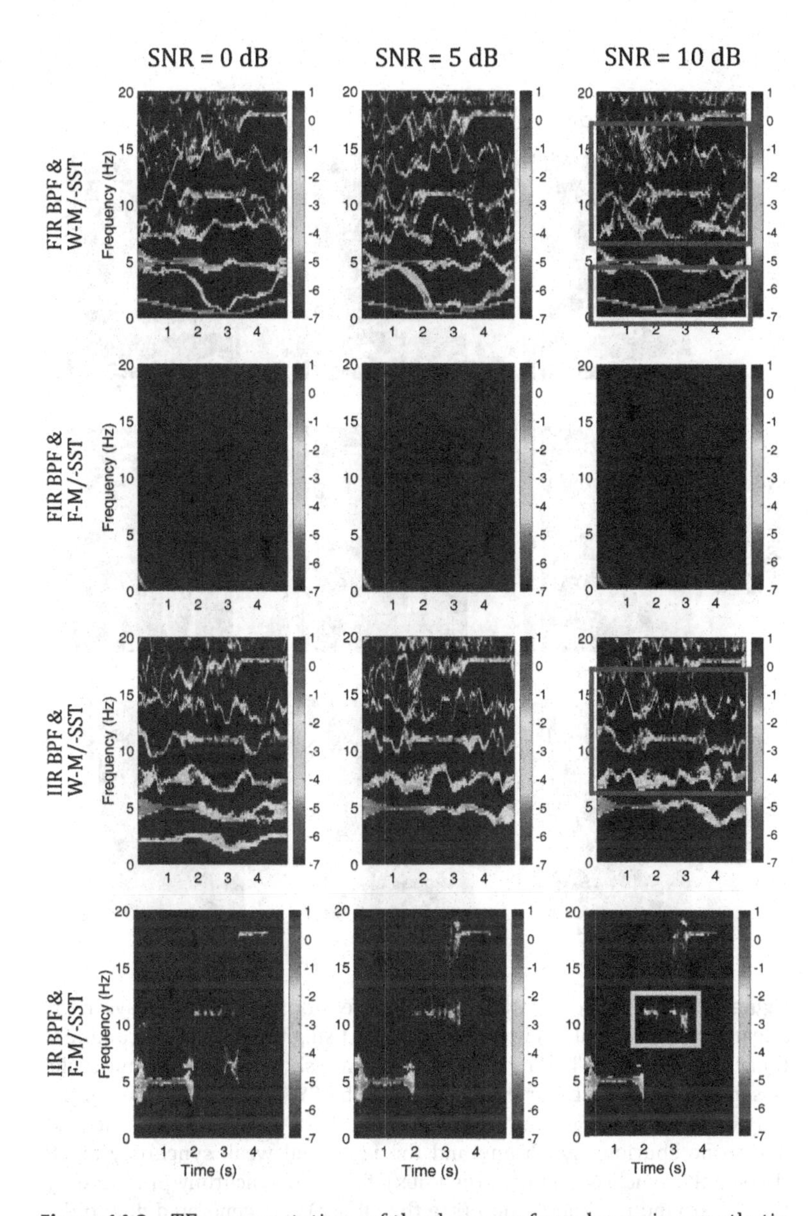

Figure 14.2 TF representations of the degrees of synchrony in a synthetic linear bivariate signal in Fig. 14.1 estimated using four different combinations of algorithms. The TF representations are shown in logarithmic scale. The SNR of the second channel governed by $1/f$ noise was set to 0 dB, 5 dB and 10 dB (left, centre and right panels, respectively). Observed mathematical artefacts produced by CWT and exemplified in the red boxes.

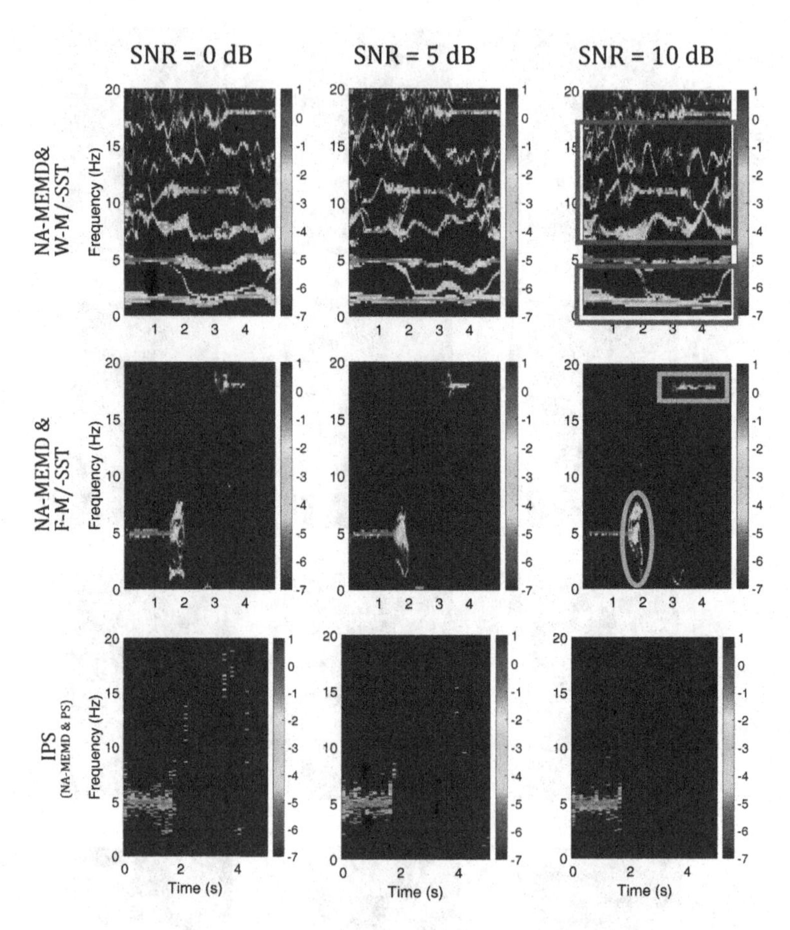

Figure 14.3 TF representations of the degrees of synchrony in a synthetic linear bivariate signal in Fig. 14.1 estimated using three different combinations of algorithms. The TF representations are shown in logarithmic scale. Observed mathematical artefacts produced by CWT and exemplified in the red boxes. The ISC algorithm, NA-MEMD and F-M/-SST, performed the best due to no spurious synchrony and the identified weak synchrony at 18 Hz (see the synchrony in the green box). Smeared synchrony produced by the ISC around the 2nd second (see the ellipse) was generated due to the frequency switches from f_1 to f_2 in the first channel, and from f_1 to f_3 in the second channel.

localised by the joint frequency estimator in W-MSST. Since the 0.01 Hz trend added to the signal was not filtered out completely using the FIR BPF and the remaining trend was localised by F-M/-MSST, the SCI values at this frequency consequently dominated the TF representations of degrees of synchrony (second row in Fig. 14.2). Undesired synchrony at f_2 was produced by the combination of IIR BPF and F-M/-SST (see fourth row in Fig. 14.2). This is because the IIR BPF introduced artefacts to the filtered second channel at f_2, thus resulting in undesired F-SST coefficients of the filtered second channel at f_2 and consequently undesired synchrony between the channels at f_2. Using IPS (last row in Fig. 14.3), the degrees of synchrony estimated at f_1 were not highly localised, spreading around f_1, and the estimation of weak synchrony at f_3 was not achieved. The proposed ISC algorithm (NA-MEMD and F-M/-SST, second row in Fig. 14.3) produced less spurious synchrony, yielded desired synchrony at f_1, and achieved the estimation of weak synchrony at f_3 at high SNR. Note that smeared synchrony produced by the ISC around the 2nd second (see the ellipse) was generated due to the frequency switches from f_1 to f_2 in the first channel, and from f_1 to f_3 in the second channel.

14.3.2 Estimating Degrees of Synchrony in a Non-Linear Narrowband Bivariate Signal

The performance of the seven combinations of algorithms was next evaluated in the task of estimating synchrony in a non-linear narrowband bivariate signal extracted from a non-linear wideband bivariate AM/FM signal corrupted by additive WGN, given by

$$\begin{aligned} x_1(t) &= ((1 + M_{AM}\cos(2\pi f_m t + \phi))A_1)\cdot \\ &\quad \sin(2\pi f_c t + (A_1 M_{FM}\sin(2\pi f_m t + \phi))) + n(t), \\ x_2(t) &= ((1 + M_{AM}\cos(2\pi f_m t + \phi))A_2)\cdot \\ &\quad \sin(2\pi f_c t + (A_2 M_{FM}\sin(2\pi f_m t + \phi))) + n(t), \end{aligned} \tag{14.17}$$

where M_{AM} denotes amplitude modulation index, M_{FM} frequency modulation index, f_c carrier frequency, f_m baseband frequency, A_1 and A_2 scalars which define signal magnitude, and $n(t)$ the added WGN. In this example, $M_{AM} = 0.5$, $M_{FM} = 20$, $f_c = 500$ Hz,

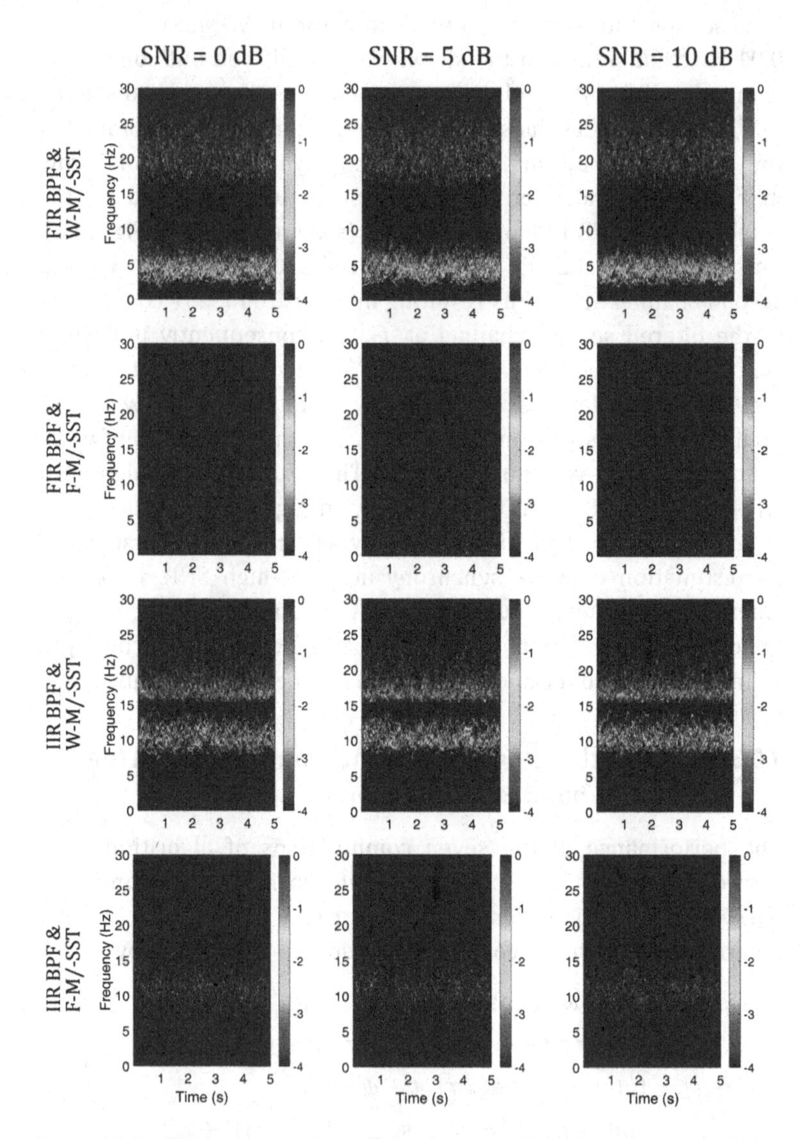

Figure 14.4 TF representations of the degrees of synchrony in a synthetic non-linear bivariate signal estimated using four different combinations of algorithms. The TF representations are shown in logarithmic scale. The SNR of the second channel governed by WGN noise was set to 0 dB, 5 dB and 10 dB (left, centre and right panels, respectively).

$f_m = 11$ Hz, $A_1 = 1$, $A_2 = 0.5$, and the sampling frequency $f_s = 1$ kHz. The SNR of both the channels was set at 3 different values (0 dB, 5 dB and 10 dB). For each combination of the algorithms 30 realisations were performed, whereby different values of phase ϕ were set in different realisations, ranging from 0 to 2π, that is, $\phi = \frac{q-1}{30}(2\pi)$, where q is the realisation number. The NA-MEMD was employed to extract a narrowband bivariate signal of 7–15 Hz ($f_m \pm 4$ Hz) from the corrupted wideband bivariate signal. The corrupted signal was also alternatively band-pass filtered to 7–15 Hz by FIR and IIR BPFs with the same orders as in the previous example. The degrees of synchrony in the filtered narrowband bivariate signals were then estimated as previously described.

Figures 14.4 and 14.5 show TF representations of degrees of synchrony in the non-linear narrowband bivariate signal estimated using the different combinations of algorithms. Using W-M/-SST combined with FIR BPF, IIR BPF, or NA-MEMD, there exist noticeable spurious synchronies outside the band of interest (see first and third rows in Fig. 14.4, and first row in Fig. 14.5). This is because there still existed a considerable (FIR BPF) and a small (IIR BPF and NA-MEMD) amount of noise—the non-linear signal outside the band of interest—after applying the FIR BPF, IIR BPF and NA-MEMD, which were consequently localised by the joint frequency estimator in the CWT-based MSST, which is sensitive to noise (Ahrabian et al., 2015). Synchrony in the band of interest was not exhibited using the combination of FIR BPF and F-M/-SST (second row in Fig. 14.4), and was poorly estimated using the combination of IIR BPF and F-M/-SST (see fourth row in Fig. 14.4). The proposed ISC algorithm (NA-MEMD & F-M/-SST, second row in Fig. 14.5) did not produce spurious synchrony outside the band of interest, and exhibited high concentration of synchrony around 11 Hz (f_m), outperforming IPS (last row in Fig. 14.5).

Figure 14.6 shows average SNRs of the TF representations of synchrony produced by the seven combinations previously shown in Figs. 14.2–14.5. The degrees of synchrony estimated at the frequencies f_1 and f_3 in the linear bivariate signal and those in the band of interest in the non-linear bivariate signal were deemed 'signals', and the degrees of synchrony outside these were deemed 'noise'. The proposed ISC algorithm had, respectively, 4.33 dB and

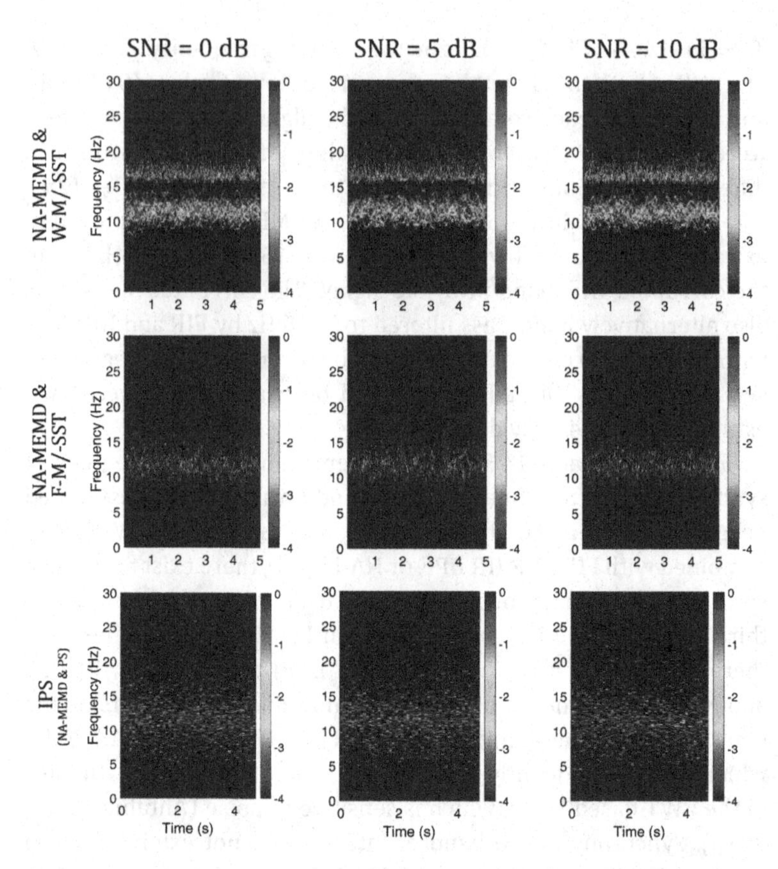

Figure 14.5 TF representations of the degrees of synchrony in a synthetic non-linear bivariate signal estimated using three different combinations of algorithms. The TF representations are shown in logarithmic scale. The ISC algorithm, NA-MEMD and F-M/-SST, performed the best due to the identified synchrony in the band of interest and no spurious synchrony outside of this band.

5.71 dB higher SNRs than the second best combination, i.e. IIR BPF and F-M/-SST, in estimating degrees of synchrony in the synthetic linear and non-linear bivariate signals. The ISC algorithm outperformed the others in both of the tasks, because essentially it did not produce spurious synchrony, its performance increased with SNR, and it exhibited highly localised synchrony at the frequencies of interest, making it the most reliable among them. It should also be

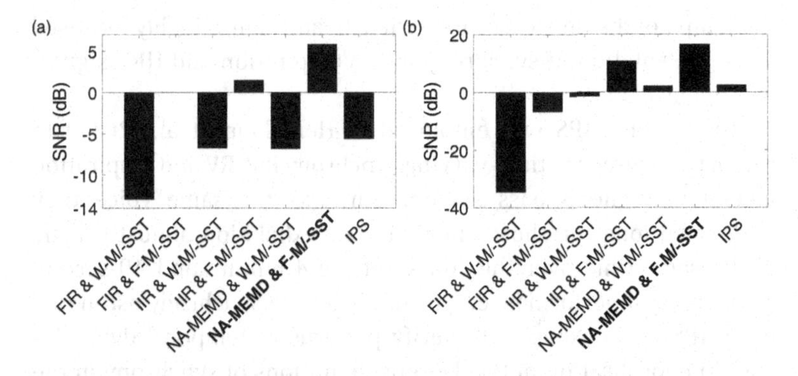

Figure 14.6 Average SNRs of TF representations of signal dependence estimated using different algorithms. (a) Average SNRs of TF representations of synchrony in the synthetic linear bivariate signal. (b) Average SNRs of TF representations of synchrony in the synthetic non-linear bivariate signal.

noted that: (i) both FIR and IIR filters generally introduce several problems, such as group delay and non-linear phase response; (ii) these filters are not data-driven and not designed specifically for non-linear and non-stationary signals as opposed to NA-MEMD; and (iii) CWT produces mathematical artefacts which degrade the performance of W-MSST.

14.3.3 Highly-Localised HRV- and Respiration-Based Choral Synchrony Analysis

Heart rate variability (HRV) represents variations of the cardiac cycle caused by the interplay between the sympathetic (SNS) and parasympathetic (PNS) nervous systems. The low frequency (LF) band of the HRV, 0.04–0.15 Hz, is linked to the interaction of the SNS and PNS, while the high frequency (HF) band, 0.15–0.4 Hz, primarily reflects the activity of PNS. The work in Hemakom et al. (2016) has shown that during choral singing, HRV and respiration of a subset of five singers synchronised during both the rehearsal and the performance. That study, however, examined the average level of cooperation by averaging the PSI of the respiratory and HRV signals of 3 different voices (1 tenor, 1 soprano and 3 bass singers) over time. This did not allow for the quantification of the level of cooperation of specific voices and at different time instants where the tempo of the musical score varied. In addition, a highly localised TF representation of synchrony in the respiration and HRV signals was not available.

To this end, IPS was employed in Hemakom et al. (2017) to reveal the degrees of time-varying synchrony in HRV and respiration among only the 3 bass singers, since as the same voice they performed long- or short- inhalation or exhalation almost exactly at the same time over the course of the 4-minute and 40-second rehearsal and performance. This time-varying synchrony estimated using IPS was next used to verify patterns of temporal dynamics and time localisation of the TF representations of synchrony in the bass singers' HRV and respiration, obtained using the proposed ISC algorithm.

The ECG and respiratory data recordings were conducted at Union Chapel, London, UK, in March 2015. A male conductor, a subset of 3 bass singers of the Eric Whitacre's choir participated in the recordings of their physiological signals during a 4-minute and 40-second rehearsal of a music piece called 'Lux Nova'. All the same subjects also participated in the recordings during a public performance of the same music piece. During both the rehearsal and the performance, ECG signals were recorded with three Ag/AgCl ECG monitoring electrodes placed on the skin, just below the collar bone; one electrode on the left hand side for the signal and the other two electrodes on the right hand side for the ground and the reference. Respiratory signals were recorded using a custom-made respiration belt placed around the chest. The respiratory signals were first low-pass filtered using an FIR low-pass filter of order 10 with the cut-off frequency set to 10 Hz, to prevent anti-aliasing, and the low-pass filtered respiratory signals were next downsampled from 1 kHz to 10 Hz by performing decimation by a factor of 100. The HRV was estimated from the ECG data using the MF-HT algorithm (Chanwimalueang et al., 2015).

The respiratory (or HRV) signals of the 3 bass singers during both the rehearsal and performance were used to form 3-channel data which was decomposed using NA-MEMD with 10 adjacent WGN channels. The IMFs with indices 3–6 produced by the NA-MEMD of the 3-channel multivariate HRV (or respiratory) signal of the bass singers contained the physically meaningful frequency range 0.04 Hz to 0.4 Hz, that is, exactly the LF/HF frequency band of HRV. The full band of interest in both the HRV and respiration data was produced by summing up the corresponding IMFs, in order to obtain the desired scale in data. The time-varying PSI values among the singers were obtained by averaging PSI values calculated from 3 combined-IMF pairs of the 3 data channels (pair 1: first and second bass singers, pair 2: first and third bass singers, pair 3: second and third bass singers), whereby the PSI values for each pair were computed using 20-second sliding windows with 18 seconds overlap (2-second increment) for the respiratory signals, and 40-second sliding windows with 36 seconds overlap (4-second increment) for the HRV signals. The PSI indices between the combined-IMFs of the noise channels were also estimated in order to provide the PSI of random signals as a baseline.

The ISC was performed on each of the same 3 combined-IMF pairs of the 3 data channels of the respiratory (or HRV) signals to obtain 3 TF representations of synchrony in each of the signals for all the pairs. The 3 TF representations were next averaged in order to obtain a TF representation of synchrony in the respiratory (or HRV) signals among the bass singers.

Figures 14.7a and 14.8a show the timings of long or short exhalation of the bass singers when they performed in unison, and when they remained silent or inhaled. Figures 14.7b and 14.8b show degrees of synchrony in, respectively, the respiratory and HRV signals during the rehearsal and the performance, estimated using IPS with 30 realisations of NA-MEMD. Observe that the respiratory synchrony in both situations (see Fig. 14.7b) reached their peaks approximately during a series of long exhalation, while the HRV synchrony during the performance (see the broken line in Fig. 14.8b) reached its first and last peaks during a series of relatively long inhalation (approximately at the 50th and 220th seconds). The increased synchrony in HRV was presumably due to synchronised acceleration of heart rates of the singers. Figures 14.7e–f and 14.8e–f) show TF representations of, respectively, respiratory and HRV synchrony during the rehearsal and the performance estimated using the proposed STFT-based ISC. In both situations, the respiratory synchrony (see Fig. 14.7e–f) was highly localised at 0.1 Hz due to a series of long exhalations they had to

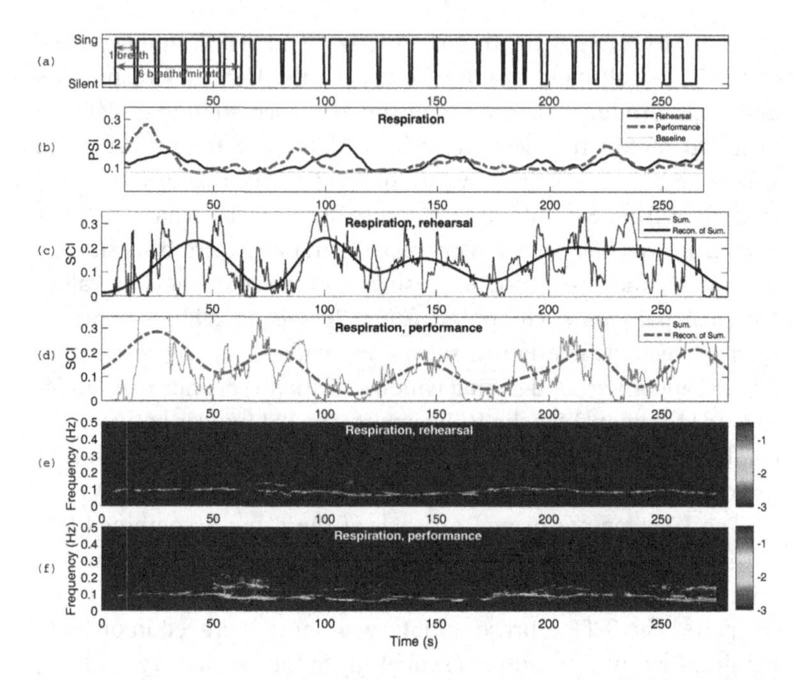

Figure 14.7 Intrinsic synchrony in respiratory signals among the bass singers during the rehearsal and the performance estimated using IPS and STFT-based ISC. (a) Singing timing of the bass singers. (b) The PSI of the respiratory signals. (c) The sum of SCI values across frequencies of the respiratory signals during the rehearsal, and its reconstruction from IMFs with indices 8–11. (d) The sum of SCI values across frequencies of the respiratory signals during the performance, and its reconstruction from IMFs with indices 8–11. (e) The SCI of the respiratory signals during the rehearsal, shown in logarithmic scale. (f) The SCI of the respiratory signals during the performance, shown in logarithmic scale. For brevity, the sum of SCI values are denoted by Sum., and the reconstruction of the sum of SCI values from IMFs with indices 8–11 by Recon. of Sum. Observe: (i) similar patterns of temporal dynamics between the Recon. of Sum. and the corresponding PSIs; and (ii) time instants of peaks in the Recon. of Sum. are close to those of the peaks in the corresponding PSIs.

perform in unison. This exemplifies that the singers were heavily demanded by the musical score to breath in unison at a very slow rate of 6 breaths per minute (breaths per minute = 60 * breathing frequency). This rate is slower than the normal breathing rate in adults which varies between 12 and 15 breaths per minutes

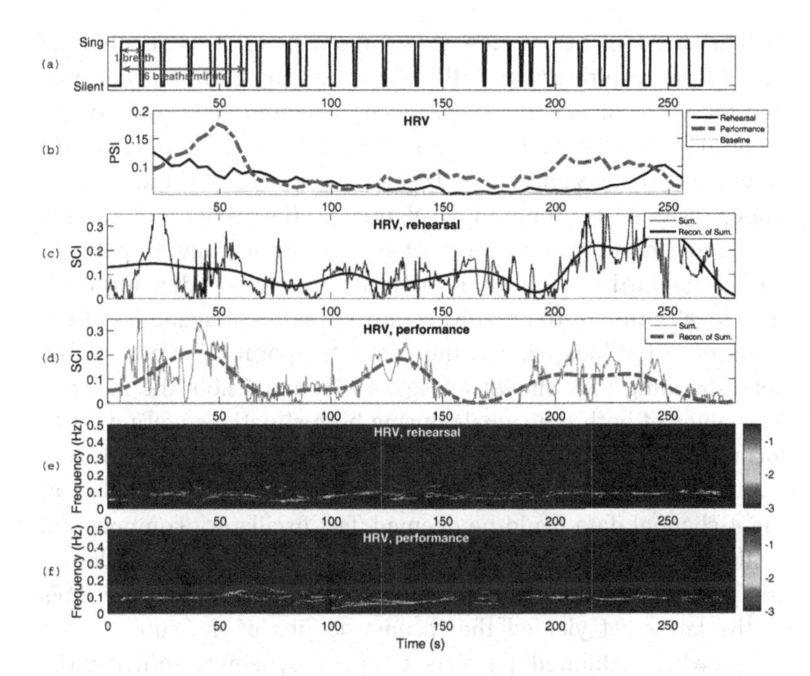

Figure 14.8 Intrinsic synchrony in HRV signals among the bass singers during the rehearsal and the performance estimated using IPS and STFT-based ISC. (a) Singing timing of the bass singers. (b) The PSI of the HRV signals. (c) The sum of SCI values across frequencies of the HRV signals during the rehearsal, and its reconstruction from IMFs with indices 8–11. (d) The sum of SCI values across frequencies of the HRV signals during the performance, and its reconstruction from IMFs with indices 8–11. (e) The SCI of the HRV signals during the rehearsal, shown in logarithmic scale. (f) The SCI of the HRV signals during the performance, shown in logarithmic scale. For brevity, the sum of SCI values are denoted by Sum., and the reconstruction of the sum of SCI values from IMFs with indices 8–11 by Recon. of Sum. Observe: (i) similar patterns of temporal dynamics between the Recon. of Sum. and the corresponding PSIs; and (ii) time instants of the peaks in the Recon. of Sum. are close to those of the peaks in the corresponding PSIs.

(0.2–0.25 Hz) (Barrett et al., 2012). The HRV synchrony (see Fig. 14.8e–f) was dominant in the LF band of HRV, 0.04–0.15 Hz. This conforms with the finding in Hemakom et al. (2016), where ICoh revealed a large proportion of coherence in the HRV signals in the LF band.

The sums of SCI values across frequencies as a function of time of the respiratory and HRV signals during both the rehearsal and the performance are, respectively, shown in Figs. 14.7c–d and 14.8c–d (see thin dotted lines). Observe that the sums of SCI values of both the signals varied dramatically during both situations, indicating that the proposed ISC algorithm effectively yielded highly localised time-varying respiratory and HRV synchrony. To perform straightforward verification of patterns of temporal dynamics and time localisation of the relatively 'faster' temporal dynamics of the sums of SCI values against the 'slow' temporal dynamics of the corresponding PSIs shown in Figs. 14.7b and 14.8b, the sums of SCI values of both the signals during both situations were used to form 4-channel data which was decomposed using MEMD. It was empirically found that the multivariate IMFs with indices 1–7 of the 4-channel data could be deemed 'fast oscillatory components', while the multivariate IMFs with indices 8–11 'slow oscillatory components', and that the combinations of the multivariate IMFs in the latter set yielded the reconstructions of the sums of SCI values which exhibited: (i) 'slow' temporal dynamics similar to the 'slow' temporal dynamics of the corresponding PSIs for both the signals and in both situations; and (ii) time instants of peaks in the reconstructions of the sums of SCI values of both the signals and in both situations being close to those of the peaks in the corresponding PSI in Figs. 14.7b and 14.8b. Observe similar patterns of temporal dynamics and close positions of the peaks between: (i) the solid lines in Figs. 14.7b and 14.7c; (ii) the thick broken lines in Figs. 14.7b and 14.7d; (iii) the solid lines in Figs. 14.8b and 14.8c; and (iv) the thick broken lines in Figs. 14.8b and 14.8d.

The degrees of synchrony in the bass singers' respiratory and HRV signals during both the situations using CWT-based ISC, which combines the conventional CWT-based SST and MSST algorithms, was also examined, as shown in Fig. 14.9a–d. Since the joint frequency estimator in the CWT-based MSST is sensitive to noise (Ahrabian et al., 2015) and CWT produced mathematical artefacts (additional noise), the TF representations produced by the CWT-based ISC exhibited noticeable spurious synchrony at several frequencies.

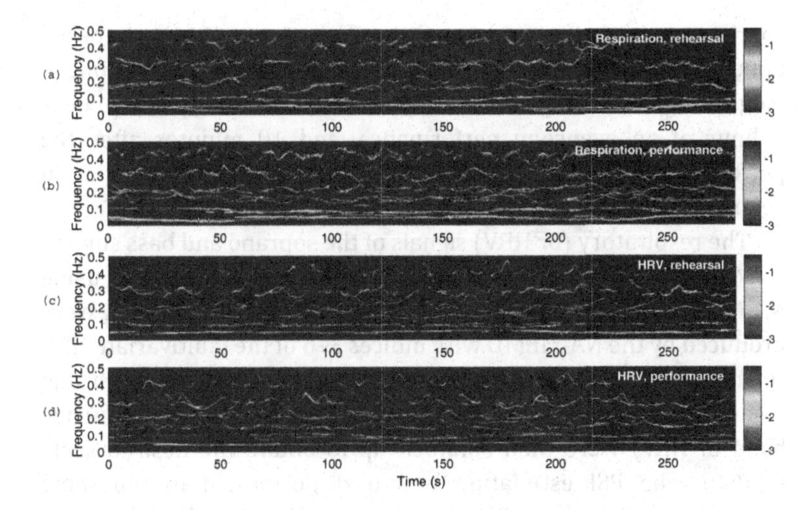

Figure 14.9 Intrinsic synchrony in respiratory and HRV signals among the bass singers during the rehearsal and performance estimated using CWT-based ISC. The TF representations are shown in logarithmic scale. (a) The SCI of the respiratory signals during the rehearsal. (b) The SCI of the respiratory signals during the performance. (c) The SCI of the HRV signals during the rehearsal. (d) The SCI of the HRV signals during the performance. Observe mathematical artefacts shown in the red boxes. They are introduced by the CWT.

14.3.4 Large Scale HRV- and Respiration-based Choral Synchrony Analysis

In addition to Section 14.3.3 where the degrees of synchronisation in the respiratory and HRV signals were highly localised in frequency and time using the proposed ISC algorithm, it is equally important to establish a measure which provides physically meaningful and straightforward interpretation of the level of cooperation in the scenarios where physiological responses are acquired from a long period of time. To this end, the proposed N-IPS algorithm was employed to empirically quantify physically meaningful and straightforward to interpret trends in the degrees of synchronisation in physiological signals of choir singers while performing an evensong.

The ECG and respiratory data recordings were conducted at St Martin-in-the-Fields church, London, UK, in July 2016. Five soprano choir singers and four bass choir singers of the Imperial College Chamber Choir participated in the recordings of their physiological signals during 10 minutes before an evensong performance, 1 hour of an evensong performance, and 10 minutes after the performance. The data were acquired in the same manner as in Section 14.3.3.

The respiratory (or HRV) signals of the soprano and bass singers were used to form multi-channel data which was decomposed using NA-MEMD with 10 adjacent WGN channels. The IMFs which were produced by the NA-MEMD with indices 3–6 of the multivariate HRV (or respiratory) signal of the singers and contained the physically meaningful frequency range 0.04 Hz to 0.4 Hz (the LF/HF frequency band of HRV) were then summed up to obtain the desired scale in data. The PSI estimation was next performed in the same manner as in Section 14.3.3. The time-varying PSI indices between the combined-IMFs of the signal and noise channels (baseline) were next used to form multi-channel data which were further decomposed using NA-MEMD with 10 adjacent WGN channels. Certain IMFs of the multivariate synchrony produced by the NA-MEMD which exhibited physically meaningful and interpretable trends in synchrony were then combined.

Figure 14.10a–d shows physically meaningful trends in synchrony in the sopranos' and basses' respiratory and HRV signals, estimated from 30 realisations of NA-MEMD. Observe that all the 'raw' synchrony obtained using the standard IPS algorithm (solid thin lines) varied dramatically during the course of events (singing and pausing), thus being less amenable to physical interpretation. Trends in the synchrony obtained using the proposed N-IPS algorithm, on the other hand, exhibited smooth variations of synchrony during the events, and smooth transitions between them. This offers more physically meaningful and interpretable results, that is, the levels of cooperation markedly increased during most of the songs (depicted by the upward pointing arrows), and decreased during the long pauses (the downward pointing arrows). Variations in the degrees of synchrony (levels of cooperation) were because the breathing rhythms, and by virtue of RSA the cardiac activities too, of the singers were modulated by the pieces of music performed, thus exhibiting stronger dynamic coupling reflected in

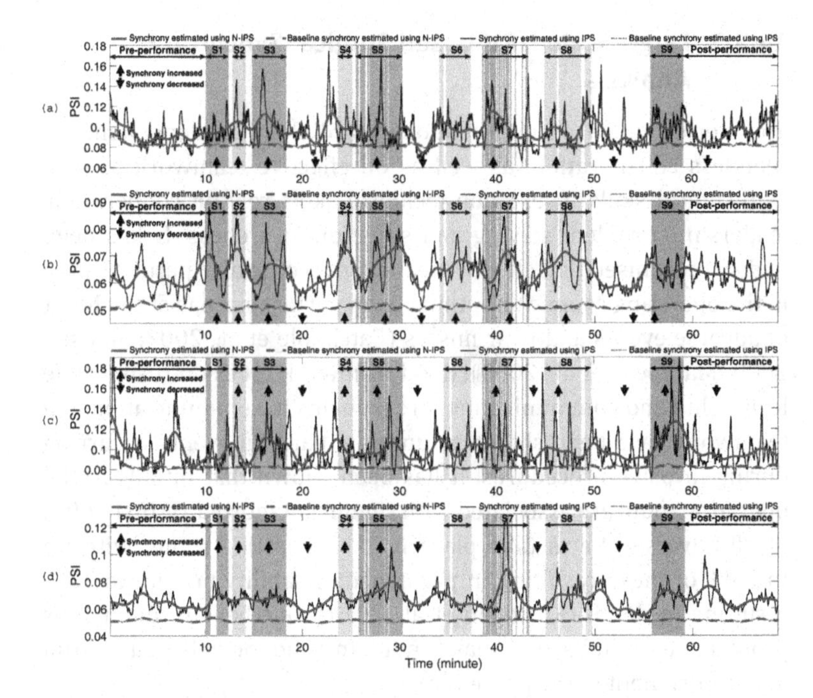

Figure 14.10 Time-varying synchrony in the sopranos' and basses' respiratory and HRV signals. (a) Respiratory PSI among the sopranos. (b) HRV PSI among the sopranos. (c) Respiratory PSI among the basses. (d) HRV PSI among the basses. Degrees of raw synchrony (thin solid lines) and the corresponding baselines (thin broken lines) were estimated using IPS. Trends of the raw synchrony (thick solid lines) and the corresponding baselines (thick broken lines) were estimated using N-IPS. Time instants when the singers sang during the 9 songs (S1–S9) are depicted by the shaded areas. Observe increases in the synchrony of their physiological responses during most of the songs depicted by the upward pointing arrows. These are due to their breathing rhythms and HRV being modulated by the demands of pieces of music.

increases in the synchrony of their physiological responses. During the pauses, however, their breathing rhythms were not dictated by any piece of music, thus resulting in decreases in the synchrony. It must be emphasised that the empirical quantification of physically meaningful and straightforward to interpret trends in synchrony is focused on here, which is a first step towards prediction.

14.3.5 Large-Scale HRV-Based Surgical Synchrony Analysis

The performance of interventional cardiology is dependent on seamless collaboration, as well as on effective teamwork as well as technical skills. These non-technical skills rely on cooperation, leadership, problem solving and situation awareness. It has been widely recognised that effective teamwork is crucial to patient safety in the operating room. Suboptimal team performance is at the heart of adverse events and near misses (Catchpole et al., 2007; Manser, 2009; Mazzocco et al., 2009). Good teamwork relies on responsible leadership and communication. Breakdowns in communication and teamwork have been shown to underlie harmful events occurring during invasive procedures. It has been shown that pitfalls in the operating room are associated with significant morbidity. Up to 60% of all adverse effects take place in the operating room, with up to 33% of these resulting in permanent disability and up to 13% resulting in death (Kurmann et al., 2014). Glitches are defined as problems and faults that may hinder, impede, or delay successful accomplishment of the procedure.

Understanding team dynamics in the operating room or interventional suite, how teams function, and the factors that facilitate or impede teamwork is critical to the achievement of high-quality care and patient safety. A large body of evidence in psychology and growing numbers of surgical studies suggest that excessively high levels of intraoperative stress are disadvantageous to performance, and, consequently, to patient care (Adjei et al., 2017; Arora et al., 2010; Hull et al., 2011). In technically challenging tasks, an appropriate level of stress is important for performance, however, greater levels of stress and mental strain may impair both non-technical skills and operative performance, potentially leading to an adverse impact on patient safety and outcomes (Catchpole et al., 2008; Mazzocco et al., 2009). Multi-disciplinary interventional cardiology catheterisation laboratory (Cath Lab) teams are composed of an interventional cardiology consultant, a cardiology registrar, a nurse, a physiologist and a radiographer, each with different background in terms of training and competencies. These team members are expected to work and function together to achieve an optimal outcome. Close cooperation between physicians performing the procedure (cardiology consultant and cardiology registrar) and physiologist outside the suite recording and monitoring pressures during coronary intervention is essential.

The ECG signals were recorded from a Cath Lab team starting 10 minutes before and during a percutaneous coronary intervention (PCI) procedure, which was 1 hour and 56 minutes long, and was carried out in the Cath Lab at Hammersmith Hospital, Imperial College Healthcare NHS Trust, London, in August 2016. During the procedure, team members were closely observed by an independent observer, who noted times and types of any glitches that occurred. Based on the published literature, possible glitches have been divided into 15 categories (Morgan et al., 2013) as follows:

(1) Absence—absence of theatre staff member when required (e.g. circulating nurse not available to obtain equipment)
(2) Communication—difficulties in communication among team members (e.g. repeat requests, incorrect terminology, and misinterpretations)
(3) Distractions—anything causing distraction from the task (e.g. phone calls or bleeps and loud music requiring to be turned down)
(4) Environment—aspects of the working environment causing difficulties (e.g. low lighting or variable temperature causing difficulties)
(5) Equipment design—issues arising from equipment design that would not otherwise be corrected with training or maintenance (e.g. compatibility problems with different implant system and equipment blockage)
(6) Maintenance—faulty or poorly maintained equipment (e.g. battery depleted during use and blunt equipment)
(7) Health and safety—any observed risk to the personnel (e.g. mask violations and food or drink in the theatre)
(8) Planning and preparations—instances that may have otherwise been avoided with appropriate prior planning and preparation (e.g. insufficient equipment resources and staffing levels and planning)

(9) Patient related—issues related to the physiological status of the patient (e.g. difficulty in extracting previous implants and unexpected anatomically related surgical difficulty and anaphylaxis)
(10) Process deviation—incomplete or reordered completion of standard task (e.g. unnecessary equipment opened)
(11) Slips—psychomotor errors (e.g. dropped instrument)
(12) Training—repetition or delay of operative steps due to training (e.g. consultant corrects assistant operating technique)
(13) Workplace—equipment or theatre layout issues (e.g. desterilising of equipment and scrubbed staff on environment)
(14) Change of the personnel during procedure
(15) All staff present during WHO check (yes/no).

The electrode placement for ECG recordings, data acquisition, HRV estimation, and PSI estimation were carried out in the same manner as in Section 14.3.3. After applying IPS to quantify degrees of synchrony in the HRV signals, the time varying PSI indices between IMFs of the HRVs and noise channels (baseline) were used to form multichannel data, which was further decomposed using NA-MEMD with 10 additional WGN channels. The IMFs produced by the NA-MEMD with indices 10–12 of the multivariate PSI values were combined to produce trends in the synchrony, since these combinations best represented changes in trends of the synchrony in response to glitches. Figure 14.11 shows trends in synchrony in HRV data during the recorded procedure, which exhibited cooperation and responses to glitches between the following pairs: (i) cardiology consultant—cardiology registrar, (ii) cardiology consultant—physiologist, and (iii) cardiology registrar—physiologist. Note that no specific types of glitches were expected to be observed during the procedure. Recorded glitches during the procedure included the following groups: (i) group 3—distractions—anything causing distraction from task (phone calls/bleeps, loud music, alarms), and (ii) group 5—equipment design—issues arising from equipment design that would not otherwise be corrected with training or maintenance (e.g. compatibility problems with different implants or wires, or equipment failure).

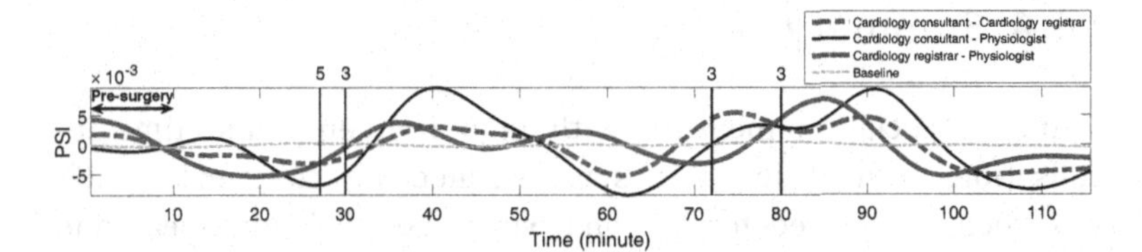

Figure 14.11 Trends in synchrony in the cardiologists' and physiologist' HRV data. Baseline of the degree of synchrony (broken thin line). Trend in the degree of synchrony between a cardio-consultant and a cardio-registrar (broken thick line), a cardio-consultant and a physiologist (solid thin line), and a cardio-registrar and a physiologist (solid thick line). The trends were estimated using the N-IPS algorithm, whereby 40-second sliding windows with 36 seconds overlap (4 second increment) were used in the PSI estimation. Two groups of glitches were recorded during the procedure: group 3 (distractions) and group 5 (equipment design). The recording was 116 minutes long.

Figure 14.11 shows that physiological responses of pairs of professionals responded to the recorded glitches with increased synchrony in HRV. Observe that the first glitch at the 27th minute caused immediate negative deflections (increases) in trends of HRV synchrony between (i) the cardiology consultant and the physiologist (thin solid line) and (ii) the cardiology consultant and the cardiology registrar (thick broken line). Also, the third glitch at the 72nd minute caused another immediate negative deflection (increase) in a trend of HRV synchrony between the cardiology registrar and the physiologist (thick solid line). These findings suggest these glitches were impeding the procedure and may affect procedural duration as well as outcome.

A growing body of literature suggests that shared knowledge and understanding of each other's roles and objectives during the invasive and high risk procedures in turn facilitates team cooperation and coordination. For a team to function efficiently, its members should share a 'mental model' of the team's tasks, objectives, means and environment (Cannon-Bowers and Salas, 2001; Klimoski and Mohammed, 1994; Mathieu et al., 2000). It is hoped that this prospective research will help to facilitate better training programme across UK hospitals in order to deliver better care to the patients.

14.4 Summary

A data association measure with high frequency and time localisation, termed intrinsic synchrosqueezing coherence (ISC), has been introduced for the analysis of coupled non-linear and non-stationary multivariate signals. This has been achieved through the combination of multivariate signal analysis using NA-MEMD and the generation of highly localised time-frequency univariate and multivariate representations of the multivariate intrinsic oscillations using the STFT-based SST and MSST algorithms. The ISC algorithm has enabled precise identification of physiological dependence in both frequency and time. The performance of the ISC algorithm has been evaluated against other combinations of algorithms in the tasks of estimating synchrony in linear and non-linear bivariate signals. The ISC algorithm has been shown to outperform the other combinations of algorithms and exhibit higher average SNR of TF representations of signal dependence. An application of the ISC algorithm to the quantification of inter-channel dependence in respiratory and HRV signals has been demonstrated. The ISC algorithm has highly exhibited localised synchrony in the respiratory and HRV signals among a subset of 3 bass singers of a choir during both a rehearsal and a performance. The STFT-based ISC algorithm has been shown to outperform CWT-based ISC, which was built upon CWT-based SST and MSST, in the localisation of synchronised frequencies in respiratory and HRV signals.

In addition to the ISC algorithm, an extension of IPS has been introduced and is referred to as nested intrinsic phase synchrony (N-IPS) as a meaningful and straightforward to interpret data association metric for trends in the level of cooperation. This is achieved by first employing the standard IPS to quantify intrinsic phase relationship between data channels, and then further decomposing time series of the multivariate degrees of phase synchrony into multiple scales, whereby certain intrinsic scales which contain physically meaningful and straightforward interpretation are then combined. This algorithm allows for empirical quantification of physically meaningful and straightforward to interpret trends in phase synchrony. Two applications of the N-IPS algorithm to the empirical quantification of physically meaningful trends in the level of cooperation through the empirical estimation of trends in synchrony in respiratory and HRV signals have been demonstrated. The N-IPS algorithm has effectively quantified physically meaningful increases and decreases in trends in levels of cooperation among subsets of soprano and bass singers of a choir during a 1-hour evensong performance. It has also revealed significant increases in trends in levels of cooperation between pairs of cardiologists and physiologist when certain types of glitches occurred during a surgery. This is a first attempt to empirical quantification of intrinsic, physically meaningful and interpretable trends in the level of cooperation of long collaborative tasks.

References

Adjei, T., von Rosenberg, W., Goverdovsky, V., Powezka, K., Jaffer, U., and Mandic, D. P. (2017). Pain prediction from ECG in vascular surgery, *IEEE J. Transl. Eng. Health Med.*, p. in print.

Ahrabian, A., Looney, D., Stanković, L., and Mandic, D. P. (2015). Synchrosqueezing-based time-frequency analysis of multivariate data, *Signal Process.* **106**, 1, pp. 331–341.

Arora, S., Tierney, T., Sevdalis, N., Aggarwal, R., Nestel, D., Woloshynowych, M., Darzi, A., and Kneebone, R. (2010). The Imperial Stress Assessment Tool (ISAT): A feasible, reliable and valid approach to measuring stress in the operating room, *World J. Surg.* **34**, 8, pp. 1756–1763.

Barrett, K. E., Barman, S. M., Boitano, S., and Brooks, H. L. (2012). *Ganong's Review of Medical Physiology*, 23rd edn. (McGraw-Hill Education/Medical).

Bhattacharyya, S., Biswas, A., Mukherjee, J., Majumdar, A. K., Majumdar, B., Mukherjee, S., and Singh, A. K. (2013). Detection of artifacts from high energy bursts in neonatal EEG, *Comput. Biol. Med.* **43**, 11, pp. 1804–1814.

Billman, G. (2013). The LF/HF ratio does not accurately measure cardiac sympatho-vagal balance, *Front. Physiol.* **4**, pp. 26–1 - 26–5.

Burger, C., and van den Heever, D. J. (2015). Removal of EOG artefacts by combining wavelet neural network and independent component analysis, *Biomed. Signal Process. Control* **15**, 1, pp. 67–79.

Cannon-Bowers, J. A., and Salas, E. (2001). Reflections on shared cognition, *J. Organ. Behav.* **22**, 2, pp. 195–202.

Catchpole, K., Mishra, A., Handa, A., and McCulloch, P. (2008). Teamwork and error in the operating room: Analysis of skills and roles, *Ann. Surg.* **247**, 4, pp. 699–706.

Catchpole, K. R., Giddings, A. E., Wilkinson, M., Hirst, G., Dale, T., and de Leval, M. R. (2007). Improving patient safety by identifying latent failures in successful operations, *Surgery* **142**, 1, pp. 102–110.

Chanwimalueang, T., von Rosenberg, W., and Mandic, D. P. (2015). Enabling R-peak detection in wearable ECG: Combining matched filtering and Hilbert transform, in *Proc. IEEE Int. Conf. on Digital Signal Processing (DSP 2015), Singapore, 21–24 July 2015* (IEEE, Piscataway, NJ), pp. 134–138.

Daubechies, I., Lu, J., and Wu, H.-T. (2011). Synchrosqueezed wavelet transforms: An empirical mode decomposition-like tool, *Appl. Comput. Harmon. Anal.* **30**, 1, pp. 243–261.

Flandrin, P., Rilling, G., and Gonçalvès, P. (2004). Empirical mode decomposition as a filter bank, *IEEE Signal Process. Lett.* **11**, 2, pp. 112–114.

Hemakom, A., Goverdovsky, V., Aufegger, L., and Mandic, D. P. (2016). Quantifying cooperation in choir singing: Respiratory and cardiac synchronisation, in *Proc. IEEE Int. Conf. on Acoustics, Speech and Signal Processing (ICASSP 2016), Shanghai, China, 20–25 March 2016* (IEEE, Piscataway, NJ), pp. 719–723.

Hemakom, A., Powezka, K., Goverdovsky, V. G., Jaffer, U., and Mandic, D. P. (2017). Quantifying team cooperation through intrinsic multi-scale measures: Respiratory and cardiac synchronisation in choir singers and surgical teams, *R. Soc. Open Sci.* **4**, 1, pp. 170853-1–170853-23.

Hlastala, M. P., and Berger, A. J. (2001). *Physiology of Respiration*, 2nd edn. (Oxford University Press, Oxford, UK).

Hu, J., Wang, C.-S., Wu, M., Du, Y.-X., He, Y., and She, J. (2015). Removal of EOG and EMG artifacts from EEG using combination of functional link neural network and adaptive neural fuzzy inference system, *Neurocomputing* **151**, 1, pp. 278–287.

Huang, N. E., Shen, Z., Long, S. R., Wu, M. C., Shih, H. H., Zheng, Q., Yen, N.-C., Tung, C. C., and Liu, H. H. (1998). The empirical mode decomposition and the Hilbert spectrum for nonlinear and non-stationary time series analysis, *Proc. R. Soc. A* **454**, 1971, pp. 903–995.

Hull, L., Arora, S., Kassab, E., Kneebone, R., and Sevdalis, N. (2011). Assessment of stress and teamwork in the operating room: An exploratory study, *Am. J. Surg.* **201**, 1, pp. 24–30.

Islam, M. K., Rastegarnia, A., and Yang, Z. (2016). Methods for artifact detection and removal from scalp EEG: A review, *Clin. Neurophysiol.* **46**, 1, pp. 287–305.

Jafarifarmand, A., and Badamchizadeh, M. A. (2013). Artifacts removal in EEG signal using a new neural network enhanced adaptive filter, *Neurocomputing* **103**, 1, pp. 222–231.

Jänig, W. (2006). *Integrative Action of the Autonomic Nervous System* (Cambridge University Press, Cambridge, UK).

Klados, M. A., Papadelis, C., Braun, C., and Bamidis, P. D. (2011). REG-ICA: A hybrid methodology combining blind source separation and regression techniques for the rejection of ocular artifacts, *Biomed. Signal Process. Control.* **6**, 3, pp. 291–300.

Klimoski, R., and Mohammed, S. (1994). Team mental model: Construct or metaphor? *J. Manag.* **20**, 2, pp. 403–437.

Kurmann, A., Keller, S., Tschan-Semmer, F., Seelandt, J., Semmer, N. K., Candinas, D., and Beldi, G. (2014). Impact of team familiarity in the operating room on surgical complications, *World J. Surg.* **38**, 12, pp. 3047–3052.

Lawhern, J., Hairston, W. D., McDowell, K., Westerfield, M., and Robbins, K. (2012). Detection and classification of subject-generated artifacts in EEG signals using autoregressive models, *J. Neurosci. Methods* **208**, 2, pp. 181–189.

Looney, D., Hemakom, A., and Mandic, D. P. (2014). Intrinsic multi-scale analysis: A multivariate EMD framework, *Proc. R. Soc. A* **471**, 2173, pp. 20140709-1 - 20140709-28.

Mandic, D. P., ur Rehman, N., Wu, Z., and Huang, N. E. (2013). Empirical mode decomposition-based time-frequency analysis of multivariate signals, *IEEE Sig. Proc. Mag.* **30**, 6, pp. 74–86.

Manser, T. (2009). Teamwork and patient safety in dynamic domains of healthcare: A review of the literature, *Acta. Anaesthesiol. Scand.* **53**, 2, pp. 143–151.

Mathieu, J. E., Heffner, T. S., Goodwin, G. F., Salas, E., and Cannon-Bowers, J. A. (2000). The influence of shared mental models on team process and performance, *J. Appl. Psychol.* **85**, 2, pp. 273–283.

Mazzocco, K., Petitti, D. B., Fong, K. T., Bonacum, D., Brookey, J., Graham, S., Lasky, R. E., Sexton, J. B., and Thomas, E. J. (2009). Surgical team behaviors and patient outcomes, *Am. J. Surg.* **197**, 5, pp. 678–685.

Morgan, L., Robertson, E., Hadi, M., Catchpole, K., Pickering, S., New, S., Collins, G., and McCulloch, P. (2013). Capturing intraoperative process deviations using a direct observational approach: The glitch method, *BMJ Open* **3**, 11, p. 003519.

Oberlin, T., Meignen, S., and Perrier, V. (2014). The Fourier-based synchrosqueezing transform, in *Proc. IEEE Int. Conf. on Acoustics, Speech and Signal Processing (ICASSP 2014), Florence, Italy, 4–9 May 2014* (IEEE, Piscataway, NJ), pp. 315–319.

Radeka, V. (1969). 1/|f|noise in physical measurements, *IEEE Trans. Nucl. Sci.* **16**, 5, pp. 17–35.

Shao, S.-Y., Shen, K.-Q., Ong, C. J., Wilder-Smith, E. P. V., and Li, X.-P. (2008). Automatic EEG artifact removal: A weighted support vector machine approach with error correction, *IEEE Trans. Biomed. Eng.* **56**, 2, pp. 336–344.

Silverthorne, D. U. (2009). *Human Physiology: An Integrated Approach*, 4th edn. (Pearson/Benjamin Cummings).

Soomro, M. H., Badruddin, N., Yusoff, M. Z., and Malik, A. S. (2013). A method for automatic removal of eye blink artifacts from EEG based on EMD-ICA, in *Proc. Int. Conf. of the IEEE 9th International Colloquium on Signal Processing and its Applications (CSPA 2013), Kuala Lumpur, Malaysia, 8–10 March 2013* (IEEE, Piscataway, NJ), pp. 129–134.

Tass, P., Rosenblum, M. G., Weule, J., Kurths, J., Pikovsky, A., Volkmann, J., Schnitzler, A., and Freund, H.-J. (1998). Detection of n:m phase locking from noisy data: Application to magnetoencephalography, *Phys. Rev. Lett.* **81**, 15, pp. 3291–3294.

ur Rehman, N., and Mandic, D. P. (2011). Filter bank property of multivariate empirical mode decomposition, *IEEE Trans. Signal Proc.* **59**, 5, pp. 2421–2426.

Vickhoff, B., Malmgren, H., Åström, R., Nyberg, G., Ekström, S.-R., Engwall, M., Snygg, J., Nilsson, M., and Jörnsten, R. (2013). Music structure determines heart rate variability of singers, *Front. Psychol.* **4**, 1, pp. 334–1 - 334–16.

von Rosenberg, W., Chanwimalueang, T., Adjei, T., Jaffer, U., Goverdovsky, V. and Mandic, D. P. (2017). Resolving ambiguities in the LF/HF ratio: LF-HF scatter plots for the categorization of mental and physical stress from HRV, *Front. Physiol.* **8**, pp. 1–12.

Wu, Z., and Huang, N. E. (2004). A study of the characteristics of white noise using the empirical mode decomposition method, *Proc. R. Soc. A* **460**, 2046, pp. 1597–1611.

Zou, Y., Nathan, V., and Jafari, R. (2014). Automatic identification of artifact-related independent components for artifact removal in EEG recordings, *IEEE J. Biomed. Health Inform.* **20**, 1, pp. 73–81.

Part VI

Language Processing, Cooperative Network, and Communications

Chapter 15

Multitask Cooperative Networks and Their Diverse Applications

Saeid Sanei,[a] Sadaf Monajemi,[b] Amir Rastegarnia,[c] Oana Geman,[d] and Ong Sim Heng[b]

[a]*School of Science & Technology, Nottingham Trent University Clifton Lane, Nottingham, NG11 8NS, UK*
[b]*School of Electrical and Electronic Engineering, National University of Singapore, Singapore*
[c]*Electrical Engineering Department, University of Malayer, Malayer, Iran*
[d]*Biomedical Engineering Department, Stefan cel Mare University of Suceava, Suceava, Romania*
saeid.sanei@ntu.ac.uk

The notion of cooperative networks and their extensions to multitask systems are visited here. The details of various network topologies and the related formulations are discussed before considering multitask systems where the groups of agents or nodes in the network can have different objectives. Then, we see how the information about the network structure particularly the strength of correspondences between the agents can enhance the solution. Despite their wide applications in communication systems, it is demonstrated that such networks can also be used in machine learning, medical engineering, security, smart grid, and mathematical modeling. In this chapter, the above

Learning Approaches in Signal Processing
Edited by Wan-Chi Siu, Lap-Pui Chau, Liang Wang, and Tieniu Tan

ISBN 978-981-4800-50-1 (Hardcover), 978-0-429-06114-1 (eBook)
www.panstanford.com

applications are further explained and the implementation results demonstrated.

15.1 Introduction

Adaptive distributed systems and optimization over networks [1–7] is an attractive research area with several applications from signal processing and optimization to information security and modeling of biological and social networks [6–11]. The performance of these self-organized networks depends on learning abilities and localized cooperation of the interconnected nodes of the network [1]. The early studies in cooperative networks consider a single objective for the whole network agents. One important advantage of these systems over centralized networks is that there wouldn't be any burden on a centralized center to fuse and process a large amount of information. The growing complexity of the networked information Systems, on the other hand, exceeds the capabilities of conventional centralized systems. Moreover, in the centralized systems since the information processing is performed in a centralized manner the entire system collapses in the event of central processor failure due to reasons such as device malfunction or communication overload. Last but not least, the nodes may be reluctant to share their sensitive data with a central processor due to privacy considerations

There are also several important cases in which different agents are interested in the estimation of different parameter vectors in a multitask network [12, 13]. In such problems each connected cluster pursues a different objective corresponding to a particular task. Distributed estimation and learning in these multitask networks is quite challenging since the agents can receive misleading information from those with different objectives. Therefore, it is crucially important to develop a fully adaptive mechanism that enables the agents to continuously learn which of their neighbors belongs to the same cluster and which ones are from a different cluster. Despite their applications in communication systems, social networks, security, smart grid, power distribution, modeling biological systems as well as brain multitask activities can be named as the areas of cooperative networking and learning research.

15.2 Cooperative Networks

Distributed information processing and learning over networks is an emerging field where a collection of agents forms an adaptive network to estimate and infer tasks from streaming data in real time. In these systems, distributed agents cooperate and interact with each other on a local level, resulting in the diffusion of information over the entire network. The single task assumption however over-simplifies the multitask problems where there are various sources of intertwined information. In multitask networks, different agents have to track and estimate different objectives simultaneously.

In the distributed adaptive networks the agents of the network process the streaming data in a cooperative manner. These networks are ubiquitous and are present in various applications such as internet-connected devices, sensor networks [23], wearable networks [24], intra-body molecular communication networks [25], biological and social networks [26], smart grids [27], game theory [28] and power control systems [29].

In the multi-agent distributed networks, the agents collaborate with each other in order to solve a global optimization problem over the network. Several strategies have been proposed in the literature to solve this problem. Three prominent approaches proposed for this purpose are incremental [30–34], consensus [35–38] and diffusion strategies [39–45]. In a general interconnected network diffusion strategy is preferred. This strategy alleviates the limitations of the incremental networks and also overcomes the instability problem of the consensus adaptive networks. Therefore, in this chapter single- and multitask diffusion adaptation (DA) methods are discussed and implemented.

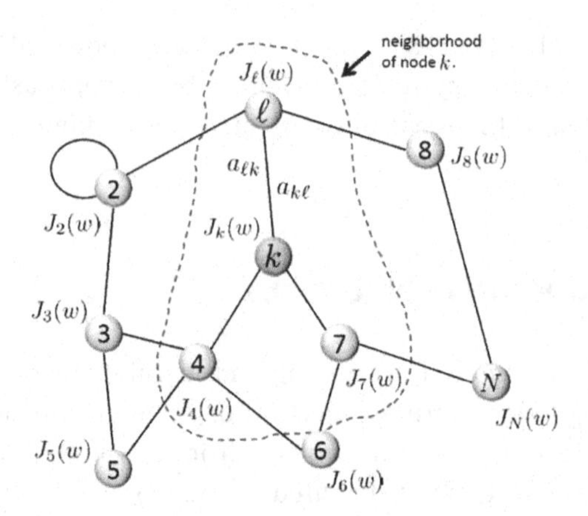

Figure 15.1 A network of nodes with a cooperation neighborhood around sensor k.

15.2.1 Diffusion Adaptation

Consider N_k as the neighbors of node k, including k itself. To model the problem, we consider a connected network consisting of N nodes. A simple network of this type can be seen in Fig. 15.1.

Each node attempts to estimate a $1 \times M$ unknown vector $\mathbf{w}_k^o$ from the collected measurements. Each node k of the network has access to scalar measurements $d_k(i)$ and an $1 \times M$ regression vector $\mathbf{x}_k(i)$ at each time instant $i \geq 0$. The data at each node is assumed to be related to the unknown parameter vector $\mathbf{w}_k^o$ via a linear regression model:

$$d_k(i) = \mathbf{x}_k^T(i)\,\mathbf{w}_k^o(i) + n_k(i), \tag{15.1}$$

where $n_k(i)$ is the measurement noise at node k and time instant i and T refers to transpose of a vector. In the context of DA, this leads to an optimization problem which minimizes a cost function, such as $J_k(\mathbf{w})$ in Fig. 15.1, proportional to the difference between an estimate and the corresponding objective, resulting in the following two-step solution including adaptation and combination [1]:

$$\begin{aligned}\psi_k(i) &= \mathbf{w}_k(i-1) + \mu_k(d_k(i) - \mathbf{x}_k^T(i)\,\mathbf{w}_k(i-1))\mathbf{x}_k(i)\\ \mathbf{w}_k(i) &= \sum_{l \in N_k} a_{lk}\psi_l(i)\end{aligned} \tag{15.2}$$

where $\mu_k \geq 0$ is the step-size parameter used by node k, $\mathbf{w}_k$ denotes the estimate of $\mathbf{w}^o$ and $\{a_{lk}\}$ are nonnegative cooperation coefficients which satisfy

$$\sum_{l \in N_k} a_{lk} = 1, \quad \text{and} \quad a_{lk} = 0 \;\text{ if }\; l \notin N_k \tag{15.3}$$

The adaptation and combination (ATC) steps can be reordered to make the combination first and then the adaptation (i.e., CTA) without any significant change in the outcome. In general, applications, there is no clue about the values of $\{a_{lk}\}$ and often for an N node neighborhood each link weight is considered equal to $1/N$. In application to multichannel electroencephalography (EEG) however, the cooperation coefficients can be estimated through brain connectivity measurement (Section III). Least mean square (LMS) optimization is often used to estimate the weights of the filter [1]. However, other methods such as Kalman filtering are also employed [54].

For the assessment of cooperative filters mean-square deviation (MSD) is often used. MSD provides the squared error in estimation of filter parameter vector which can be calculated in each iteration.

15.2.2 Multitask Diffusion Adaptation

Unlike in single task where there is a single objective for the network, in a multitask scenario, the nodes are divided into groups or clusters each following a particular objective. Figure 15.2 shows a network in which each of the three sets of nodes has a different target to achieve. Solving the optimization problem leads to the following solution [12]:

$$\begin{aligned}\psi_k(i) &= \mathbf{w}_k(i-1) + \mu_k\Bigg(\left[d_k(i) - \mathbf{x}_k^T(i)\,\mathbf{w}_k(i-1)\right]\mathbf{x}_k(i) \\ &\qquad + \beta \sum_{l=N_k} \gamma_{lk}(i)\left[\mathbf{w}_l(i-1) - \mathbf{w}_k(i-1)\right]\Bigg) \\ \mathbf{w}_k(i) &= \sum_{l=N_k} a_{lk}(i)\psi_l(i),\end{aligned} \tag{15.4}$$

where it is assumed that the combination, a_{lk}, and the regularization parameters, γ_{lk}, change versus time. Selection of the regularization coefficients has a significant impact on the network performance. These coefficients are estimated in an adaptive manner so that the

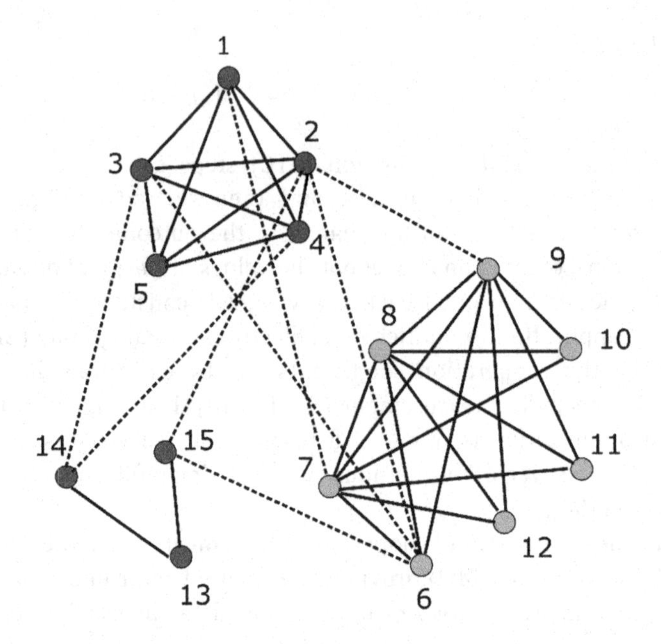

Figure 15.2 A three class/task cooperative network.

agents can be clustered more accurately, i.e., they allocate higher weights to neighbors seeking similar objectives. In other words, the penalty term in Eq. (15.4) must be omitted when the objectives of node k and l are not similar. As an empirically good choice:

$$\gamma_{lk}(i) = \begin{cases} \frac{\exp\left(-\|w_k(i-1)-\psi_l(i)\|^2/h\right)}{\sum_{n\in N_k}\exp\left(-\|w_k(i-1)-\psi_l(i)\|^2/h\right)} & l \in N_k \\ 0 & \text{otherwise} \end{cases} \quad (15.5)$$

where h is a constant which can be changed to adjust the width of the distribution. The multitask strategy has potential for many applications. As an example, different sub-networks within the brain may be responsible for different mental and physical tasks which may happen simultaneously. Counting numbers while cycling is a good example of the brain two-task scenario.

More accurate estimation of the parameter vector, **w**, in the single or multitask scenarios is highly dependent on the accuracy in estimation of the combination coefficients, a_{lk}. In some applications, these parameters can be estimated or approximated either off-line or during the adaptation iterations. For the latter case:

$$a_{lk}^{(MT)}(i) = \begin{cases} \frac{\|w_k(i-1)-\psi_l(i)\|^{-2}}{\sum_{n\in N_k}\|w_k(i-1)-\psi_n(i)\|^{-2}} & l \in N_k \\ 0 & \text{otherwise} \end{cases} \quad (15.6)$$

There are many approaches, such as in [63], and applications, such as those described in the later sections for the multitask DA.

15.3 Brain Connectivity

In the case of brain application, the cooperation parameters can be approximated through connectivity measures of the brain zones under EEG electrodes. A set of EEG electrodes with some causality between their signals, making a cooperative network, is shown in Fig. 15.3. Besides the basic Granger Causality which is popular but perhaps less efficient for estimation of brain connectivity, the directed transfer function (DTF) as an extension of Granger causality has become an established approach to detect and quantify the coupling between the nodes and their directions. A time-varying DTF can also be generated to track the source signals by calculating the DTF over short windows to achieve a short time DTF (SDTF) [14].

Considering $\mathbf{z}_i = [x_1(i), x_2(i), \ldots, x_N(i)]$ the vector signal is assumed to satisfy a multivariate auto-regressive model (MVAR) of the form

$$\mathbf{z}_i = \sum_{m=1}^{p} \Lambda_m \mathbf{z}_{i-m} + \mathbf{e}_i, \quad (15.7)$$

where p denotes the model order, $\mathbf{e}_i$ is the error term, and Λ_m is the $N \times N$ matrix of MVAR coefficients at lag m. By transforming to frequency domain we have

$$\mathbf{H}(\omega) = \left[\mathbf{I}_N - \sum_{m=1}^{p} \Lambda_m e^{-j\omega m}\right]^{-1}. \quad (15.8)$$

Then, as an effective measure of the link weights between the nodes with indices (l, k), for each data segment the DTF coefficients are

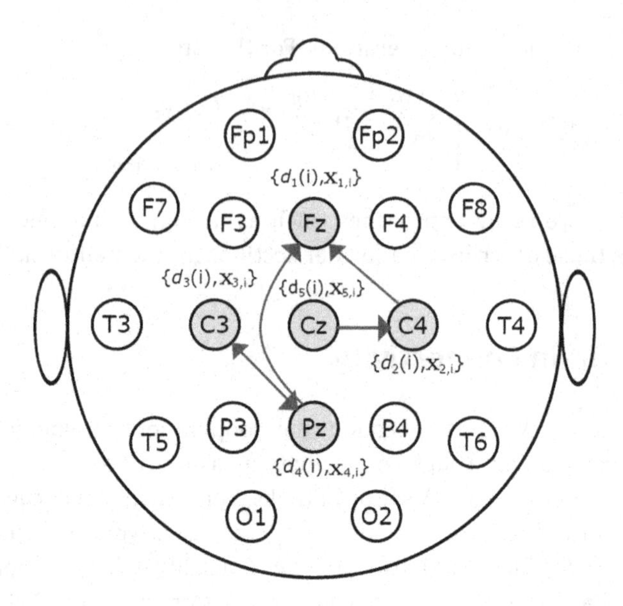

Figure 15.3 A symbolic group of electrodes cooperate in performing a brain task.

defined as

$$\xi_{lk}^2(\omega) = \frac{|H_{lk}(\omega)|^2}{\sum_{q=1}^{N} \left|H_{qk}(\omega)\right|^2}. \tag{15.9}$$

Then, the combination coefficient between l and k is estimated by averaging this value over frequency ω.

In very recent practices however, another effective method for the brain connectivity estimation has been proposed [14–18]. In order to estimate the combination weights in each time interval for more accurate estimation of cooperative filter parameters, $\mathbf{w}_k$, the Stockwell time-frequency transform (S-transform) [47] is employed here. This method is more accurate and less sensitive to the changes in time-frequency parameters compared to the autoregressive based methods [2]. It is defined as

$$\mathbf{X}_k(\tau, f) = \int_{-\infty}^{\infty} \mathbf{x}_k(t) \frac{|f|}{\sqrt{2\pi}} e^{-\frac{(\tau-t)^2 f^2}{2}} e^{-i2\pi f t} dt \tag{15.10}$$

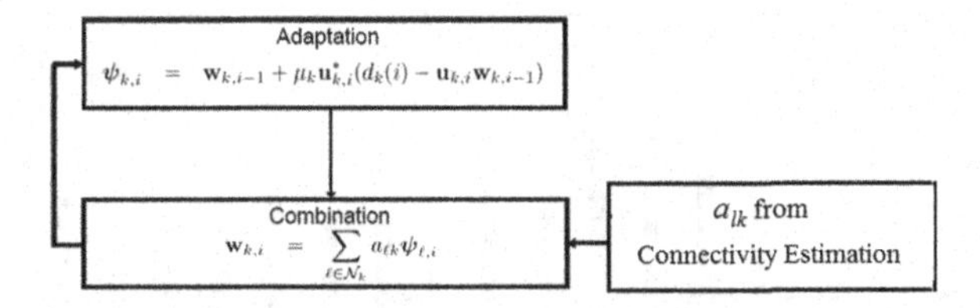

Figure 15.4 A model for the brain thought-to-action where the cooperative learning is informed by the estimates of brain connectivity.

Then, the cross-spectrum of the signal is defined as

$$C_{kl}^{(ST)}(t, f) = \frac{S_{kl}^{(ST)}(t, f)}{\sqrt{S_{kk}^{(ST)}(t, f)S_{ll}^{(ST)}(t, f)}}, \tag{15.11}$$

where

$$S_{kl}^{(ST)}(t, f) = \langle X_k(t, f)X_l^*(t, f)\rangle \tag{15.12}$$

is complex-valued. The imaginary part of S-coherency (ImSCoh) is related to the phase difference between the signals $\mathbf{x}_k(t)$ and $\mathbf{x}_l(t)$ at each frequency f; e.g., if ImSCoh is positive then, $\mathbf{x}_k(t)$ and $\mathbf{x}_l(t)$ are interacting and $\mathbf{x}_k(t)$ leads $\mathbf{x}_l(t)$. The cross-spectra values are then used in estimation of the combination weights as

$$a_{kl}^{(ST)} = \frac{\max\left(\mathrm{Im}(C_{kl}^{(ST)}(t, f), 0\right)}{\sum_{l \in N_k} \max\left(\mathrm{Im}(C_{kl}^{(ST)}(t, f), 0\right)}. \tag{15.13}$$

15.4 Exploiting Brain Connectivity into the DA Model

Figure 15.4 demonstrates how the connectivity estimates are used in the DA formulation. The estimates, measured for each data segment, are plugged into the adaptation equation for estimation of the DA weights $\mathbf{w}_k$. The system can be trained to classify any brain action particularly those originated from motor cortex. In the following sections, we review some of the applications to brain computer interfacing (BCI) and restoration of movement-related brain activity from tremor for Parkinsonian patients, and for tracking the ERPs.

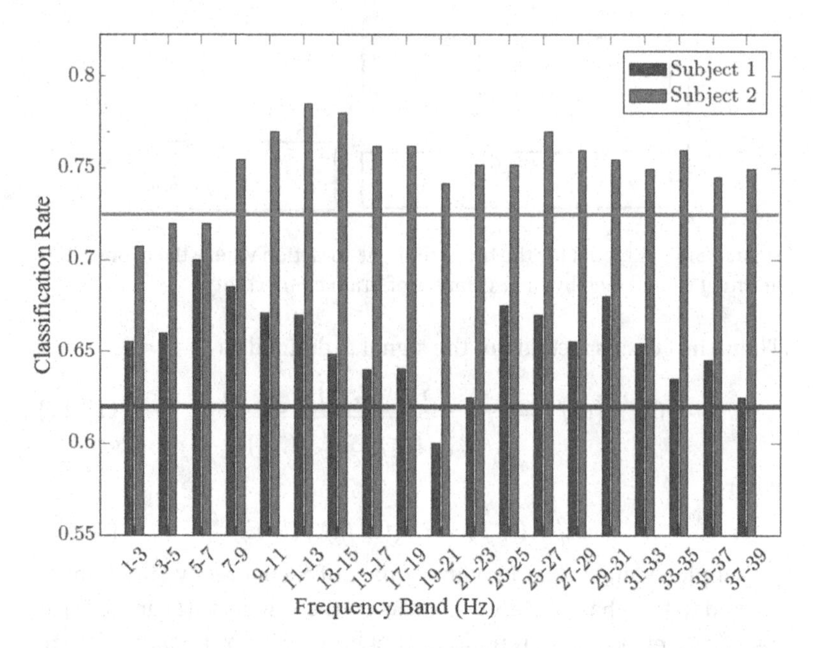

Figure 15.5 Clockwise and anti-clockwise classification success rates for different frequency bands using both diffusion and non-diffusion based classifiers. The blue and red horizontal lines represent the scores of the non-diffusion case for subjects 1 and 2, respectively.

15.4.1 BCI Using Single Task Diffusion Adaptation

In an application to EEG signals the algorithm in Fig. 15.4 has been used to model the brain activity during clockwise and anti-clockwise hand-drawing of a circle. The EEG artifacts were removed using a combined blind source separation and support vector machines (BSS-SVM) [16, 46].

After designing the model using a set of training EEG data, it was used as a classifier to distinguish between the two directions of hand rotation [14]. The results for two subjects performing the above tasks are given in Fig. 15.5. This is a very challenging BCI problem and the crisp classification techniques fail to identify the direction of hand movement. Therefore, this method opens a new direction in BCI research as well as rehabilitation engineering.

15.4.2 Parkinson's and Multitask Diffusion Learning

There are many processes in nature that can be modeled using multitask DA strategy, e.g., in [8–11]. In these approaches the groups of objects or events with different objectives are clustered iteratively and during the adaptation process. In an application for restoration of movement related brain activity from Parkinsonian tremor [17]. In an experiment 20 patients with Parkinson's disease and different levels of tremor have been considered. Two systems work in synchrony; first, an EEG system is used with 28 electrodes that are positioned based on the 10–20 EEG system. Active electrodes have been used and the sampling rate was 256 Hz. The EEG signals are filtered between 2 and 45 Hz. Eye blink artifacts are removed using BSS-SVM [46]. The second system is a motion capture system manufactured by Xsens. One sensor was attached to the subjects hand to capture its movement. The sensor includes an accelerometer, a gyroscope and a magnetometer. This is used to segment the EEG data. We can group the subjects into two categories based on the level of their hand's tremor. Among the 20 patient's with Parkinson's disease, 11 have high tremor intensity and 9 patients have low tremor intensity. The patients were asked to perform a simple motor task consisting of four stages: (1) the patient is asked to pick up a cylindrical object with medium weight from a table and place it on a box with approximately 15 cm height located on the same table; (2) the patient then brings back the hand to its initial position and rest for some seconds; (3) the patient raises his hand again and moves it towards the object; (4) the patient grabs the object from top of the box and returns it to its initial position. the movement segments are then: the grasping of the cylindrical object from the table until the placement on the box (stage 1) and the similar task when the bottle is already on the box (stage 3). The DA is expected to cluster movement related brain activity and tremor based artifact into two different classes.

The multitask DA is performed by combining automatic estimation of cooperation parameters $a_{lk}^{(MT)}(t)$ as estimated in Eq. (15.6) and the brain connectivity using S-coherency $a_{lk}^{(SC)}(t)$ as given in

Eq. (15.13) through [12, 17]:

$$a_{lk}(t) = \begin{cases} \frac{\lambda a_{lk}^{(MT)}(t)+(1-\lambda)a_{lk}^{(SC)}(t)}{\sum_{n\in N_k}\lambda a_{nk}^{(MT)}(t)+(1-\lambda)a_{nk}^{(SC)}(t)}, & l \in N_k \\ 0, & \text{otherwise} \end{cases}, \tag{15.14}$$

where λ is a penalty term which can be set manually for the best performance. To assess the system performance the weighted degree centrality for each node, $\text{WDC}_k(t)$, defined as

$$\text{WDC}_k(t) = \sum_{l=1}^{N} a_{kl}(t), \tag{15.15}$$

is used. In Fig. 15.6 the between-group differences, i.e., the combination values of high tremor group minus the combination values of the low tremor subjects for $\lambda = 0$ and $\lambda = 1$ are provided. The significance of brain connectivity estimate is clear in this figure. It is also shown that [17] using $\lambda = 0.7$ the WDC is significantly higher for higher tremor groups around frontal electrodes and lower for posterior electrodes, whereas for $\lambda = 0$, there is almost no differences in the WDC levels. Also, the method is able to distinguish between high and low tremor intensities using the combination weights and centrality information. Moreover, multitask learning methods can be used to enhance the connectivity patterns of the brain as distinctive features to distinguish normal and abnormal brain functions.

15.5 Cooperative Smart Grid

In a smart grid scenario the consumer appliances are connected through communication links. There are often, fixed (non-shiftable), shiftable, and curtailable appliances for each household. Also, the overall energy consumption is bounded by the available energy from the retailer. On the other hand, a factor of consumer convenience governs the adaptation process. The traditional centralized schemes, where a central controller collects all the necessary information from the consumers and computes the control solution [56, 57], is becoming less attractive, if not infeasible, due to the limited capacity in computation and communication resources. On the other hand, due to reduction in the communication load and

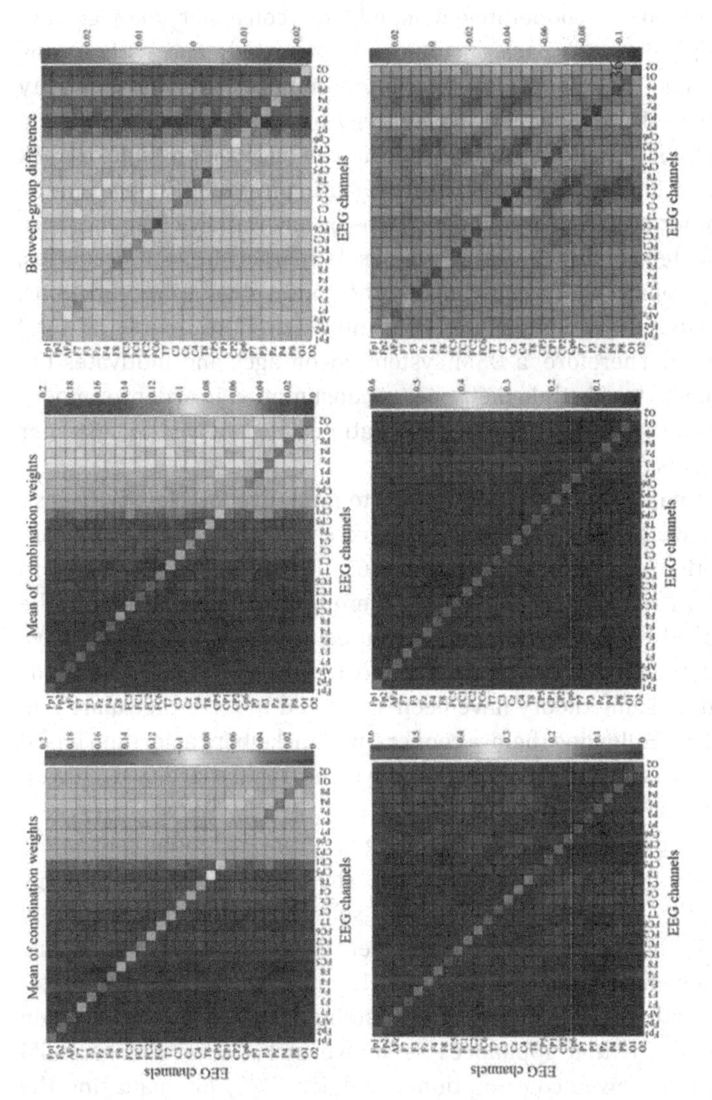

Figure 15.6 Between group differences for $\lambda = 0.7$ (top) and (b) $\lambda = 0$ (bottom).

robustness to failures, fully distributed demand response (DR) programs have become more attractive in the smart grids.

Unlike the centralized power distribution systems, in a fully distributed and cooperative demand-side (consumer) management (DSM) framework each customer autonomously and without any need for the global information, minimizes his incommodity function. Using a cooperative strategy based on adaptive diffusion and providing a decentralized mechanism makes it possible to track and alleviate the effects the drifts resulting from the changes in the customer preferences and conditions or any rapidly changing price parameter coming from the wholesale market. In such scenarios, the customers aim at maximizing their individual utility functions, while the utility company aims at minimizing the smart grid total payment. Therefore, a DSM system encourages and motivates the customers to schedule their energy consumption in order to smooth out the total energy demand fluctuations and reduce the customer payments.

A smart grid may be optimized to minimize the electricity cost, maximize the social welfare, minimize the aggregate power consumption, or combination of them [61]. Therefore, the optimization system has to be regularized as there are many constraints to be complied with. Constrained distributed convex optimizations have been studied recently [58]. Around the same time, the methods based on game theory have been suggested to solve the smart grid problem. Following them, a cooperative Stackelberg algorithm based on DA has been suggested to solve the regularized, yet convex, problems [55].

Consider a smart grid including several residential consumers and a service provider (retailer). The retailer buys the energy from the wholesale market and sells it to the consumers in a way to ensure that the retailer is not scathed under any circumstance. Also, each consumer is equipped with a smart meter (SM) with the capability of scheduling the power consumption of the residential appliances as shown in Fig. 15.7. For each SM there is a power consumption scheduler (PCS) for managing the power consumption of its appliances, using a price signal and other necessary information. The PCS is also able to inform the retailer and neighboring consumers about the users' power consumption profile.

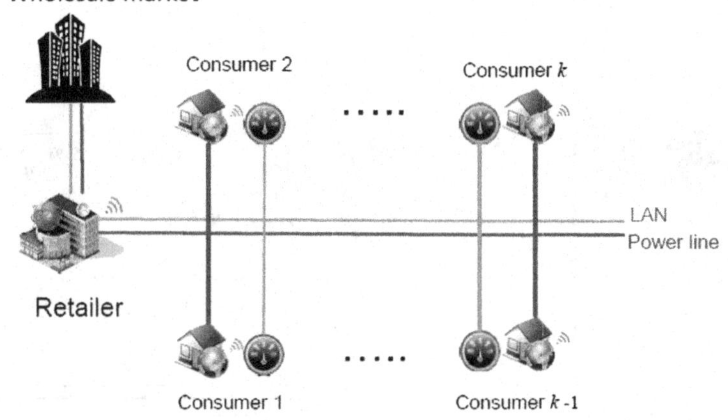

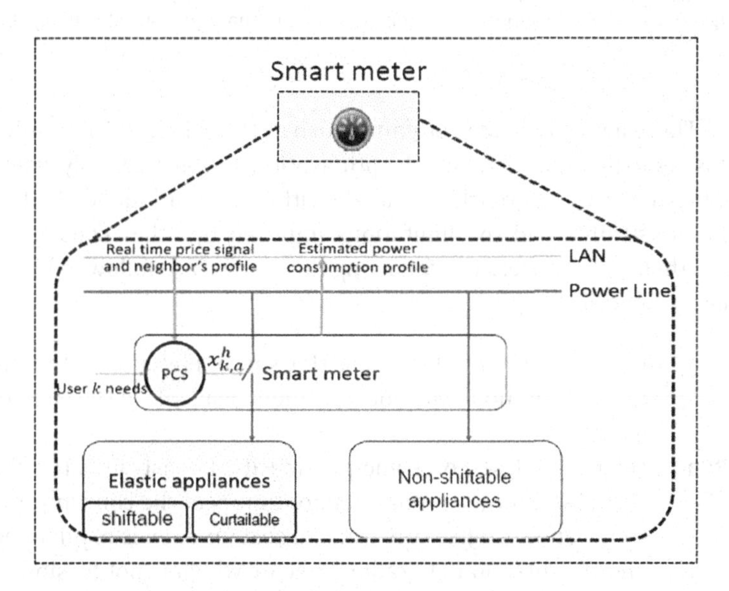

Figure 15.7 The model of cooperative smart grid; smart cooperative grid (top) and the components of the smart meter (bottom) [55].

In addition, it is assumed that the SM is able to monitor and collect all the data of its appliances for turning on/off and choosing the level of power consumption for these appliances in different situations. This is through a local area network (LAN) that connects the consumers to the retailer as well as their neighbors.

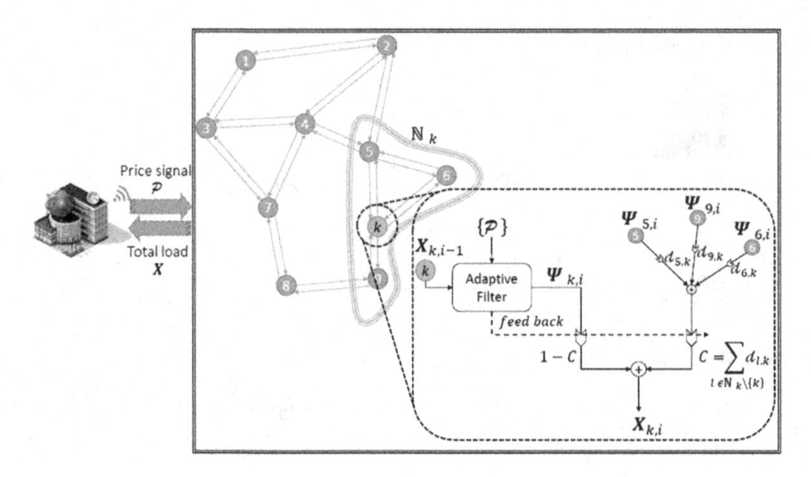

Figure 15.8 The cooperative (decentralized) smart grid model using DA [55].

There have been some attempts such as those in [59, 60] to solve this demand-side distributed optimization problem mainly using game theoretic approaches. The algorithm is then independent of the constraints and the limitations imposed by other consumers and their preferences. The following three steps are followed in the algorithm [55]:

Step 1. (Adaptation): At this stage the smart meter updates the current solution using local gradient available at the current iteration.

Step 2. (Penalty): Based on some pre-specified constraints (on the available energy and price, smoothness of the consumption trend, and such like) the original cost function is regularized and the direction of gradient vector which is not feasible is penalized.

Step 3. (Combination): At this stage, each consumer combines his estimate of the optimum $\boldsymbol{X}_k$ with estimates in his neighborhood N_k.

Application of this method results in smoothing the price trend as well as instantaneous demand fluctuations. Figure 15.9 shows a typical outcome of application of the method.

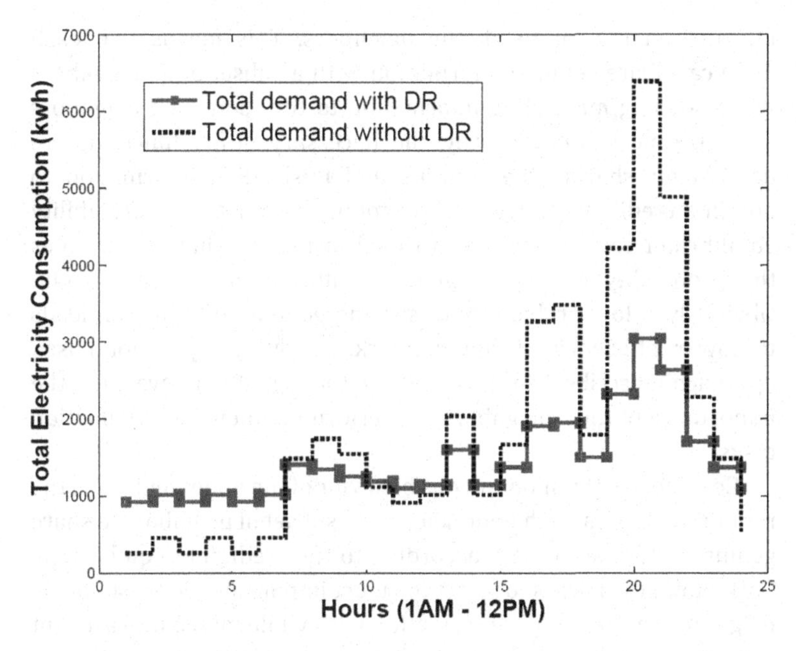

Figure 15.9 The global power consumption pattern before and after applying DR.

15.6 Information Reliability in Multitask Networks

The network agents should accept and process the reliable information only. Therefore, Information reliability in multitask networks is essential and there is need for establishing a measure of trustworthiness among the agents. This requires an adaptive mechanism to decide between reliable and unreliable information in a multi-agent communication system. In this study, initially multiclass multitask networks are considered without any prior assumption about the clusters. Then, the case where sharing genuine information is subject to a cost and each node can decide whether to share genuine or misleading information is studied. To consider this problem in a more generalized way, the nodes are allowed to make this decision in a pairwise manner rather than using holistic approaches where the genuine or false information is sent to all the neighbors. This means that each node can share genuine information with a subset of its neighbors while sending misleading information to the remaining neighbors at each time instant. Thirdly, the necessary utility functions are defined for sharing the genuine and misleading information to obtain a credibility model for the cooperative network. Credibility equilibrium can be achieved through a model which determines the probability of sharing genuine information for each agent. Obviously, a low probability of sharing genuine information leads to lower efficiency of the network. Lastly, a reputation-based approach described in [48] allows the agents to evaluate the importance of their neighbors for performing their own estimation tasks.

Considering the spontaneous behavior of the nodes and using the reputation scores, each node selects the subset of neighbors to share genuine information with according to the credibility equilibrium. This allows the agents to select the most important and trustworthy neighbors to share their information with while taking into account their own privacy and cost budgets.

Unlike the assumption in Eqs. (15.2) and (15.4) where all the agents within a neighborhood share their information, in a more realistic and challenging scenario some agents are reluctant to share their genuine information with all their neighbors. This can be due to the cost or security of the information. Therefore, in such cases the received information from a neighboring node l might be different from its true estimate $\psi_l(i)$. In our model, genuine information of each node k refers to its true estimate $\psi_k(i)$, which is obtained according to Eq. (15.2) and sharing anything rather than $\psi_k(i)$ is considered as fabricated information. There is therefore a need to integrate the information credibility and multitask diffusion concepts. This leads to a more general formulation as follows:

$$
\begin{aligned}
\psi_k(i) &= \mathbf{w}_k(i-1) + \mu_k \left[d_k(i) - \mathbf{x}_k^T(i)\,\mathbf{w}_k(i-1) \right] \mathbf{x}_k(i) \\
\tilde{\psi}_l^k(i) &= \begin{cases} \psi_l(i), & k \in N_G^l(i) \\ \chi_l(i), & \text{otherwise} \end{cases} \\
\mathbf{w}_k(i) &= \sum_{l=N_k} a_{lk}(i)\tilde{\psi}_l^k(i),
\end{aligned}
\qquad (15.16)
$$

where $\tilde{\psi}_l^k(i)$ is the information received by node k from node l at time instant i, and $\chi_l(i)$ is the $M \times 1$ fabricated information shared by node l at time instant i.

This fake or fabricated data can be any information different from the true estimate $\psi_k(i)$ and can have different forms and distributions. Moreover, $N_G^l(i)$ represents the subset of node l's neighbors, N_l, that node l shares its genuine information with at time instant i. Since node k is not aware whether the received information is genuine or fabricated, it combines all the received information $\tilde{\psi}_\mathrm{l}^\mathrm{k}(\mathrm{i})$ from its neighbors using the combination weights to estimate $\mathbf{w}_k(i)$. The estimation of combination weights iteratively and automatically also follows Eq. (15.6) in a multitask scenario.

15.6.1 Information Credibility Modeling

To model this behavior, the approach in [49] is followed in [48] assuming that at each time instant i, each agent can decide whether to share information truthfully or manipulate and share fabricated misleading information. These two strategies can be represented by G and F, respectively. Clearly, this decision by an agent k depends on the cost that it has to pay for sharing the genuine information, which can be represented by c_k. Moreover, the probability of sharing genuine information for agent k at time instant i, represented by $p_{k,\mathrm{G}}^i$, depends on this cost.

Since the nodes are connected and also collaborate with each other the utility of each node depends on the actions and behaviors of other nodes in the network. Hence, it can be assumed that the benefit of utilizing shared information increases as the number of genuine agents in the network grows. Using a commonly used utility function [48, 50], we have

$$u_\mathrm{G}^k(n) = \eta - e^{-\delta n} - c_k \qquad (15.17)$$

where $u_\mathrm{G}^k(n)$ is the utility function for agent k sharing genuine information with n agents. Moreover, η and δ are positive constants where η is the maximum utility of the user and δ represents the speed of saturation for genuine information. It is important to note that parameters $0 \leq \eta \leq 1$ and c_k are normalized in our model and can be directly compared with each other. Although the cost of sharing genuine information c_k can vary among the nodes of the network, we assume that $c_k = c$ is the same for all the agents without loss of generality. On the other hand, the utility function of an agent k sharing false information, $u_{\mathrm{F}}^k(n)$, approximately fits the following equation [49, 50]:

$$u_{\mathrm{F}}^k(n) = p_a^k\left(\eta - e^{-\delta(n+1)}\right) + \left(1 - p_a^k\right)\left(\eta - e^{-\delta n)}\right) \qquad (15.18)$$

Here, p_a^k is the probability of acquiring new information for each agent k, which decreases as the information acquisition cost increases. Although this probability can be different among the agents, we assume that it is the same for all the agents for simplicity and replace it with p_a. Moreover, the term $n+1$ in the first exponent is due to the n genuine agents in the network plus node k's own genuine data that it has access to.

Using the above two equations, the average utility for sharing genuine information $\bar{u}_G(p_G)$ over the network is defined as [49, 50]:

$$\bar{u}_G(p_G) = \sum_{n=1}^{N-1}\binom{N-1}{n} p_{\mathrm{G}}^n (1 - p_{\mathrm{G}})^{N-1-n}\left(\eta - e^{-\delta(n+1)} - c\right) \qquad (15.19)$$

Similarly for the average utility of sharing false information, $\bar{u}_{\mathrm{F}}(p_G)$:

$$\bar{u}_{\mathrm{F}}(p_G) = \sum_{n=1}^{N-1}\binom{N-1}{n} p_{\mathrm{G}}^n (1 - p_{\mathrm{G}})^{N-1-n}\left(p_a^k\left(\eta - e^{-\delta(n+1)}\right) + \left(1 - p_a^k\right)\left(\eta - e^{-\delta n)}\right)\right) \qquad (15.20)$$

Hence, the average utility function of the entire network can be given as

$$\bar{u}(p_{\mathrm{G}}) = p_{\mathrm{G}}\bar{u}_G(p_G) + (1 - p_{\mathrm{G}})\,\bar{u}_{\mathrm{F}}(p_G) \qquad (15.21)$$

After defining the utility functions for the network, the probability of sharing genuine information p_G at each time instant i can be obtained by [49, 50]

$$p_{\mathrm{G}}^{i+1} = p_{\mathrm{G}}^i + \alpha\left[\bar{u}_{\mathrm{G}}\left(p_{\mathrm{G}}^i\right) - \bar{u}(p_{\mathrm{G}}^i)\right] \qquad (15.22)$$

where α is a constant positive variable. It can be observed from Eq. (15.22) that in the cases where the utility of sharing genuine information is higher than the average utility, the probability of sharing genuine information increases over the agents, thus encouraging them to be more truthful. Also, increasing the information sharing cost c results in decreasing the probability of sharing genuine information between the network agents. This means that there is a trade-off between the cost for sharing genuine information and the chance of receiving truthful and useful data by the nodes. Therefore, the value of c can be different based on the undertaken case and is set by the designer. Since a low probability of sharing genuine information can result in a degraded benefit of cooperation and the poor performance of the network, a reputation protocol is required to enhance the performance of the adaptive network.

15.6.2 Adaptive Distributed Reputation

The agents of an adaptive distributed network might be inclined to share false and misleading information rather than their genuine data due to several reasons. A model of the dynamics of probability of sharing genuine information among the agents was presented in previous section where factors such as the high cost of information sharing can contribute to a low probability of sharing reliable information. Therefore, it is important to tackle this barrier by an adaptive and distributed strategy. Therefore, a reputation scoring method is proposed which enables the agents to evaluate the importance of their neighbors for their own estimation task. With the help of this reputation scheme, each agent can summarize the past actions of its neighbors as a reputation score. This score is then used to evaluate the importance and truthfulness of each particular neighbor. Using this score, the agents are then able to select a subset of their neighbors that send more relevant and useful information to them for their own estimation task. As a result, each agent shares its genuine information with this subset of neighbors while sharing false information with the remaining neighbors according to the probability of sharing genuine information, p_G^i.

15.6.2.1 Reputation score

According to the reputation protocols, cooperative and truthful agents have higher reputation scores among the agents while non-cooperative and deceptive agents suffer from a lower reputation score [51]. Among several mechanisms to design reputation scores (such as average and cumulative scores), an exponentially weighted moving average scheme is more robust to cheating. This is because they allocate higher weights to recent observations compared to past actions [52, 53]. Monajemi et al. [48] proposed to utilize such reputation scores in multitask diffusion strategies to help the agents overcome the negative and adverse effects of misleading information. Unlike in some previous approaches [53], the reputation scores here are based on the dynamics of sharing genuine information and the importance and similarity of the received data. The cooperative reputation score is therefore modeled as a smooth exponential moving average which involves the past communications as

$$R_{lk}(i) = \beta_k R_{lk}(i-1) + (1-\beta_k)\, a_{lk}(i) \tag{15.23}$$

where $\beta_k \in (0, 1)$ is the smoothing factor. From (15.23) a higher value of β_k results in a higher weight and influence of the past actions. Each node k of the network can then utilize the obtained reputation scores $R_{lk}(i)$ of its neighbors at each time instant to evaluate the importance of each particular neighbor for its own estimation task. Using this score, each node k can sort its neighbors so that the nodes with higher reputation scores have a higher rank. Based on this ranking and by considering the probability of sharing the genuine information, node k forms a subset of its neighbors to share genuine information with. In particular, node k shares its own genuine information with a subset of its neighbors with higher reputation scores. The dynamics of this subset can change over time according to the dynamics of the probability of sharing genuine information, p_{G}^{i}, estimated in Eq. (15.22). To make sure that none of the agents violates p_{G}^{i} the maximum number of agents for each node k to share the genuine information with at each time instant should be less than or equal to:

$$N_{\mathrm{G},k}^{\max} = \left\lfloor p_{\mathrm{G}}^{i} \times |N_k| \right\rfloor \tag{15.24}$$

where $N_{\mathrm{G},k}^{\max}$ represents the maximum number of neighbors that node k at time instant i can share genuine information with $|N_k|$ is the cardinality of the neighbor set N_k and $\lfloor . \rfloor$ is the integer floor operator.

Now, at each time instant i each agent k ranks its neighbors according to their reputation scores $R_{lk}(i)$ obtained by (15.23) and

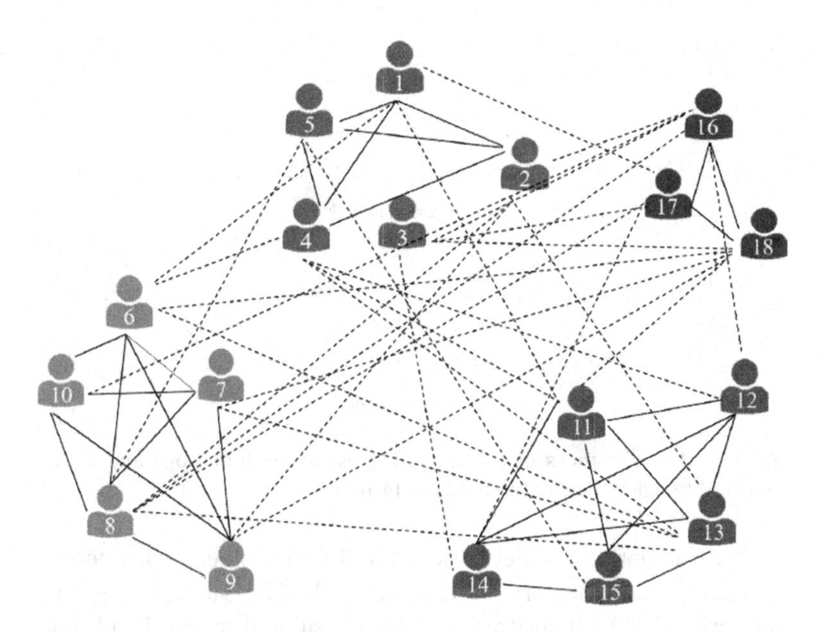

Figure 15.10 Topology of the multitask network used in the simulation. The network consists of $N = 18$ nodes and $Q = 4$ clusters. The solid lines show the connections between the nodes of the same cluster while the dashed lines represent the connections between the nodes of different clusters [48].

share its genuine information with the first $N_{G,k}^{\max}$ neighbors that have higher reputation scores. For example, if the probability of sharing genuine information at time i is $p_G^i = 0.5$, agent k shares its genuine information with at most half of its neighbors that have higher reputation scores.

Finally, the most important information in the set of DA equations in (15.16), i.e., the reliable set of neighbors, $N_G^l(i)$, is re-estimated, using (15.24) as [48]:

$$N_G^k(i+1) = [\text{first } N_{G,k}^{\max}(i) \text{ neighbors of node } k \text{ with higher reputation scores } R_{lk}(i)] \quad (15.25)$$

In summary, in each iteration of the algorithm the probabilities p_G^i estimated using (15.22) is plugged into (15.24) and then used in (15.25) to estimate the set of neighbors whom the genuine information can be shared with before applying the DA algorithm in (15.16).

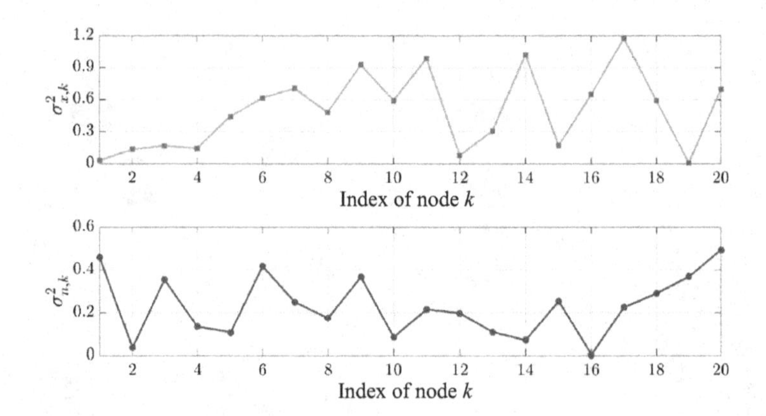

Figure 15.11 Variances of the input regression vectors (top) and noise (bottom) for each node of the network [48].

In a simulation the network of Fig. 15.10 has been considered. There are four clusters in this network. The variables of the networks and the algorithm are also considered in Fig. 15.11 and Table 15.1.

Figure 15.12 shows the evolution of node 1 selecting a subset of neighbors to share genuine information at 500 time intervals. In this figure, the green arrows represent the nodes that receive genuine information while the red ones show those who receive fabricated data from node 1. Additionally, the color of neighboring nodes at each iteration shows whether they are sending genuine information to node 1 or not. For instance, we observe that at iteration 1000 nodes {2, 4, 5} send genuine information to node 1, while nodes {6, 11, 17} transmit fabricated data. The latter information is not available to node 1 and this node selects the subset $N^1_G(i)$ using the reputation scores according to (15.23).

In the previous simulation setup, it is assumed that the fabricated data follows a Gaussian distribution. However, in the

Table 15.1 The parameters used in the simulation

Network variables				Info. credibility			Reputation			
N	Q	M	μ	c	p^0_G	p_a	η	δ	α	β
18	4	2	0.02	0.01	0.5	0.8	1	0.1	1	0

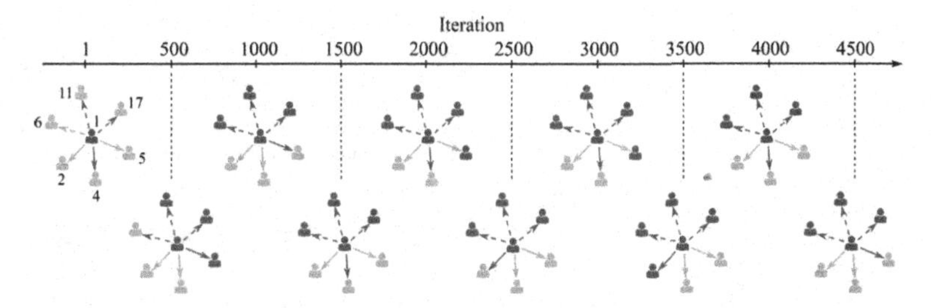

Figure 15.12 Evolution of node 1 selecting a subset of neighbors with whom to share genuine information. The green arrows represent the nodes receiving genuine information from node 1, while red ones show fabricated data from node 1. The color of the neighboring nodes at each iteration shows whether they are sending genuine information to node 1 or not.

real-world scenarios the fabricated or misleading data $\chi_k(i)$ might not be drawn from any specific distribution such as the Gaussian distribution. Therefore, it is important to evaluate the performance of different methods for more general cases where the nodes deal with other types of fabricated data. To do so, we generate the misleading information $\chi_k(i)$ using chi-square distributions with various degrees of freedom ν.

Figure 15.13 represents the results where $\chi_k(i)$ is drawn from a chi-square distribution with a degree of freedom $\nu = 3$. The other parameters such as the sharing cost c remain the same.

The results in Fig. 15.13 reveal that the proposed reputation-based method is robust to various forms of fabricated data while the performance of other learning strategies might be affected by it. In particular, it can be observed from this figure that the performance of the adaptive multitask algorithm where the subset $N_{\text{G}}^k(\text{i})$ is randomly selected is adversely affected and is almost the same as the non-cooperative case.

15.7 Other Applications

Distributed dictionary learning and its application in classification of anti-bodies in optical images [9–11], modeling biological systems [8, 62], distributed power control [64], cooperative adaptive line

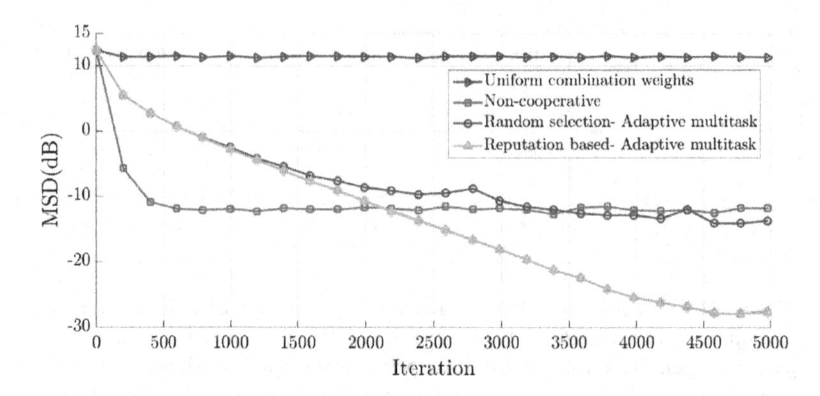

Figure 15.13 The MSD curves representing the performance of different learning strategies over the distributed network. Here the cost c is equal to 0.01 and the misleading data shared by the nodes follow a chi-square distribution. Lower values of MSD show lower estimation error and better performance.

enhancer, etc. are out of many other applications of cooperative networks.

On the other hand distributed tracking has become a new tool in distributed probability density estimation [65] and particle filtering [18].

Biological systems can be modeled using cooperative networks. In such models the collection of bacteria forms an adaptive network, where the agents cooperate with each other by propagating signals that stem from biochemical reactions. In the model presented by Monajemi et al. [8], the bacteria motility including birth and death, at the presence of nutrition and antibacterial chemical has been formulated through diffusion adaptation.

In the model, at every time instant, each bacterium in the network has a noisy estimate of the nutrition intensity. This estimate is a scalar measure which depends on the distance of the bacterium from the nutrition source. Additionally, each bacterium has access to the information emitted by other bacteria in the network. The nature of this information arises from the chemicals bacteria are able to release in their surrounding environment. Similar to the information released by the bacteria about the nutrition intensity, the estimated chemical (used to kill the bacteria) intensity by each bacterium is

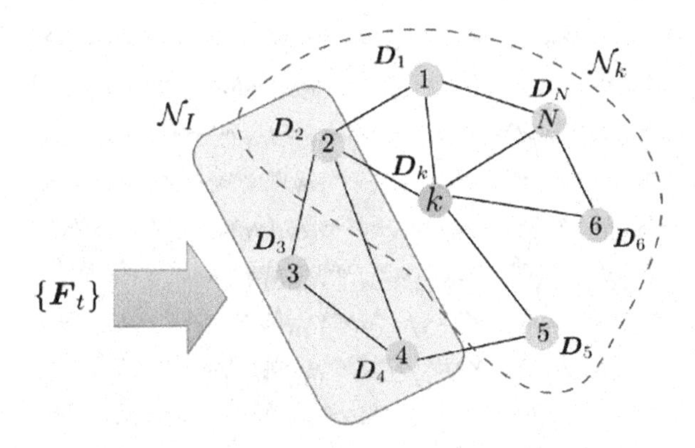

Figure 15.14 Sample of a connected network where each agent k is responsible for learning a sub-dictionary $\boldsymbol{D}_k$. The input data (feature vectors) F_t can be presented to a subset of agents represented by N_I.

considered noisy. Based on these two items of information, i.e., the nutrition and chemical intensities, each bacterium decides whether to diffuse information to the network or not.

In the simulation results it has been shown that the diffusion of information helps them adapt their performance, survive, and locate the food sources more efficiently.

Distributed dictionary learning is a fundamental approach for reconstruction or classification of data recorded in various conditions or using different modality recording systems [9]. Multitask diffusion adaptation strategy can be extended to designing a dictionary learning (DL) system [10, 11]. It is then used for diagnosis of autoimmune diseases by classification of human epithelial type-2 (HEp-2) cells and antinuclear antibodies (ANA) captured through indirect immunofluorescence (IIF) [11].

It is possible to form a connected network of N nodes where each node k is in charge of updating its own sub-dictionary D_k. In this way, the dictionary matrix D is distributed over the network. The network in Fig. 15.14 represents the concept. Similar to the other distributed adaptive networks, each node can share information with and receive information from its neighbors. Moreover, the input features F_t might be only presented to a subset of the nodes (shown by N_I) rather than all of them. In such cases the

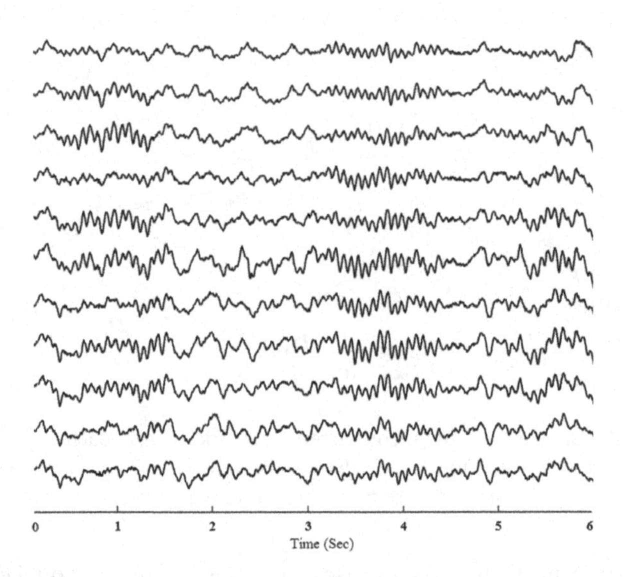

Figure 15.15 A multichannel EEG segment including sleep spindles. It is clear that the spindles are not spread uniformly over the EEG electrodes.

learning system is computationally more efficient while retaining comparable performance with other methods [11]. The approach called adaptive distributed dictionary learning (ADDL), has been tested using two datasets of HEp-2 cell images. The outcome verifies the superiority of the method in terms of both classification rate and computational cost to a number of benchmark methods [11].

Cooperative adaptive line enhancer (CoALE) is another application of DA. A CoALE can detect or restore bursts of periodic signals over a network of connecting agents. One application can be in detection of a jamming signal in communication networks. This can be advantageous in detection and prevention of the cyber-attacks and keeping the communication safe. Another application can be on detection and restoration of 1–2 sec burst of alpha-range spindles from noisy sleep EEG data. A set of such data can be seen in Fig. 15.15. The spindles often appear during the second and third stages of sleep.

The design follows the concept of ALE where the error between the data and its shifted version is minimized. The shift, Δ, is equivalent to an integer number of cycles of the periodic signal, is shorter than the correlation length of the signal and longer that correlation length of the existing noise (or background artifact). Therefore, the DA formulation takes the form

$$\psi(i) = \mathbf{w}_k \ (i-1) + \mu(\mathbf{x}_k(i-\Delta) - \mathbf{x}_k^T(i)\,\mathbf{w}_k \ (i-1))\mathbf{x}_k(i)$$
$$\mathbf{w}_k(i) = \sum_{l \in N_k} a_{lk}\psi_l(i), \tag{15.26}$$

where $\mathbf{x}_k$s are the EEG signals. For this particular application the combination weights, a_{lk} are estimated off-line through measuring the brain connectivity.

Distributive tracking using DA has been designed and employed for tracking of event related potentials (ERPs). ERPs directly measure the brain responses to specific sensory, cognitive, or motor stimuli [18]. Often averaging over multiple trials is used for assessing ERPs assuming that the ERP components are constant over different trials and therefore, averaging over single trials attenuates the background EEG signals. However, the averaging process can result in loss of information related to trial-to-trial variability. This is crucial in scenarios where the ERP waveform changes due to habituation, fatigue, or the level of attention. Moreover, several brain abnormalities such as schizophrenia and depression can be identified by ERP variation over different trials. The ERP parameters such as amplitude and latency change from trial to trial.

The most important ERP called P300 wave consists of two main overlapping subcomponents known as P3a and P3b. P3a has a fronto-central distribution since it is generated within the prefrontal, frontal, and anterior temporal brain regions. The P3a subcomponent is an unintentional response reflecting an automatic orientation of attention to a salient stimulus which is independent of the task. On the other hand, P3b is mainly biased toward the centro-parietal region since it is generated principally by the posterior temporal, parietal, and posterior cingulate cortex regions. Compared to P3b, P3a has a shorter latency and a more rapid habitation.

The inherent lateral dependencies and similarities as well as spatial overlaps have been exploited in designing a cooperative particle filter for tracking the variabilities [18]. A network similar

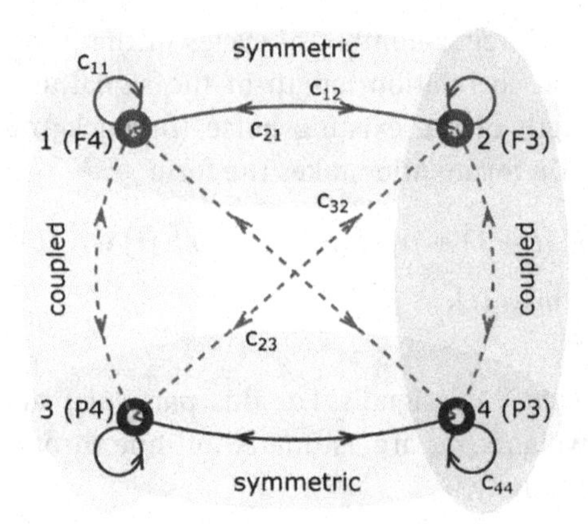

Figure 15.16 Cooperation between the electrodes for P3a and P3b tracking [18]; F3 and F4 are the left and right frontal electrodes, respectively, where p3a elicits, and P3 and P4 are the posterior left and right electrodes, respectively, where p3b elicits.

to that in Fig. 15.16 has been modeled and used for the tracking purposes.

In terms of the cooperation gains, it is assumed that the lateral nodes equivalently contribute to each other's information, whereas the frontal and posterior electrodes share a third of their information mainly due to the spatial overlap.

It has been shown that the amplitudes, latencies, and the width of both p3a and p3b can be tracked over multiple trials much better compared to the traditional single channel or non-cooperative techniques.

15.8 Conclusions

Single-task and multitask groups of agents can be represented by cooperative networks and their actions against various objectives and targets can be modeled and optimized by diffusion adaptation. In addition, various constraints can be incorporated into the formulation to enhance the adaptation process. There are tremendous applications for the diffusion adaptation of cooperative networks including communications, modeling physiological system, power and water control, security, robotics, and social networks. In places where there is some information about the communication intensities between the agents, the system can be further enhanced for better performance. This could be seen in the brain-related applications including BCI and Parkinson's problems. The multitask diffusion adaptation scenario has even more applications due to its ability to group the agents into those with different objectives. Distribution of power in a smart grid, incorporating variety of demands (constraints) by consumers and retailer, is an impressive application of the approach. Diffusion adaptation may be coupled with other signal processing techniques (such as mathematical modeling or game theory) to more accurately solve many practical problems. The concept can also be used in better estimation of data distribution and tracking its variabilities through cooperation.

References

1. Sayed, A. H. (2014). Adaptive networks, *Proceedings of the IEEE*, 102(4), pp. 460–497.
2. Khalili, A., Rastegarnia, A., Bazzi, W. M., and Sanei, S. (2016). Maximum correntropy based distributed estimation of adaptive networks, in *Advances in Computer Communications and Networks: From Green, Mobile, Pervasive Networking to Big Data Computing*, River Publisher, Eds. Aaron Striegel, Min Song, and Kewei Sha.
3. Khalili, A., Rastegarnia, A., Bazzi, W. M., and Sanei, S. (2017). Analysis of incremental augmented affine projection algorithm for distributed estimation of complex-valued signals, *Journal of Circuits, Systems & Signal Processing*, 36 (1), pp. 119–136.
4. Khalili, A., Rastegarnia, A., and Sanei, S. (2016). Tracking performance of incremental augmented complex least mean square adaptive network in the presence of model non-stationarity, *IET Signal Processing* 10 (7), pp. 798–804.
5. Khalili, A., Rastegarnia, A., and Sanei, S. (2015). Quantised augmented complex least mean-square algorithm: Derivation and performance, *Journal of Signal Processing*, 121(C), pp. 54–59.
6. Khalili, A., Rastegarnia, A., and Sanei, S. (2015). Robust Frequency Estimation in Three-Phase Power Systems Using Correntropy Based

Adaptive Filter, *IET Journal of Science, Measurement & Technology*, 8 pages, DOI: 10.1049/iet-smt.2015.0018.

7. Khalili, A. Rastegarnia, A. Bazzi, W. M., and Sanei, S. (2014). Analysis of cooperation gain for adaptive networks in different communication scenarios, *Journal of Electronics and Communications*, 2014, doi.org/10.1016/j.aeue.2014.03.014.
8. Monajemi, S., Ong, S.-H., and Sanei, S. (2014). Advances in bacteria motility modelling via diffusion adaptation, *Proceedings of the European Signal Processing Conference, EUSIPCO, Portugal.*
9. Ensafi, S., Monajemi, S. Ong, S.-H., and Sanei, S. (2015). Adaptive Distributed Dictionary Learning Method for HEP-2 Cell Image Classification, *Proceedings of the 18th International Conference on Medical Image Computing and Computer Assisted Intervention (MICCAI), Germany.*
10. Monajemi, S., Ensafi, S. Lu, S., Kassim, A. A., Tan, C. L., Sanei, S., and Ong, S.-H. (2016). Classification of HEp-2 cells using distributed dictionary learning, *Proceedings of the 24th European Signal Processing Conference, EUSIPCO Budapest, Hungary.*
11. Monajemi, S., Ensafi, S., Lu, S., Kassim, A. A., Tan, C. L., Sanei, S., and Ong, S.-H. (2017). *Cooperative Dictionary Learning for HEp-2 Cell classification, Biomed. Signal Processing in Big Data*, Ed. E. Sajdic, CRC Press.
12. Monajemi, S., Sanei, S., Ong, S.-H., and Sayed, A. H. (2015). Adaptive regularized diffusion adaptation over multitask networks, *Proceedings of the of IEEE Workshop on Machine Learning for Signal Processing*, Boston.
13. Chen, J., Richard, C., and Sayed, A. H. (2015). Diffusion LMS over multitask networks. *IEEE Transactions Signal Processing*, 63(11), pp. 2733–2748.
14. Eftaxias, K., Sanei, S., and Sayed, A. (2013). A new approach to evaluation and modeling of brain connectivity using diffusion adaptation. *Proceedings of the IEEE International Conference on Acoustics, Speech and Signal Processing, ICASSP, Vancouver, Canada.*
15. Eftaxias, K., and Sanei, S. (2013). Diffusion adaptive filtering for modeling brain responses to motor tasks, *18th International Conference on Digital Signal Processing, DSP*, Santorini, Greece.
16. Eftaxias, K., and Sanei, S. (2014). Discrimination of task-based EEG signals using diffusion adaptation and S-transform coherency, *Proceedings of IEEE Workshop on Machine Learning for Signal Processing, France.*

17. Monajemi, S., Eftaxias, K., Ong, S.-H., and Sanei, S. (2016). An informed multitask diffusion adaptation approach to study tremor in Parkinson's disease, *IEEE Journal of Selected Topics in Signal Processing; Special Issue on Advanced Signal Processing in Brain Networks*, 10(7), pp. 1306–1314.
18. Monajemi, S., Jarchi, D., Ong, S.-H., and Sanei, S. (2017). Cooperative particle filtering for detection and tracking of ERP subcomponents from multichannel EEG, *Entropy, Special Issue on Entropy and Electroencephalography*, 19(5), 199.
19. Sanei, S. (2013). *Adaptive Processing of Brain Signals* (John Wiley & Sons).
20. Enshaeifar, S., Spyrou, L., Sanei, S., and Took, C. C. (2016). A regularised EEG informed Kalman filtering algorithm, *Biomedical Signal Processing and Control*, 25, pp. 196–200.
21. Jarchi, D., Sanei, S., and Lorist, M. M. (2011). Coupled particle filtering: A new approach for P300-based analysis of mental fatigue, *Biomedical Signal Processing and Control*, 6(2), pp. 175–185.
22. Mohseni, H. R.,Wilding, E., and Sanei, S. (2010). Variational Bayes for spatiotemporal identification of event-related potential subcomponents, *IEEE Transactions Biomedical Engineering*, 57(10), pp. 2413–2428.
23. Akyildiz, I. F., Pompili, D., and Melodia, T. (2005). Underwater acoustic sensor networks: Research challenges, *Ad Hoc Networks*, 3(3), pp. 257–279.
24. Ashok, R. L., and Agrawal, D. P. (2003). Next-generation wearable networks, *Computer*, 36(11), pp. 31–39.
25. Atakan, B., and Akan, O. B. (2007). An information theoretical approach for molecular communication, *Proceedings of the 2nd Conference on Bio-Inspired Models of Network, Information and Computing Systems*, pp. 33–40, Budapest, Hungary.
26. Di Lorenzo, P. (2012). Bio-inspired dynamic radio access in cognitive networks based on social foraging swarms, Ph.D. dissertation, Sapienza University of Rome.
27. Latifi, M., Khalili, A., Rastegarnia, A., Zandi, S., and Bazzi, W. M. (2017). A distributed algorithm for demand-side management: Selling back to the grid, *Heliyon*, 3(11) pp. 1–28.
28. Latifi, M., Khalili, A., Rastegarnia, A., and Sanei, S. (2017). Fully distributed demand response using adaptive diffusion Stackelberg algorithm, *IEEE Transactions on Industrial Informatics*, 13(5), pp. 2291–2301.

29. de Azevedo, R., Cintuglu, M. H., Ma, T., and Mohammed, O. A. (2017). Multiagent-based optimal microgrid control using fully distributed diffusion strategy, *IEEE Transactions on Smart Grid*, 8(4), pp. 1997–2008.
30. Bertsekas, D. P. (1997). A new class of incremental gradient methods for least squares problems, *SIAM Journal on Optimization*, 7(4), pp. 913–926.
31. Tsitsiklis, J. N., and Athans, M. (1984). Convergence and asymptotic agreement in distributed decision problems, *IEEE Transactions on Automatic Control*, 29(1), pp. 42–50.
32. Lopes, C. G., and Sayed, A. H. (2007). Incremental adaptive strategies over distributed networks, *IEEE Transactions on Signal Processing*, 55(8), pp. 4064–4077.
33. Nedic, A., and Bertsekas, D. P. (2001) Incremental subgradient methods for nondifferentiable optimization, SIAM Journal on Optimization, 12(1), pp. 109–138.
34. Blatt, D., Hero, A. O., and Gauchman, H. (2007). A convergent incremental gradient method with a constant step size, *SIAM Journal on Optimization*, 18(1), pp. 29–51.
35. DeGroot, M. H. (1974). Reaching a consensus, *Journal of the American Statistical Association*, 69(345), pp. 118–121.
36. Xiao, L., and Boyd, S. (2004). Fast linear iterations for distributed averaging, *Systems & Control Letters*, 53(1), pp. 65–78.
37. Nedic, A., and Ozdaglar, A. (2009). Distributed subgradient methods for multi-agent optimization, *IEEE Transactions on Automatic Control*, 54(1), pp. 48–61.
38. Srivastava, K., and Nedic, A. (2011). Distributed asynchronous constrained stochastic optimization, *IEEE Journal of Selected Topics in Signal Processing*, 5(4), pp. 772–790.
39. Sayed, A. H. (2014). Adaptation, learning, and optimization over networks, *Foundations and Trends in Machine Learning*, 7(4–5), pp. 311–801.
40. Sayed, A. H., Tu, S.-Y., Chen, J., Zhao, X., and Towfic, Z. J. (2013). Diffusion strategies for adaptation and learning over networks: An examination of distributed strategies and network behavior, *IEEE Signal Processing Magazine*, 30(3), pp. 155–171.
41. Chen, J., and Sayed, A. H. (2012). Diffusion adaptation strategies for distributed optimization and learning over networks, *IEEE Transactions on Signal Processing*, 60(8), pp. 4289–4305.

42. Cattivelli, F. S., and Sayed, A. H. (2010). Diffusion LMS strategies for distributed estimation, *IEEE Transactions on Signal Processing*, 58(3), pp. 1035–1048.
43. Chen, J., and Sayed, A. H. (2013). Distributed pareto optimization via diffusion strategies, *IEEE Journal of Selected Topics in Signal Processing*, 7(2), pp. 205–220.
44. Lopes, C. G., and Sayed, A. H. (2008). Diffusion least-mean squares over adaptive networks: Formulation and performance analysis, *IEEE Transactions on Signal Processing*, 56(7), pp. 3122–3136.
45. Chen, J., and Sayed, A. H. (2012). Distributed pareto-optimal solutions via diffusion adaptation, *Proceedings of the IEEE Statistical Signal Processing Workshop, SSP*, pp. 648–651.
46. Shoker, L., Sanei, S., and Chambers, J. A. (2006). Artifact removal from electroencephalograms using a hybrid BSS-SVM algorithm, *IEEE Signal Processing Letters*, 12(10), pp. 721–724.
47. Stockwell, R., Mansinha, L., and Lowe, R. (1996). Localization of the complex spectrum: The s-transform, *IEEE Transactions on Signal Processing*, 44(4), pp. 998–1001.
48. Monajemi, S., Sanei, S., and Ong, S.-H. (2018). Information credibility over multitask distributed networks, *Future Generation Computer Systems, Special Issue on Measurements and Security of Complex Networks and Systems*, 83(C), pp. 485–495.
49. Jiang, C., Han, Z., Ren, Y., and Hanzo, L. (2016). Information credibility equilibrium of cooperative networks, *IEEE Wireless Communications and Networking Conference*, WCNC, pp. 1–6.
50. Bouakiz, M., and Sobel, M. J. (1992). Inventory control with an exponential utility criterion, *Operations Research*, 40(3), pp. 603–608.
51. Artz, D., and Gil, Y. (2007). A survey of trust in computer science and the semantic web, *Journal of Web Semantics: Science, Services and Agents on the World Wide Web*, 5(2), pp. 58–71.
52. Macnab, R. M., and Koshland, D. E. (1972). The gradient-sensing mechanism in bacterial chemotaxis, *Proceedings of the National Academy of Sciences*, 69(9), pp. 2509–2512.
53. Kutalik, Z., Razaz, M., and Baranyi, J. (2005). Connection between stochastic and deterministic modelling of microbial growth, *Journal of Theoretical Biology*, 232(2), pp. 285–299.
54. Vahidpour, V., Rastegarnia, A., Khalili, A., Bazzi, W., and Sanei, S. (2018). Partial diffusion Kalman filtering for distributed state estimation in multi-agent networks, *IEEE Transactions on Automatic Control* (arXiv preprint arXiv:1705.08920).

55. Latifi, M., Khalili, A., Rastegarnia, A., and Sanei, S. (2017). Fully distributed demand response using adaptive diffusion Stackelberg algorithm, *IEEE Transactions on Industrial Informatics*, 13(5), pp. 2291–2301.
56. Paterakis, N. G., Erdin, O., Bakirtzis, A. G., and Catalo, J. P. S. (2015). Optimal household appliances scheduling under day-ahead pricing and loadshaping demand response strategies, *IEEE Transactions on Industrial Informatics*, 11(6), pp. 1509–1519.
57. Deng, R., Xiao, G., Lu, R., and Chen, J. (2015). Fast distributed demand response with spatially and temporally coupled constraints in smart grid, *IEEE Transactions on Industrial Informatics*, 11(6), pp. 1597–1606.
58. Towfic, Z. J., and Sayed, A. H. (2014). Adaptive penalty-based distributed stochastic convex optimization, *IEEE Transactions on Signal Processing*, 62(15), pp. 3924–3938.
59. Mohsenian-Rad, A. H., Wong, V. W. S., Jatskevich, J., Schober, R., and Leon-Garcia, A. (2010). Autonomous demand-side management based on game-theoretic energy consumption scheduling for the future smart grid, *IEEE Transactions on Smart Grid*, 1(3), pp. 320–331.
60. Soliman, H. M., and Leon-Garcia, A. (2014). Game-theoretic demand-side management with storage devices for the future smart grid, *IEEE Transactions on Smart Grid*, 5(3), pp. 1475–1485.
61. Vardakas, J. S., Zorba, N., and Verikoukis, C. V. (2015). A survey on demand response programs in smart grids: Pricing methods and optimization algorithms, *IEEE Communication Surveys & Tutorials*, 17(1), pp. 152–178.
62. Jianshu, C., Xiaochuan, Z., and Sayed, A. H. (2010). Bacterial motility via diffusion adaptation, *Proceedings of the Asilomar Conference on Signals, Systems and Computers*, Pacific Grove, CA, USA.
63. Nassif, R., Richard, C., Ferrari, A., and Sayed, A. H. (2016). Multitask diffusion adaptation over asynchronous networks, *IEEE Transactions on Signal Processing* 64(11), pp. 2835–2850.
64. Khelladi, L., and Badache, N. (2017). Revisiting directed diffusion in the era of IoT-WSNs: Power control for adaptation to high density, *Proceedings of the 8th International Conference on Information, Intelligence, Systems & Applications, IISA*.
65. Hlinka, O., Hlawatsch, F., and Djuric, P. M. (2013). Distributed particle filtering in agent networks: A survey, classification, and comparison, *IEEE Signal Processing Magazine*, 30, pp. 61–81.

Chapter 16

Spoken Language Processing: From Isolated Word Recognition to Neural Representation of Syntactical Structures, Based upon Kernel Memory

Tetsuya Hoya
Department of Mathematics, College of Science & Technology, Nihon University, 1-8-14, Kanda-Surugadai, Chiyoda-Ku, Tokyo 101-8308, Japan
hoya@math.cst.nihon-u.ac.jp

Language is an essential faculty, equipped with human beings for enabling smooth communication. Modeling the language faculty of humans has become more intriguing, considering the recent overwhelming advancements made in artificial intelligence, and is expected to be inevitable for its further success. In the signal processing context, "spoken language processing" embraces multiple stages of processing, i.e., the feature extraction from raw speech waveform data, followed by recognition, association, and other higher-order processing of linguistically meaningful patterns. This chapter deals with some of such processing, from the recognition of individually spoken words to the formation of a composite network representing syntactical structures, under a unified principle of kernel memory. Kernel memory is a connectionist model, comprised

Learning Approaches in Signal Processing
Edited by Wan-Chi Siu, Lap-Pui Chau, Liang Wang, and Tieniu Tan

ISBN 978-981-4800-50-1 (Hardcover), 978-0-429-06114-1 (eBook)
www.panstanford.com

by a set of units and their connections, as in the ordinary artificial neural network models, but is heavily based upon localistic representation, unlike the well-known multi-layered perceptrons. Within the context of kernel memory, while the model is represented by a network structure, the notion of nodes, as well as their connections, is extended in various aspects, and was originally proposed to provide a basis for modeling various cognitive modalities. In this chapter, with a review of some relevant works, a holistic model of the spoken language processing is proposed based upon the kernel memory concept.

16.1 Introduction

Spoken language processing is an interdisciplinary field of study. In the field, speech recognition has long been a central topic of interest in a wide range of studies and is considered to be of fundamental for any further language related processing. While to pursue an engineering solution for effectively recognizing speech signals is the main purpose of the automatic speech recognition (ASR) area of studies, many of psycholinguistic studies put a large emphasis on elucidating how humans recognize spoken language (i.e., human speech recognition; HSR). In the HSR, it is generally conceived that speech is processed through a hierarchical structure having both explicit pre-lexical (i.e., sub-word) and lexical (i.e., word) layers, the latter of which dynamically integrates sub-word information gathered from the former (i.e., pre-lexical layer) during the course of an utterance.

An illustrative connectionist model, known as TRACE II (McClelland and Elman, 1986), was proposed within this context. The connectionist model is based upon a three-layered structure, incorporating a feature, pre-lexical, and lexical layer. Although the TRACE II model gives a reasonable account for the recognition process of individual words segmented from a continuous speech stream, it is completely pre-specified and thus does not learn by itself. Moreover, the validity of the model has been verified through the simulation study using only artificial speech segments (i.e., "wickelphones"), and, therefore, the applicability to real speech data is still not clear.

The lack of handling real speech inputs in many of the connectionist models proposed so far in the HSR domain contrasts with the success of the ASR approaches, as represented by the models based upon hidden Markov models (HMMs; Young et al., 2005). In recent years, a further success has been made by way of deep learning (DL) approaches combined with the HMMs (for a comprehensive survey, see Yu and Li, 2017); in the DL context, neural network models with many layers, such as deep neural networks (DNNs), are used for robust estimation of the posteriori probabilities of HMM (see e.g., Hinton et al., 2012), in place of conventional Gaussian mixture models (GMMs). A DNN is a feedforward, multi-layered perceptron (MLP) type network (Rosenblatt, 1958; Widrow, 1962) and has multiple hidden layers (typically, 2–7 layers, each with hundreds to thousands of units). Training such a large-scale network requires a massive amount of computational resource, as it normally involves long and iterative parameter tuning due to the utility of back-propagation (BP) algorithm (Amari, 1967; Rumelhart et al., 1986). Moreover, since BP algorithm is based upon a gradient-descent parameter optimization, training DNN/MLP inherently suffers from numerical instability, i.e., becoming stuck in local minima or slow convergence/divergence. In practice, exploiting the hybrid scheme of DL and HMM is not straightforward and requires a deep understanding and expertise, in order to achieve its optimal performance (cf. Shinozaki and Watanabe, 2015).

Bearing these advancements in the ASR and HSR area of studies in mind, this chapter first describes another approach of speech recognition proposed within the context of kernel memory; a review of the author's previous works relevant to the neural models of spoken word recognition is given, followed by the description of kernel memory. Then, a composite model representing syntactical structures is eventually proposed, as a post-processing model of word recognition.

16.2 Kernel Memory

Kernel memory is a connectionist model, consisting of a set of kernel units and their connections (for the detail, see Hoya, 2005). A kernel unit (or "a kernel," in short) generates its output via the response (or kernel) function assigned, given the input. Unlike typical artificial neural network (ANN) models, both the input to and output from a single kernel can have multiple values (e.g., in either a vector or a matrix form). Moreover, depending upon the situation, a network comprising of kernel units, each with different domain input data is possible (i.e., the input to Kernel 1 is $\mathbf{x}^1$, Kernel 2 $\mathbf{x}^2$, and so forth, each with different modality; e.g., $\mathbf{x}^1$ represents a set of feature data points obtained from an image, while $\mathbf{x}^2$ that obtained from a raw speech sample). In the kernel memory context, units with different kernel functions may also coexist in a network structure, and they can be connected to each other, via the feed-forward/feedback (and even their uni-/bi-directional) connections in between (i.e., such connections as these are called "link weight connections"). These manners of connection essentially remove any topological constraints in designing a network structure. From a neural modeling perspective, a kernel does not always correspond to a single neuron or cell but rather is compared to a functional unit represented by a population of neurons (i.e., a "functional web"; Pulvermüller, 2002) or a column structure as in the cerebral cortex of brain.

A kernel function is generally expressed in a form

$$K(\mathbf{x}) = K(\mathbf{x}, \mathbf{t}), \tag{16.1}$$

where $\mathbf{x}$ and $\mathbf{t}$ are the input and *template* data, respectively (without loss of generality, here, both $\mathbf{x}$ and $\mathbf{t}$ are given in a vector form). The latter corresponds to the inner-held data within a kernel and can thus be compared to a locally stored "memory" item (so the term "kernel memory"). A kernel is said to be "activated" (or excited) when the output of the function (16.1) is greater than or equal to a certain threshold, given the input data $\mathbf{x}$

$$K(\mathbf{x}) \geq \theta_K \tag{16.2}$$

For the application to pattern recognition (or function approximation), for instance, a kernel function that yields a similarity metric between $\boldsymbol{x}$ and a prototype $\boldsymbol{t}$ may be chosen for a particular set of kernel units, where $\boldsymbol{t}$ is set during the construction (i.e., learning) phase of kernel memory.

16.2.1 Network Models Based upon Radial Basis Function Kernels

For measuring the similarity between the input to a kernel unit $\boldsymbol{x}$ and its template data $\boldsymbol{t}$, a standard choice would be a radial basis function (RBF):

$$K(\boldsymbol{x}, \boldsymbol{t}) = \exp\left(\frac{-||\boldsymbol{x} - \boldsymbol{t}||_2^2}{\sigma^2}\right) \qquad (16.3)$$

where σ is the radius, $\boldsymbol{t}$ is called the centroid vector, and $\|\cdot\|_2$ denotes $L2$-norm.

A number of studies have suggested that RBFs are more appealing in view of neural modeling, compared to sigmoidal functions as used in conventional MLP type networks: Hopfield's analysis on sensory modalities (Hopfield, 1995) indicates the resemblance between neurobiological computation of analogue match and RBF, while Poggio and Edelman (1990) proposed a biologically plausible model of 3D object recognition using RBFs. Moreover, RBFs can also yield an intuitive interpretation of the so-called "hand-cells" found within the inferior cortex of macaque that selectively respond only to visual stimuli of hand-images (Gross et al., 1972). On the other hand, from the modeling point of view, the theoretical work (Gori and Scarselli, 1998) concluded that neural networks based upon RBFs appear more suitable for pattern recognition and verification tasks than MLPs, in the case where a reliable rejection is required. Along with these supportive studies, RBF neural networks (RBF-NNs) have been successfully applied in practice to a variety of pattern recognition problems, since the early work due to Broomhead and Lowe (1988).

Specht (1990) proposed a variant of the RBF-NNs for general pattern classification purposes, termed Probabilistic Neural Network (PNN). Like an ordinary RBF-NN, a PNN has a three-layered structure; each unit in the first layer corresponds to an element of the input vector, the second layer consists of multiple RBFs (i.e., each given as (16.3) with an independent template vector, while a unique value of the radius is used), and each unit in the third layer outputs a linear sum value of the second layer unit activations. In pattern classification tasks, a final result can be generally obtained via a winner-takes-all competition of the third layer unit activations. The difference between RBF-NNs and PNN resides in the setting of the weight values between the second and third layer. For instance, when the i-th RBF unit that falls into Class j is allocated in the second layer, the unit will have no connection but that to the j-th third layer unit; the third layer unit eventually outputs a linear sum value over all the RBF units that belong to Class j. To put it another way, the output of the j-th third layer unit $o(j)$ is given in the form of a kernel function with an inner-product operator

$$\begin{aligned} o(j) &= K(\boldsymbol{h}_j, \boldsymbol{t}_j) = \boldsymbol{h}_j \cdot \boldsymbol{t}_j, \\ \boldsymbol{h}_j &= [h_j(1), h_j(2), \ldots, h_j(N_j)] \quad (1 \times N_j), \\ \boldsymbol{t}_j &= [1, 1, \ldots, 1]^T \quad (N_j \times 1), \end{aligned} \tag{16.4}$$

where $h_j(i)$ are the respective activation values of the RBF units that belong to Class j. In the structural sense, a PNN can, therefore, be interpreted as a particular instance of kernel memory. Network structures constructed on the basis of kernel units with template data, such as RBF kernels above, are said to be *transparent*, in that the inner-held data (i.e., denoted as the template data $\boldsymbol{t}$ in (16.3)) are readily available, as well as accessible from other units, where required. For instance, provided that a kernel K_α is activated first (i.e., when its activation exceeds a certain threshold, as given by (16.2)), it then feed-forwards the template data $\boldsymbol{t}$, instead of its activation value, to another kernel K_β connected to it, and K_β takes the forwarded data as input for a further processing. Such a data-forwarding principle can be exploited for the embodiment of e.g., information retrieval or associative memory, but is generally not considered in the family of RBF-NNs or PNNs, nor possible by conventional MLPs where the stored data is represented in a distributed manner by their node parameters and inter-layer weight connections.

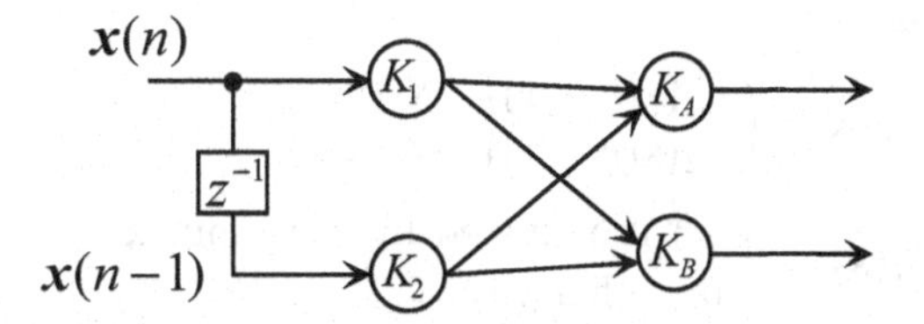

Figure 16.1 A simple kernel network that can perform serial order detection; K_A detects the activation sequence of $K_1 \rightarrow K_2$, whereas K_B does $K_2 \rightarrow K_1$. Adapted from Hoya (2005).

16.2.2 Representing Temporal Data Structures by Kernel Memory

One of the main difficulties in applying directly conventional ANN models to speech recognition resides in handling temporally varying sequential data (Morgan and Bourlard, 1995); in speech processing, raw speech waveforms are converted into a sequence of the feature data (or a "frame"), each typically obtained in the form of approximately 10–40 Mel-frequency cepstral coefficients (MFCCs, i.e., each computed over a single speech window in the order of 10 ms; Furui, 1981), in which the number of the frames is varying from utterance to utterance. Unlike conventional ANNs, HMMs are good at modeling the temporal structures of speech. The recent dramatic progress made in speech recognition is ascribed mostly to the hybrid HMM-DL approaches, where the DL-based neural networks play a role in the feature extraction of raw speech signals.

In the kernel memory context, such temporal data structures can be handled straightforwardly by exploiting another interpretation of the template data. As an example, consider a simple network in Fig. 16.1. In the figure, let us assume that both the kernels K_A and K_B on the right hand side hold a kernel function that performs a pattern matching between the input and template data, the latter given by the following matrices $\boldsymbol{T}_A$ and $\boldsymbol{T}_B$ as their inner-held data (instead of the template vectors as in (16.1)), respectively,

$$\mathbf{T}_A = \begin{pmatrix} 1 & 0 \\ 0 & 1 \end{pmatrix}, \mathbf{T}_B = \begin{pmatrix} 0 & 1 \\ 1 & 0 \end{pmatrix}, \tag{16.5}$$

and the input

$$\mathbf{X} = \begin{pmatrix} K_1(\mathbf{x}(n-1)) & K_1(\mathbf{x}(n)) \\ K_2(\mathbf{x}(n-1)) & K_2(\mathbf{x}(n)) \end{pmatrix},$$

$$K_i(\mathbf{x}) = \begin{cases} 1 \ ; \text{ if activated by the input } \mathbf{x} \\ 0 \ ; \text{ otherwise.} \end{cases} \qquad (16.6)$$

Now, with the settings (16.5) and (16.6), the activation states (i.e., given as binary values) of the kernels K_1 and K_2 on the left hand side in the figure, at both time n and n–1, are forwarded to the two kernels K_A and K_B. Then, K_A can be activated, e.g., if and only if the input $\boldsymbol{X}$ is identical to $\boldsymbol{T}_A$; K_1 is activated at time n–1 followed by that of K_2 at time n. In this case, the kernel K_A detects the sequential activations of $K_1 \rightarrow K_2$. Likewise, K_B acts as an activation sequence detector of the opposite direction, i.e., $K_2 \rightarrow K_1$.

By generalizing the notion of the template matrix representation as in (16.5) to incorporate more than two kernels, a network structure that is able to perform the matching between more complex patterns of spike trains can be considered. Thereby, such a network can be exploited in modeling a more general temporal coding mechanism of neural activity.

16.3 A Neuro-Computational Model of Isolated Spoken Word Recognition

Within the kernel memory principle, Hoya and van Leeuwen (2010) proposed a neuro-computational model of isolated spoken word recognition, the Cascaded Neuro-Computational (CNC) model. The connectionist model is composed of the three cascading layers; the first layer consists of multiple RBF units, the second layer the so-called "word-candidate" units, and the third layer the units, each representing an isolated word. Upon a completion of the utterance, each of the third layer eventually yields the recognition score for a particular class (i.e., word) by integrating the activations of the second units that fall into the word.

While CNC processes real speech input, as ordinary ASR models do, it conforms to several psycholinguistic viewpoints. First, the network has an explicit pre-lexical layer (i.e., corresponding to the first layer) and lexical layers (i.e., the second and third layers); the former processes the sub-word information of a word, upon the receival of a short segment of speech (i.e., represented by a frame) by the receptive fields (i.e., each modeled by an RBF), and is explicitly separated from the succeeding layers (i.e., the second and third layers) representing the lexical information. Such a hierarchical manner of the information processing also corresponds to the observation in human-beings that the analysis of auditory signals is performed by a layered architecture of the temporal cortex (cf. Allen, 1994). Second, word competition is performed within the lexical part (i.e., the third layer), on the basis of the sub-word information collected within the pre-lexical layer.

Third, each of the second layer units exploits and models a neural mechanism of serial-order detection, the evidence of which is supported by a wealth of neuropsychological/psycholinguistic findings (for the detail, see Pulvermüller, 2002); each unit in the second layer performs a pattern matching between its inner-held template and a series of labels, each label representing a winning unit in the first layer, obtained during the course of the utterance of a word. To be more specific, suppose that a second layer unit has a kernel function of the form (16.1) with the template vector (instead of using a matrix expression as in (16.5))

$$\mathbf{t} = [1, 1, 1, 2, 3, 3, 4], \qquad (16.7)$$

where each element represents the index of a maximally activated RBF kernel in the first layer of CNC, given an MFCC frame as the input $\boldsymbol{x}$ of all the RBF kernels. Then, the template $\boldsymbol{t}$ in (16.7) represents an activation pattern of the receptive units given the 7 consecutive frames of speech data, i.e., where the RBF kernel h_1 maximally activated three times, followed by the maximal activation of h_2 once, h_3 twice, and h_4 once. (Here, note that the second layer unit has the inter-layer connections only to these four RBF kernels). The second layer unit then performs a pattern matching between $\boldsymbol{x}$ and $\boldsymbol{t}$, the actual computation of which can be done by applying an "edit-distance" (Duda et al., 2001) type matching algorithm between two strings; all the elements in $\boldsymbol{t}$, as well as those in $\boldsymbol{x}$, are concatenated to form a string "1112334", prior to performing the matching.

Note that this manner of string matching allows not only to cope with the input vector of varying length but also to yield a midway score of the matching, given partially the input data; the former enables the neural recognizer to handle the temporal variability in speech, while the latter provides it a capability of generating an intermediate result of lexical competition before the completion of the utterance, the capability of which is considered to be important for giving a connectionist account of the lexical analysis within the psycholinguistic context. (Digging into this issue further is, however, beyond the scope of the chapter and should be discussed more in the future study.)

16.3.1 Construction of the CNC model

The construction of the CNC model is totally date-driven; there are initially no units in the layers but is only given a *prescription* that the resultant network yields a three-layered structure. The entire network structure (i.e., both the assignment of units in each of the three layers, as well as their inner-held template data, and the inter-layer connections) is constructed via an incremental learning scheme, given a set of the MFCC frame data for training.

The scheme involves two independent construction stages. In the first stage, only the first layer units, i.e., each represented by an RBF kernel, (as well as their template data) are assigned. The assignment is performed on a frame-to-frame basis; each of the MFCC frame data in the training set is given one-by-one as the input to the CNC, and, where appropriate (to be described later in this subsection), it is chosen as the template vector of an RBF kernel. In the second stage, both the second and third layer units, as well as the inter-layer connections (i.e., the connections between the first and second layer and those between the second and third layer, respectively), are allocated. In contrast to the first stage, the allocation of units in both the second and layers, as well as the establishment of the inter-layer connections, are carried out on a whole-word basis; the input to the second layer is, as aforementioned, represented by a series of labels (i.e., as in (16.7)) obtained using all the MFCC frame data spanning over the whole utterance of a word.

For both the construction stages, a simple "one-pass" (i.e., whole training data are presented to the network only once) incremental learning algorithm similar to that for self-structuring kernel memory (SSKM) networks (Hoya, 2003; Hoya and Washizawa, 2007), which is originally motivated from the learning of resource allocation network (RAN; Platt, 1991), is exploited; in the first stage, the learning is done in an unsupervised mode using the one-pass algorithm once, whereas a supervised learning procedure using essentially the same one-pass algorithm as that for the first stage is applied to the second stage. (Note that there is essentially only a single parameter to set for the construction of the entire CNC structure using the one-pass algorithm, i.e., the unique radius value of the RBF units, the degrees of freedom of which equal those of the PNN model; for the detail, see Hoya, 2016.)

In the first stage, a new RBF unit will be assigned to the first layer, whenever the pattern space spanned by the existing RBFs fails to cover the local region corresponding to the input frame data (i.e., this implies that a set of the first MFCC frame data of the very first sample in the training set is initially set as the template vector of the first RBF unit). Such an assignment is, therefore, done, regardless of any attributes the frame data hold, i.e., the class (word) to which the input belongs.

In contrast to this, a new unit will be assigned in the second/third layer during the second stage of learning, either (i) if there exists no other unit in the third layer, representing the same word as that corresponding to the input series of labels or (ii) when the third layer obtained by applying an winner-takes-all strategy yields a wrong recognition result, given the input series. For both the cases, a new unit will be allocated in the second layer; a string matching unit with the template vector identical to the input series will be newly added in the second layer. For the case (i), a new unit representing the same class (word) as the input will also be assigned in the third layer. Upon the allocation of a new second layer unit, both the connections to the first layer units, the indices of which are represented by the respective labels (i.e., elements) in the template vector, and a connection to the third layer unit representing the same word as the input will be established.

16.3.2 Summary of the Simulation Results

In the simulation study (Hoya and van Leeuwen, 2010), the model was empirically validated using three databases for Japanese spoken digit recognition tasks (i.e., the number of digits: 12/13), each containing around a total of 840–1600 speech samples collected from a single/seven speakers for speaker-dependent and 88 speakers for speaker-independent tasks, recorded in an ordinary room with a moderate level of background noise. In summary, the overall recognition performance of around 90.0–98.5% was obtained using the CNC model (with the network size of around 100-3600 L1, 90-310 L2, and 12/13 L3 units), the performance of which was almost comparable to that of a whole-word based HMM.

16.4 An Augmented CNC Model for Connected Word Recognition

In the recent work (Hoya and van Leeuwen, 2016), an augmented version of the CNC model is proposed for performing connected word recognition tasks. Connected word recognition (CWR) refers to a task of recognizing individually a varying number of words (i.e., the number of which is unknown) contained in a continuous speech stream and eventually yielding a final result by concatenating each of the word recognition scores. While the same approaches as used for the isolated word recognition tasks described in the previous section can be applied to recognize each word, to perform an CWR task requires to solve an additional problem of segmenting a speech stream into its sub-parts, each representing a word. However, the segmentation into word-tokens is a non-trivial task, as the locations of speech/non-speech boundaries are not generally known in advance, nor the word-boundaries. Moreover, the detection of speech-to-non-speech boundaries can be unreliable, since similar transitions in signal energy often occur within words. Boundary detection between two adjacent words (i.e., detection of speech-to-speech boundaries) can be falsified, due to the co-articulation effect (cf. Rabiner and Juang, 1993); around the boundary, some speech portions may be shared by adjacent words, which causes the difference in sound between words uttered in a continuous manner and those separately. Compared to the isolated spoken word recognition, to solve the CWR problem is, therefore, generally considered to be much harder.

16.4.1 Segmentation of Training Samples into Word-Tokens by the Augmented CNC

In the augmented model of CNC, speech/non-speech portions are identified via the processing using the first layer (L1) units. First, prior to performing the identification, the L1 units that respond selectively to the frame data containing non-speech portions, such as speech pause or silent periods, of continuous speech streams (hereafter, denoted L1S units) are identified among all the L1 units. (Recall that given the training data set, the L1 units are generated without taking any properties of the frame data into account, as described in Section 16.3.1.) This identification is carried out after the assignment of all the L1 units during the first stage of the CNC construction (Section 16.3.1), using a separate training data set that contains non-speech portions only; in practice, such a separate set can be obtained conveniently, e.g., via applying a simple energy-based signal splitting scheme (e.g., Deller et al., 1993) to both the very first and last parts of each training sample.

Then, using both the L1 and L1S units, the following segmentation procedure is performed for each training sample:

(1) Obtain a label sequence by presenting all the MFCC frame data of a training sample to the first layer; each label in the sequence corresponds to the index of a winning L1/L1S unit;
(2) Trim the sequence by deleting the L1S unit labels (i.e., corresponding to the non-speech parts) appeared, if any, in both the beginning and ending portions; thereby, an overall location of the entire speech episode can be identified;
(3) Remove all but a single L1S unit label from the trimmed sequence obtained in (2); this process corresponds to shrinking the inter-/intra-word non-speech portions, if any;

(4) Split the edited label sequence obtained from (3) into the subsequences, each then corresponding to a word-token for training.

Note that step (3) above is intended to cope with the temporal variability in terms of the non-speech portions around/in word utterances.

Gathering all the subsequences obtained in (4) above constitutes a word-based training data set for constructing both the second (L2) and third (L3) layers of CNC. Then, using the word-based training set, the same supervised incremental learning procedure as in the second stage for the construction of the original CNC model is applied to build these two layers.

16.4.2 Performing Connected Word Recognition Using the Augmented CNC

In the recognition mode, a testing sample is first converted into the corresponding trimmed version of the label sequence as obtained in step (2) above using both the L1 and L1S units. Then, connected word recognition is performed based upon the label sequence using the augmented CNC model (for the detail, see Hoya and van Leeuwen, 2016):

(1) Starting from the very first label of the sequence, consecutive labels are fed-forwarded to each of the L2 units up to the same lengths of the respective template vectors, while skipping all but a single L1S unit label appeared, if any. When all the L2 units that belong to a particular word generate the final outputs, the L3 unit representing the corresponding word emits its final activation value;

(2) Isolated word recognition is performed to each of the subsequences obtained in (1), when all the L3 units emit their final activations;

(3) The result of the connected word recognition is obtained by concatenating all the scores of isolated word recognition; i.e., this is done by repeating steps (1) and (2) above till the end of the input label sequence.

16.4.3 Summary of the Simulation Study Using the Augmented CNC Model for Connected Digit Recognition Tasks

The augmented CNC model was applied to two speech data sets for connected digit recognition tasks (Hoya and van Leeuwen, 2016). Both the two data sets used for the simulation study (i.e., one for the multi-speaker dependent and the other for speaker-independent tasks) contain the speech samples of digit strings, where the duration of the utterance for each digit in a string was assumed as roughly equal. Therefore, in step (4) of the segmentation procedure for the training mode (Section 16.4.1), the length of the subsequence per word-token was uniformly set to each value obtained by simply dividing the length of the edited label sequence by the number of words it contains, rather than resting on a statistical estimation (i.e., thus avoiding an arduous computation).

For the simulation study, speech samples uttered in Japanese from seven native male speakers were obtained to constitute a connected digits data set for performing the multi-speaker dependent tasks (i.e., denoted "7-speaker data set"). A subset of publicly available database Aurora-2J (Yamamoto et al., 2003) was used for the speaker-independent tasks, in which the samples were obtained from a total of 41 Japanese native speakers (i.e., 22 females and 19 males). Each data set contains a total of around 1300-1500 connected digit strings; each of the connected digit strings in the 7-speaker data set was chosen as one of the prespecified 21 digit combinations; the total number of times each digit appeared in the 21 combinations was almost equal. In contrast, the appearance of each digit was not specified in advance for the Aurora-2J data set, and the number of digits appeared in each string of the Aurora-2J varied from two to seven.

In summary, the overall recognition performance of around 88-93% in terms of word accuracy was obtained by the augmented CNC (with the network size: around 2870-3430 L1, 310-540 L2, and 9/11 L3 units), the performance of which was, again, almost comparable to a state-of-the-art approach based upon the HMM (i.e., the HMM with embedded training; for the detail, see Young et al., 2005), similar to the isolated word recognition cases as described in Section 16.3.2.

As we have seen so far, the CNC models proposed for isolated/connected word recognition tasks can be viewed as particular examples of kernel memory: first, different types of kernel functions coexist in a single network: i.e., RBF kernels in the first, non-linear ones in the second, each performing a pattern matching of label-series, and the linear sum kernels in the third layer, each integrating the second layer units' activation for a particular class (word). Second, a variation of the temporal coding scheme (Section 16.2.2) is exploited in each second layer unit for the pattern matching. Third, the whole network is constructed by applying a simple one-pass incremental learning algorithm, similar to that for SSKM networks (Hoya, 2003; Hoya and Washizawa, 2007).

16.5 Neural Representation of Syntactical Structures by Way of Kernel Memory

One of the salient features the CNC models exhibit is that the input data given to the network are processed at different time scales within the respective layers; the lowest level of the layered structure (i.e., the respective RBF units in the first layer) handles the data varying at a faster time-scale (the order of ten-milliseconds, say) than those within the higher levels (i.e., the second and third layers). For the case of word recognition tasks, as described, these correspond to the processing of the sub-word (pre-lexical) and word (lexical) levels of linguistic information, respectively.

Such a hierarchical principle can, then, be extended seamlessly to encompass those beyond the single-word level; i.e., the phrases, each comprised by an ordered sequence of multiple words, or even syntactical level, where each constituent in turn represents a lexical category and their ordered sequences form the respective syntactical structures of a language.

To begin with, let us revisit the simple network structure depicted in Fig. 16.1: given a word for testing, a word recognition result can be obtained via the winning L3 unit of a CNC model, as described in the previous Sections 16.3 and 16.4. Now, suppose that the word recognition result is conveyed to both K_1 and K_2 where K_i

($i = 1, 2$) emits a binary value

$$K_i = \begin{cases} 1 \; ; & \text{if the word recognized is Word } i, \\ 0 \; ; & \text{otherwise.} \end{cases} \tag{16.8}$$

Then, given the template vectors $\boldsymbol{t}_A = [1, 2]$ and $\boldsymbol{t}_B = [2, 1]$ for K_A and K_B (or, alternatively, the matrix ones $\boldsymbol{T}_A$ and $\boldsymbol{T}_B$ as in (16.5)), respectively, the network in Fig. 16.1 can detect the word-sequence Word 1 $\rightarrow$ Word 2 by the activation of K_A, or the reversed one i.e., Word 2 $\rightarrow$ Word 1 by that of K_B. Similar to the L2 units of the CNC model given in the previous section, the sequence detection is essentially done by performing a pattern matching between the input given to the kernel unit K_A/K_B and the template vector $\boldsymbol{t}_A/\boldsymbol{t}_B$. By augmenting the simple kernel network in Fig. 16.1, word-sequence patterns, or phrases, containing more than two words can be straightforwardly detected.

16.5.1 Kernel memory representation of elementary grammatical structures

Chomsky (1957; Chapter 3) argued that any approach representing the grammar of a language must have the capability of generating infinite number of sentences, with a finite set of elements. For this aim, he denied any approaches based exclusively on conventional left-to-right Markov processes, including the HMM, by showing their failure to represent some rudimentary languages.

Chomsky (1957) then carried on to examine the following three primitive languages for giving his rebuttals to such Markovian approaches, each consisting only of two letters α and β (i.e., these two letters essentially can be substituted for certain respective words):

(1) $\alpha\beta, \alpha\alpha\beta\beta, \alpha\alpha\alpha\beta\beta\beta, \ldots$, i.e., sentences only with $n(\geq 1)$-times repetitions of α followed by those of β;

(2) $\alpha\alpha, \beta\beta, \alpha\beta\beta\alpha, \beta\alpha\alpha\beta, \alpha\alpha\alpha\alpha, \beta\beta\beta\beta, \alpha\alpha\beta\beta\alpha\alpha, \alpha\beta\beta\beta\beta\alpha, \ldots$, i.e., sentences comprised by one half with a string S, containing either both or only one of the letters α and β, and the other with the string reversing the appearance order of each letter in the first half;

(3) $\alpha\alpha, \beta\beta, \alpha\beta\alpha\beta, \beta\alpha\beta\alpha, \alpha\alpha\alpha\alpha, \beta\beta\beta\beta, \alpha\alpha\beta\alpha\alpha\beta, \alpha\beta\beta\alpha\beta\beta, \ldots$, i.e., sentences comprised by one half with a string S, containing either both or only one of the letters α and β, and the other identical to S.

Now, we consider the kernel memory representation of the three primitive languages above: let us assume that in Fig. 16.1 the kernels K_1 and K_2 are activated upon the receival of the letter α and β, respectively (i.e., these two kernels can regarded as the "post-processing" units of spoken word recognition, and their activation functions can be given in the form as (16.8)), and K_A has the template

$$\mathbf{t}_A = [1, 1, \ldots, 1, \ 2, 2, \ldots 2] \ ,$$

or, in a more concise form

$$\mathbf{t}_A = [1 : n, \ 2 : n] \quad (n \geq 1), \tag{16.9}$$

where in the latter in each element p:n is read as "the kernel K_p is consecutively activated n times." Then, with an appropriate setting of the value n, the kernel K_A can be activated, when it detects the corresponding sentence pattern in (1) above (i.e., this detection can be actually done by performing a pattern matching between the input $\boldsymbol{x}$ to the kernel K_A and $\boldsymbol{t}_A$, as described in Section 16.3).

For the patterns in (2) above, if the templates of K_A and K_B in Fig. 16.1 are given as

$$\mathbf{t}_A = [1 : n_1, \ 2 : n_2, 2 : n_2, 1 : n_1],$$
$$\mathbf{t}_B = [2 : n_1, \ 1 : n_2, 1 : n_2, 2 : n_1] \ (n_1 \geq 1, n_2 \geq 0), \tag{16.10}$$

respectively, the kernel K_A can detect such patterns as $\alpha\alpha$ ($n_1 = 1$, $n_2 = 0$), $\alpha\alpha\alpha\alpha$ ($n_1 = 2, n_2 = 0$), $\alpha\beta\beta\alpha$ ($n_1 = 1, n_2 = 1$), $\alpha\alpha\beta\beta\alpha\alpha$ ($n_1 = 2, n_2 = 1$), and the like, whereas K_B that of $\beta\beta$ ($n_1 = 0, n_2 = 1$), $\beta\alpha\alpha\beta$ ($n_1 = 1, n_2 = 1$), etc. Similarly, the kernel units K_A and K_B, with the template (and the appropriate settings of both n_1 and n_2)

$$\mathbf{t}_A = [1 : n_1, \ 2 : n_2, 1 : n_1, 2 : n_2],$$
$$\mathbf{t}_B = [2 : n_1, \ 1 : n_2, 2 : n_1, 1 : n_2] \ (n_1 \geq 0, n_2 \geq 0), \tag{16.11}$$

can detect the sentence patterns as in (3) above.

For all the three languages (1–3), extension to the case where more than two letters are involved in the sentence generation is then straightforward.

16.5.2 Representation of the sentences with inter-dependent words by a kernel network

Chomsky (1957) continued his arguments to the cases where there is a dependency between words, i.e., "if-then" or "either-or", in a sentence and within the inter-dependent words where there is embedded another sentence, i.e.

(1) If S1, then S2.
(2) Either S3 or S4.
(3) The boy who told that S5 smiles again.

In the above, S1-S5 are embedded sentences and can even be one of the three sentences, e.g., S3 in (2) is another sentence of (1) or again (2), or S1 and S2 in (1) are other sentences of (3) and (2), respectively, and so on.

Now, before tackling the generalized forms (1)–(3) above, let us first consider less complicated ones below (i.e., using a limited vocabulary, consisting of fourteen words only):

(1a) If the boy smiles happily, then the girl smiles too.
(2a) Either the boy or the girl.
(3a) The boy who cried yesterday smiles again.

In order to represent the three sentences (1a)–(3a), a two-layer kernel network as illustrated in Fig. 16.2 may be conceived, where in the figure each of the first layer kernels K_{11}-$K_{1/14}$ (here, a slash "/" is used to distinguish the layer number from that of unit, for convenience) has the activation function of the form (16.8) and can thus be activated upon recognizing one of the words in the vocabulary set $\{w_i\}$ = {"if", "then", "either", "or", "who", "the", "boy", "girl", "smiles", "happily", "too", "cried", "yesterday", "again"} and where each of the second layer kernels K_{21}-K_{23} can be activated, when a particular activation sequence pattern of the first layer units is detected. In this case, the inner-held template of each second layer kernel K_{21}-K_{23} is set, respectively, as

$$\begin{aligned} t_{21} &= [\mathbf{11}, 16, 17, 19, 1/10, \mathbf{12}, 16, 18, 19, 1/11] \;, \\ t_{22} &= [\mathbf{13}, 16, 17, \mathbf{14}, 16, 18] \;, \\ t_{23} &= [16, 17, \mathbf{15}, 1/12, 1/13, 19, 1/14] \;, \end{aligned} \qquad (16.12)$$

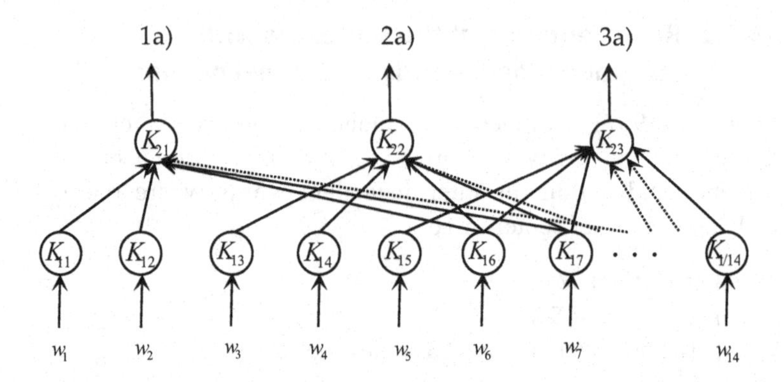

Figure 16.2 A two-layer kernel network representing the sentences (1a)-(3a).

where each index i appeared in the respective template vectors denotes the activation of the first layer kernel K_{1i} (here, an index number in bold indicates the activation of a kernel representing one of the "functional" words w_1-w_5, in order to distinguish it from the non-functional ones). Then, the kernels K_{21}-K_{23} can be activated, if and only if the respective word-sequences (1a)–(3a) are detected.

While the kernel network in Fig. 16.2 embodies the interdependency in terms of a particular set of the functional words w_1-w_5, it is apparent that the number of the second layer units soon becomes prohibitively large, as the number of words used in the vocabulary increases, or a combinatorial explosion eventually occurs. This violates the aforementioned principle of modeling a language with a finite set of elements (i.e., in the beginning of Section 16.5.1), and the network representation in Fig. 16.2, therefore, appears to be inappropriate for the purpose of the language modeling.

16.5.3 Sequence Detection of Lexical Categories by Means Of Kernel Memory

Next, let us consider the following three sentence patterns:

(1b) If NP1 VP1, then NP2 VP2;
(2b) Either NP1 or NP2;
(3b) NP1 who VP1 VP2,

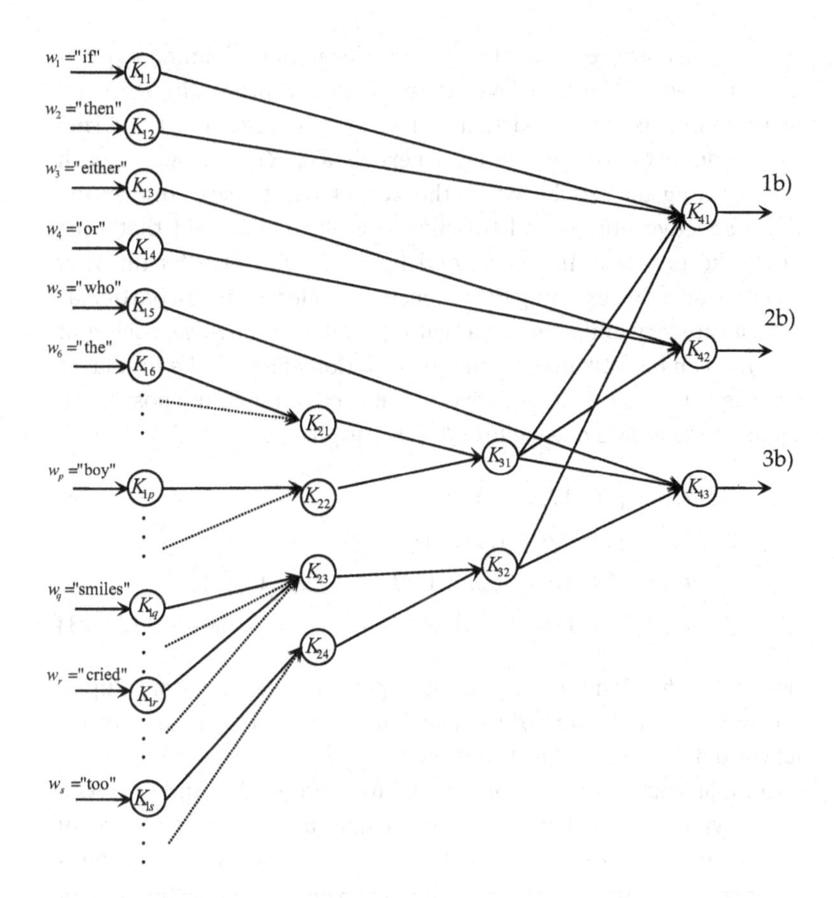

Figure 16.3 A four-layer kernel network that processes the three-sentence patterns 1b–3b).

where NP*i* and VP*i* stand for a noun-phrase and verb-phrase, respectively. Without loss of generality, let us assume here that each NP*i*/VP*i* is restricted only to a specific phrase pattern; NP*i* is composed of a single noun followed by an article (i.e., the boy, a girl, that car, etc.), whereas each VP*i* consists of no more than a single verb followed by an adverb (i.e., smiles again, cried yesterday, etc.).

Figure 16.3 shows a four-layer kernel network which is capable of processing the three sentence patterns (1b)–(3b), where each pattern now contains some varying constituents (i.e., NP*i* and VP*i*).

As in the figure, each of the second layer kernel units K_{21}-K_{24} is connected to the first layer kernels, each representing a single word that falls into a particular lexical category; i.e., the kernel K_{21} is connected to the first layer kernels $K_{16}, K_{17}, \ldots, K_{1p-1}$, each representing an article, while the kernel K_{22} those representing the respective nouns, and the like. Therefore, it is said that each of the kernel units in the second layer K_{21}-K_{24} *bundles* the first layer kernels representing the respective single words of a particular lexical category and acts like a "gating" unit, or a *variable*; each unit K_{21}-K_{24} can be activated, upon the activation of *any* of the first layer kernels connected to it, and forwards its activation to the post-layer. Then, the template of each unit K_{21}-K_{24} is given as

$$
\begin{aligned}
t_{21} &= \{16, 17, \ldots, 1p-1\} \ , \\
t_{22} &= \{1p, 1p+1, \ldots, 1q-1\} \ , \\
t_{23} &= \{1q, 1q+1, \ldots, 1r, 1r+1, \ldots, 1s-1\} \ , \\
t_{24} &= \{1s, 1s+1, \ldots\} \ , \qquad (16.13)
\end{aligned}
$$

where $\{a, b\}$ denotes an OR-like operator, meaning that "upon receiving the activation of the previous layer unit K_a *or* that of K_b, forward it to the post-layer units connected."

In contrast, each of the third layer units K_{31} and K_{32} can be activated upon detecting a particular activation sequence of the second layer units connected to it, i.e., each representing a completion of the noun-/verb-phrase, whereas the activation of the fourth-layer unit K_{41}-K_{43} occurs, when it detects an activation sequence pattern, consisting of the five first-layer units K_{11}-K_{15} and the two third layer ones K_{31} and K_{32}. Then, the fourth-layer kernels K_{41}-K_{43} eventually represent the sentence patterns (1b)–(3b), respectively, with the template settings of (16.13) and

$$
\begin{aligned}
t_{31} &= [21, 22] \, , \\
t_{32} &= [23, 24] \, , \\
t_{41} &= [11, 31, 32, 12, 31, 32] \, , \\
t_{42} &= [13, 31, 14, 31] \, , \\
t_{43} &= [31, 15, 32 : 2] \, . \qquad (16.14)
\end{aligned}
$$

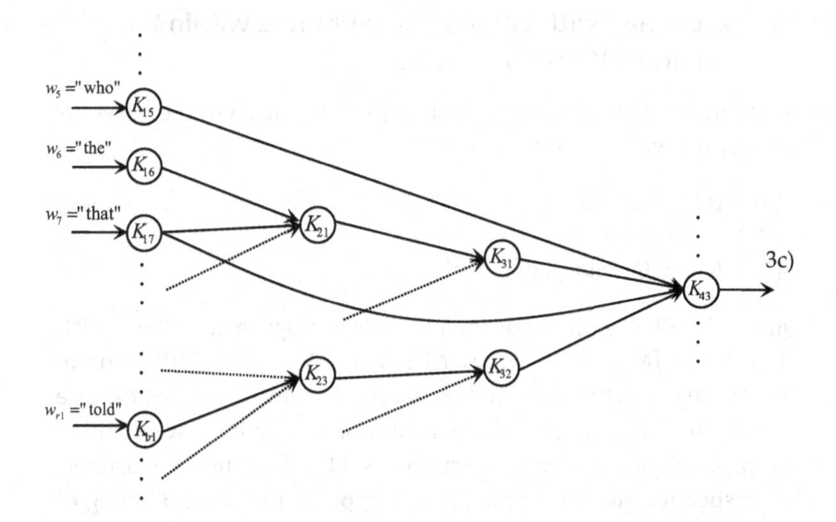

Figure 16.4 A part of the modified version of the four-layer kernel network shown in Fig. 16.3; the fourth-layer kernel K_{43} now has the connections to the kernels different from those in Fig. 16.3, in order to represent the sentence pattern 1c).

Therefore, the kernel network in Fig. 16.3, together with the template settings of (16.13) and (16.14), also allows to generalize the sentences such as

(1b-1) If the girl stops suddenly, then the cat runs quickly; where NP1 = "the girl", VP1 = "stops suddenly", NP2 = "the cat", and VP2 = "runs quickly";

(2b-1) Either this car or that train; where NP1 = "this car" and NP2 = "that car";

(3b-1) The teacher who sings sadly leaves here; where NP1 = "the teacher," VP1 = "sings sadly", and VP2 = "leaves here."

It is also said that the kernel network in Fig. 16.3 can alleviate the problem of "combinatorial explosion" to a great extent, due to the introduction of lexical category-wise sequence detection (cf. Pulvermüller, 2002), rather than performing word-to-word detection of the sequences.

16.5.4 Dealing with Embedded Sentences within the Kernel Memory Principle

Now, let us revisit the patterns (1b)–(3b) appeared in the previous subsection, with a modification:

(1c) If S1, then S2;
(2c) Either S3 or S4;
(3c) NP1 who told that S5 VP1,

where S1–S4 is either one of (1c)–(3c) or given as S1 = [NP1, VP1], S2 = [NP2, VP2], S3 = NP1, and S4 = NP2 (NPi: a noun followed by an article; VPi: a verb followed by an adverb) and where S5 is either (1b) or (2b), without loss of generality. The modified setup, therefore, yields the sentences (1c)–(3c) much closer to the respective general forms (1)–(3) appeared in the beginning of Section 16.5.2.

In order to cope with such forms as those with a nested structure, it is considered that a mechanism separate from the primary processing part is necessary, i.e., temporal memory to store a subpart of the sentence that is being nested, that is, the storage representing short-term memory (STM) or working memory (WM).

The STM/WM can also be comprised by a set of the kernel units $K_{\mathrm{STM},j}$ ($j = 1, 2, \ldots, N_{\mathrm{STM}}$). Initially, each of the STM units $K_{\mathrm{STM},j}$ has no connections to other existing units in the network, nor is set its template.

16.5.4.1 Introduction of the STM kernel units to the kernel network in Fig. 16.3

In accordance with the introduction of the STM kernel units $K_{\mathrm{STM},j}$, the template vectors of the two fourth-layer units K_{41} and K_{42} in Fig. 16.3 are modified:

$$\mathbf{t}_{41} = [11, \mathrm{STM}, 12, \mathrm{STM}],$$
$$\mathbf{t}_{42} = [13, \mathrm{STM}, 14, \mathrm{STM}], \qquad (16.15)$$

where the symbol "STM" that appears in the second/fourth elements in (16.15) indicates the use of the STM/WM represented by $K_{\mathrm{STM},j}$; once a kernel unit representing a functional word (i.e., if the word is "if", say, then it corresponds to the kernel K_{11}, the index of which appears as the first element of $\boldsymbol{t}_{41}$ in (16.15)) is activated, the accumulation process of the subsequent activation followed by that of the kernel so as to construct the input data, by K_{41} as well as the detection of the activation sequence upon receival of the whole input, is temporarily suspended; these processes are superseded by the j-th STM kernel $K_{\mathrm{STM},j}$.

As specified above, for instance, both S1 and S2, appeared in the second and fourth positions of the sentence pattern (1c), respectively, are allowed to be one of the four patterns of (1a), (1b), (1c), or [NP, VP], each represented by the kernel K_{41}-K_{44}, respectively. The processing of S1 and S2 in (1c) thus involves a sub-processing by the STM kernels (i.e., as indicated by the second and fourth elements of $\boldsymbol{t}_{41}$ in (16.15)). In order to perform such a sub-processing, the following template is temporarily assigned to $K_{\mathrm{STM},j}$

$$\boldsymbol{t}_{STM,j} = \{41, 42, 43, 44\}, \tag{16.16}$$

and thereby are established the connections between K_{41}-K_{44} and $K_{\mathrm{STM},j}$. This setting then specifies how the activation of $K_{\mathrm{STM},j}$ can occur; upon completing either the four sentence patterns (i.e., each represented by the activation of K_{41}-K_{44},), the j-th STM kernel $K_{\mathrm{STM},j}$ will be activated. Besides the template setting (16.16), the STM kernel is also assigned a "pointer" $p_{\mathrm{STM},j}$ to the kernel unit which has initiated its sub-process; in this example, $p_{\mathrm{STM},j} = 41@2$, where the right-hand side $p@q$ is read as "the sub-process by $K_{\mathrm{STM},j}$ has been initiated by the kernel K_p, at the q-th element of its template $\boldsymbol{t}_p$." This pointer is used to store the auxiliary information of where to resume in the upper-level processing.

Then, the input data accumulation process for the kernel $K_{\mathrm{STM},j}$ starts from scratch, instead of K_{41}, and continues, till the activation of one of the four kernels specified in $\boldsymbol{t}_{\mathrm{STM},j}$. Upon the termination of the sub-processing by $K_{\mathrm{STM},j}$, the primary processing by K_{41} is resumed from the position the pointer $p_{\mathrm{STM},j}$ specifies, whereas $K_{\mathrm{STM},j}$ is back to the initial state; the contents of the template $\boldsymbol{t}_{\mathrm{STM},j}$, as well as the pointer $p_{\mathrm{STM},j}$, are all cleared.

16.5.4.2 A further modification to the network structure in Fig. 16.3

Next, in order to represent the sentence pattern (1c) by the fourth-layer unit K_{43}, the following modification to the template vector setting is made:

$$t_{43} = [31, 15, 1r1, 17, \text{STM},32]. \quad (16.17)$$

In Fig. 16.4, a part of the augmented four-layer network that reflects the modification to K_{43} is also shown.

In addition to the modification above, a fourth-layer unit K_{44} is newly added to the network, with the template

$$t_{44} = [31, 32], \quad (16.18)$$

in order to represent the (basic) sentence form [NP, VP].

16.5.4.3 Some illustrative examples of processing the embedded sentences by way of the augmented kernel network

Now, to be more concrete, let us first consider again the sentence 1b-1) "If the girl stops suddenly, then the cat runs quickly" appeared in the previous subsection, but here in terms of the modified form (1c) above, and observe step-wise its processing by way of the network representation in Fig. 16.3, which is now augmented with a set of the STM/WM kernel units $K_{\text{STM},j}$ and the aforementioned modifications using (16.15)–(16.18):

Step (1) "if": K_{11} is activated by the word recognition, and thereby the STM kernel $K_{\text{STM},1}$ is exerted (i.e., indicated by the second element of $\boldsymbol{t}_{41}$ in (16.15)). Then, $K_{\text{STM},1}$ is assigned the template $\boldsymbol{t}_{\text{STM},1} = \{41, 42, 43, 44\}$ and waits for one of the four sentence patterns (1c)–(3c) and [NP1, VP1] to be completed, each completion of which is detected in terms of the activation of the fourth-layer unit K_{41}-K_{44}, respectively. Also is set the pointer $p_{STM,1} = 41@2$.

Step (2) "the girl stops suddenly": since the activation sequence (Young et al., 2005; Yu et al., 2017) is eventually detected by K_{44} (i.e., due to the template setting (16.18)), $K_{\text{STM},1}$ is co-activated, and, accordingly, the sub-process by $K_{\text{STM},1}$ is terminated. Flush the contents of $\boldsymbol{t}_{\text{STM},1}$ and $p_{\text{STM},1}$, then return to the primary process specified by $\boldsymbol{t}_{\text{STM},1}$, i.e., the position to resume the processing is specified by the third element in $\boldsymbol{t}_{41}$ in (16.15).

Step (3) "then": K_{12} is activated by the word recognition, and thereby the STM kernel $K_{\text{STM},1}$ is exerted again. Set $\boldsymbol{t}_{\text{STM},1} = \{41, 42, 43, 44\}$ and $p_{\text{STM},1} = 41@4$.

Step (4) "the cat runs quickly": similar to Step (2) above, both K_{44} and $K_{\text{STM},1}$ are co-activated by the sentence processing, and, accordingly, the sub-process by $K_{\text{STM},1}$ is terminated. Return to the primary process, while flushing the contents of $\boldsymbol{t}_{\text{STM},1}$ and $p_{\text{STM},1}$; at this point, since there is no process remaining, the primary process is also terminated here. Stop.

Then, let us consider an example of the sentence of the form (3c) with an embedded structure:

(3c-1) The boy who told that if the girl stopped suddenly then the cat ran quickly smiles again.

In this example, the sentence pattern (2c) is embedded. Now, we start the observation of each process by the augmented version of the kernel network:

Step (1) "the boy who told": the activation sequence of K_{44}, K_{15}, and K_{1r1} occurs.

Step (2) "that": K_{17} is activated by the word recognition, and thereby is exerted the STM kernel $K_{\text{STM},1}$. Set $\boldsymbol{t}_{\text{STM},1} = \{41, 42\}$ and $p_{\text{STM},1} = 43@5$ (i.e., for the former, it is assumed that S5 is restricted only to the form of either (1b) or (1c), without loss of generality).

Step (3) "if": K_{11} is activated by the word recognition, and thereby is exerted $K_{\text{STM},2}$. Set $\boldsymbol{t}_{\text{STM},2} = \{41, 42, 43, 44\}$ and $p_{\text{STM},2} = 41@2$.

Step (4) "the girl stopped suddenly": the activation of K_{44} eventually occurs. This causes the co-activation of $K_{\text{STM},2}$, and, accordingly, the sub-process by $K_{\text{STM},2}$ is terminated. Return to the sub-process by $K_{\text{STM},2}$, while flushing the contents of $\boldsymbol{t}_{\text{STM},2}$ and $p_{\text{STM},2}$, i.e., the position to resume the processing is specified by the third element in $\boldsymbol{t}_{41}$ in (16.15).

Step (5) "then": K_{12} is activated by the word recognition, and thereby the STM kernel $K_{\text{STM},2}$ is exerted again. Set $\boldsymbol{t}_{\text{STM},2} = \{41, 42, 43, 44\}$ and $p_{\text{STM},2} = 41@4$.

Step (6) "the cat ran quickly": both K_{44} and $K_{\text{STM},1}$ are co-activated by the processing, and, accordingly, the sub-process by $K_{\text{STM},2}$ is terminated; this also terminates the processing by $K_{\text{STM},1}$, since it causes the activation of the kernel K_{41} (i.e., specified by the first element in $\boldsymbol{t}_{\text{STM},1}$ above). Therefore, return to the primary process (i.e., 43@5), while flushing the contents of $\boldsymbol{t}_{\text{STM},2}$ and $p_{\text{STM},2}$, as well as those of $\boldsymbol{t}_{\text{STM},1}$ and $p_{\text{STM},1}$.

Step (7) "smiles again": the activation of K_{32} eventually occurs. Since this activation is described by the last element in $\boldsymbol{t}_{43}$, there is no more process to be done. Stop.

16.6 Conclusion

This chapter has been devoted to the topic of spoken language processing, dealt as uniformly as possible within the context of kernel memory. In this chapter, it has been described that the concept of kernel memory offers neurally versatile representations and thereby various features of the language faculty can be modeled. Then, a unified account within this connectionist principle has been given for the multitude levels of auditory perceptual processing relevant to the language, i.e., from the lower level recognition of isolated spoken words to that of the connected ones and the higher-level processing of syntactical structures, the latter of which has not so far been discussed much within the signal processing context but rather abundant in the psycholinguistic area of studies. The ultimate aim of the present chapter is, therefore, to fill in the gap between these two disciplines, within the principle of kernel memory.

The kernel networks appeared in this chapter can, by and large, be subdivided into two parts: (i) one part comprised by a group of kernel units, each acting as a "receptive field" upon receiving a short segment of sensory data (i.e., each MFCC frame data, in the case of speech samples) and eventually emitting its output in a more compact form of representation (i.e., its unique identification number or "spike" signal corresponding to it) for enabling a rapid and effective post-processing, and (ii) the other part being a hierarchical network structure, where there are multiple layers of kernel units, each unit integrating the activation of the previous layer kernel units connected to it in a certain manner. Each unit in the other part then processes the activation of the previous layer units at a slower time scale, and eventually forwards its activation to the post-layer units for a further processing. More specifically, the former (i) can be represented by the RBF kernels in the first layer of CNC (Section 16.3) or that in its augmented model (Section 16.4), whereas the latter (ii) by the second/third layer units of the CNC models or the units within the network structures appeared in Section 16.5. For the latter (ii), each of the kernel units either functions as a gating unit (i.e., the unit emitting its activation upon when one of the previous layer units connected to it is activated, in Section 16.5.3) or detects a particular activation sequence pattern of previous layer kernel units. Moreover, such an activation sequence detection performed in a hierarchical fashion has been exploited in Section 16.5, in order to provide a basis for the problem of representing syntactical structures of a language (Chomsky, 1957) via the neural modeling. Note that, unlike the representation of the syntactical structures by the simple recurrent network (SRN) model (a.k.a. the Elman's net; Elman, 1990), the network representations shown in Section 16.5 are based entirely upon a feed-forward architecture, i.e., without any feedback connections established between the units. Such a feed-forward architecture then ensures the transparency in terms of the network dynamics and provides an analytically tractable model of language grammars.

The main thrust of this chapter is to provide a new way of connectionist's representation that enables a seamless integration of the network responsible for a lower level (i.e., spoken word) recognition, given the real speech input, and the hierarchical one in the upper level acting as a post-processor of language modeling. For the latter, however, only a fragment of the language modeling within the connectionist principle of kernel memory has been given by examining a few examples in Section 16.5, which is far from complete. Also, the issue of how such kernel networks as shown in Section 16.5, each representing a certain syntactical structure, are actually formed, given a set of the speech samples for training, i.e., the learning aspect of the networks, has not yet been discussed; it is considered that giving an eloquent account for the construction of a psycho-linguistically plausible neural network model representing syntactical structures, while accepting real speech input, is not a simple story. Even so, the pursuit in this direction is quite beneficial, as it would eventually lead to the answer to one of the most interesting and important questions of how language is acquired, i.e., embodiment of the "language acquisition device (LAD)." In contrast to the network models for syntactical structures in Section 16.5, the construction (i.e., learning) of the CNC models, capable of performing isolated/connected word recognition tasks using real speech data as described in Sections 16.3 and 16.4, has been carried out, based primarily upon the acoustic properties of word, as in the currently de facto standard approaches of HMM-DL. However, since the CNC models was empirically validated only for the case of a small set of vocabulary (i.e., consisting of less than 15 digits) under a controlled environment, a further investigation is still necessary, where a larger size of lexicon is given and/or assumed a less controlled situation.

The issues mentioned above remained to be addressed and should be elaborated more, which is, however, beyond the scope of this chapter and left to the future study.

References

Allen, J. B. (1994). How do humans process and recognize speech? *IEEE Transactions on Speech and Audio Processing*, **2–4**, pp. 567–577.

Amari, S. (1967). Theory of adaptive pattern classifiers, *IEEE Transactions on Electronic Computers*, **EC-16**, pp. 299–307.

Broomhead, D. S., and Lowe, D. (1988). Multivariate functional interpolation and adaptive networks, *Complex Systems*, **2**, pp. 321–355.

Chomsky, N. (1957). *Syntactic Structures* (Mouton de Gruyter, Berlin/New York).

Dellar, Jr., J. R., Proakis, J. G., and Hansen, J. H. L. (1993). *Discrete-Time Processing of Speech Signals* (Macmillan Publishing, New York).

Duda, R. O., Hart, P. E., and Stork, D. G. (2001). *Pattern Classification* (2nd ed., Wiley, New York).

Elman, J. L. (1990). Finding structure in time, *Cognitive Science*, **14**, pp. 179–211.

Furui, S. (1981). Cepstral analysis technique for automatic speaker verification, *IEEE Transactions on Acoustic Speech and Signal Processing*, **29**, pp. 254–272.

Gori, M., and Scarselli, F. (1998). Are multilayer perceptrons adequate for pattern recognition and verification? *IEEE Transactions on Pattern Analysis and Machine Intelligence*, **20–11**, pp. 1121–1132.

Gross, C. G., Rocha-Miranda, C. E., and Bender, D. B. (1972). Visual properties of neurons in inferotemporal cortex of the macaque, *J. Neurophysiology*, **35**, pp. 96–111.

Hinton, G., Deng, L., Yu, D., Dahl, G., Mohamed, A., Jaitly, A. Senior, N., Vanhoucke, V., Nguyen, P., Sainath, T., and Kingsbury, B. (2012). Deep neural networks for acoustic modeling in speech recognition: The shared views of four research groups, *IEEE Signal Processing Mag.*, **29–6**, pp. 82–97.

Hopfield, J. J. (1995). Pattern recognition computation using action potential timing for stimulus representation, *Nature*, **376(6)**, pp. 33–36.

Hoya, T. (2003). On the capability of accommodating new classes within probabilistic neural networks, *IEEE Transactions on Neural Networks*, **14–2**, pp. 450–453.

Hoya, T. (2005). *Artificial Mind System: Kernel Memory Approach* (Series: Studies in Computational Intelligence, Vol. 1: Springer-Verlag, Heidelberg).

Hoya, T. (2016). On the parameter setting of a network-growing algorithm for radial basis kernel networks, *Proceedings of the 2016 Joint 8th International Conference on Soft Computing and Intelligent Systems and 2016 17th International Symposium on Advanced Intelligent Systems*, pp. 355–359.

Hoya, T., and van Leeuwen, C. (2010). A cascaded neuro-computational model for spoken word recognition, *Connection Science*, **22–1**, pp. 87–101.

Hoya, T., and van Leeuwen, C. (2016). Connected word recognition using a cascaded neuro-computational model, *Connection Science*, **28–4**, pp. 332–345.

Hoya, T., and Washizawa, Y. (2007). Simultaneous pattern classification and multidomain association using self-structuring kernel memory networks, *IEEE Transactions on Neural Networks*, **18–3**, pp. 732–744.

McClelland, J. L., and Elman, J. L. (1986). The TRACE model of speech perception, *Cognitive Psychology*, **18**, pp. 1–86.

Morgan, N., and Bourlard, H. A. (1995). Neural networks for statistical recognition of continuous speech, *Proceedings of the IEEE*, **83–5**, pp. 742–770.

Platt, J. (1991). A resource-allocating network for function interpolation, *Neural Computation*, **3–2**, pp. 213–225.

Poggio, T., and Edelman, S. (1990). A network that learns to recognize three-dimensional objects, *Nature*, **343(18)**, pp. 263–266.

Pulvermüller, F. (2002). *The Neuroscience of Language: on Brain Circuits of Words and Serial Order* (Cambridge University Press, New York).

Rabiner, L., and Juang, B.-H. (1993). *Fundamentals of Speech Recognition* (Prentice Hall, Englewood Cliffs, NJ).

Rosenblatt, F. (1958). The perceptron: A probabilistic model for information storage and organization in the brain, *Psychological Review*, **65**, pp. 386–408.

Rumelhart, D. E., McClleland, J. L., and the PDP Research Group. (1986). *Parallel Distributed Processing: Explorations in the Microstructure of Cognition – Vol. 1. Foundations* (MIT Press, Cambridge).

Shinozaki, T., and Watanabe, S. (2015). Structure discovery of deep neural network based on evolutionary algorithms, *Proceedings International Conference on Acoustics, Speech and Signal Proceedings (ICASSP-2015)*, pp. 4979–4983.

Specht, D. F. (1990). Probabilistic neural networks, *Neural Networks*, **2**, pp. 568–576.

Widrow, B. (1962). Generalization and information storage in networks of adaline "neurons," in M. C. Yovitz, G. T. Jacobi, and G. D. Goldstein (Eds.), *Self-Organizing Systems*, pp. 435–461 (Sparta, Washington, D.C.).

Yamamoto, K., Nakamura, S., Takeda, K., Kuroiwa, S., Kitaoka, N., Yamada, T., Mizumachi, M., Nishimura, T., and Fujimoto, M. (2003). Aurora-2J/Aurora-2J corpus and evaluation baseline, *Technical Report, Information Processing Society of Japan*, **2003-SLP–47(19)**, pp. 101–106.

Young, S., Evermann, G., Gales, M., Hain, T., Kershaw, D., Moore, G., Odell, J., Ollason, D., Povey, D., Valtchev, V., and Woodland, P. (2005). *The HTK Book (Version 3.3)* (Dept. Engineering, Cambridge University).

Yu, D., and Li, J. (2017). Recent progresses in deep learning based acoustic models, *IEEE/CAA J. Automatica Sinica*, 4–3, pp. 396–409.

Part VII

Discussion on AI for Healthcare

Chapter 17

The Brave New World of Machine Learning in AI and Medicine

Paulina Y. Chan[a] and Stephen K. Ng[b]
[a]*Chartered Management Institute, United Kingdom*
[b]*Mailman School of Public Health, Columbia University, New York City, New York USA*
paulinaue@aol.com

In 50 short years, DSP has progressed from quality signal processing to entirely new fields of computing, machine learning, and artificial intelligence (AI). It has also penetrated into almost every area of human endeavour, from manufacturing to commerce to healthcare, pointing to new directions with profound impact on world economy and human welfare. With machine learning and AI on par or surpassing human performances in areas such as pattern recognition and complex decision making (in Chess and Go), we are on the brink of a brave new era when even the most personal human services such as medicine are increasing dependent on AI.

This chapter reviews some recent advances in AI with special attention to its emergent use in healthcare. Medicine has always had a strong idiosyncratic component, with the best experts disagreeing on various matters. By incorporating the logic and discipline of

Learning Approaches in Signal Processing
Edited by Wan-Chi Siu, Lap-Pui Chau, Liang Wang, and Tieniu Tan

ISBN 978-981-4800-50-1 (Hardcover), 978-0-429-06114-1 (eBook)
www.panstanford.com

AI in medicine, we can expect great strides in clinical medicine and public health in areas of disease diagnosis, treatment, and prevention.

17.1 DSP Platform

Signal Processing is the historic technology platform for moving information to transform communications landscape. In the digital age precise bits and bytes of vast amount of data are transmitted meticulously across computer and communications networks for quality improvement. We celebrated the 50th Anniversary of Digital Signal Processing (DSP) in 2017. DSP technology is deployed through sensors and devices, machines and systems to serve designated human recipients and robotic objects. Methodologies, algorithms and modelling tools of DSP do not stop with ensuring data accuracy, efficiency and quality. The technology facilitates breakthroughs that will be important in the years ahead, whether developing machine learning approaches to process big data for artificial intelligence (AI), wrestling with the implications of robotics, pursuing in medical diagnoses, enabling facial and language recognition, or social media and initiatives.

This chapter is a succession of the finale of *Trends in Digital Signal Processing: A Festschrift in Honour of A. G. Constantinides* (2016), 'Our World is Better Served by DSP Technologies and Their Innovative Solutions'. Much progress has been made in the innovative landscape since then. Enthusiasts look forward to the advent of super-intelligent artificial brains from intelligent machines, discovering solutions to intractable scientific problems in disciplines from cell biology to cosmology and quantum physics, and leading the way to clean energy. All are current and future applications of artificial intelligence (Fig. 17.1). This chapter focuses on some significant impact of AI and machine learning that are causing global interest and discussion, with an emphasis on medicine and healthcare.

Figure 17.1

17.2 From Human Behavioural Learning to Machine Learning

What is learning? How does learning occur? How do we learn? Learning requires cognitive, linguistic, motor and social skills and can take many forms. It involves the acquisition and modification of knowledge, skills, strategies, beliefs, attitudes and behaviours [1]. Aristotle (384–322 BC) focused on cognitive areas pertaining to memory, perception and mental imagery. The philosopher ensured that his studies were based on empirical evidence that was gathered through observation and experimentation. Behaviouristic learning with cognitive approaches emerged in the 15th century when it meant 'thinking and awareness'. Learning since has involved experimental techniques, utilizing information gathering, computation and comprehension, management and analyses, and judgement and decision making (Fig. 17.2). This process is applied to the nature of knowledge, the formation and reconstruction of knowledge, and reflection and self-assessment.

The optimism and perils generated by machine intelligence have played out for decades in the works of fictional writers such as Philip K Dick and Isaac Asimov. Machine learning is developed by cross-disciplinary professionals from many fields such as mathematics,

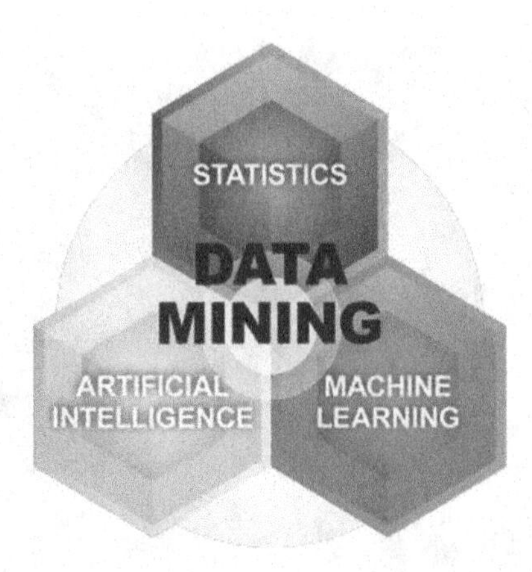

Figure 17.2

Figure 17.3

neurotechnology, bioengineering, computer science, and electrical & electronic engineering. They use human learning techniques and behavioural capabilities to program computer systems so that a machine would imitate intelligent human behaviour and use human reasoning as a model (Fig. 17.3).

Machines are given the ability to 'learn', process cognitive skills, progressively improve performance on a specific task from big data. Machines are also taught deep learning: to learn from datasets and perform functions that they are not specifically programmed to do (Fig. 17.4). As machines are taught to imitate human behaviours

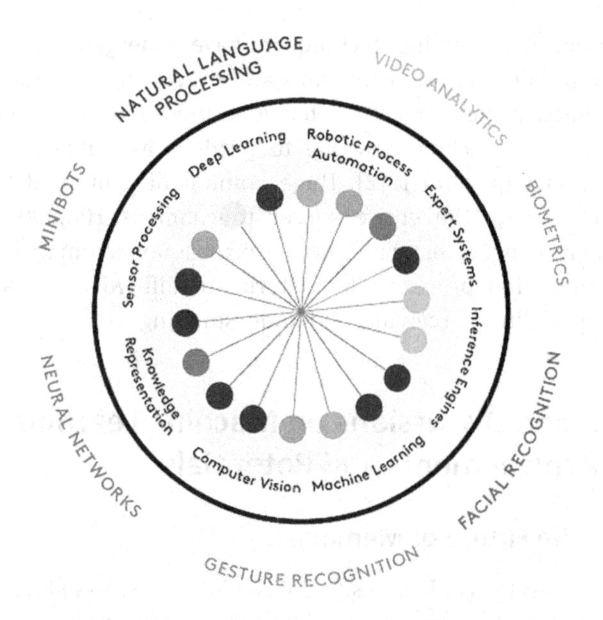

Figure 17.4

Figure 17.5

in exponential speed, capacities and memories, machine learning is outperforming conventional statistical approaches, achieving human parity in speech recognition and being Digital Assistants to human, e.g. Siri and Alexa (Fig. 17.5).

Random data mining techniques have emerged with some common pitfalls such as distortions and overfitting of parameters. MIT Technology Review (2018) has identified how researchers in Germany used machine learning to predict the outcome of the 2018 World Cup winner [2]. Huge amount of training-data was used to simulate the entire soccer tournament 100,000 times, factoring ranking, priority and other socioeconomic impacts in the AI determination process. The criteria to fulfil AI are: learning, reasoning, feeling, perceiving, and understanding.

17.3 Some Discussions on Machine Learning Achievements and Potentials

17.3.1 The Future of Memory

Cognitive scientists at University College London stressed that most acts of human memory are exercises in reconstruction, or creative imagination. Human memory is not the retrieval of an identical 'bit' of information from a database in a computer-like brain. Human memory is a dynamic act, imaginative reconstruction rather than recovering an inert photographic image.

Replicating human brain cell groups in neural net machines, Schacter et al. (2012) [3] predicted that new generations of AI machines will mimic the human processes not only for the construction of memory but of future scenarios. This would involve

Figure 17.6

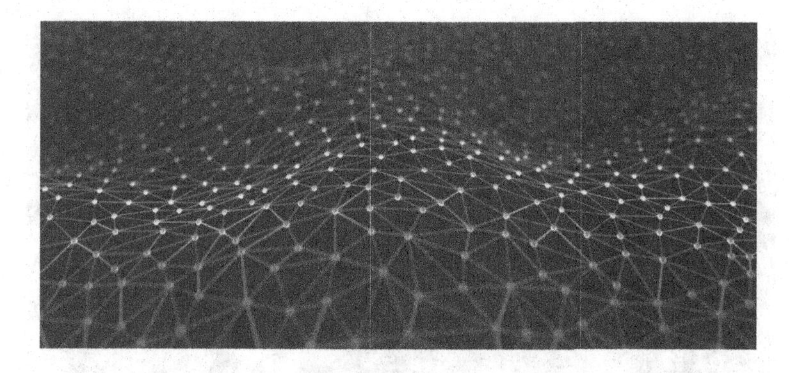

Figure 17.7

many layers of 'neural nets' replicating the functions of groups of brain cells (Fig. 17.7). Learning is achieved in the nets by their active capacity to reconstruct past 'experience' and to project possible future situations. A value guide for priorities and decision-making is established by an algorithm, a governing principle within the system.

DeepMind (cofounded by Hassabis) was sold to Google in 2014. Employing 700 engineers and technicians DeepMind is developing the ultimate machine known as Singularity, or Artificial General Intelligence (AGI). The AGI will multitask across a vast range of activities at prodigious speeds and capacities. The future of AI is the future of artificial imagination: the capacity of the machine to predict and assess the consequences of a range of different scenarios ahead of 'acting'. The board game Go, 'with an astonishing 10 to the power of 170 possible board configurations', played for 3000 years in China, requires extraordinary levels of intelligence and intuition. DeepMind's system AlphaGo was initially trained on thousands of past professional games of Go and beat the world's top Go player in 2016.

The current development of AGI is that machines are learning by playing against itself, something unique to human beings. This may generate the next set of questions: what are the differences between intelligence and wisdom? (Fig. 17.8). The connections between unlike and opposites? Future machines may even have the capacity to make moral decisions.

Figure 17.8

17.3.2 Supply and Demand for the New Champions in the Labour Market

Artificial intelligence and automation have been identified as the key primary drivers of social changes, particularly in the labour market. AI could both create and limit economic opportunities. An analysis from PricewaterhouseCooper (PwC) predicted that AI, robotics and other forms of 'smart automation' technologies could boost productivity and create better products and services. While some jobs will be displaced or fundamentally changed in nature, new jobs will also be created and the long-term net effect should be positive for the economy as a whole. The global audit firm reported at the World Economic Forum 2018 that AI and robotics will create almost 60 million more jobs than the 75 million jobs they displaced by 2022 (World Economic Forum, 2018).

LinkedIn's platform, which has 575 million members globally, has provided a vantage point to examine the trends that will shape the future of work (Fig. 17.9). LinkedIn uncovered two concurrent trends: the continued rise of tech jobs and skills and, in parallel, a growth in 'human-centric' jobs and skills [4] (Fig. 17.10).

The growing presence of AI skills was 190% (2015–2017) in the workforce. Expertise in areas like neural networks, deep learning and machine learning, tech jobs like software engineers and data analysts, technical skills such as cloud computing, mobile

Figure 17.9

application and machine learning system development are on the rise and in great demand in most industries and across all regions. Jobs that have been the largest decreases in share of hiring include administrative assistants, customer service representatives, accountants, electrical/mechanical technicians. The latter jobs depend on more repetitive tasks.

All types of technical AI skills are spreading beyond just the tech industry. The top 10 industries are: software and IT services, education, hardware and networking, finance, manufacturing, consumer goods, healthcare, corporate services, entertainment and design. Industries with growing AI skills penetration are also changing fast. 'Change' is a proxy for innovation. The records indicate that the presence of AI skills correlates strongly with innovation within an industry. It also means there is an opportunity for many industries to invest more heavily in their AI capabilities.

There are significant shifts in the quality, location, and format of new roles. Machines will overtake humans and perform more tasks at the workplace by 2025 (Fig. 17.11). The amount of work done by machines will jump from 29% to more than 50%. However, another 133 million new roles may emerge as companies shake up their division of labour between humans and machines, translating to 58 million net new jobs being created by 2022. The rapid shift will be accompanied by new labour market demands that may result in more, rather than fewer jobs. Uncertainty remains as to the types of jobs created, their permanence, and the training required. Preparing

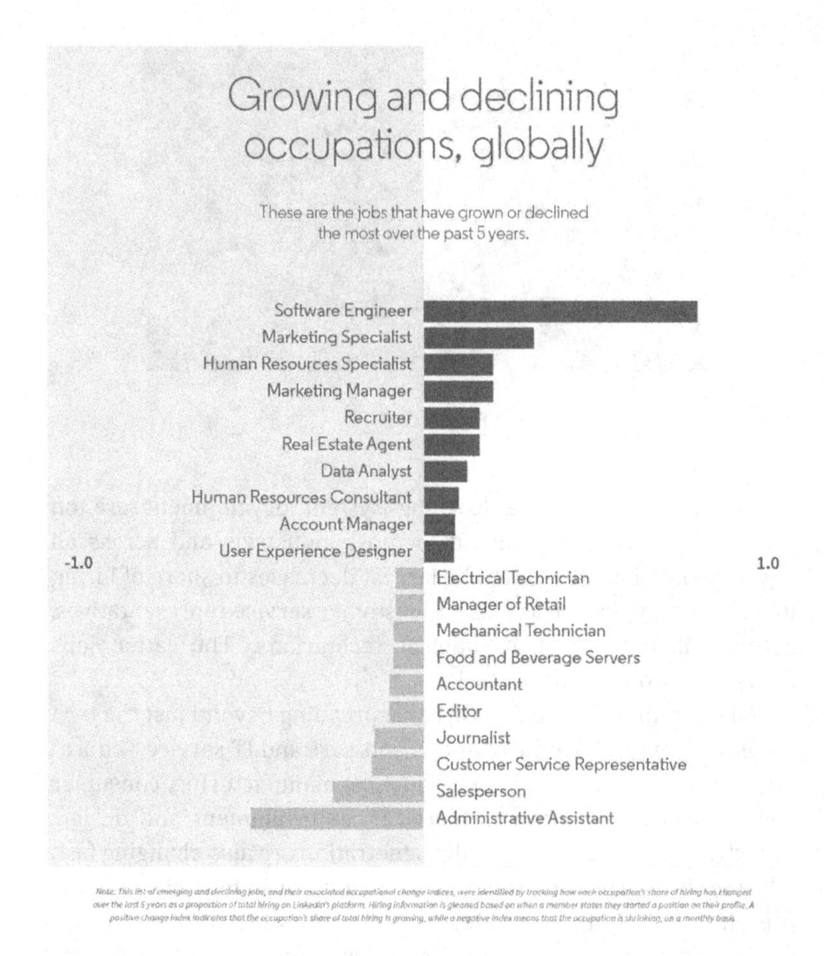

Figure 17.10

the workforce for these changes will depend on a data-driven approach to understand the trends that are shaping the future of the labour market, and a commitment to invest in lifelong learning opportunities that can help workers adapt to rapid economic shifts.

As new skills emerge, governments, educational institutions and employers should consider how they can most effectively develop learning programmes that equip people with the skills they will

Figure 17.11

need to keep up with the modern economy. AI skills are global. The countries with the highest penetration of AI skills are the United States, China, India, Israel and Germany.

17.3.3 AI at the Fourth Industrial Revolution

Artificial Intelligence and robots are transforming how we work and live (Fig. 17.12). Just as electricity transformed the way industries functioned in the past century, AI being the science of programming cognitive abilities into machines, has the power to substantially change society as the fourth industrial revolution. AI is being harnessed to enable robots. The robotics business is booming. Worldwide sales of robots increased by 31% in 2017 over 2016 (International Federation of Robotics) [5]. Worldwide manufacturers are trying to automate more tasks to make up for shortages of human labour. E-commerce industries are exploring new approaches to automate the picking, packing and processing of goods. However, the vast majority of robots are not very smart nor adaptive. They can do things precisely and tirelessly, but they are easily 'confused' by real-world complexity and are mostly painful to program. They require cutting-edge machine vision, and it means balancing precision with compliance in designs.

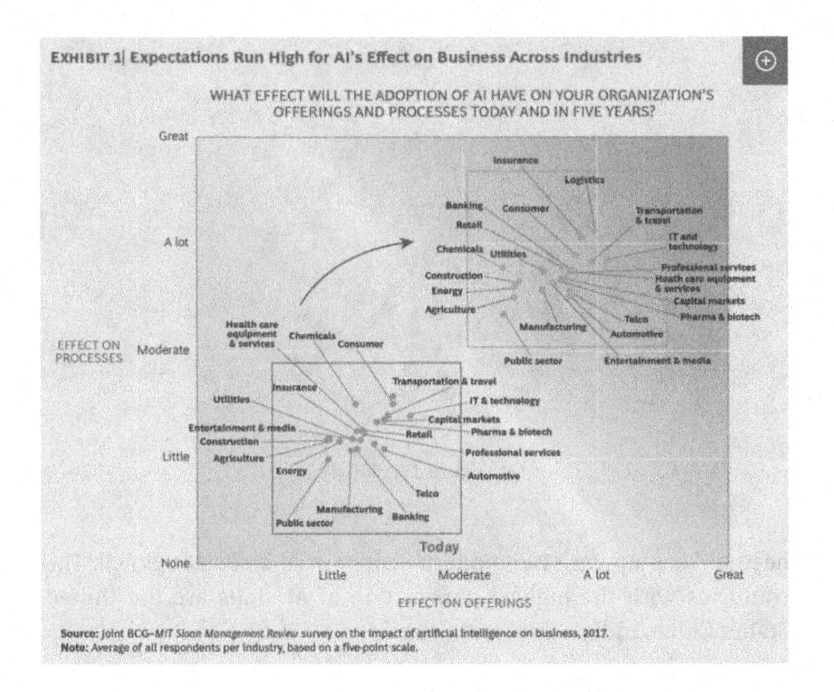

Figure 17.12

17.3.4 Outsource Fact-Checking to a Machine to Detect 'Fake News'

Mark Zuckerberg, Facebook's chief executive, said it would require a vast reservoir of big data to automatically identify dubious news (Fig. 17.13). He promised the US Congress that AI would help solve the problem. Researchers from MIT, Sofia University in Bulgaria, and Qatar Computing Research Institute [6] tested over 700 possible variables to predict a media outlet's trustworthiness using a machine learning model. The best model accurately labelled news outlets with factuality just 65% of the time! These experiments reveal the challenges of outsourcing fact checking to a machine in close to real time. Researchers are using variables that could be tabulated independent of human fact checkers. These include analyses of the content, sentence structure of headlines and the word diversity in articles, URL site indicators, website traffic, and measures of the outlet's influence, etc. The critical factors to resolve

Figure 17.13

the fake-news epidemic are an increased availability of training data and the improvement of the performance of the machine training models. 'Fake news will never stop completely, but we can put them under control'.

17.3.5 Human–Robot Collaboration

Partnering between human and robots is a current trend (Fig. 17.14). The aim is to build collaborative robots, i.e. cobots, so that robots can be used alongside human workers [7].

One emerging application of human-robot collaboration is autonomous farming in which robots are the majority of the 'farm-workers'. 'Robots in agriculture' is a promising strategy to increase food production and alleviate the shortage of farm workers in developed countries. Farmers share their work space with robots.

Figure 17.14

Figure 17.15

Tech companies are developing robots to pick strawberries in the UK (BBC Money) and leafy green lettuce in the USA (Iron Ox in San Carlos, California). Farmers and robots are producing leafy greens at 26,000 heads a year in the 8000 square feet indoor hydroponic facility. That is the production level of a typical outdoor farm which is five times bigger. The robotic workers are a series of robotic arms and movers. The arms individually pluck the plants from their hydroponic trays and transfer them to new trays to maximize their health and output. The mechanic movers carry the 800-pound water-filled trays around the facility. Deep learning software (The Brain) keeps the arms and movers synchronized and monitors nitrogen levels, temperature, and robot locations. It also alerts both robot and human attention wherever they are needed.

The next step would be a fully autonomous farm where software and robotic replace human agricultural workers. The upfront investment in this new technology is a challenge to smaller family-owned operations. However, bringing automation to farming is necessary to help a wider swatch of the agricultural industry in developed countries (Fig. 17.15).

17.4 Machine Learning and Medicine

Computer decision-making algorithms in medicine can generally be classified as either 'expert systems' or 'machine learning systems.' Expert systems are didactic algorithms with rigid input, boundaries,

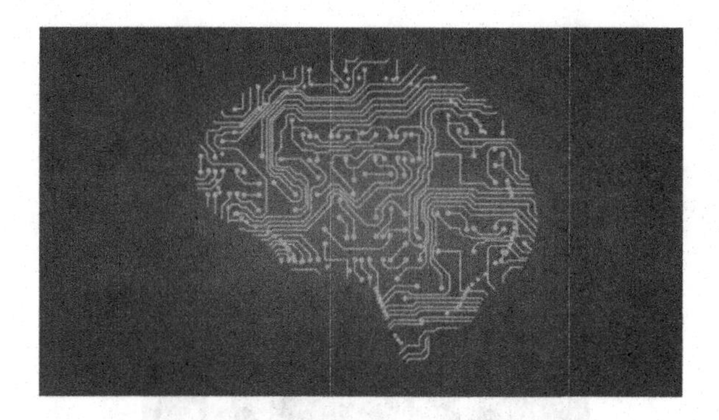

Figure 17.16

and output criteria. They arise from the consensus of medical experts, follow strict decision-making rules, and do not allow deviations. They are subjectively formulated and do not require large data sets or modelling. The use of these systems is analogous to colouring exercises of grade school students when colours cannot go beyond the outline of the drawings. Machine learning on the other hand allow more flexibility and the algorithm can improve its performance with more data input. The modelling is interactive and dependent on the training sets provided to the system. Programming a machine to learn is analogous to lecturing to university students, where thinking out-of-the box and dissent are tolerated or even encouraged. Machine learning can be supervised, semi-supervised or unsupervised. Training data sets, loose decision-making rules, and output endpoints are given to the computer in supervised or semi-supervised machine learning. In unsupervised machine learning the computer is allowed a free hand to determine the variables to be included and the complexity of the model. Another name for unsupervised machine learning is 'data mining (Fig. 17.16).

Machine learning in medicine is a relatively unexplored field. For a long time the extensive training required of physicians, the vast amount of knowledge and experience to be acquired, stored, and retrieved are believed to be the epitome of human intelligence and beyond the capabilities of machines. The empathy which makes physicians true healers is also not conducive to machine learning.

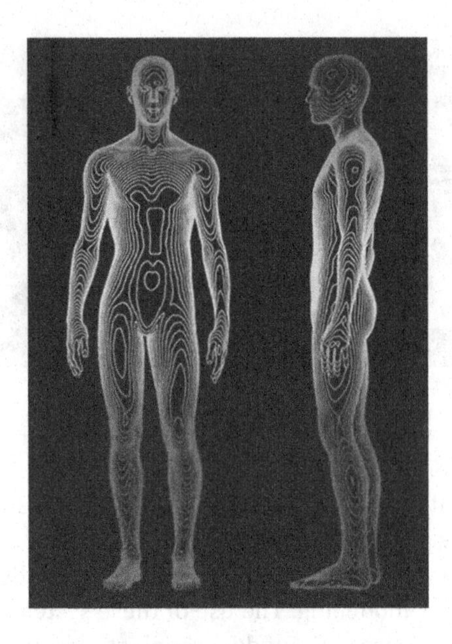

Figure 17.17

However, as our understanding of the human body increases in complexity, the amount of knowledge in even the smallest of subspecialties can be daunting to the brightest human minds. Physicians simply cannot keep abreast of all the different aspects of human physiology and pathology and their interactions with therapeutic agents pertaining to their own field.

17.4.1 The Complete Patient

In clinical medicine, we are always trying to draw conclusions from incomplete information. We do standard blood tests, radiologic examinations, scans etc. that look at a small part of the whole patient at a time. As in the Chinese proverb of 'The blind feeling the Elephant,' a person's impression of the elephant depends on which part of the animal he touches. But every part of the patient is important in diagnosis and treatment (Fig. 17.17). Up till now we do not have the ability to look at the complete patient because the data would be too expensive to gather and too massive to be modelled and analysed. With our increased ability to gather information, the

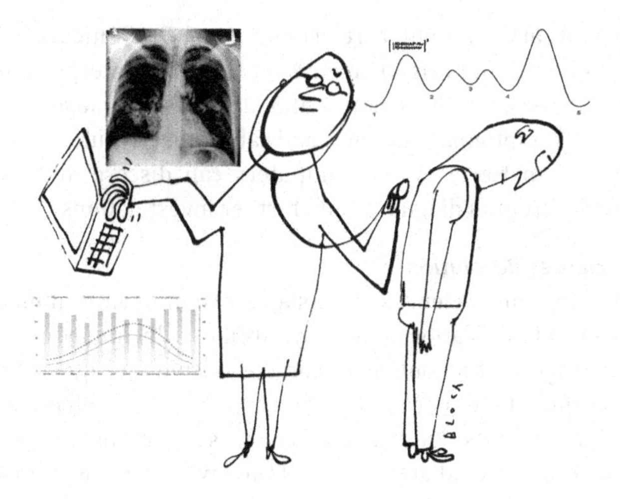

Figure 17.18

challenge is to synthesize and draw better conclusions from the data collected. Advances in machine learning enable us to look at the complete patient for the first time using all the information at our disposal. The opportunity is huge but so is the challenge. With the veil lifted from our eyes we may not be seeing more clearly because the richness of the data can defy human interpretation. Unless we have better machine learning algorithms, better models, and more data sets to verify our models, we may not be able to benefit from the new information we can gather.

From the perspective of a brave new world in medicine where sophisticated machine learning is at the disposal of clinicians and researchers, the following areas may receive the most benefits.

17.4.2 Clinical Medicine

17.4.2.1 Diagnosis

Pattern Recognition

Many diagnostic tests depend on the acquisition and interpretation of digital visual images, such as mammograms in radiology or cellular morphology in anatomical pathology (Fig. 17.18). Machine learning in pattern recognition can sometimes outperform human eyes. Re-screening of mammograms by computers is now routinely carried out and computer re-screening of Pap smears is FDA approved with results equal or superior to human interpretations.

Pattern recognition is not confined to visual images. Combinations of symptoms and signs and laboratory results may form recognisable patterns which computers can discern and suggest possible differential diagnoses and further investigations.

Omics Sciences Revolution

Advances in omics sciences and its application in clinical medicine can lead to identification of new syndromes which may otherwise be missed without a complete characterization and quantification of the myriad of biological molecules in genomics, proteomics, and metabolomics. Tests that use to take weeks and months to perform can now be completed almost in real time with the help of machine learning. An example is rapid genomic tests, which can identify pathogens quickly and accurately.

Reality Mining

The mobile smartphone and other wearable gadgets can monitor physical location and vital body functions such as heart rate, respiratory rate, movements and speech patterns. Some models can also measure blood pressure. The various implantables such as pacemaker and insulin pump also generate continuous signals of heart rate and blood sugar level. Other medical appliances which people use in their daily life such as supplemental oxygen, C-pap for sleep apnoea are also potential sources of digital signals.

The processing and analysis of background data generated automatically and continuously has been called Reality Mining [8] (Fig. 17.19). It is a form of data mining used to discover unknown patterns and generate hypotheses. Already it has been shown to be useful in detecting psychiatric illness such as depression (by changes in speech pattern) and neurologic conditions such as Parkinson's Disease or Multiple Sclerosis (by changes in movement and gait).

17.4.2.2 Treatment

Personalised Medicine

Advances in genomics and rapid genotyping enable individuals to receive medications tailored to their needs (Fig. 17.20). Instead

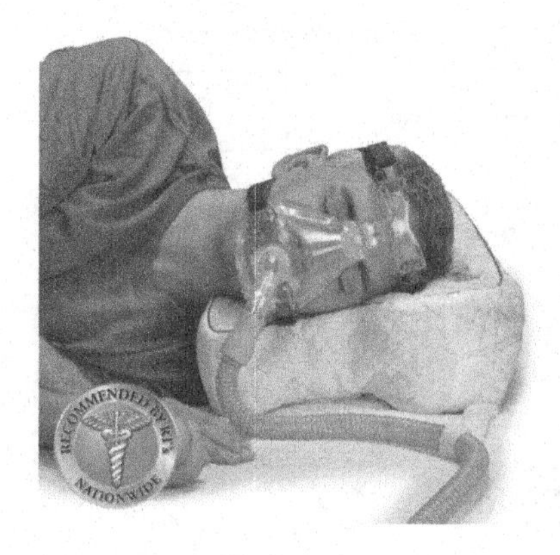

Figure 17.19

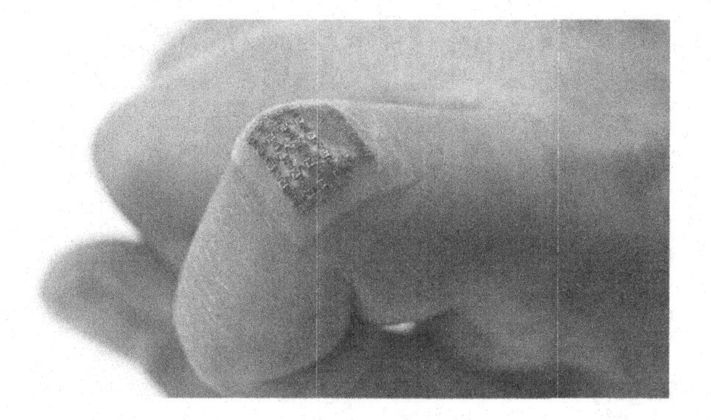

Figure 17.20

of a hit and miss approach, each person will receive drugs that are sure to work for him/her. Idiosyncratic drug reactions and interactions will be known before the start of therapy. Patients can thus avoid expensive, painful, and unnecessary treatments. Using Reinforcement Learning (RL) researchers were able to optimize chemotherapy regimens for glioblastoma. There were also successful attempts to predict the chance of neutropenia in breast cancer patients on chemotherapy using AI algorithms on large databases which include genetics, cancer mutations, physical environment, and lifestyle. The goal of personalized and precision medicine is in sight.

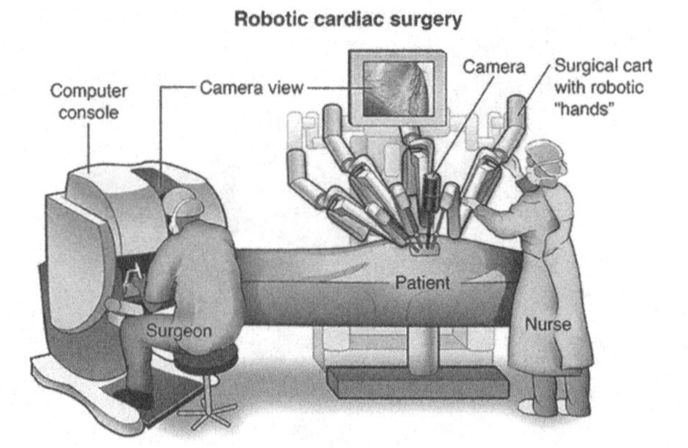

Figure 17.21

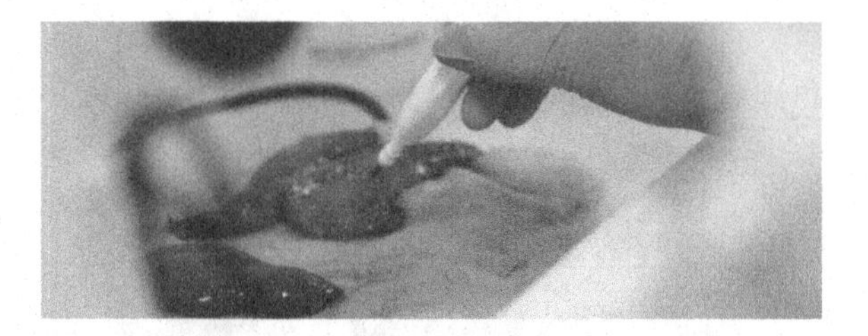

Figure 17.22

Robotic Surgery

Many delicate surgical procedures are now performed using robotic systems. Robotic surgery can offer better fields of vision, perform movements not possible with human hands, and reach spaces too small for human fingers (Fig. 17.21). Robotic surgery is often less invasive than traditional surgery, has fewer complications and faster recovery time. An exciting new field in robotic surgery is robotic catheterization systems in endovascular procedures. These robotic catheters have steerable tips which can change in shape and stiffness and can bend up to 180 degrees and rotate 360 degrees distally without affecting the proximal end.

Intelligent Knife (iKnife)

Researchers at Imperial College led by Professor Jeremy Nicholson have recently developed an intelligent surgical knife which vaporises

Figure 17.23

tissues on contact (Fig. 17.22). The smoke is immediately analysed by a mass spectrometer to tell whether a tumour is benign or malignant.

Bionic Limbs

Electrodes implanted on the motor cortex of quadriplegic patients can now collect brain signals and use them to direct movements of either the patient's own limbs or a prosthesis (Fig. 17.23). A patient just has to think about performing a particular task and the brain signals will be translated into movements. This is made possible by real-time collection and analysis of neuronal signals from the motor cortex. At present, only a small amount of the billions of signals generated in the brain can be analysed due to limitations in computing power. With better sensors and analytic algorithms, the movements can be more refine and seamless. In patients with intact spinal cord, electrodes implanted in the stump of an amputated limb can pick up weak signals from the brain and after amplification by a machine learning computer can direct movements of a prosthesis.

Metabonomics (Metabolic Profiling)

Collection and analysis of real-time chemical metabolic data can be very helpful in monitoring ICU patients and patients undergoing anaesthesia. It can also be used for prognostic predictions as well as monitoring treatment compliance.

Figure 17.24

Telemedicine

Long-distance diagnosis using radiologic and cardiologic data are now routine (Fig. 17.24). Tele-robotic surgery is also commonplace. One serious limitation to the widespread use of telemedicine is the lack of sensors for the physician to touch and feel the patient in a physical examination. There are exciting new developments in tactile sensors. When such sensors become inexpensive and widely available, even the remotest patient can have the service of specialists living hundreds of miles away.

17.4.3 Public Health

Machine learning and related technologies are also extremely useful in public health practice. Surveillance and contact tracing of infectious diseases are the obvious applications.

17.4.3.1 Surveillance

Detection and control of infectious disease outbreaks depend on vigilant surveillance, which in turn depend on the rapid identification of the responsible pathogens. Machine learning has made point of need rapid DNA analysis possible by comparing collected

Figure 17.25

DNA samples with pre-existing database. The same principles apply to surveillance of environmental pollution and hazards: point of need molecular and spectrometric analyses will greatly facilitate the detection and control of environmental hazards (Fig. 17.25).

17.4.3.2 Contact tracing

In an epidemic outbreak investigation, contact tracing using Reality Mining can help identify mode and route of spread. Ordinary people usually have incomplete recall of where they have been and with whom they have come into contact. GPS data of the index patients and their social network data would be invaluable in establishing contact patterns and potential infection routes.

17.4.3.3 Predicting an epidemic

Recurring epidemics depend on the accumulation of newly susceptible individuals (Fig. 17.26). In childhood diseases such as enterovirus infection epidemics are intimately linked to birth rates. In a recent paper published in Science by Imperial College researchers, it was possible to predict enterovirus epidemics two years in advance using mathematical modelling [9].

In monitoring epidemics of infectious diseases such as Ebola, knowing the physical location and movement of the human population in endemic areas is critical. Since Ebola has no known human reservoir, outbreaks must start with encroachment into

Figure 17.26

animal habitats by surrounding human populations. Such contacts with animals can be monitored and stopped more effectively if they can be recorded in real time. When an outbreak has occurred, knowing the whereabouts of individuals in the endemic area is again of vital importance in controlling its spread. Such data will be available if smartphone ownership is ubiquitous.

17.4.3.4 Resource and service allocation

Allocation of scarce resources and services, as well as public health forecasting and planning can be done better with AI and big data analysis (Fig. 17.27). Adequate prioritization can only be achieved if public health officials look at the complete picture using all the data available.

17.4.3.5 Disaster prediction and relief

Many natural disasters have harbingers which may not be recognizable until after the event. With AI and sufficient data, such warning patterns may be discernible beforehand.

Figure 17.27

17.4.3.6 Research

Nutritional epidemiologic studies are notorious for difficulties in obtaining food intake data. Few people can accurately remember or record their daily food intake without help. Smartphone applications can go a long way to remedy the situation. Food can be photographed and recorded, and portion sizes estimated instantaneously using specially designed food intake applications. Calories can be calculated quickly with immediate feedback to the study subject if necessary.

17.4.4 Future Development

Before you can process digital signals, you need to collect them; therefore novel ways of signal collection with existing and future sensors are imperative.

17.4.4.1 Existing sensors

The mobile smartphone and wearable gadgets such as Fitbit can monitor physical locations and vital body functions (Fig. 17.28). Various other implantables such as pacemaker and insulin pump also generate continuous vital information. Besides automatic collection of background signals, the cell phone and other devices can also be used to record information voluntarily to supplement human memory.

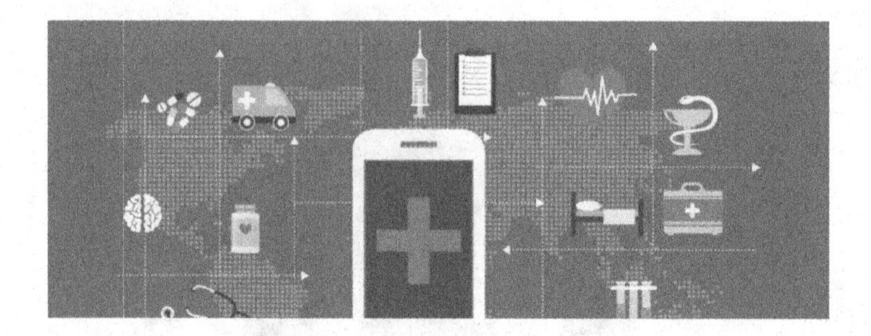

Figure 17.28

17.4.4.2 Future sensors

Environmental pollutants

Air pollution is a leading cause of human diseases. We now monitor air quality regularly with roadside sensors. But these sporadic samples do not give a full picture of the source, severity, and distribution of environmental pollution. What if we can invent inexpensive wearable sensors for individuals to detect pollutants? With hundreds of thousands of these sensors deployed we will have a more complete picture of the extent of pollution. These sensors can also be used to monitor indoor air pollution.

Environmental pathogens

Sensors to detect pathogens such as swine flu or bird flu viruses quickly and inexpensively can be extremely useful to recognise and stop the spread of these diseases. Before an outbreak such sensors can be deployed in farms and slaughter houses. After an outbreak these sensors can be deployed in hospitals, clinics, and public arenas.

Disaster sensors

Natural disasters such as tornadoes and earthquakes are usually preceded by changes in the atmosphere and earth movements. We now have sensors to detect these changes, but they are bulky and expensive and can only be placed in a few selected locations. Wearable or portable sensors of atmospheric pressure and seismic activities can be deployed to detect warning signs before disasters occur. They are also useful in monitoring the aftermath of disasters.

17.4.5 Social Implications

With cutting-edge advances of DSP in clinical medicine in developed economies, the disparities of healthcare in different populations are going to be exaggerated at least in the short term. It is hard to image getting a bionic limb when most amputees in third world countries have trouble getting a regular proper-fitting prosthesis. Drugs are also often unaffordable that it is unrealistic to talk about individualised medicine. In many third world hospitals having a qualified surgeon is a luxury and giving them the most advanced robotic surgical system is a cruel joke.

Does it mean that resources spent on these new frontiers are a misallocation and waste? This question is similar to the ones raised about space exploration in the 1960s when people questioned the wisdom of sending humans to the moon. But to explore the unknown and what is possible in the future is human nature. This is what differentiates us from other animals. We cannot stop scientific progress until we have attained a totally egalitarian society on earth. What seems in the beginning to be wasteful and meaningless can eventually be extremely useful as developments bring about unforeseen dividends. The space program is an excellent example: so many of the technologies developed are now being used in ordinary civilian life. We can expect the same with machine learning in medicine.

17.4.6 Privacy Issues

Invasion of personal privacy is a big concern of reality mining as well as wearable sensors. Legal and ethical issues have to be explored and dealt with. All such sensors should have an opt-out switch for individuals who want to keep their data private.

17.5 Business Response to AI in US Dollars

The U.S. and China lead the world in investments in AI (James Manyika, chairman and director of the McKinsey Global Institute). In 2017 AI investment in North America ranged from $15 billion to $23 billion, Asia (mainly China) was $8 billion to $12 billion, and Europe lagged at $3 billion to $4 billion. Tech giants have been the primary investors in AI (between $20 billion and $30 billion), venture capitalists and private equity firms invested $6 billion to $9 billion.

Machine learning received 56% of the investments with computer vision second at 28%. Natural language garnered 7%, autonomous vehicles was at 6% and virtual assistants made up the rest.

Reference to the McKinsey survey: sectors leading in AI are telecom and tech companies, financial institutions and automakers. These early adopters tend to be larger and digitally mature companies that incorporate AI into core activities, focus on growth and innovation over cost savings and enjoy the support of C-suite level executives. The slowest adopters are companies in health care, travel, professional services, education and construction.

McKinsey believes that AI can deliver more than double the impact of other analytics and has the potential to materially raise corporate performance.

17.6 Conclusion: Can Intelligence Ever Be Artificial?

Artificial intelligence is an old concept to highlight man-machine symbiosis (Fig. 17.29). Both cooperate in making decisions and

Figure 17.29

Figure 17.30 Celebratory gathering of world experts and scholars at the 50th Anniversary of DSP, 2017, Imperial College London.

controlling complex situation flexibly. Scientists aim to maximize complementarity: that man to think as no human brain ever thought, and machine to process data in new ways not approached by machine we know now. The whole is greater than the sum of its parts! The current challenge in machine-learning is notably in the human factors of sentiments and imagination.

Acknowledgement

Alex Tam is acknowledged for graphics designs and support. Photos and graphics subject to third-party copyright used with permission, or from the Internet or © Imperial College London.

References

1. Schunk, D. *Learning Theories: An Educational Perspective*, Pearson Education, Inc, 2000.
2. Emerging Technology from the ar Xiv Machine learning predicts World Cup winner. *MIT Technology Review*, 2018.
3. Schacter, D., et al. The future of memory: Remembering, imagining, and the brain. *Neuron*, 2012, 76(4), 677–694.

4. Perisic, I. *The Economic Graph Research Program*, Microsoft Research Podcast Transcript, Episode 11, February 11, 2018.
5. Ransbotham, S., et al. Reshaping business with artificial intelligence. MIT Sloan Management Review in collaboration with The Boston Consulting Group 2017.
6. Nakov, P. Predicting factuality of reporting and bias of news media sources, *Empirical Methods in Natural Language Processing Conference*, November, 2017.
7. Brooks R. Why we will rely on robots. *MIT Technology Review*, August 2014.
8. Pentland, A. et al. Using reality mining to improve public health and medicine. In Bushko, R. G. (ed.): *Strategy for the Future of Health*, IOS Press, 2009, pp. 93–102.
9. Pons-Salort, M, Grassly, N. Serotype-specific immunity explains the incidence of diseases caused by human enteroviruses. *Science* 2018, 361, pp. 800–803.

Index

Printed in the United States
by Baker & Taylor Publisher Services

Printed in the United States
by Baker & Taylor Publisher Services